T0202659

Lecture Notes in Computer Science 14159

Founding Editors

Gerhard Goos
Juris Hartmanis

Editorial Board Members

Elisa Bertino, *Purdue University, West Lafayette, IN, USA*
Wen Gao, *Peking University, Beijing, China*
Bernhard Steffen ⓘ, *TU Dortmund University, Dortmund, Germany*
Moti Yung ⓘ, *Columbia University, New York, NY, USA*

The series Lecture Notes in Computer Science (LNCS), including its subseries Lecture Notes in Artificial Intelligence (LNAI) and Lecture Notes in Bioinformatics (LNBI), has established itself as a medium for the publication of new developments in computer science and information technology research, teaching, and education.

LNCS enjoys close cooperation with the computer science R & D community, the series counts many renowned academics among its volume editors and paper authors, and collaborates with prestigious societies. Its mission is to serve this international community by providing an invaluable service, mainly focused on the publication of conference and workshop proceedings and postproceedings. LNCS commenced publication in 1973.

Chiara Di Francescomarino ·
Andrea Burattin · Christian Janiesch ·
Shazia Sadiq
Editors

Business Process Management

21st International Conference, BPM 2023
Utrecht, The Netherlands, September 11–15, 2023
Proceedings

 Springer

Editors
Chiara Di Francescomarino 🆔
University of Trento
Trento, Italy

Andrea Burattin 🆔
Technical University of Denmark
Kgs. Lyngby, Denmark

Christian Janiesch 🆔
TU Dortmund University
Dortmund, Germany

Shazia Sadiq 🆔
The University of Queensland
Brisbane, QLD, Australia

ISSN 0302-9743   ISSN 1611-3349   (electronic)
Lecture Notes in Computer Science
ISBN 978-3-031-41619-4   ISBN 978-3-031-41620-0   (eBook)
https://doi.org/10.1007/978-3-031-41620-0

© The Editor(s) (if applicable) and The Author(s), under exclusive license
to Springer Nature Switzerland AG 2023
Chapter "Can I Trust My Simulation Model? Measuring the Quality of Business Process Simulation Models"
is licensed under the terms of the Creative Commons Attribution 4.0 International License (http://
creativecommons.org/licenses/by/4.0/). For further details see license information in the chapter.

This work is subject to copyright. All rights are reserved by the Publisher, whether the whole or part of the
material is concerned, specifically the rights of translation, reprinting, reuse of illustrations, recitation,
broadcasting, reproduction on microfilms or in any other physical way, and transmission or information
storage and retrieval, electronic adaptation, computer software, or by similar or dissimilar methodology now
known or hereafter developed.
The use of general descriptive names, registered names, trademarks, service marks, etc. in this publication
does not imply, even in the absence of a specific statement, that such names are exempt from the relevant
protective laws and regulations and therefore free for general use.
The publisher, the authors, and the editors are safe to assume that the advice and information in this book are
believed to be true and accurate at the date of publication. Neither the publisher nor the authors or the editors
give a warranty, expressed or implied, with respect to the material contained herein or for any errors or
omissions that may have been made. The publisher remains neutral with regard to jurisdictional claims in
published maps and institutional affiliations.

This Springer imprint is published by the registered company Springer Nature Switzerland AG
The registered company address is: Gewerbestrasse 11, 6330 Cham, Switzerland

Preface

This volume comprises all papers presented at the 21st International Conference on Business Process Management (BPM), held during September 11–15, 2023 in Utrecht, the Netherlands. Emerging from the pandemic and returning to the normalcy absent since BPM 2019 in Vienna, the BPM community has continued as a determined and flexible community, evident from the following excellent conference program.

BPM 2022 in Münster, Germany marked the cautious and successful return to a full in-person conference. In light of BPM 2023's submission and attendance numbers, the appreciation and importance of BPM as a physical venue for the BPM community stands unquestioned. The conference was flanked by a multitude of events, such as the Blockchain, Educators, and RPA Fora, 11 workshops, tutorials, a doctoral consortium, and wonderful social events, which gave rise to the opportunity for networking and exchanging the latest research ideas.

BPM 2023 followed the history and philosophy of previous editions with respect to the three main research tracks, *Foundations* (Track I), *Engineering* (Track II), and *Management* (Track III), reflecting the different communities of the conference series. Track I (chaired by Chiara Di Francescomarino) addressed computer science research methods for researching the underlying principles of BPM, computational theories, algorithms, semantics, conceptual models, identification of novel problems, languages, and architectures. Track II (chaired by Andrea Burattin) dealt with engineering aspects of information systems research, including business process intelligence, process mining, process modelling, and process enactment, and employed rigorous and repeatable empirical evaluations. Track III (chaired by Christian Janiesch) aimed at advancing the understanding of socio-technical, cognitive or psychological aspects of BPM techniques, tools, and methods as well as managerial aspects of BPM in and across organisations. Shazia Sadiq served as the Consolidation Chair.

This year, the conference received a total of 167 submissions, out of which 151 entered the review phase for full papers. Out of these 151 full paper submissions, 40 were submitted to Track I, 62 to Track II, and 49 to Track III. The review process followed the high-quality standards of the BPM conference series. Each paper was reviewed by at least three Program Committee members of the respective track; reviews were single blind. Then, an extensive discussion phase between the reviewers and a Senior Program Committee member followed. As a result, the discussion was summarised in a meta-review by the Senior Program Committee member who also offered their recommendation to the Track Chairs. This thorough review process resulted in 7 accepted papers for Track I, 11 accepted papers for Track II, and 9 accepted papers for Track III, totalling 27 contributions included in the main research track (16.17% overall acceptance rate). Moreover, the review process resulted in the inclusion of 23 papers in the BPM Forum program, published in a separate volume of the Springer LNBIP series. These papers aim at presenting highly innovative research and ideas.

The program also included three invited keynote talks about timely topics in BPM: Marta Kwiatkowska, Professor of Computing Systems and Fellow at Trinity College, University of Oxford, delved into the ways in which formal methods and robust machine learning can be applied in the context of BPM; Matthias Weidlich, Professor with the Department of Computer Science at Humboldt-Universität zu Berlin, focused on the link between database systems and BPM and its opportunities; and Marc Kaptein, Medical Director at Pfizer, talked about his instrumental role in the roll-out process of Pfizer/BioNTech's corona vaccine. In addition to keynote abstracts, this volume features abstracts accompanying the tutorials.

The research presented in this volume shows the wide variety of topics and methods that characterise the BPM community across the three tracks. These proceedings report on diverse insights obtained via behavioural-science thinking (e.g., case studies) and design-science research (e.g., method development). Topics range from process modelling and mining, over conformance checking, to stakeholder engagement and digital process innovation. The main topics of this volume are reflected in the session themes, including among others design patterns and languages, resource and task management, cognitive aspects of BPM, and real-world applications as well as the process-mining-related topics of anomaly detection, conformance and alignment, event log manipulation, performance metrics, process discovery, and simulation.

Open Science continued to be a major principle for the BPM community, aiming at reproducibility and replicability of the research results. Following the tradition started in 2020, the authors were explicitly requested to link one or more repositories with additional artefacts such as data sets, prototypes, and interview protocols alongside implemented prototypes to their papers. Furthermore, Shazia Sadiq, Jens Gulden, and Adela del Río Ortega acted as inaugural Diversity & Inclusion Chairs and encouraged all participants to consider diversity, equity, and inclusion (DEI) in their writing, reviews, presentations, and all other interactions related to the BPM conference.

We would like to thank all authors, both regular and senior members of the Program Committees, and the external reviewers of the three tracks: foundations, engineering, and management. They made a rigorous, extensive, and timely review procedure possible and enabled the high-quality research output reflected by the papers in this volume. In addition to the committees of the BPM 2023 main track and BPM 2023 Forum, committees for the workshops, the tutorials, the RPA Forum, the Educators Forum, the Blockchain Forum, the Industry Day, the Demonstration and Resources Track, the Doctoral Consortium, the BPM Dissertation Award, and the Journal-First Track did an outstanding job in reviewing and selecting high-quality contributions to the different tracks and fora.

We acknowledge our sponsors for their support in making BPM 2023 happen: Celonis and Software AG as platinum sponsors; BPM Consult as bronze sponsor; and Hogeschool Utrecht, the Netherlands Research School for Information and Knowledge Systems, Springer, and Utrecht University as academic sponsors. We also appreciated the use of EasyChair for streamlining an intensive reviewing process and CEUR Workshop Proceedings for supporting the publication of all other formats as well as all other sponsors.

Finally, we would like to express our special thanks to Hajo Reijers as the General Chair of BPM 2023, together with the Organizing Committee Chairs Inge van de

Weerd, Jan Martijn van der Werf, and Pascal Ravesteijn, and their staff. The Utrecht team did an impeccable job in planning and organising an unforgettable conference.

Last but not least, we thank you as the readers of this volume and wish you a great experience in examining the latest in BPM research.

September 2023

Chiara Di Francescomarino
Andrea Burattin
Christian Janiesch
Shazia Sadiq

Organization

Steering Committee

Jan Mendling (Chair)	Humboldt-Universität zu Berlin, Germany
Marlon Dumas	University of Tartu, Estonia
Avigdor Gal	Technion – Israel Institute of Technology, Israel
Chiara Ghidini	Fondazione Bruno Kessler, Italy
Manfred Reichert	University of Ulm, Germany
Hajo A. Reijers	University of Utrecht, The Netherlands
Stefanie Rinderle-Ma	Technical University of Munich, Germany
Adela del Río Ortega	University of Seville, Spain
Michael Rosemann	Queensland University of Technology, Australia
Shazia Sadiq	The University of Queensland, Australia
Barbara Weber	University of St. Gallen, Switzerland
Matthias Weidlich	Humboldt-Universität zu Berlin, Germany
Mathias Weske	University of Potsdam, Germany

Executive Committee

General Chair

Hajo A. Reijers	University of Utrecht, The Netherlands

Main Conference Program Committee Chairs

Chiara Di Francescomarino (Track I Chair)	University of Trento, Italy
Andrea Burattin (Track II Chair)	Technical University of Denmark, Denmark
Christian Janiesch (Track III Chair)	TU Dortmund University, Germany
Shazia Sadiq (Consolidation Chair)	The University of Queensland, Australia

Workshop Chairs

Jochen De Weerdt	KU Leuven, Belgium
Luise Pufahl	Technical University of Munich, Germany

Demonstration and Resources Chairs

Tijs Slaats	University of Copenhagen, Denmark
Andrés Jiménez Ramírez	University of Seville, Spain
Karolin Winter	Eindhoven University of Technology, The Netherlands

Industry Day Chairs

Arjen Maris	University of Twente, The Netherlands
Willemijn van Haeften	HU University of Applied Sciences Utrecht, The Netherlands

Blockchain Forum Chairs

Julius Köpke	University of Klagenfurt, Austria
Orlenys López Pintado	University of Tartu, Estonia

RPA Forum Chairs

Ralf Plattfaut	University of Duisburg-Essen, Germany
Jana-Rebecca Rehse	University of Mannheim, Germany

Educators Forum Chairs

Jorge Munoz-Gama	Pontificia Universidad Católica de Chile, Chile
Katarzyna Gdowska	AGH University of Science and Technology, Poland
Fernanda Gonzalez-Lopez†	Pontificia Universidad Católica de Chile/Human and Process Data Science Research Lab, Chile
Koen Smit	HU University of Applied Sciences Utrecht, The Netherlands
Jan Martijn van der Werf	Utrecht University, The Netherlands

Tutorial Chairs

Michael Rosemann	Queensland University of Technology, Australia
Sietse Overbeek	Utrecht University, The Netherlands

Journal-First Track Chairs

Mieke Jans	Hasselt University, Belgium
Henrik Leopold	Kühne Logistics University, Germany

Doctoral Consortium Chairs

Dirk Fahland	Eindhoven University of Technology, The Netherlands
Barbara Weber	University of St. Gallen, Switzerland
Johan Versendaal	HU University of Applied Sciences Utrecht, The Netherlands

Best BPM Dissertation Award Chair

Jan Mendling	Humboldt-Universität zu Berlin, Germany

Publicity Chairs

Iris Beerepoot	Utrecht University, The Netherlands
Andrea Delgado	Universidad de la República, Uruguay
Mahendrawathi ER	Institut Teknologi Sepuluh Nopember, Indonesia

Diversity and Inclusion Chairs

Shazia Sadiq   The University of Queensland, Australia
Jens Gulden   Utrecht University, The Netherlands
Adela del Río Ortega   University of Seville, Spain

Proceedings Chairs

Xixi Lu   Utrecht University, The Netherlands
Felix Mannhardt   Eindhoven University of Technology, The Netherlands

Organizing Committee Chairs

Inge van de Weerd   Utrecht University, The Netherlands
Jan Martijn van der Werf   Utrecht University, The Netherlands
Pascal Ravesteijn   HU University of Applied Sciences Utrecht,
                                    The Netherlands

Track I: Foundations

Senior Program Committee

Wil van der Aalst   RWTH Aachen University, Germany
Jörg Desel   FernUniversität in Hagen, Germany
Claudio Di Ciccio   Sapienza University of Rome, Italy
Dirk Fahland   Eindhoven University of Technology, The Netherlands
Chiara Ghidini   Fondazione Bruno Kessler, Italy
Thomas Hildebrandt   University of Copenhagen, Denmark
Arthur ter Hofstede   Queensland University of Technology, Australia
Rick Hull   University of California, USA
Sander J. J. Leemans   RWTH Aachen University, Germany
Fabrizio Maria Maggi   Free University of Bozen-Bolzano, Italy
Andrea Marrella   Sapienza University of Rome, Italy
Marco Montali   Free University of Bozen-Bolzano, Italy
Oscar Pastor   Universidad Politécnica de Valencia, Spain
Artem Polyvyanyy   The University of Melbourne, Australia
Manfred Reichert   University of Ulm, Germany
Hagen Voelzer   University of St. Gallen, Switzerland
Matthias Weidlich   Humboldt-Universität zu Berlin, Germany
Jan Martijn van der Werf   Utrecht University, The Netherlands
Mathias Weske   Hasso Plattner Institute/University of Potsdam,
                                    Germany

Program Committee

Lars Ackermann   University of Bayreuth, Germany
Ahmed Awad   University of Tartu, Estonia
Patrick Delfmann   Universität Koblenz, Germany
Rik Eshuis   Eindhoven University of Technology, The Netherlands

Peter Fettke	German Research Center for Artificial Intelligence (DFKI)/Saarland University, Germany
Valeria Fionda	University of Calabria, Italy
Francesco Folino	ICAR-CNR, Italy
Maria Teresa Gómez López	University of Seville, Spain
Alessandro Gianola	Free University of Bozen-Bolzano, Italy
Guido Governatori	Independent Researcher, Australia
Giancarlo Guizzardi	University of Twente, The Netherlands
Ekkart Kindler	Technical University of Denmark, Denmark
Akhil Kumar	The Pennsylvania State University, USA
Irina Lomazova	National Research University Higher School of Economics, Russia
Hugo Andrés López	Technical University of Denmark, Denmark
Qinghua Lu	CSIRO, Australia
Xixi Lu	Utrecht University, The Netherlands
Felix Mannhardt	Eindhoven University of Technology, The Netherlands
Werner Nutt	Free University of Bozen-Bolzano, Italy
Chun Ouyang	Queensland University of Technology, Australia
Luigi Pontieri	ICAR-CNR, Italy
Daniel Ritter	SAP SE, Germany
Andrey Rivkin	Technical University of Denmark, Denmark
Arik Senderovich	York University, Canada
Tijs Slaats	University of Copenhagen, Denmark
Monique Snoeck	KU Leuven, Belgium
Ernest Teniente	Universitat Politècnica de Catalunya, Spain
Eric Verbeek	Eindhoven University of Technology, The Netherlands
Karsten Wolf	Universität Rostock, Germany
Francesca Zerbato	University of St. Gallen, Switzerland

Track II: Engineering

Senior Program Committee

Boualem Benatallah	University of New South Wales, Australia
Jochen De Weerdt	KU Leuven, Belgium
Remco Dijkman	Eindhoven University of Technology, The Netherlands
Boudewijn van Dongen	Eindhoven University of Technology, The Netherlands
Marlon Dumas	University of Tartu, Estonia
Avigdor Gal	Technion – Israel Institute of Technology, Israel
Massimo Mecella	Sapienza University of Rome, Italy
Jorge Munoz-Gama	Pontificia Universidad Católica de Chile, Chile
Luise Pufahl	Technical University of Munich, Germany
Hajo A. Reijers	Utrecht University, The Netherlands
Stefanie Rinderle-Ma	Technical University of Munich, Germany
Pnina Soffer	University of Haifa, Israel
Barbara Weber	University of St. Gallen, Switzerland

Ingo Weber   Technical University of Munich, Germany
Moe Thandar Wynn   Queensland University of Technology, Australia

Program Committee

Han van der Aa   University of Mannheim, Germany
Robert Andrews   Queensland University of Technology, Australia
Abel Armas Cervantes   The University of Melbourne, Australia
Nick van Beest   CSIRO, Australia
Cristina Cabanillas   University of Seville, Spain
Fabio Casati   ServiceNow, USA/University of Trento, Italy
Johannes De Smedt   KU Leuven, Belgium
Benoît Depaire   Hasselt University, Belgium
Joerg Evermann   Memorial University of Newfoundland, Canada
Walid Gaaloul   Télécom SudParis, France
Luciano García-Bañuelos   Tecnológico de Monterrey, Mexico
Laura Genga   Eindhoven University of Technology, The Netherlands
Daniela Grigori   Université Paris-Dauphine - PSL, France
Georg Grossmann   University of South Australia, Australia
Mieke Jans   Hasselt University, Belgium
Anna Kalenkova   The University of Adelaide, Australia
Dimka Karastoyanova   University of Groningen, The Netherlands
Agnes Koschmider   University of Bayreuth, Germany
Massimiliano de Leoni   University of Padua, Italy
Henrik Leopold   Kühne Logistics University, Germany
Francesco Leotta   Sapienza University of Rome, Italy
Elisa Marengo   Free University of Bozen-Bolzano, Italy
Rabeb Mizouni   Khalifa University, United Arab Emirates
Timo Nolle   Technical University of Darmstadt, Germany
Hye-Young Paik   University of New South Wales, Australia
Cesare Pautasso   University of Lugano, Switzerland
Pierluigi Plebani   Politecnico di Milano, Italy
Pascal Poizat   Université Paris Nanterre/LIP6, France
Simon Poon   The University of Sydney, Australia
Barbara Re   University of Camerino, Italy
Manuel Resinas   University of Seville, Spain
Stefan Schönig   University of Regensburg, Germany
Marcos Sepúlveda   Pontificia Universidad Católica de Chile, Chile
Natalia Sidorova   Eindhoven University of Technology, The Netherlands
Renuka Sindhgatta   IBM Research, India
Minseok Song   Pohang University of Science and Technology,
                              South Korea
Niek Tax   Meta, UK
Seppe vanden Broucke   Ghent University, Belgium
Karolin Winter   Eindhoven University of Technology, The Netherlands

| Nicola Zannone | Eindhoven University of Technology, The Netherlands |
| Sebastiaan van Zelst | RWTH Aachen University/Fraunhofer Institute for Applied Information Technology FIT, Germany |

Track III: Management

Senior Program Committee

Daniel Beverungen	Paderborn University, Germany
Adela del Río Ortega	University of Seville, Spain
Paul Grefen	Eindhoven University of Technology, The Netherlands
Mojca Indihar Štemberger	University of Ljubljana, Slovenia
Marta Indulska	The University of Queensland, Australia
Peter Loos	German Research Center for Artificial Intelligence (DFKI)/Saarland University, Germany
Jan Mendling	Humboldt-Universität zu Berlin, Germany
Jan Recker	University of Hamburg, Germany
Maximilian Röglinger	FIM Research Center Finance & Information Management/University of Bayreuth, Germany
Michael Rosemann	Queensland University of Technology, Australia
Flavia Santoro	Universidade do Estado do Rio de Janeiro, Brazil
Peter Trkman	University of Ljubljana, Slovenia
Amy Van Looy	Ghent University, Belgium
Jan vom Brocke	University of Münster, Germany/University of Liechtenstein, Liechtenstein

Program Committee

Amine Abbad-Andaloussi	University of St. Gallen, Switzerland
Banu Aysolmaz	Eindhoven University of Technology, The Netherlands
Christian Bartelheiner	Paderborn University, Germany
Iris Beerepoot	Utrecht University, The Netherlands
Marco Comuzzi	Ulsan National Institute of Science and Technology, South Korea
Irene Bedilia Estrada Torres	University of Seville, Spain
Renata Gabryelczyk	University of Warsaw, Poland
Kanika Goel	Queensland University of Technology, Australia
Fernanda Gonzalez-Lopez†	Pontificia Universidad Católica de Chile/Human and Process Data Science Research Lab, Chile
Thomas Grisold	University of St. Gallen, Switzerland
Tomislav Hernaus	University of Zagreb, Croatia
John Krogstie	Norwegian University of Science and Technology, Norway
Michael Leyer	University of Marburg, Germany/Queensland University of Technology, Australia
Alexander Mädche	Karlsruhe Institute of Technology, Germany
Niels Martin	Hasselt University, Belgium

Martin Matzner	Friedrich-Alexander-Universität Erlangen-Nürnberg, Germany
Alexander Nolte	University of Tartu, Estonia
Nadine Ostern	Queensland University of Technology, Australia
Ralf Plattfaut	University of Duisburg-Essen, Germany
Geert Poels	Ghent University, Belgium
Gregor Polančič	University of Maribor, Slovenia
Pascal Ravesteyn	Utrecht University of Applied Sciences, The Netherlands
Jana-Rebecca Rehse	University of Mannheim, Germany
Kate Revoredo	Vienna University of Economics and Business, Austria
Dennis M. Riehle	Universität Koblenz, Germany
Estefanía Serral	KU Leuven, Belgium
Rehan Syed	Queensland University of Technology, Australia
Oktay Turetken	Eindhoven University of Technology, The Netherlands
Irene Vanderfeesten	KU Leuven, The Netherlands
Inge van de Weerd	Utrecht University, The Netherlands
Sven Weinzierl	Friedrich-Alexander-Universität Erlangen-Nürnberg, Germany
Axel Winkelmann	University of Würzburg, Germany
Bastian Wurm	Ludwig-Maximilians-Universität Munich, Germany
Patrick Zschech	Friedrich-Alexander-Universität Erlangen-Nürnberg, Germany
Michael zur Muehlen	Stevens Institute of Technology, USA

Additional Reviewers

Nour Assy	Heerko Groefsema
Charlotte Bahr	Michael Grohs
Luca Barbaro	Jonas Gunklach
Marisol Barrientos	Nico Hambauer
Janik-Vasily Benzin	Tiphaine Henry
Jonas Blatt	Mubashar Iqbal
Lasse Bohlen	Adrian Joas
Katharina Brennig	Merlin Knaeble
Qifan Chen	Sven Kruschel
Minsu Cho	Annina Liessmann
Axel Christfort	Yue Liu
Carl Corea	Yang Lu
Paul Cosma	Bernd Löhr
Sebastian Dunzer	Edoardo Marangone
Peter A. François	Alessandro Marcelletti
Yuhong Fu	Rahma Mukta
Alexandre Goossens	Roman Nesterov
Valerio Goretti	Xavier Oriol

Philippe Queinnec
Luis Quesada
Kristo Raun
Julian Rosenberger
Gwen Salaün
Daniel Schloß
Julia Seitz
Nafiseh Soveizi
Klara Steflova
Vinicius Dani Stein

Willi Tang
Niek Tax
Wouter van der Waal
Ignacio Velásquez
Charlotte Verbruggen
Jia Wei
Bemali Wickramanayake
Boming Xia
Arash Yadegari Ghahderijani
Sandra Zilker

Keynotes

Database Systems and Process Management – A Call for a Closer Look

Matthias Weidlich🆔

Humboldt-Universität zu Berlin, Berlin, Germany
matthias.weidlich@hu-berlin.de

In many application scenarios, the handling of data and the execution of processes is inherently intertwined. Data may trigger the instantiation of business processes and potentially influences their behaviour in terms of branching and performance characteristics. At the same time, the effects of process execution typically materialize in the form of data being created or updated. As such, it is just natural that database systems play a fundamental role in the automation and analysis of processes and a plethora of supporting mechanisms have been developed. Those reach from models for data handling in processes, through connectors between database systems and process engines, to querying mechanisms that facilitate process mining.

In this talk, however, I argue for a closer look at the intersection of database systems and process management, which reveals opportunities to improve the interplay of data and processes that may not be obvious at first glance. Insights on processes are valuable for the design and operation of database systems, for instance in the area of transaction management. The automation and analysis of processes, in turn, may benefit from models and algorithms that have been developed for database systems, including notification mechanisms, query models for pattern detection, and indexing structures. Drawing on existing work in the area, I will point to several promising directions for future research.

Robust Decision Pipelines: Opportunities and Challenges for AI in Business Process Modelling

Marta Kwiatkowska(iD)

Department of Computer Science, University of Oxford, UK
marta.kwiatkowska@cs.ox.ac.uk

Keywords: Modelling and verification · Adversarial robustness · Optimality guarantees

1 Extended Abstract

Traditional business process modelling techniques, which leverage handcrafted pipelines and expert knowledge, are being revolutionised by artificial intelligence (AI). Deep learning (DL), in particular, has been successfully employed in process mining and discovery to build predictive process models from event logs [1], and reinforcement learning (RL) can be utilised for policy synthesis. Data-driven decision pipelines are now commonly deployed in application domains such as financial services, and rigorous modelling of the associated processes can aid in their stress testing, optimisation and 'what if' analysis.

However, a known concern about DL is that it lacks robustness; more specifically, DL systems such as neural networks are susceptible to so called adversarial attacks, i.e., minor modifications to inputs, often imperceptible, which can catastrophically change the decision of the network. Before they can be deployed within decision pipelines, DL components require certifiable guarantees not just for accuracy and performance, but also properties such as safety and robustness [2].

Fortunately, much progress has been made in formal modelling and verification techniques, especially model checking, with which rigorous models of software systems can be built and automatically checked against specifications expressed in temporal logic [3]. Building on existing verification technologies, a fast growing research effort is tackling the problem of computing robustness guarantees for deep learning; examples include search-based safety verification using SMT (Satisfiability Modulo Theory) [2] for DL, guaranteed robust explanations for DL [4], provable robustness to

---
Supported by the EPSRC Prosperity Partnership FAIR (grant number EP/V056883/1). MK receives funding from the ERC under the European Union's Horizon 2020 research and innovation programme (FUN2MODEL, grant agreement No. 834115).

causal interventions for DL decisions [6], and optimality guarantees for RL policies from temporal logic specifications [5].

However, while data-rich scenarios and deep learning enable ease of automation for business processes, they also present significant new challenges due to their complexity, as well as their black-box and adaptive nature. Achieving robust decision pipelines will require concerted effort to develop integrated methods for certifiable training, robust explainability, certification guarantees, robustness to distribution shift and interventions, optimal policy synthesis and real-time monitoring.

References

1. Evermann, J., Rehse, J.R., Fettke, P.: Predicting process behaviour using deep learning. Decis. Support Syst. **100**, 129–140 (2016)
2. Huang, X., Kwiatkowska, M., Wang, S., Wu, M.: Safety verification of deep neural networks. In: Majumdar, R., Kunčak, V. (eds.) CAV 2017. LNCS, vol. 10426, pp. 3–21. Springer, Cham (2017). https://doi.org/10.1007/978-3-319-63387-9_1
3. Kwiatkowska, M., Norman, G., Parker, D.: PRISM 4.0: verification of probabilistic real-time systems. In: Gopalakrishnan, G., Qadeer, S. (eds.) CAV 2011. LNCS, vol. 6806, pp. 585–591. Springer, Heidelberg (2011). https://doi.org/10.1007/978-3-642-22110-1_47
4. La Malfa, E., Zbrzezny, A., Michelmore, R., Paoletti, N., Kwiatkowska, M.: On guaranteed optimal robust explanations for NLP models. In: Proceedings of the International Joint Conference on Artificial Intelligence (IJCAI-21) (2021)
5. Shao, D., Kwiatkowska, M.: Sample efficient model-free reinforcement learning from LTL specifications with optimality guarantees. In: Proceedings of the International Joint Conference on Artificial Intelligence (IJCAI) (2023)
6. Wang, B., Lyle, C., Kwiatkowska, M.: Provable guarantees on the robustness of decision rules to causal interventions. In: Proceedings of the International Joint Conference on Artificial Intelligence (IJCAI-21) (2021)

Tutorials

Task Support for Process Mining: From Formulating Questions to Evaluating Results

Iris Beerepoot[1], Francesca Zerbato[2], Barbara Weber[2]
and Pnina Soffer[3]

[1] Utrecht University, Utrecht, The Netherlands
i.m.beerepoot@uu.nl
[2] University of St. Gallen, St. Gallen, Switzerland
[3] University of Haifa, Mount Carmel, 3498838 Haifa, Israel

Abstract. Despite the rising popularity of process mining in practice, executing a process mining project is a daunting task that requires significant expertise. As such, there is a pressing need to provide comprehensive support for *process analysts*. This tutorial aims to provide participants with an overview of state-of-the-art practices followed by process analysts at each stage of a typical process mining project, from defining questions to evaluating results. Based on empirical evidence and experience from several projects, we go over concrete strategies to support analysts, with a focus on specific tasks and areas that require extra attention. This sets the stage for further research in developing support for process analysts and allows identifying blind spots that future research might address.

Tutorial Content. This tutorial aims at providing participants with an overview of state-of-the-art practices followed by process analysts throughout a process mining (PM) project. The tutorial is organized as follows.

Part 1: Introduction. The tutorial starts by setting the scene and providing an overview of the objectives. Then, we specify the perspective that we will take: Rather than the enterprise level, we specifically focus on the *individual* and *team* levels. In particular, we choose to look at PM projects from the eyes of process analysts since individual support for process mining analysts is still lacking [5]. Therefore, we leave organizational activities such as obtaining project support and change management out of scope. We conclude this part with an agenda.

Part 2: Four Stages of Process Mining Projects. During the core part of the tutorial, we systematically walk through the different stages of a PM project. Existing PM methodologies focus on providing high-level guidance, often not dwelling on task-specific support. We use existing methodologies as a skeleton and enhance them with evidence gathered from practice to help make the guidance within high-level stages more tangible and relatable for the participants. Here, we build on our experience from interview, think-aloud and action research studies conducted in several research and applied projects. For each of the following stages, we discuss key tasks and areas where support might be needed using a running example based on a real project conducted in Dutch hospitals and give pointers on where to find more.

1. Define questions: How to develop questions for process mining? We draw on interviews with PM experts and present concrete examples of how process analysts develop questions, closing with recommendations for question formulation and refinement [4].
2. Data collection and preparation: How to decrease the effort of event log extraction? We draw on a structured literature review of process mining case studies to present a taxonomy of human tasks in event log extraction [2] and illustrate how tasks can be automated through matching [3].
3. Mining and analysis: What strategies to adopt when analysing the data? From interviews with process analysts, we discuss strategies to structure a PM analysis and factors that influence their practical application [5].
4. Results: How should insights from the analysis be evaluated to be translated into concrete improvements? We draw on a structured literature review and action research to describe how artefacts and insights are currently evaluated with domain experts and outline concrete validation strategies [1].

Closing. We end by discussing takeaways and limitations before collecting additional suggestions for supporting process analysts from the audience.

Intended Audience. We invite academics and practitioners with some basic knowledge of process mining, such as: (1) researchers interested in state-of-the-art process mining practices, with a particular focus on developing support for process analysts, (2) students aiming to get an overview of strategies that process analysts apply in practice or understand where support for process analysts is lacking, and (3) practitioners wishing to gain knowledge in existing support for individuals involved in process mining projects.

Acknowledgements. Special thanks to our advisors Xixi Lu, Niels Martin, Vinicius Stein Dani, and Lisa Zimmermann, who have helped with their research to fill some of the gaps in supporting process analyst. F. Zerbato and B. Weber are supported by the ProMiSE project funded by the SNSF under Grant No.: 200021_197032}

References

1. Koorn, J.J., et al.: Bringing rigor to the qualitative evaluation of process mining findings: an analysis and a proposal. In: International Conference on Process Mining (ICPM), pp. 120–127 (2021). https://doi.org/10.1109/ICPM53251.2021.9576877
2. Stein Dani, V. et al.: Towards understanding the role of the human in event log extraction. In: Marrella, A., Weber, B. (eds.) BPM 2021. LNBIP, vol. 436, pp 86–98. Springer, Cham (2022). https://doi.org/10.1007/978-3-030-94343-1_7

3. Stein Dani, V., Leopold, H., van der Werf, J.M.E.M., Reijers, H.A.: Supporting event log extraction based on matching. In: Cabanillas, C., Garmann-Johnsen, N.F., Koschmider, A. (eds.) BPM 2022. LNBIP, vol. 460, pp. 322–333. Springer, Cham (2023). https://doi.org/10.1007/978-3-031-25383-6_24
4. Zerbato, F., Koorn, J.J., Beerepoot, I., Weber, B., Reijers, H.A.: On the origin of questions in process mining projects. In: Almeida, J.P.A., et al. (eds.) EDOC 2022. LNCS, vol. 13585, pp. 165–181. Springer, Cham (2022). https://doi.org/10.1007/978-3-031-17604-3_10
5. Zerbato, F., Soffer, P., Weber, B.: Process mining practices: evidence from interviews. In: Di Ciccio, C., Dijkman, R., del Río Ortega, A., Rinderle-Ma, S. (eds.) BPM 2022. LNCS, vol. 13420, pp. 268–285. Springer, Cham (2022).https://doi.org/10.1007/978-3-031-16103-2_19

Using Large Language Models in Business Processes

Thomas Grisold[1], Jan vom Brocke[2], Wolfgang Kratsch[3],
Jan Mendling[4,5,6], and Maxim Vidgof[7]

[1] University of St. Gallen
thomas.grisold@unisg.ch
[2] ERCIS - European Research Center for Information Systems, University of
Münster, University of Liechtenstein
jan.vom.brocke@uni-muenster.de
[3] University of Applied Sciences Augsburg, Fraunhofer FIT
wolfgang.kratsch@fim-rc.de
[4] Humboldt-Universität zu Berlin, Berlin, Germany
jan.mendling@hu-berlin.de
[5] Weizenbaum Institute, Berlin, Germany
[6] WU Vienna, Vienna, Austria
[7] Vienna University of Economics and Business, Austria
maxim.vidgof@wu.ac.at

Abstract. Large language models, such as ChatGPT, provide ample opportunities for organizational work. These models are capable of collecting, integrating, and generating information with no or little human supervision [1]. Despite their wide and rapid uptake, we lack systematic knowledge about how large language models can be used in business processes. Our tutorial sheds light on the organizational, managerial and design-related implications of using large language models in business processes. We present a theoretical framework that integrates and synthesizes research from relevant streams, including task complexity [2], task automation [5], and human-AI delegation [1]. We specify potential opportunities and threats in relation to various forms of tasks, such as decision tasks and judgment tasks. Along these lines, we also explore how the use of large language models may affect the overall outcome of a process, for example, by providing new value propositions. We use, reflect, and discuss the implications of our framework based on real-world examples. Our conceptual framework is relevant to guide future research [e.g. 1] but also inform managerial decisions in organizations [e.g. 3].

Keywords: Large language models · BPM · ChatGPT

The research by Jan Mendling was supported by the Einstein Foundation Berlin under grant EPP-2019-524 and by the German Federal Ministry of Education and Research under grant 16DII133.

1   Aims and Contents of the Tutorial

Our tutorial will entail two parts. In the first part, we present a theoretical perspective to explore how large language models can be used in business process. We draw from relevant research streams, such as task complexity [2], task automation [4] and human-AI delegation [1], to specify opportunities and potential threats of using large language models on the levels of specific tasks and the levels of process outcomes. We also consider implications that arise with regards to process outcomes. We present a conceptual framework that is relevant both for future research [e.g. 1] as well as managerial decision-making in organizations [e.g. 3]. In the second part, we will conduct an interactive session where participants will learn about real-world examples before they gather in smaller groups to use our conceptual framework for business-process related tasks. Subsequently, we will have a joint discussion to reflect on the strengths and weaknesses of using large language models in business processes, as well as their implications for BPM in more general terms. To this end, we will discuss implications that arise with regard to the analysis, design and implementation of business processes [4].

2   Intended Audience

Our tutorial focuses on managerial and organizational implications of using large language models in business processes. In that regard, we discuss various aspects that pertain to the analysis, design and performance of certain process-related tasks. Furthermore, we shed light on implications that arise on the level of the overall business process. Against this backdrop, our tutorial addresses researchers who are studying, or planning to study, the role of large language models in business processes. The tutorial speaks to practitioners who are dealing with large language models, and want to know more about their applications, threats and opportunities. A technical focus or background is not required.

References

1. Baird, A., Maruping, L.M.: The next generation of research on is use: a theoretical framework of delegation to and from agentic IS artifacts. MIS Q. **45**, 315–341 (2021)
2. Campbell, D.J.: Task complexity: a review and analysis. Acad. Manage. Rev. **13**, 40–52 (1988)
3. Cromwell, J.R., Harvey, J.-F., Haase, J. et al.: Discovering where ChatGPT can create value for your company. In: Harvard Business Review (2023)

4. Dumas, M., La Rosa, M., Mendling, J. et al.: Fundamentals of business process management. Springer, Heidelberg (2018). https://doi.org/10.1007/978-3-662-56509-4
5. Parasuraman, R., Sheridan, T.B., Wickens, C.D.: A model for types and levels of human interaction with automation. IEEE Trans. Syst. Man Cybern. Syst. Humans **30**, 286–297 (2000)

Natural Language Processing for Business Process Analysis

Han van der Aa[1], Henrik Leopold[2,3], Kiran Busch[2],
and Adrian Rebmann[1]

[1] University of Mannheim, Mannheim, Germany
{han.van.der.aa, rebmann}@uni-mannheim.de
[2] Kühne Logistics University, Hamburg, Germany
{henrik.leopold, kiran.busch}@the-klu.org
[3] Hasso Plattner Institute, Potsdam, Germany

Abstract. Natural Language Processing (NLP) has become an essential tool for many organizations aiming to analyze and understand the vast amounts of text data they generate. The latest developments related to language models have significantly boosted the analytical capabilities of NLP tools and have created completely new use cases. In this tutorial, we will focus on the intersection of NLP and Business Process Management (BPM) and explore how NLP can support various BPM analysis tasks. We first introduce fundamentals of NLP and explore how language models work. Then, we focus on the automated analysis of textual descriptions, after which we turn to the analysis of process-oriented artifacts, where we show how NLP can be used to obtain novel insights from process models and event logs. These parts are followed by a hands-on exercise session, in which participants will learn how to use general and process-specific NLP libraries and techniques. Finally, we conclude the tutorial with a discussion of future directions.

After the tutorial, participants will have learned about the fundamentals of NLP, the potential of using NLP in the context of BPM, and how to apply NLP to their own BPM research and analyses.

1 Content Outline

The content of this tutorial will consist of the following parts:

Introduction to NLP and Language Models. (15 minutes)
In this part of the tutorial, we will provide the audience with the prerequisite knowledge necessary to understand state-of-the-art NLP tools. Besides introducing basic concepts, such as word embeddings, we will provide a gentle introduction to language models. Afterwards, we explain the core idea behind the transformer architecture and show they can be easily integrated using Python.

Extracting Process Information from Texts. (15 minutes)
This part of the tutorial will focus on the extraction of process information from textual process descriptions. We will show the challenges involved in this task, such as the large degree of variety, describe how this task has so far been tackled, and

briefly demonstrate how the extracted information can be used for downstream tasks such as process model extraction [3] and conformance checking [1].

Using NLP on Process-oriented Data. (15 minutes)
Next, we turn to the use of NLP for the analysis of process-oriented artifacts, focusing on process models and event logs. We will show how NLP can be used to extract semantic information from these artifacts, such as the actions and business objects contained in them. Subsequently, we will show how this extracted information can be used for purposes such as anomaly detection [2] and event abstraction [4], bringing a new layer of insight to process mining and analysis.

Hands-on Session. (30 minutes)
In this part, we will provide the attendees with practical demonstrations and exercises to extract and analyze process information from both textual descriptions and real-world event logs. To support this part, we will establish dedicated Jupyter notebooks that guide the attendees through the different steps, so that they can quickly gain insights. The notebooks will be designed to encourage audience members to try out different prompts (for ChatGPT) and also apply them on their own examples. During the exercises, we will provide guidance and answer any questions that may arise. To participate, audience members just need a laptop with Python, for which we will provide installation instructions.

Future of NLP Applications in BPM. (15 minutes)
In the final part of the tutorial, we will discuss how we expect that the role of NLP will further develop in the context of BPM in the future and engage the audience in a discussion on this matter.

2   Intended Audience

Our tutorial targets academics, practitioners, and students who are interested in learning about the application and potential of NLP for BPM and process mining. Audience members are not expected to have any prior experience with NLP. However we do expect them to be familiar with the fundamentals of BPM and process mining.

References

1. Van der Aa, H., Leopold, H., Reijers, H.A.: Checking process compliance against natural language specifications using behavioral spaces. Inf. Syst. **78**, 83–95 (2018)
2. Van der Aa, H., Rebmann, A., Leopold, H.: Natural language-based detection of semantic execution anomalies in event logs. Inf. Syst. **102**, 101824 (2021)
3. Bellan, P., Dragoni, M., Ghidini, C.: Process extraction from text: state of the art and challenges for the future. arXiv preprint arXiv:2110.03754 (2021)
4. Rebmann, A., van der Aa, H.: Enabling semantics-aware process mining through the automatic annotation of event logs. Inf. Syst. **110**, 102111 (2022)

Contents

Management

Foundations

Efficient Optimal Alignment Between Dynamic Condition Response Graphs and Traces

Axel Kjeld Fjelrad Christfort[(✉)] and Tijs Slaats

Department of Computer Science, University of Copenhagen, Copenhagen, Denmark
{axel,slaats}@di.ku.dk

Abstract. Dynamic Condition Response (DCR) Graphs is a popular declarative process modelling notation which is supported by commercial modelling tools and has seen significant industrial adoption. The problem of aligning traces with DCR Graphs, with it's multitude of applications such as conformance checking and log repair, has surprisingly not been solved yet. In this paper we address this open gap in the research by developing an algorithm for efficiently computing the optimal alignment of a DCR Graph and a trace. We evaluate the algorithm on the PDC 2022 dataset, showing that even for large models and traces alignment problems can be solved within milliseconds, and present a case study based on test-driven modelling.

Keywords: Trace alignment · DCR graphs · Conformance checking · Declarative process models · Test-driven Modelling

1 Introduction

Traditionally processes have been modelled using flow-based, imperative notations. The key property of these notations is that the edges, usually arrows, in a model capture the flow of the process. For example, an arrow between two activities may indicate that after executing the first, one moves on to executing the second. Constraint-based declarative notations [1] on the other hand forego this notion of flow. Edges in declarative models capture the logical relation between activities. For example, an arrow between two activities may capture that they are mutually exclusive and can never occur together in the same process instance. Whereas imperative notations are well-suited to capture straightforward structured processes with few variations, declarative notations have been argued to be better suited to capturing flexible knowledge-intensive processes, which allow for many different variants of the same process [20].

The last twenty years have seen the development of a multitude of declarative notations. In this paper we focus on Dynamic Condition Response (DCR) Graphs [13], which have seen significant industrial adoption, in particular in Denmark. This includes a commercial online modelling tool, the integration of a DCR process engine in commercial case management and workflow tools already

© The Author(s), under exclusive license to Springer Nature Switzerland AG 2023
C. Di Francesomarino et al. (Eds.): BPM 2023, LNCS 14159, pp. 3–19, 2023.
https://doi.org/10.1007/978-3-031-41620-0_1

used widely in local and central government institutions in Denmark, and a wide array of published use cases in various domains such as health care, governmental processes, and financial services.

Alignment algorithms compare a process trace and process model to determine to what degree they deviate from each other. Instead of making the binary distinction of whether a trace is a valid run of the model or not, they compute minimal changes, based on some cost function, that can be made to the trace so that it becomes a valid run. This approach has many applications within the fields of process modelling and mining. In conformance checking [6] alignment is used to provide a fine-grained measure for the degree to which a trace conforms to a model. In noise-filtering alignment can be used both to filter only the worst non-aligning traces out of a log, but also to inspect in more detail why traces misalign with a (potentially discovered) model.

Interestingly, even though there are many potential applications of computing (optimal) alignments between traces and DCR Graphs, this problem has yet to be solved. This leads us to the primary contributions of this paper:

(1) We develop an algorithm for computing the optimal alignment of a DCR Graph and a trace.
(2) We develop a method for reducing the search space of the algorithm, important for making the computation feasible in time-sensitive applications, and include a mechanically checked proof of correctness for the search space reduction.
(3) We evaluate the algorithm on the dataset from the Process Discovery Contest at ICPM 2022 (PDC2022).
(4) We present a case study based on checking open tests [23, 26] for the real-life process model of the Dreyers Foundation [9, 10].

In the remainder of this paper we first discuss related work (Sect. 2) and follow by recalling the formal definitions of DCR Graphs (Sect. 3). We then introduce alignment for DCR Graphs (Sect. 4), our algorithm for finding an optimal alignment (Sect. 5) and show how we can significantly reduce the search space of the algorithm in order to improve its efficiency (Sect. 6). Afterwards we evaluate the performance of the algorithm (Sect. 7) and as a case study show how it can be applied to test-driven modelling (Sect. 8).

2   Related Work

The term trace alignment was originally coined in [14] and introduced as a preprocessing step to other process mining techniques, where traces where aligned against one another in order to distinguish common from exceptional behaviour.

Since then, trace alignment has been extended to not only cover the alignment of traces against traces, but the alignment of traces against model behaviour and has become a cornerstone of conformance checking [6]. It is covered comprehensibly in [2], which includes defining alignments between traces and models, computing trace alignment for petri nets, and showing general applications, namely regarding conformance checking and process enhancement.

Trace alignment is, however, a computationally expensive task and as such, much work has been done on improving the performance of these computations. Lee et al. [16] recomposes conformance results of sub-problems and present a divide-and-conquer based alignment framework and Reißner et al. [21] propose a method of computing an combining the automatons corresponding to both log and model. Multiple approaches have also been suggested for approximating alignments, *i.e.* using simulation [22], using subset selection and edit distance [11], and using Trie data structures [3].

Other efforts have been made in extending alignments, namely computing online alignments of event streams [25], defining partial alignments over partial traces [19], and computing alignments of data aware processes by decomposition [18] and SMT-encoding respectively [12].

Finally, efficient computation of these conformance checking artefacts has been addressed in [5] by optimized SAT-encodings.

Trace alignment of models has since been extended to cover many process notations, including hybrid [24] and declarative notations. Namely two approaches have been suggested for computing trace alignments for declare models. De Leoni et al. [8,17] defines trace alignment for declare models, shows a mapping to the A^* algorithm, elaborates on conformance checking using alignments, before showing applications regarding log repair and -cleaning. De Giacomo et al. [7] extends this work by showing another mapping from alignment of declare models to a planning problem.

In this paper we extend the previous work by defining alignments for DCR graphs, providing a novel algorithm for computing these that is efficient in time and space, before showing applications of alignments as not just tools of conformance checking, but also tools of model checking.

3   Dynamic Condition Response Graphs

We first recall the basic syntax and semantics of DCR graphs [4]. The executable nodes of DCR Graphs are known as *events*. These DCR events can be repeatedly executed, unlike the concept of events in event logs, that represent a particular instance of the execution of an activity in time. Events have a *marking*, a boolean triple, indicating if the event has been executed (at least once) in the past, if the event is currently pending (and therefore must either be executed or excluded for the process to be finished successfully), and if the event is currently included. Events are labelled with the activities that they represent, and multiple events may have the same label. The edges between the events are known as *relations*, of which there are 5 types: the condition relation ($\rightarrow\bullet$) captures that, as long as the source event is included, it must be executed at least once before the target event can be executed, the milestone relation ($\rightarrow\diamond$) captures that, as long as the source event is included and pending, it blocks the target event from executing, the response relation ($\bullet\rightarrow$) captures that the source event makes the target pending, and the exclusion ($\rightarrow\%$) and inclusion ($\rightarrow+$) relations respectively remove events from and add them back into the current graph.

Definition 1. *A* DCR *graph* *is* *a* *tuple* $(E, M, \Sigma, \ell, \rightarrow\bullet, \bullet\rightarrow, \rightarrow\diamond, \rightarrow+, \rightarrow\%)$, *where*

- *E is the set of events*
- *M* $= (\mathsf{Ex}, \mathsf{Re}, \mathsf{In}) \in \mathcal{P}(E) \times \mathcal{P}(E) \times \mathcal{P}(E)$ *is the* marking *of the graph*
- *Σ is the set of* activities
- *$\ell : E \rightarrow \Sigma$ is the* labelling function, *assigning activities to events*
- *$\rightarrow\subseteq E \times E$ for $\rightarrow\in \{\rightarrow\bullet, \bullet\rightarrow, \rightarrow\diamond, \rightarrow+, \rightarrow\%\}$ are respectively the condition, milestone, response, inclusion and exclusion relations between events*

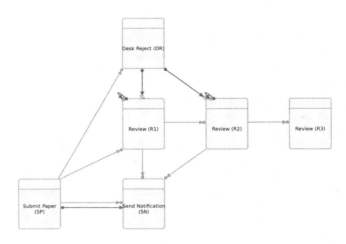

Fig. 1. A DCR graph for a simple paper submission process. Events are named as follows: "Label (EventID)".

For a DCR graph G with events E and an event $e \in E$, and for $\rightarrow\in \{\rightarrow\bullet, \bullet\rightarrow, \rightarrow\diamond, \rightarrow+, \rightarrow\%\}$, we write $(\rightarrow e)$ for the set $\{e' \in E \mid e' \rightarrow e\}$ and $(e\rightarrow)$ for the set $\{e' \in E \mid e \rightarrow e'\}$. We write $\ell^{-1} : \Sigma \rightarrow \mathcal{P}(E)$ for the inverse labelling function, returning for each activity the set of events to which it is assigned.

Example 1. *In Fig. 1 we show an example of a DCR graph modelling a simple paper submission system. When a paper is submitted, a notification is eventually required (modelled through a response relation). Any number of reviews can be written, but to send a notification we need at least two. To model this counting of the reviews we've created three separate review events, the first is a condition for the second and excludes itself, similarly the second excludes itself and is a condition for the third event, which is otherwise unconstrained. The first two review events are conditions for sending a notification. A paper can also be desk rejected, this removes the first and second review event of the process, thereby allowing a notification to be send immediately after. If desired, reviews can still be added under a desk reject through the third review event.*

Now that we have defined the syntax of DCR Graphs, we consider their semantics. First of all, we define that an event of a DCR graph is *enabled*, when it is included, all included events that are a condition for it have been executed, and all included events that are a milestone to it are not pending.

Definition 2 (Enabled events). *Let $G = (E, M, L, \ell, \to\bullet, \bullet\to, \to\diamond, \to+, \to\%)$ be a DCR graph, with marking $M = (\mathsf{Ex}, \mathsf{Re}, \mathsf{In})$. An event $e \in E$ is enabled, written $e \in \mathsf{enabled}(G)$, iff (a) $e \in \mathsf{In}$, (b) $\mathsf{In} \cap (\to\bullet e) \subseteq \mathsf{Ex}$, and (c) $\mathsf{In} \cap (\to\diamond e) \subseteq E \setminus \mathsf{Re}$.*

If an event e is enabled in a graph with marking $(\mathsf{Ex}, \mathsf{Re}, \mathsf{In})$, then it can be executed. Executing e will (a) add it to the set of executed events, (b) set all events in $(e\bullet\to)$ to pending, and (c) include and exclude all events from $(e\to+)$ and $(e\to\%)$ respectively.

Definition 3 (Execution). *Let $G = (E, M, L, \ell, \to\bullet, \bullet\to, \to\diamond, \to+, \to\%)$ be a DCR graph, with marking $M = (\mathsf{Ex}, \mathsf{Re}, \mathsf{In})$. When $e \in \mathsf{enabled}(G)$, the result of executing e, written $\mathsf{execute}(G, e)$ is a new DCR graph G' with the same events, labels, labelling function and relations, but a new marking $M' = (\mathsf{Ex}', \mathsf{Re}', \mathsf{In}')$, where (a) $\mathsf{Ex}' = \mathsf{Ex} \cup \{e\}$ (b) $\mathsf{Re}' = (\mathsf{Re} \setminus \{e\}) \cup (e\bullet\to)$, and (c) $\mathsf{In}' = (\mathsf{In} \setminus (e\to\%)) \cup (e\to+)$.*

We say that a graph is *accepting* if there are no included pending events.

Definition 4 (Accepting). *Let G be a DCR graph, with marking $M = (\mathsf{Ex}, \mathsf{Re}, \mathsf{In})$. We say that G is accepting, written $\mathsf{accepting}(G)$, iff $\mathsf{In} \cap \mathsf{Re} = \{\}$.*

Example 2. *In the DCR graph of Fig. 1 only the event Submit Paper is initially enabled, all other events are blocked by conditions. Executing this event enables Desk Reject and the first Review Event. Send Notification also becomes pending, which makes the graph no longer accepting. We can now execute the first Review event (R1), which removes itself from the workflow, after which the second Review event (R2) becomes enabled. Doing another review excludes the second Review event and enables the third Review event (R3) and Send Notification. The third Review event can be repeated an unbounded number of times to add additional reviews to the paper. Executing Send Notification removes the pending response on this activity and makes the graph accepting again.*

Finally we define the (accepting) runs and traces of a DCR Graph. We write $G \to_e G'$ if $e \in \mathsf{enabled}(G)$ and $G' = \mathsf{execute}(G, e)$.

Definition 5 (Runs). *A sequence of events $\phi = \langle e_1...e_n \rangle$ is a run of a DCR Graph iff $G \to_{e_1} ... \to_{e_n} G'$. It is an accepting run iff $\mathsf{accepting}(G')$.*

Definition 6 (Traces). *Let $\phi = \langle e_1...e_n \rangle$ be a run of a DCR Graph G. Then $\sigma = \langle \ell(e_1)...\ell(e_n) \rangle$ is a trace of G. It is an accepting trace iff ϕ is an accepting run.*

4   Trace Alignment for DCR Graphs

In order to quantify the misalignment between a log and a DCR graph, we can compute alignments which relate moves in a trace to moves in a model. In order to define such alignments we follow the style of [8] and introduce the symbol $\gg$, denoting 'no move', and the notation $S_\gg = S \cup \{\gg\}$ for $S \in \{E, \Sigma\}$.

Definition 7 (Alignment and complete alignment). *Let Σ be a set of activity names, $\sigma \in \Sigma^*$ a trace, and $G = (E, M, \Sigma, \ell, \rightarrow\bullet, \bullet\rightarrow, \rightarrow\diamond, \rightarrow+, \rightarrow\%)$ a DCR graph.*
  A pair $(l, e) \in (\Sigma_\gg \times E_\gg) \setminus \{\gg, \gg\}$ is

- *a move* in log *if $l \in \Sigma$ and $e = \gg$;*
- *a move* in model *if $l = \gg$ and $e \in E$;*
- *a move* in both log and model *if $l \in \Sigma$, $e \in E$ and $l = \ell(e)$.*

  Let $\Sigma_A = (\Sigma_\gg \times E_\gg) \setminus \{\gg, \gg\}$ be the set of legal moves. The alignment of a trace σ' and run $\phi \in E^*$ is a sequence $\gamma = \langle (l_1, e_1)...(l_n, e_n) \rangle \in \Sigma_A^*$, s.t. $\sigma' = \langle l_1...l_n \rangle$ and $\phi = \langle e_1...e_n \rangle$ (ignoring $\gg$). We say that γ is a complete *alignment of σ and G iff $\sigma' = \sigma$ and ϕ is an accepting run of G.*

As some moves may be more costly than others from a conformance point of view, we will need to introduce a cost function $\mathcal{K} : \Sigma_A \to \mathbb{R}_0^+$. Using $\mathcal{K}$, we can now define an optimal alignment between a trace and a model.

Definition 8 (Optimal alignment). *Let σ be a trace and G a DCR graph. Let $\Gamma_{(\sigma,G)}$ be the set of all complete alignments of σ and G. Given a cost function $\mathcal{K}$, we now define an alignment $\gamma \in \Gamma_{(\sigma,G)}$ to be* optimal, *iff $\forall \gamma' \in \Gamma_{(\sigma,G)}.\mathcal{K}(\gamma) \leq \mathcal{K}(\gamma')$.*

Example 3. *Consider the DCR graph in Fig. 1, the trace $\sigma = \langle SubmitPaper, Review, SendNotification \rangle$, and the alignment:*

$$\gamma_1 = \frac{\sigma' : |SubmitPaper|Review|\gg|Send\ Notification}{\phi : |SP \qquad\qquad |R1 \qquad|R2|SN}$$

γ_1 is a complete alignment that is optimal with a uniform cost function that assigns cost 0 to synchronous moves and 1 to all log and model moves. It leads to an accepting trace by inserting a second review. Note that another optimal alignment could be found by inserting a desk reject instead.

5   Computing an Optimal Alignment for DCR Graphs

There are several ways to compute an optimal alignment for DCR Graphs. The most straightforward would be to build the underlying automaton or transition system of the DCR Graph and apply out-of-the-box alignment methods. However, such an approach is hampered by the fact that the state space of DCR Graphs grows exponentially in the number of events. To be precise, the states of the automaton are the reachable markings of the graph and in the worst case

Algorithm 1. Align(G, σ, cost **default** 0)

---

global min $= \infty$.

```
 1: if cost ≥ min then
 2:    return ∞
 3: end if
 4: if accepting(G) ∧ |σ| = 0 then
 5:    return cost
 6: end if
 7: if |σ| > 0 then
 8:    l:σ' = σ
       // Synchronous move
 9:    for e ∈ ℓ⁻¹(l) do
10:       if e ∈ enabled(G) then
11:          G' = execute(e, G)
12:          cost' = Align(G', σ', cost + 𝒦(l,e))
13:          if cost' < min then
14:             min = cost'
15:          end if
16:       end if
17:    end for
       // Log move
18:    cost' = Align(G, σ', cost + 𝒦(l,≫))
19:    if cost' < min then
20:       min = cost'
21:    end if
22: end if
    // Model move
23: for e ∈ enabled(G) do
24:    G' = execute(e, G)
25:    cost' = Align(G', σ, cost + 𝒦(≫,e)))
26:    if cost' < min then
27:       min = cost'
28:    end if
29: end for
30: return min
```

---

a DCR Graph may have $2^{3|E|}$ reachable markings. While not all markings are reachable, and similar optimizations as we discuss below can be directly applied to the state space, real-life DCR models usually contain enough reachable markings that one will quickly run into memory allocation problems. Similarly a breadth-first approach tends to reach a deep enough depth in real-world applications that the exponential space complexity becomes a practical limitation.

A more efficient approach builds only the parts of the automaton that is needed for solving the alignment computation and keeps only the strictly required parts in memory at any point in time. Therefore we propose a simple depth first search of the alignment state space, which bounds the search efficiently enough for DCR graphs to not severely affect performance and also is fairly intuitive to implement and extend.

The algorithm, as seen in Fig. 1, simply performs synchronous-, log- and model-moves until an alignment is found. Once an alignment is found, it checks each node in the search-space that has a lesser cost, updating the optimal alignment cost if necessary.

As this is a depth first algorithm, the order in which it performs the moves matter, since reaching a more optimal solution earlier, will allow it to bound the search-space more efficiently. We believe that for most practical uses, an optimal alignment will be dominated by synchronous moves, which is therefore the path explored first. Starting with synchronous moves also gives the benefit of running in linear time for accepting traces. Next, log-moves are explored when possible, leaving the highly exponential model-moves until absolutely necessary.

Before computing an alignment, we can, however, pre-compute a trivial alignment that often results in a quite effective bound. In Fig. 1 we note that the graph is already accepting until a paper is submitted, meaning an empty run will be accepting. Having the empty run be accepting is the case for many modelled DCR processes, and is in particular the case for any model discovered with the DisCoveR algorithm [4]. Even if that is not the case, the state-space for a DCR graph will necessarily be simpler than that of a full alignment. We therefore propose the trivial alignment composed of only log and model moves, resulting in the bound of

$$\sum_{l \in \sigma} \mathcal{K}(l, \gg) + Align(G, \langle \rangle)$$

We note that as the alignment is of the empty trace and G, if G is accepting, the alignment will finish in constant time. As we have no initial bound for this alignment, we instead incrementally increase the bound from 0 in an iteratively deepening depth first search. This approach once again avoids exponential space complexity while yielding comparable performance to a breadth-first search [15].

6   Search Space Reduction

From Definition 3, it should be apparent that if an event has been executed and is still enabled afterwards, any immediately repeated executions will not further alter the marking. Thus execution of events that leave themselves enabled is idempotent, and any such idempotent model-moves will necessarily result in an overlapping sub-problem. However, a cache along all three parameter dimensions of the algorithm can be employed to avoid visiting duplicate nodes in the search space. The remainder of this section will thus focus on reducing two out of three of these dimensions.

One of these dimensions is the currently accumulated cost of aligning the prefix of the trace. Since the algorithm is clearly deterministic, we know that if two recursive calls have the same marking and the same trace suffix, they must necessarily find the same alignment. As the cost of aligning the suffix is simply added to the cost of aligning the prefix, we know that given these two traces with the same marking and trace suffix, only the one with the lowest accumulated alignment cost can yield an optimal alignment.

The other, more involved, dimension is the state, or marking, of the DCR graph. As this marking consists of three sets, each of which can potentially contain any permutation of events, a bound for the complexity of the state-space could be as high as $O(2^{3|E|})$. However, if we observe Definition 3, we notice that for a marking $M = (\mathsf{Ex}, \mathsf{Re}, \mathsf{In})$, neither Re nor In can be freely permuted. Both of these sets have limited states they can occur in based on the response and exclude/include relations respectively. Ex can, however, occur in significantly more permutations, as these will only be bounded by which events can be executed at any given time.

We therefore wish to reason about which permutations of Ex, and thus which markings, will behave the same during execution and therefore alignment. We call these markings *execution equivalent* with the formal definition given below.

Definition 9 (Execution equivalent markings). *Let $G = (E, M, L, \ell, \rightarrow\bullet, \bullet\rightarrow, \rightarrow+, \rightarrow\%)$ be a DCR graph, with marking $M = (\mathsf{Ex}, \mathsf{Re}, \mathsf{In})$. Let $M' = (\mathsf{Ex}', \mathsf{Re}', \mathsf{In}')$ be another possible marking of G. We say that M and M' are* execution equivalent *within G, if we have (1) $\mathsf{Re} = \mathsf{Re}'$, (2) $\mathsf{In} = \mathsf{In}'$, and (3) $\forall e \in E.\mathsf{Ex} \cap (\rightarrow\bullet e) = \mathsf{Ex}' \cap (\rightarrow\bullet e)$.*

Example 4 (Execution equivalent markings). *Consider graph G with marking $M = (\mathsf{Ex}, \mathsf{Re}, \mathsf{In})$ from Fig. 1. As only R1 and R2 can be excluded $(\rightarrow\%)$ and the ordering of their execution is fixed, there are only 2 permutations of In. Likewise, as there is only one response relation $(\bullet\rightarrow)$, the only two permutations of Re will be the one where no event is pending, or where only SN is pending.*

Now consider the number of permutations of Ex. The permutations are already slightly limited, since SP, R1, R2, R3 is a linear chain of conditions, meaning Ex over these events can't permute freely, but only increase linearly. When we also consider SN and DR, these can be intertwined much more freely, with executing DR leaving R3 enabled, yielding even more permutations.

However, since neither DR nor SO condition for any event, the markings that have these in Ex will be execution equivalent to the markings that do not. Therefore, if ignoring execution equivalent markings, the only permutations of Ex left are $\emptyset$, $\{SP\}$, $\{SP, R1\}$, and $\{SP, R1, R2\}$.

In order to safely ignore execution equivalent markings during alignment, we must ensure that the set of accepting runs will be identical between two equivalent markings within a graph, thus ensuring the same optimal alignments will be found. First we will prove that enabledness will be identical between execution equivalent markings within the same graph.

Lemma 1. *Let G be a graph with marking M and have that M and M' are execution equivalent within G. Let G' be G with M' substituted for M. For any event e, given $e \in \mathsf{enabled}(G)$ we also have $e \in \mathsf{enabled}(G')$.*

Proof. Looking at Definition 2, in order to have $e \in \mathsf{enabled}(G')$, we must show that (a) $e \in \mathsf{In}'$, (b) $\mathsf{In}' \cap (\rightarrow\bullet e) \subseteq \mathsf{Ex}'$, and (c) $\mathsf{In}' \cap (\rightarrow\diamond e) \subseteq E \setminus \mathsf{Re}'$.

As M and M' are execution equivalent within G we have (1) $\mathsf{In} = \mathsf{In}'$, (2) $\mathsf{Re} = \mathsf{Re}'$, as well as (3) $\mathsf{Ex} \cap (\rightarrow\bullet e) = \mathsf{Ex}' \cap (\rightarrow\bullet e)$.

As $e \in \mathsf{enabled}(G)$ we have (a) and (c) by trivial substitution with (1) and (2). Furthermore we have $In \cap (\rightarrow\bullet e) \subseteq \mathsf{Ex}$ which by set algebra and substitutions using (1) and (3) gives (b):

$$\mathsf{In} \cap (\rightarrow\bullet e) \subseteq \mathsf{Ex} \Longrightarrow$$
$$\mathsf{In} \cap (\rightarrow\bullet e) \cap (\rightarrow\bullet e) \subseteq \mathsf{Ex} \cap (\rightarrow\bullet e) \Longrightarrow$$
$$\mathsf{In} \cap (\rightarrow\bullet e) \subseteq \mathsf{Ex} \cap (\rightarrow\bullet e) \Longrightarrow$$
$$\mathsf{In}' \cap (\rightarrow\bullet e) \subseteq \mathsf{Ex}' \cap (\rightarrow\bullet e) \Longrightarrow$$
$$\mathsf{In}' \cap (\rightarrow\bullet e) \subseteq \mathsf{Ex}'$$

Thus showing $e \in \mathsf{enabled}(G')$. □

Next, we must also reason that execution equivalence persists during event execution.

Lemma 2. *Let G be a graph with marking M and have that M and M' are execution equivalent markings within G. Let G' be G with M' substituted for M. For any event e with $\mathsf{execute}(e, G) = G_e$ and $\mathsf{execute}(e, G') = G'_e$, we have that the markings M_e of G_e and M'_e of G'_e are execution equivalent withing G_e.*

Proof. As M and M' are execution equivalent within G we have (1) $\mathsf{In} = \mathsf{In}'$, (2) $\mathsf{Re} = \mathsf{Re}'$ and (3) $\mathsf{Ex} \cap (\rightarrow\bullet e) = \mathsf{Ex}' \cap (\rightarrow\bullet e)$.

Looking at Definition 3, we must trivially have $\mathsf{In}_e = \mathsf{In}'_e$ and $\mathsf{Re}_e = \mathsf{Re}'_e$ by (1) and (2).

From (3) we must have $(\mathsf{Ex} \cup \{e\}) \cap (\rightarrow\bullet e) = (\mathsf{Ex}' \cup \{e\}) \cap (\rightarrow\bullet e)$, as adding the same element to both executed sets can never alter the equivalence.

By definition of execution we have $\mathsf{Ex}_e = \mathsf{Ex} \cup \{e\}$ and $\mathsf{Ex}'_e = \mathsf{Ex}' \cup \{e\}$, hence $\mathsf{Ex}_e \cap (\rightarrow\bullet e) = \mathsf{Ex}'_e \cap (\rightarrow\bullet e)$, thus giving us that M_e and M'_e are execution equivalent within G_e. □

For our final lemma, we must prove that acceptance is also identical between execution equivalent markings within the same graph.

Lemma 3. *Let G be a graph with marking M and have that M and M' are execution equivalent markings within G. Let G' be G with M' substituted for M. Given $\mathsf{accepting}(G)$, we have $\mathsf{accepting}(G')$.*

Proof. As M and M' are execution equivalent within G we have $\mathsf{In} = \mathsf{In}'$, $\mathsf{Re} = \mathsf{Re}'$. If $\mathsf{accepting}(G)$, we have $\mathsf{In} \cap \mathsf{Re} = \emptyset$, thus $\mathsf{In}' \cap \mathsf{Re}' = \emptyset$, giving $\mathsf{accepting}(G')$. □

Now we can state and prove our theorem, allowing us to disregard markings that are execution equivalent to any already visited marking.

Table 1. Results from doing the 960000 trace alignments for the PDC2022 dataset. All results shown in ms.

Avg. time	Avg. time (cost 0)	Avg. time (rest)	Min time	Max time
0.5985	0.1013	1.3244	0.0078	268.7772

Theorem 1. *Let G be a graph with marking M and have that M and M' are execution equivalent markings within G. Let G' be G with M' substituted for M. Any run ϕ will be accepting in G if and only if it is accepting in G'.*

Proof. By recursion over ϕ in direction G to G'. The proof for direction G' to G is identical.

By Definition 6 we have that $e_1 \in$ enabled(G) and for $G \rightarrow_{e_1} G_1$, the run suffix $\langle e_2...e_n \rangle$ will be an accepting run of G_1. Using Lemma 1 we have $e_1 \in$ enabled(G'), and using Lemma 2 we have that for $G' \rightarrow_{e_1} G_1'$, the markings of G_1 and G_1' are execution equivalent withing G_1.

Applying these lemmas recursively, we get $G \rightarrow ... \rightarrow G_n$ and $G' \rightarrow ... \rightarrow G_n'$ with the markings of G_n and G_n' being execution equivalent within G_n. As ϕ is an accepting run of G we have accepting(G_n), allowing us to apply Lemma 3 yielding accepting(G_n'), thus showing that ϕ must also be an accepting run of G'.   □

Definition 9, Lemmas 1, 2 & 3 as well as **Theorem 1** have been defined and proven using the proof assistant Isabelle and can be found online[1].

7   Evaluation of Run-Time Performance

The algorithm and all optimizations as described in Sect. 6 has been implemented in Typescript[2] and benchmarked on the dataset from the Process Discovery Contest at ICPM 2022 (PDC2022). All experiments have been run on an Intel(R) Core(TM) i7-10700 CPU @ 2.90 GHz.

This dataset consists of 480 training logs generated from 96 underlying models with either no noise or 4 varying degrees of noise added. For each of these training logs there were given a test log and a base log to use for classification, each with 1000 traces.

For our experiments we mined a model from each training log and performed alignment of the mined model and each trace of the corresponding test and base logs. In total this yielded 960000 alignments across 480 models with varying model size and degree of non-conformance. We have defined model size as the number of constraints in the DCR graph, and degree of non-conformance as the resulting cost of the found optimal alignment. The aggregated results of these runs can be seen in Table 1. Notably, we can present an average alignment time of 0.6 ms, which only increases to 1.3ms if we disregard perfectly fitting traces.

---

[1] https://www.isa-afp.org/entries/DCR-ExecutionEquivalence.html.
[2] https://github.com/Axel0087/DCR-Alignment.

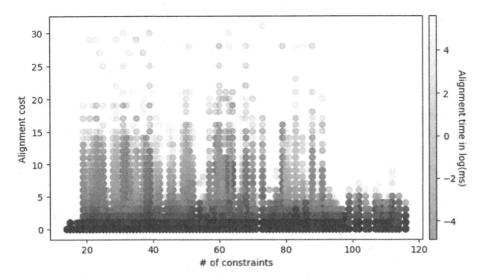

Fig. 2. A heatmap showing alignment time in ms as a function of alignment cost and model size measured in number of constraints. Shown on a logarithmic scale, due to outliers as seen in Fig. 3

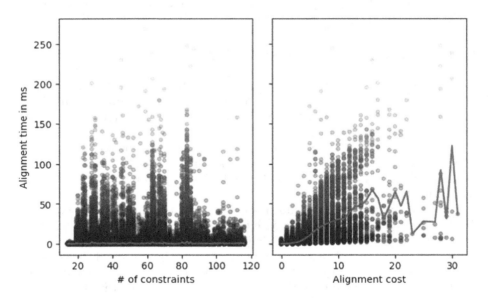

Fig. 3. Alignment times shown in ms as functions of alignment cost and model size measured in number of constraints. The average values are plotted as red lines. (Color figure online)

The full results can be seen in Fig. 2, where the timing of each run has been plotted as a function of model size and alignment cost respectively. From this figure we can draw a few key points: (1) it seems that the running time is more dependent on the degree of non-conformance than on model size. (2) even though the alignment cost seems to be the main factor in determining the running time of the algorithm, we still see quite some variance. Notably, this variance seems to manifest in vertical stripes, indicating that some models are much faster to align, independently of their size. From this we can draw that the composition of the particular model may say more than purely model size.

We can further inspect the effect of these parameters by examining Fig. 3. We see that it does indeed appear that there is a clear linear tendency between alignment cost and running time. We can, however, note that while there is this linear tendency, there is an extreme variance that increases with alignment cost. Looking at model-size, we once again see this tendency of streaks appearing. For any model size, however, the average running time remains low, confirming that model size in itself is not a determining factor of the running time of alignment computation, but that some model compositions may lead to higher variance in alignment running times.

8 Applying Alignment to Test Driven Modeling for the Dreyers Foundation Application Process

In this section we present a case study demonstrating the application of the alignment algorithm to a large DCR Graph in use as the executable process model in a real-life electronic case management system. We focus on solving the computational problem of checking open tests in test driven modelling, mapped to an alignment problem. This allows us to show how the algorithm performs when confronted with unusual cost functions that significantly increase the difficulty of the search. We start by introducing the case and model, continue by giving an informal introduction to open tests, and finish by presenting and discussing the results of checking open tests on the model.

8.1 The Dreyers Foundation Application Process

The Dreyers Foundation is a Danish funding agency specializing in grants for projects and activities promoting the development of the lawyer and architect professions[3], and their interaction with society. As of 2023 the foundation aims to award an approximate yearly total of 50M Danish kroner (6.71M euros).

As part of an IT overhaul in 2013 their internal processes regarding both application assessment, rejections, grants and payouts were modelled as DCR Graphs [9,10]. In broad terms, applications are accepted in rounds. In each round a caseworker first pre-screens applications. The remaining applications are independently reviewed by 2–4 reviewers, at least one of which must be an architect or

---

[3] https://dreyersfond.dk/fonden/oekonomi/.

a lawyer, depending on the type of application. Once all reviews are in, the Foundation's board decides on which applications to accept during a board meeting. Payouts can happen as a single payment or in several tranches, until the grant period expires and an end-report is produced. In practice, board members and reviewers overlap, and reviews might happen or be amended at the board meeting. Figure 4 shows the DCR Graph modelling their main application process.

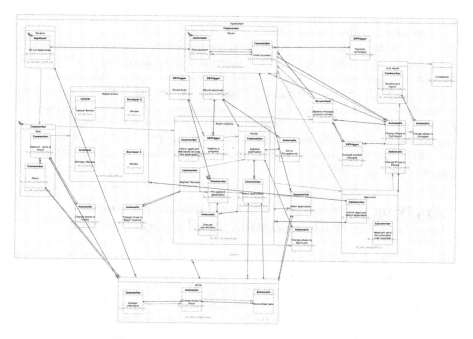

Fig. 4. Original DCR Graph modelling the application process of the Dreyers Foundation. Time- and data-perspectives have been removed by the authors.

8.2   Test Driven Modelling

Test driven modelling deals with the complexity of declarative modelling notations by first asking the modeller to define a number of (desired) or negative (undesired) example runs (tests) of a process. During the modelling of the process these tests are checked against the evolving model, showing if they pass or fail. When all positive tests pass and all negative tests fail, the model can be seen as satisfying the tests. Allowing the modeller to think in terms of simple traces and checking these automatically against the model helps untangle complex interactions between the constraints.

Traditional test driven modelling [26] defines tests as full runs of the process, thereby acting as a kind of system tests. Open tests [23] on the other hand are essentially unit tests for process models. They allow the tester to define a partial trace of the system and the *context* of relevant activities for that trace.

Checking if a system passes an open test is a model checking problem. For a positive test we check if there exists a trace in the whole system which corresponds to the test case when projected over the context. For negative tests we check that no such trace exists. This model checking task maps neatly to an alignment checking problem when using the cost function that assigns cost ∞ to log-moves and model-moves inside the context, and cost 0 to synchronous-moves and model-moves outside the context. This cost function effectively implements the projection of the context, and fully disallows deviations from the model.

8.3   Results and Discussion

To evaluate the practical applicability of our alignment algorithm to checking open tests, we defined a set of 20 tests (10 positive and 10 negative). We checked alignment for each test twice: once with no depth limit (exploring the entire state space if no alignment was found) and once with the depth limit set to 100. Both settings yielded the same outcome for each test. Table 2 shows an example of one of these tests and which alignment was returned by the algorithm. This was the fastest test to complete, rounding to 0 ms. For the details of all tests we refer to the GitHub repo[4].

Table 2. A test case and it's alignment, with the activity abbreviations: **Re**ject application, **An**søger informeret, **F**ill **o**ut application, **Ap**proved - to board, **L**awyer **R**eview, **R**egister **D**ecision, **C**hange phase to **A**bort, **An**onymiser data.

trace:	$\langle Re, Ai \rangle$
context:	$\{Ap, Re, Ai\}$

σ':	$\gg$	$\gg$	$\gg$	$\gg$	Re	Ai	$\gg$	$\gg$
ϕ:	Foa	Ap	LR	RD	Re	Ai	CA	An

(alignment)

Table 3. Aggregated results for running all open tests

Outcome	Count	Time for depth: 100 (s)	Time for depth: inf (s)
Alignment found	8	0.005	0.003
No alignment found	12	3.946	106.163

The aggregated results are shown in Table 3. Tests are grouped based on if an alignment was found or not. For each group we show the number of tests in the group and the average running time in seconds for both runs. Tests that did find an alignment completed on average in 5 and 3 ms. Tests that did not find an alignment were significantly slower: 4 s on average when limited in depth, 106 s on average when forced to explore the entire state space. It is notable however that the depth-limited search completed in a reasonable time and resulted in the correct answer for each test. While in general bounding the search means

---

[4] https://github.com/Axel0087/DCR-Alignment/blob/main/runDreyers.ts.

sacrificing the guarantee of soundness of a true negative or false positive result, we posit that in most cases it is unlikely that the only counter examples are significantly longer than the total number of activities in the model. In practice this means that a modelling tool can check tests up-to a configurable level of confidence (given as max depth) in real-time and be prompted to do complete checks by the user when convenient.

9   Conclusion

We introduced an algorithm for computing the optimal alignment between a DCR Graph and a trace. The provided algorithm has been implemented in Typescript and evaluated on the dataset for the process discovery contest at ICPM 2022 consisting of 480 logs. The experiment showed that the algorithm is highly efficient, making it a useful tool for the conformance checking and noise filtering of declarative process models. We also presented a case study where we used alignment to check open tests on the Dreyers Foundation case. In future work we expect to further pursue this angle and explore the development of improved alignment algorithms for DCR Graphs that can handle unusual cost functions.

Acknowledgements. This work is supported by Digital Research Centre Denmark (DIREC). We acknowledge Søren Debois for sharing his Isabelle/HOL DCR Graph formalization.

References

1. van der Aalst, W.M., Pesic, M.: DecSerFlow: towards a truly declarative service flow language. In: Bravetti, M., Nunez, M., Zavattaro, G. (eds.) Proceedings of Web Services and Formal Methods (WS-FM 2006), vol. 4184, pp. 1–23 (2006)
2. Adriansyah, A.: Aligning observed and modeled behavior. Ph.D. thesis, Mathematics and Computer Science (2014)
3. Awad, A., Raun, K., Weidlich, M.: Efficient approximate conformance checking using trie data structures. In: 2021 3rd International Conference on Process Mining (ICPM), pp. 1–8 (2021)
4. Back, C.O., Slaats, T., Hildebrandt, T.T., Marquard, M.: Discover: accurate and efficient discovery of declarative process models. Int. J. Softw. Tools Technol. Transf. **24**, 1–25 (2021)
5. Boltenhagen, M., Chatain, T., Carmona, J.: Optimized sat encoding of conformance checking artefacts. Computing **103**(1), 29–50 (2021)
6. Carmona, J., van Dongen, B., Solti, A., Weidlich, M.: Conformance Checking. Springer, Cham (2018). https://doi.org/10.1007/978-3-319-99414-7
7. De Giacomo, G., Maggi, F.M., Marrella, A., Sardina, S.: Computing trace alignment against declarative process models through planning. In: Proceedings of the International Conference on Automated Planning and Scheduling, vol. 26, no. 1, pp. 367–375 (2016)
8. de Leoni, M., Maggi, F.M., van der Aalst, W.M.: An alignment-based framework to check the conformance of declarative process models and to preprocess event-log data. Inf. Syst. **47**, 258–277 (2015)

9. Debois, S., Hildebrandt, T., Slaats, T., Marquard, M.: A case for declarative process modelling: Agile development of a grant application system. In: 2014 IEEE 18th International Enterprise Distributed Object Computing Conference Workshops and Demonstrations, pp. 126–133. IEEE (2014)
10. Debois, S., Slaats, T.: The analysis of a real life declarative process. In: IEEE Symposium Series on Computational Intelligence, SSCI 2015, pp. 1374–1382 (2015)
11. Fani Sani, M., van Zelst, S.J., van der Aalst, W.M.P.: Conformance checking approximation using subset selection and edit distance. In: Advanced Information Systems Engineering, pp. 234–251 (2020)
12. Felli, P., Gianola, A., Montali, M., Rivkin, A., Winkler, S.: Cocomot: conformance checking of multi-perspective processes via SMT. In: Business Process Management, pp. 217–234 (2021)
13. Hildebrandt, T.T., Normann, H., Marquard, M., Debois, S., Slaats, T.: Decision modelling in timed dynamic condition response graphs with data. In: Business Process Management Workshops, pp. 362–374 (2022)
14. Jagadeesh Chandra Bose, R., van der Aalst, W.M.: Process diagnostics using trace alignment: opportunities, issues, and challenges. Inf. Syst. 37(2), 117–141 (2012). Management and Engineering of Process-Aware Information Systems
15. Korf, R.E.: Depth-first iterative-deepening: an optimal admissible tree search. Artif. Intell. 27(1), 97–109 (1985)
16. Lee, W.L.J., Verbeek, H., Munoz-Gama, J., van der Aalst, W.M., Sepúlveda, M.: Recomposing conformance: closing the circle on decomposed alignment-based conformance checking in process mining. Inf. Sci. 466, 55–91 (2018)
17. de Leoni, M., Maggi, F.M., van der Aalst, W.M.P.: Aligning event logs and declarative process models for conformance checking. In: Business Process Management, pp. 82–97 (2012)
18. de Leoni, M., Munoz-Gama, J., Carmona, J., van der Aalst, W.M.P.: Decomposing alignment-based conformance checking of data-aware process models. In: On the Move to Meaningful Internet Systems: OTM 2014 Conferences, pp. 3–20 (2014)
19. Lu, X., Fahland, D., van der Aalst, W.M.P.: Conformance checking based on partially ordered event data. In: Business Process Management Workshops, pp. 75–88 (2015)
20. Reijers, H.A., Slaats, T., Stahl, C.: Declarative modeling-an academic dream or the future for BPM? In: Business Process Management, pp. 307–322 (2013)
21. Reißner, D., Conforti, R., Dumas, M., La Rosa, M., Armas-Cervantes, A.: Scalable conformance checking of business processes. In: On the Move to Meaningful Internet Systems, OTM 2017 Conferences, pp. 607–627 (2017)
22. Sani, M.F., Gonzalez, J.J.G., van Zelst, S.J., van der Aalst, W.M.: Conformance checking approximation using simulation. In: 2020 2nd International Conference on Process Mining (ICPM), pp. 105–112 (2020)
23. Slaats, T., Debois, S., Hildebrandt, T.: Open to change: a theory for iterative test-driven modelling. In: Business Process Management, pp. 31–47 (2018)
24. van Dongen, B.F., De Smedt, J., Di Ciccio, C., Mendling, J.: Conformance checking of mixed-paradigm process models. Inf. Syst. 102, 101685 (2021)
25. van Zelst, S.J., Bolt, A., Hassani, M., van Dongen, B.F., van der Aalst, W.M.P.: Online conformance checking: relating event streams to process models using prefix-alignments. Int. J. Data Sci. Anal. 8(3), 269–284 (2019)
26. Zugal, S., Pinggera, J., Weber, B.: Creating declarative process models using test driven modeling suite. In: IS Olympics: Information Systems in a Diverse World, pp. 16–32 (2012)

Can I Trust My Simulation Model? Measuring the Quality of Business Process Simulation Models

David Chapela-Campa[1]([envelope]), Ismail Benchekroun[2], Opher Baron[3], Marlon Dumas[1], Dmitry Krass[3], and Arik Senderovich[4]

[1] University of Tartu, Tartu, Estonia
{david.chapela,marlon.dumas}@ut.ee
[2] Department of Statistics, Faculty of Arts and Science, University of Toronto, Toronto, Canada
ismail.benchekroun@mail.utoronto.ca
[3] Rotman School of Management, University of Toronto, Toronto, Canada
{opher.baron,dmitry.krass}@rotman.utoronto.ca
[4] School of Information Technology, York University, Toronto, Canada
sariks@yorku.ca

Abstract. Business Process Simulation (BPS) is an approach to analyze the performance of business processes under different scenarios. For example, BPS allows us to estimate what would be the cycle time of a process if one or more resources became unavailable. The starting point of BPS is a process model annotated with simulation parameters (a BPS model). BPS models may be manually designed, based on information collected from stakeholders and empirical observations, or automatically discovered from execution data. Regardless of its origin, a key question when using a BPS model is how to assess its quality. In this paper, we propose a collection of measures to evaluate the quality of a BPS model w.r.t. its ability to replicate the observed behavior of the process. We advocate an approach whereby different measures tackle different process perspectives. We evaluate the ability of the proposed measures to discern the impact of modifications to a BPS model, and their ability to uncover the relative strengths and weaknesses of two approaches for automated discovery of BPS models. The evaluation shows that the measures not only capture how close a BPS model is to the observed behavior, but they also help us to identify sources of discrepancies.

Keywords: Business process simulation · Process mining

1 Introduction

Business Process Simulation (BPS) is a technique for estimating the performance of business processes under different scenarios [9]. BPS enables analysts to address questions such as "what would be the cycle time of a process if one or more resources became unavailable?" or "what would be the impact of automating an activity on the waiting times of other activities in the process?".

© The Author(s) 2023
C. Di Francescomarino et al. (Eds.): BPM 2023, LNCS 14159, pp. 20–37, 2023.
https://doi.org/10.1007/978-3-031-41620-0_2

The starting point of BPS is a process model, e.g. in the Business Process Model and Notation (BPMN)[1], enhanced with simulation parameters [21] (herein, a BPS model). These simulation parameters capture, for example, the processing times of each activity or the rate at which new process instances (cases) are created.

BPS models may be manually created based on information collected via interviews or empirical observations, or they may be automatically discovered from execution data recorded in process-aware information systems (event logs) [6,7,17,22]. Regardless of the origin, a key question when using a BPS model is how to assess its quality. This question is particularly relevant when tuning the simulation parameters. Several approaches have been proposed to address this problem. However, these approaches are either manual and qualitative [22] or they produce a single number that does not allow one to identify the source(s) of deviations between the BPS model and the observed reality [6,10].

In this paper, we study the problem of automatically measuring the quality of a BPS model w.r.t. its ability to replicate the observed behavior of a process as recorded in an event log. We advocate a multi-perspective approach to this problem, thus proposing a set of quality measures that address different perspectives of process performance. The starting point is the idea that a good BPS model is one that generates traces consisting of events similar to the observed data. Accordingly, the proposed approach maps an event log produced by the BPS model and an event log recording the observed behavior into histograms or time series capturing a given perspective, and then compares the resulting histograms or time series using a distance metrics.

We conduct a two-fold evaluation of the measures using synthetic and real-life datasets. In the synthetic evaluation, we study the ability of the proposed measures to discern the impact of modifications to a BPS model, whereas in the real-life evaluation, we analyze their ability to uncover the relative strengths and weaknesses of two approaches for automated discovery of BPS models. Our results show that the measures not only capture how close a BPS model is to the observed behavior, but also help us identify sources of discrepancies.

The rest of the paper is structured as follows. Section 2 gives an overview of prior research related to the discovery and evaluation of BPS models. Section 3 introduces relevant process mining concepts and distance measures. Section 4 analyzes the problem and proposes a set of measures of quality of BPS models. Section 5 discusses the empirical evaluation, and Sect. 6 draws conclusions and sketches future work.

2 Background

2.1 Business Process Simulation Models

A BPS model consists of *i)* a stochastic control-flow model, *ii)* an activity performance model, and *iii)* an arrival and congestion model. The stochastic model

---

[1] https://www.bpmn.org/.

is composed of a process model (e.g., a BPMN model or a Petri net) and a stochastic component capturing the probability of occurrence of each path in the model (branching probabilities). In a BPS model, the stochastic model is enhanced by adding an activity performance model, which determines the duration of the activity instances (e.g., by associating a parametric distribution to each activity in the model). Finally, in a BPS model, an arrival and congestion model determines when new cases arrive in the system, and when the execution of each enabled activity instance starts, given the available resource capacity.

Traditionally, BPS models are constructed manually by experts. Recent approaches advocate for the automated discovery of BPS models from event logs. Below, we consider two such approaches. The first one, namely *SIMOD* [6], starts by constructing a stochastic process model by applying the SplitMiner algorithm [3] to discover a BPMN model from the input log, and replaying the traces of the log to calculate the branching probabilities. Next, SIMOD discovers the activity performance model (activity duration distributions) and a congestion model consisting of: *i)* a case inter-arrival time distribution; *ii)* a set of resources, their availability timetables, and the activities they perform; and *iii)* the distribution of extraneous waiting times between activities (i.e. waiting times not attributable to congestion) [8]. Once a BPS model is discovered, its parameters are tuned to fit the data using a Bayesian hyper-parameter optimizer.

The second BPS model discovery technique we consider is *ServiceMiner*©. ServiceMiner operates in three steps: *i)* data preprocessing, where techniques for data cleaning and categorical feature encoding are applied; *ii)* data enhancement, where new data attributes that capture trend, seasonality, and system congestion are created using methods described in [23]; and *iii)* model learning, where the BPS model is created by combining process discovery, queue mining (learning of queueing building blocks from data), and machine learning (to boost the accuracy of arrival and activity time generation). For process discovery, ServiceMiner mines a Markov chain, estimating the case routing probabilities between consecutive activity pairs. An abstraction mechanism allows for filtering out rare activities, paths, and transitions. Next, using queue mining, the various queueing building blocks are fitted from data, by using techniques described in [24]. Lastly, ServiceMiner applies a machine learning technique that uses congestion features that come from queueing theory, which, via cross-validation, leads to accuracy improvements when generating inter-arrival times and activity durations.

While the evaluation reported below focuses on BPS models discovered by SIMOD and ServiceMiner, the proposed measures can be used to assess the quality of any model that generates event logs. For example, the proposal can also be used to evaluate generative deep learning models of business processes [7]. On the other hand, it cannot be used to assess coarse-grained BPS models, e.g. based on system dynamics [20], unless these are refined to generate event logs.

2.2   Quality Measures for Business Process Simulation Models

Leemans et al. [12,13] and Burke et al. [5] studied the evaluation of stochastic models using, among other measures, Earth Movers' Distance. However, they focus solely on the control-flow perspective, and their purpose is mainly conformance checking. In this paper, we focus on the assessment of BPS model quality considering both temporal and control-flow dimensions.

Prior studies have considered the evaluation of BPS models. Rozinat et al. [22] perform an evaluation of BPS models following manual comparisons. However, they do not propose concrete and automatable evaluation measures. Camargo et al. [7] study the performance of data-driven simulation and deep learning techniques, proposing measures that combine the control-flow and the temporal perspectives. The latter measures are not scalable, and they do not identify the sources of discrepancies between BPS models. To overcome these shortcomings, we propose an approach that views the process from different perspectives and provides a separation of concerns between the three BPS model components (the stochastic model, the activity performance model, and the congestion model). We then propose efficient measures for each component.

3 Preliminaries

3.1 Event Logs

Modern enterprise systems maintain records of business process executions, which can be used to extract *event logs*: sets of timestamped events capturing the execution of the activities in a process [9]. We assume that each event record in the log relates to a case, an activity, and an activity start and end timestamp (as in Table 1). However, the proposal of this paper can be generalized to include other life-cycle events (e.g., activity enablement or cancellation). We shall refer to events and activity instances interchangeably, even though they could mean different things in other contexts. Let $\mathcal{E}$ be the universe of events, C be the universe of case identifiers, A be the set of possible activity labels, and T be the time domain.

Definition 1 (Event Log). An *event log* (denoted by $\mathcal{L}$) is a set of executed activity instances, $E \subseteq \mathcal{E}$, with each event having a schema $\sigma_\mathcal{E} = \{\xi, \alpha, \tau_{start}, \tau_{end}\}$, that assigns the following attribute values to events:

- $\xi : \mathcal{E} \to C$ assigns a case identifier,
- $\alpha : \mathcal{E} \to A$ assigns an activity label,
- $\tau_{start} : \mathcal{E} \to T$ assigns the start timestamp of the executed activity, and,
- $\tau_{end} : \mathcal{E} \to T$ assigns the end timestamp of the executed activity.

Note that the transformation from a traditional event log that contains only a single timestamp to our notion of an event log is straightforward (see [18]).

3.2 Measures for Time-Series and Histogram Comparison

To analyze the temporal performance of a process, an event log can be mapped to a variety of time series (e.g. activity starts, activity ends). Accordingly, we

Table 1. Example of 6 events of an event log from a Procure-to-Pay Process

Case	Activity	Start	End	Case	Activity	Start	End
111	CreatePO	7:12:00	7:20:00	222	CreatePO	10:12:00	10:47:00
111	ApprovePO	9:30:00	10:12:00	222	PO_Rejected	10:47:00	11:26:00
111	GoodsReceived	10:12:00	10:44:00	333	CreatePO	9:26:00	10:32:00

consider the use of techniques to quantify the distance between two time series, $x = (x_1, \ldots, x_n)$, and $y = (y_1, \ldots, y_m)$, of (potentially different) lengths n and m, respectively. To this end, one may employ various measures, such as computing $||x - y||_l$ in any of the standard norms (i.e., $l = 1, 2, \infty$).[2] These comparisons would only be possible after padding the shorter time series. In addition, standard norms do not capture the temporal differences between the two time series. For example, a temporal shift in x vs y may produce $l1$ or $l2$ norm, but represents a significant failure in the model to capture time-series patterns properly. To overcome the two limitations, namely the need for padding, and ignoring temporal differences, a natural measure is the Wasserstein Distance (WD) [19]; in this work, we consider two variations of WD.

- *Earth Mover's Distance (EMD)* [13] computes the effort it takes to balance two vectors x and y of different lengths, treating each entry x_i, y_j as 'masses' to move from location to location until the two time series are equal. EMD does not assume that $\sum_i x_i = \sum_j y_j$, i.e., the sum of the 'earth mass' to be moved can be different; in such cases, we add a penalty for creating redundant mass to fill in gaps. Herein, we consider the EMD problem with absolute distance measure [14].
- *1st Wasserstein Distance (1WD)* [14] is a computationally efficient variation of the EMD. It introduces the constraint that the sum of masses must be the same in x and y (i.e., the constraint $\sum_i x_i = \sum_j y_j$ is enforced). 1WD is suitable for comparing empirical distribution functions (histograms), since the sum of the mass in each is 1.

When comparing two histograms, we let $f = (f_1, \ldots, f_n)$ be the n normalized frequency values of the first histogram, and let $g = (g_1, \ldots, g_m)$ be the m normalized frequencies of the second histogram. We treat the two histograms f and g similarly to the two time series x and y, and employ 1WD distance, since the sum of masses is 1 (EMD and 1WD lead to the same results).

4   Framework for Measuring BPS Model Quality

In this part, we develop an approach for measuring the quality of BPS models. There are two main reasons why directly evaluating a BPS model would be impractical: *i)* typically, the 'true' BPS model of the process is not available (and

---

[2] See Sect. 2.2 in [15] for a survey on time-series comparison measures.

often does not exist), thus, we cannot perform a model-to-model comparison; *ii)* different simulation engines (e.g., BIMP [1], Prosimos [16]) support different BPS model formats, hindering a generic comparison of BPS models. Therefore, we propose to generate a collection of logs simulated with the BPS model under evaluation, and compare them to event logs of the actual system (i.e., the system that the model aims to mimic). Consequently, one can apply a 'transitive argument': the 'closer' the simulated logs are to the actual data, the better is the model. In other words, we treat the (test) data as our 'ground truth', since useful models are supposed to be faithful generators of 'reality'.

Two challenges arise when measuring the quality of a BPS model: *i)* a model can be very close to the data in one aspect (e.g., control-flow), yet very different in another (e.g., in inter-arrival times), and, *ii)* a model can generate many realities as it is probabilistic in nature (durations and routing are stochastic), while the data consists of a single realization. To overcome the first limitation, we propose a collection of measures to quantify the distance across multiple process perspectives. Specifically, we shall consider control-flow, temporal, and congestion distance measures. As for the second limitation, our approach is to generate multiple event logs simulating the 'ground-truth' event log (i.e., with the same number of cases, and starting from the same instant in time), and use the generated logs to construct confidence intervals for each of our measures.

For all measures, we consider a collection of K generated logs (GLogs) that came from K simulation runs, and compare these K GLogs to the actual test event log (ALog) that, importantly, was not used to construct the BPS model. Below, we outline control-flow, temporal, and congestion, discuss their rationale, and briefly provide their computation by comparing GLogs and an ALog.[3]

4.1   Control-Flow Measures

To evaluate the quality w.r.t. the control-flow perspective (i.e., the capability of the model to represent the event sequences in the actual event log), we propose two measures. The first one, namely control-flow log distance (CFLD), is a variation of a measure introduced by Camargo et al. in [7]. CFLD precisely penalizes the differences in the control-flow by pairing each case in GLog to the case in the ALog that minimizes the sum of their distances. However, due to its steep computational complexity, we propose an additional measure, the n-gram distance (NGD), that approaches the problem in a more efficient way.

Control-Flow Log Distance (CFLD). Given two logs $\mathcal{L}_1$ and $\mathcal{L}_2$ with the same number of cases, we compute the average distance to transform each case in $\mathcal{L}_1$ to another case in $\mathcal{L}_2$ (see [7] for a description of a similarity version of this measure). To compute this measure, we first transform each process case of $\mathcal{L}_1$ and $\mathcal{L}_2$ to their corresponding activity sequences, abstracting from temporal information. Then, we compute the Damerau-Levenshtein (DL) distance [26] between each pair of cases i, j belonging to $\mathcal{L}_1$ and $\mathcal{L}_2$, respectively, normalizing

---

[3] For clarity, each measure is described as a distance between two event logs. However, we propose to report the average of K individual comparisons (GLog against ALog).

them by the maximum of their lengths (obtaining a value in $[0, 1]$). Subsequently, we compute the matching between the cases of both logs (such that each i is matched to a different j, and vice versa) minimizing the sum of distances using the Hungarian algorithm for optimal alignment. The CFLD is the average of the normalized distance values.

CFLD requires pairing each case in the simulated log with a case in the original log, minimizing the total sum of distances. The computational complexity of computing the DL-distance for all possible pairings is $O(N^2 \times MTL^3)$ where N is the number of traces in the logs (assuming both logs have an equal number of cases, which holds in our setting) and MTL is the maximum trace length. Since all pairings are put into a matrix to compute the optimal alignment of cases (the one that minimizes the total sum of distances), CFLD's memory complexity is quadratic on the number of cases. The optimal alignment of traces using the Hungarian algorithm has a cubic complexity on the number of cases.

N-Gram Distance (NGD). Leemans et al. [13] measure the quality of a stochastic process model by mapping the model and a log to their Directly-Follows Graph (DFG), viewing each DFG as a histogram, and measuring the distance between these histograms. We note that the histogram of 2-grams of a log is equal to the histogram of its DFG.[4] Given this observation, we generalize the approach of [13] to n-grams, noting that the histogram of n-grams of a log is equal to the $(n-1)^{th}$-Markovian abstraction of the log [2]. In other words, the histogram of 2-grams is the 1st-order Markovian abstraction (the DFG), the histogram of 3-grams is the 2nd-order Markovian abstraction, and so on.

Given two logs $\mathcal{L}_1$ and $\mathcal{L}_2$, and a positive integer n, we compute the difference in the frequencies of the n-grams observed in $\mathcal{L}_1 \bigcup \mathcal{L}_2$. To compute this measure, we transform each case of $\mathcal{L}_1$ and $\mathcal{L}_2$ to its corresponding activity sequences, abstracting temporal information, and adding $n-1$ dummy activities to both start and end of the case (e.g., 0-A-B-C-0 for case A-B-C and $n = 2$). Then, we compute all sequences of n activities (n-grams) observed in each log, and measure their frequency. Finally, we compute the sum of absolute differences between the frequencies of each computed n-gram, and normalize the total distance by the sum of frequencies of all n-grams in both logs (obtaining a value in $[0, 1]$).

For example, consider $\mathcal{L}_1$ having three cases A-B-C-D, and $\mathcal{L}_2$ having three cases A-B-E-D. Given $n = 2$, the observed n-grams are 0-A, A-B, B-C, C-D, and D-0 in $\mathcal{L}_1$; and 0-A, A-B, B-E, E-D, and D-0 in $\mathcal{L}_2$ (each one with a frequency of three). The n-grams B-C, C-D, B-E, and E-D have a frequency of 3 in one log, and 0 in the other, thus, the NGD between $\mathcal{L}_1$ and $\mathcal{L}_2$ is 0.4 (12 divided by 30). By adding dummy activities, all activity instances have the same weight in the measure, as each of them is present in n n-grams. Otherwise, the first and last activity instances of each trace would be present only in one n-gram. Note that we do not use the EMD to compute the NGD, because the order of the n-grams in the histogram is irrelevant and EMD would take this order into account.

---

[4] An n-gram is a vector of n consecutive activities in a trace of a log. A 2-gram is a pair of consecutive activities in a log. Every arc in the DFG of a log is a 2-gram of the log and vice-versa.

NGD is considerably more efficient than CFLD, as the construction of the histogram of n-grams is linear on the number of events in the log, and the same goes for computing the differences between the n-gram histograms.

4.2   Temporal Measures

We propose three measures that assess the ability of a BPS model to capture the temporal performance perspective, based on the idea that the time series of events generated by a BPS model should be similar to the time series of the test data, with respect to seasonality, trend, and time-to-event.

The first two measures come from time-series analysis, where most approaches in the literature (e.g., SARIMA) decompose the time series into components of trend, seasonality, and noise [4]. We follow a similar path by analyzing the trend (comparing the absolute distribution of events), and the seasonality (comparing the circadian distribution of events). The third measure comes from time-to-event (or survival) analysis [11], a field in statistics that analyzes the behavior of individuals from some point in time until an event of interest occurs. Specifically, we are interested in analyzing the capability of the simulator to correctly reconstruct the occurrence of events (and their timestamps) from the beginning of the corresponding case to its end. Below, we provide the details of the three aforementioned measures.

Absolute Event Distribution (AED). Given two event logs $\mathcal{L}_1$ and $\mathcal{L}_2$, we transform the events into a time series by binning the timestamps in the event log (both start and end timestamps) by date and hour of the day (e.g., timestamps between '02/05/2022 10:00:00' and '02/05/2022 10:59:59' will be placed into the same bin). Let $i = 1, \ldots, B$ be the hours from the first until the last timestamp in $\mathcal{L}_1 \bigcup \mathcal{L}_2$ (i.e., the timeline of both logs), and $dh(\tau(e))$ a function returning the i corresponding to the date and hour of the day of a timestamp of event e (for brevity, we refer to both τ_{start} and τ_{end} as τ), the binning procedure is as follows,

$$x_i = |\{e \in \mathcal{L}_1 \mid dh(\tau(e)) = i\}|, \quad y_i = |\{e \in \mathcal{L}_2 \mid dh(\tau(e)) = i\}| \qquad (1)$$

Finally, the AED distance between $\mathcal{L}_1$ and $\mathcal{L}_2$ corresponds to the EMD between $x_1, \ldots, x_B$ and $y_1, \ldots, y_B$.

Circadian Event Distribution (CED). Given two event logs $\mathcal{L}_1$ and $\mathcal{L}_2$, we partition each log into sub-logs by the day of the week (Mon-Sun). Let $wd(\tau(e))$ be a function that returns the day of the week for timestamp $\tau(e)$. Then, for $i = 1, \ldots, 7$, we obtain the corresponding sub-logs as follows,

$$\mathcal{L}_{1,i} = \{e \in \mathcal{L}_1 \mid wd(\tau(e)) = i\}, \quad \mathcal{L}_{2,i} = \{e \in \mathcal{L}_2 \mid wd(\tau(e)) = i\} \qquad (2)$$

Subsequently, we bin each sub-log into hours with Eq. (1) using $h(\tau(e))$, a function returning the hour of the day of a timestamp of event e, instead of $dh(\tau(e))$. In this way, all the timestamps recorded on any Monday between '10:00:00' and '10:59:59' will be placed in the same bin), obtaining $x_{1,d}, \ldots, x_{B,d}$ and

$y_{1,d}, \ldots, y_{B,d}$ with $d \in \{1, \ldots, 7\}$. Finally, the CED distance between $\mathcal{L}_1$ and $\mathcal{L}_2$ corresponds to the average of the EMD between $x_{1,d}, \ldots, x_{B,d}$ and $y_{1,d}, \ldots, y_{B,d}$ with $d \in \{1, \ldots, 7\}$.

Relative Event Distribution (RED). Here, we wish to analyze the ability of the simulator to mimic the temporal distribution of events w.r.t. the origin of the case (i.e., the case arrival). To this end, given two event logs $\mathcal{L}_1$ and $\mathcal{L}_2$, we offset all log timestamps from their corresponding case arrival time (the first timestamp in a case is set to time 0, the second one is set to the inter-event time from the first, etc.). Formally, let $a(\xi(e)) = \min_{t'} \{t' \mid t' = \tau_{start}(e') \wedge e' \in \mathcal{L} \wedge \xi(e') = \xi(e)\}$ be the arrival time of a case associated with an event in the log. Then, the relative event times $\rho(e)$ are defined as,

$$\rho(e) = \tau(e) - a(\xi(e)), \tag{3}$$

with $\tau(e)$ being $\tau_{start}(e)$ for start times, and $\tau_{end}(e)$ denoting end times. We apply Eq. (3) to the timestamps in $\mathcal{L}_1$ and $\mathcal{L}_2$ and, for each log, discretize the resulting $\rho(e)$ into hourly bins (e.g., those durations between 0 and 3,599 s go to the same bin). Finally, the RED distance between $\mathcal{L}_1$ and $\mathcal{L}_2$ corresponds to the EMD between the discretized $\rho(e)$ of each log.

4.3   Congestion Measures

To measure the capability of a model to represent congestion, we rely on queueing theory, a field in applied probability that studies the behavior of congested systems [25]. The workload in a queueing system is dominated by two factors: the *arrival rate* of cases over time, and the *cycle time*, which is the length-of-stay of a case in the system. Below, we propose two measures to compare the two workload components over pairs of event logs by comparing the time series of the arrivals, and the distribution of the cycle times (assuming that its variability is captured by the arrivals time-series comparison).

Case Arrival Rate (CAR). This measure compares case arrival patterns (shape) and counts (number of arrival per bin). Given two event logs $\mathcal{L}_1$ and $\mathcal{L}_2$, we use the function $a(c), c \in C$ to obtain the sets of arrival timestamps of each log. Subsequently, we bin them using Eq. (1) (timestamps between '02/05/2022 10:00:00' and '02/05/2022 10:59:59' are placed in the same bin), obtaining two vectors $x_1, \ldots, x_B$ and $y_1, \ldots, y_B$ corresponding to the binned arrival timestamps of $\mathcal{L}_1$ and $\mathcal{L}_2$, respectively. Finally, the CAR distance between $\mathcal{L}_1$ and $\mathcal{L}_2$ corresponds to the EMD between $x_1, \ldots, x_B$ and $y_1, \ldots, y_B$.

Cycle Time Distribution (CTD). Here, we seek to measure the ability of the BPS model to capture the end-to-end cycle time of the process. Given two event logs $\mathcal{L}_1$ and $\mathcal{L}_2$, we collect all cycle times into a single histogram per log, which depicts their empirical probability distribution functions (PDF). The CTD distance between $\mathcal{L}_1$ and $\mathcal{L}_2$ corresponds to the 1WD between both histograms.[5]

---

[5] Since we are comparing two distributions, 1WD and EMD yield the same result.

5   Evaluation

We report on a two-fold experimental evaluation. The first part aims to validate the applicability of the proposed measures by testing the following evaluation question: *are the proposed measures able to discern the impact of different known modifications to a BPS model?* (**EQ1**). Given the potential efficiency issues of CFLD, the first part of the evaluation also aims to answer the question: *is the N-Gram Distance's performance significantly different from the CFLD's performance?* (**EQ2**). The second part of the evaluation is designed to test if: *given two BPS models discovered by existing automated BPS model discovery techniques in real-life scenarios, are the proposed measures able to identify the strengths and weaknesses of each technique?* (**EQ3**). Given the complexity of the EMD (cf. Sect. 3), the second part of this evaluation also focuses on answering: *does the 1-WD report the same insights in real-life scenarios as the EMD?* (**EQ4**).

In the case of the NGD, we report on this measure for a size $N = 2$.[6] The distance computed by the EMD is not directly interpretable, as it is an absolute number on a scale that depends on the range of values of the input time series. Accordingly, we divide the raw EMD by the number of observations in the original log. In this way, we can interpret the resulting scaled-down EMD as the average number of bins that each observation of the original log must be moved to transform it into the simulated log. For example, a value of 10 implies that, on average, each observation had to be moved 10 bins.

5.1   Synthetic Evaluation

Datasets. To assess EQ1 and EQ2, we manually created the BPS model of a loan application process based on the examples from [9, Chapter 10.8]. The process comprises 12 activities (with one loop, a 3-branch parallel structure, 3 exclusive split gateways, and 3 possible endings) and 6 different resource types (performing different activities with a working schedule from Monday to Friday, from 9am to 5pm). We simulated a log of 1,000 cases as the log recording the process (i.e., the ALog). We created 7 modifications of the original BPS model: *i)* altering the control-flow by arranging the parallel activities as a sequence (Loan$_{SEQ}$); *ii)* altering, on top of the previous modification, the branching probabilities (Loan$_{S\text{-}G}$); *iii)* modifying the rate of case arrivals (Loan$_{ARR}$); *iv)* increasing the duration of the activities of the process (Loan$_{DUR}$); *v)* halving the available resources to create resource contention (Loan$_{RC}$); *vi)* changing the resource working schedules from 9 am–5 pm to 2 pm–10 pm (Loan$_{CAL}$); and *vii)* adding timer events to simulate extraneous waiting time [8] delaying the start of 4 of the activities (Loan$_{EXT}$).

We simulated $K = 10$ logs (as the GLogs) with 1,000 cases for each altered BPS model. Table 2 shows the results of the proposed measures for each modified scenario, and for the original BPS model as ground truth (Loan$_{GT}$) to measure the distance associated with the stochastic nature of the simulation.

---

[6] Augusto et al. [2] found that, for models with no duplicate activities, the size of $N = 2$ captures enough information to compare processes from a control-flow perspective.

Table 2. Results (average and 95% confidence interval) of the proposed measures for the original and modified BPS models of a loan application process.

	NGD	CFLD	AED	CED
Loan$_{GT}$	0.02 (±0.00)	0.02 (±0.00)	2.39 (± 0.51)	0.05 (±0.01)
Loan$_{SEQ}$	0.34 (±0.00)	0.20 (±0.00)	2.58 (± 0.34)	0.05 (±0.01)
Loan$_{S-G}$	0.46 (±0.00)	0.32 (±0.00)	44.17 (± 9.59)	0.24 (±0.01)
Loan$_{ARR}$	0.04 (±0.01)	0.03 (±0.00)	35.66 (±14.46)	0.08 (±0.01)
Loan$_{DUR}$	0.03 (±0.01)	0.02 (±0.00)	4.29 (± 1.34)	0.05 (±0.01)
Loan$_{RC}$	0.23 (±0.03)	0.15 (±0.01)	23.67 (± 8.81)	0.06 (±0.01)
Loan$_{CAL}$	0.02 (±0.00)	0.02 (±0.00)	6.27 (± 0.37)	3.51 (±0.02)
Loan$_{EXT}$	0.02 (±0.01)	0.02 (±0.00)	5.43 (± 1.64)	0.09 (±0.01)

	RED	CAR	CTD
Loan$_{GT}$	0.22 (± 0.05)	0.00 (± 0.00)	7.06 (± 1.41)
Loan$_{SEQ}$	1.91 (± 0.26)	0.00 (± 0.00)	42.38 (± 3.83)
Loan$_{S-G}$	235.36 (±12.93)	0.00 (± 0.00)	7,667.13 (±443.38)
Loan$_{ARR}$	0.62 (± 0.16)	42.39 (±14.12)	11.89 (± 2.42)
Loan$_{DUR}$	7.09 (± 0.51)	0.09 (± 0.04)	200.53 (± 15.66)
Loan$_{RC}$	31.66 (± 8.62)	0.03 (± 0.02)	759.45 (±210.93)
Loan$_{CAL}$	0.26 (± 0.06)	6.24 (± 0.00)	7.51 (± 1.94)
Loan$_{EXT}$	8.02 (± 0.18)	0.00 (± 0.00)	262.30 (± 6.19)

Results and Discussion. Regarding EQ1, Table 2 shows how the proposed measures appropriately penalize the BPS models for the modifications affecting their corresponding perspectives. In the control-flow measures, the BPS models showing significant differences w.r.t. the ground truth are those with control-flow modifications. The distances of Loan$_{RC}$ and Loan$_{SEQ}$ are explained by the parallel activities being executed more frequently in a specific order. In the first BPS model, due to resource contention, which delays the execution of one of the parallel activities in some cases. In the second one, due to the control-flow modification. Finally, Loan$_{S-G}$ reports the highest distance as, in addition to the modification in Loan$_{SEQ}$, it also alters the frequency of each process variant.

For temporal measures, the AED distance captures the difference in the distribution of events along the entire process. However, to identify the sources of these differences. We require a combination of the penalties incurred by CED, RED, and CAR. Thus, we must analyze them to find the root-causes for the discrepancies in AED. Starting from the seasonal aspects captured by CED, only Loan$_{S-G}$ and Loan$_{CAL}$ report significant differences, being the latter the only BPS model altering seasonal aspects. Loan$_{S-G}$'s distance is due to the change in the gateway probabilities, which in turns impacts the overall distribution of executed events. As expected, Loan$_{CAL}$ presents the highest CED distance due to the change in schedules that displaces executed events from morning to evening.

Moving to RED, which reports the distance in the distribution of events over time within each case, we observe that all modifications except Loan$_{CAL}$ should affect this perspective. The slightly higher penalization of Loan$_{ARR}$ is due to the higher case arrival rates, which delay the start activities due to resource contention. Loan$_{SEQ}$ presents a higher distance (close to a displacement of 2 h per event) as the three parallel activities are executed as a sequence, delaying subsequent activities. Similarly, in Loan$_{DUR}$, Loan$_{EXT}$, and Loan$_{RC}$, activity delays are caused by longer durations, extraneous delays, and resource contention waiting times, respectively. Finally, Loan$_{S\text{-}G}$ presents the highest RED distance due to the high-frequency differences in each process variant.

Switching to CAR, we do not observe significant differences in BPS models that exhibit the same arrival rate, except for Loan$_{CAL}$. The latter is explained by the change in schedules, as cases cannot start until the resources start their working period (which skews effective start times). Unsurprisingly, for Loan$_{ARR}$, the difference in CAR is due to the change in the arrival model.

Finally, the last proposed measure is CTD, which reports the distance in case duration among all the cases. The results of CTD follow a similar to RED (yet, with different values), since cycle times correspond to the time distance between the first and last events of the case. However, this correlation might not hold across all scenarios. Specifically, if the distribution of executed activities in the middle of each case is different, but the last event does not change, RED would detect discrepancies that CTD would not (as the cycle time would remain the same). Thus, CTD is most relevant when the analysis revolves around total cycle times, while disregarding the temporal distribution of events within the case.

To answer EQ2, we computed the Kendall rank correlation coefficient between NGD and CFLD, and we obtained a correlation of 1.0. Thus, in light of the complexity of CFLD (cf. Sect. 4), we recommend using NGD to assess the quality of a BPS model from the control-flow perspective.

5.2   Real-Life Evaluation

Datasets. To evaluate EQ3 and EQ4, we selected four real-life logs of different complexities: *i)* a log from an academic credentials' management process (AC_CRE), containing a high number of resources exhibiting low participation in the process. *ii)* a log of a loan application process from the Business Process Intelligence Challenge (BPIC) of 2012[7] – we preprocessed this log by retaining only the events corresponding to activities performed by human resources (i.e., only activity instances that have a duration). *iii)* the log from the BPIC of 2017[8] – we pre-processed this log by following the recommendations reported by the winning teams participating in the competition.[9] And *iv)* a log from a call centre process (CALL) containing numerous cases of short duration – on average, two activities per case. To avoid data leakage, we split the log of each dataset into two sets (*training* and *testing*). These datasets correspond to disjoint (non-overlapping) intervals in time with similar case and event intensity. The training dataset contains cases that are fully contained in the training period, and same

---

[7] https://doi.org/10.4121/uuid:3926db30-f712-4394-aebc-75976070e91f.

[8] https://doi.org/10.4121/uuid:5f3067df-f10b-45da-b98b-86ae4c7a310b.

[9] https://www.win.tue.nl/bpi/doku.php?id=2017:challenge.

Table 3. Characteristics of the real-life logs used in the evaluation.

Event log	Cases	Activity instances	Variants	Activities	Resources
AC_CRE_TR	398	1,945	54	16	306
AC_CRE_TE	398	1,788	35	16	281
BPIC12_TR	3,030	16,338	735	6	47
BPIC12_TE	2,976	18,568	868	6	53
BPIC17_TR	7,402	53,332	1,843	7	105
BPIC17_TE	7,376	52,010	1,830	7	113
CALL_TR	260,889	445,567	1,689	19	2,712
CALL_TE	260,890	454,807	1,573	19	2,711

Table 4. Distance measures for the BPS models discovered by SIMOD and ServiceMiner on the logs in Table 3. The CFLD ran out of memory (48 GB of allocated memory) on the CALL dataset after $> 2\,\text{h}$, thus no values are reported in those cells.

		AC_CRE	BPIC12	BPIC17	CALL
NGD	SIMOD	0.24 (±0.01)	0.56 (±0.00)	0.37 (±0.00)	0.08 (±0.00)
	ServiceMiner	0.13 (±0.01)	0.13 (±0.01)	0.06 (±0.00)	0.04 (±0.00)
CFLD	SIMOD	0.21 (±0.00)	0.55 (±0.00)	0.34 (±0.00)	-
	ServiceMiner	0.18 (±0.01)	0.16 (±0.00)	0.06 (±0.00)	-
AED	SIMOD	91.72 (± 16.66)	61.57 (±10.80)	192.18 (±33.03)	48.13 (±0.11)
	ServiceMiner	298.17 (±105.87)	29.22 (± 8.33)	51.24 (±11.90)	1.67 (±0.13)
CAR	SIMOD	110.38 (± 16.94)	336.42 (±42.98)	390.04 (±43.39)	61.68 (±0.00)
	ServiceMiner	327.85 (±118.33)	153.25 (±24.76)	121.90 (±23.60)	3.57 (±0.20)
CED	SIMOD	2.22 (±0.09)	20.55 (±0.82)	10.72 (±0.50)	18.17 (±0.04)
	ServiceMiner	1.63 (±0.19)	28.52 (±1.24)	9.87 (±0.42)	0.70 (±0.01)
RED	SIMOD	9.96 (±6.46)	3.99 (±0.95)	62.80 (±2.16)	0.10 (±0.00)
	ServiceMiner	70.49 (±4.08)	45.91 (±2.09)	149.60 (±0.42)	0.03 (±0.01)
CTD	SIMOD	62.23 (±1.65)	93.45 (±0.71)	102.85 (±0.84)	8.18 (±0.08)
	ServiceMiner	99.88 (±0.30)	124.21 (±0.33)	112.83 (±0.16)	12.04 (±0.07)

for the testing dataset. Table 3 shows the characteristics of the four training and four testing- event logs. For each dataset, we ran two automated BPS model discovery techniques (SIMOD and ServiceMiner) on the training log, and evaluated the quality of the discovered BPS models on the test log.

Results and Discussion. Regarding EQ3, Table 4 shows the results of the proposed measures for the BPS models automatically discovered by SIMOD and ServiceMiner (henceforth M_{SIMOD} and M_{SerMin}, respectively). From the control-flow perspective, M_{SerMin} performs closer to the original log than M_{SIMOD} for all four datasets. The reason lies in the methods that the approaches use to model the control-flow. SIMOD is designed to discover an interpretable process model to support modification for *what-if* analyses. To this end, SIMOD uses a model discovery algorithm that applies multiple pruning techniques to simplify the discovered model. Conversely, ServiceMiner discovers a Markov chain, which yields more accurate results, yet can lead to complex 'spaghetti models'.

Two main differences are reported w.r.t. the temporal and congestion aspects. First, for Case Arrival Rate (CAR), M_{SerMin} presents better results in BPIC12, BPIC17, and CALL, while M_{SIMOD} outperforms it in AC_CRE. To model the arrival of new cases, ServiceMiner splits the timeline into one-hour windows, and bootstraps the arrivals per time window. SIMOD computes the inter-arrival times (i.e., the time between each arrival and the next one) and estimates a parametrized distribution to model them. The complexity of ServiceMiner's arrival model allows it to capture better the arrival rate in scenarios where the density of case arrivals per hour is high, and/or the rate of arrivals varies through time (BPIC12, BPIC17, and CALL). On the contrary, if cases are scattered over time (AC_CRE), SIMOD's approach presents a better result.

The second main difference lies in the Relative Event Distribution (RED) and the Cycle Time Distribution (CTD) distances. Here, M_{SIMOD} obtains better results in both measures except in one case. In the CALL dataset, M_{SerMin} obtains a smaller RED distance (both methods perform well w.r.t. the original log). SIMOD outperforms ServiceMiner due to a high amount of extraneous activity delays (i.e., waiting times not related to the resource allocation or activity performance) exhibited in these processes. Specifically, SIMOD includes a component to discover extraneous delays, which improves the distribution of the events within the case. Both techniques perform close to the original log in the CALL dataset because extraneous delays are rare for the call centre process.

For seasonality, the Circadian Event Distribution (CED) reports slight differences between the two methods for AC_CRE and BPIC17, and a moderately better result for M_{SIMOD} in BPIC12. The CALL dataset presents the highest difference, where M_{SerMin} obtains better results, which can be attributed to its highly accurate arrival model. The CALL dataset has mostly cases with one or two events. Hence, case execution depends more on the arrival time of the case, than on the activity performance and congestion models.

Combining all the temporal perspectives in one measure, the results of Absolute Event Distribution (AED) follow the same distribution as CAR, where M_{SerMin} presents better results in BPIC12, BPIC17, and CALL, while M_{SIMOD} performing better in AC_CRE. Although this measure summarizes all the temporal performance in one, it is highly affected by the performance of the arrival model. A wrong arrival rate propagates the error to all the events per case, displacing them even if their relative distribution is accurate.

The proposed measures detected key differences between the considered BPS model discovery techniques. Additionally, our results can help to identify potential improvements in these techniques. SIMOD's inferior performance in the control-flow perspective is expected, given that it takes a simplified process model as input. Moreover, there is a natural fit between the control-flow measures (e.g., NGD) and the Markovian approach of ServiceMiner – as a Markov chain is, in essence, a generative 2-gram model. The results also highlight the benefits of SIMOD's extraneous waiting time discovery component (a feature

Table 5. Results of the proposed measures for the BPS models discovered by SIMOD and ServiceMiner with the real-life logs in Table 3.

		AC_CRE	BPIC12	BPIC17	CALL
AED	SIMOD	117.32 (± 18.85)	313.30 (±41.50)	314.92 (±43.02)	61.76 (±0.08)
	ServiceMiner	315.88 (±115.60)	79.19 (±16.12)	65.56 (±12.55)	3.47 (±0.19)
CAR	SIMOD	110.38 (± 16.94)	336.42 (±42.98)	390.04 (±43.39)	61.68 (±0.00)
	ServiceMiner	327.85 (±118.33)	153.25 (±24.76)	121.89 (±23.60)	3.57 (±0.20)
CED	SIMOD	3.11 (±0.18)	2.10 (±0.07)	1.65 (±0.03)	4.72 (±0.00)
	ServiceMiner	2.50 (±0.27)	2.34 (±0.03)	1.65 (±0.02)	1.03 (±0.03)
RED	SIMOD	48.19 (±1.72)	96.82 (±1.61)	132.31 (±1.68)	0.00 (±0.00)
	ServiceMiner	74.50 (±0.18)	150.83 (±0.57)	150.26 (±0.28)	0.00 (±0.00)

that ServiceMiner does not have). Finally, although ServiceMiner's arrival model achieved the best results in most of the scenarios, the evaluation in AC_CRE point towards an improvement opportunity in the situation where cases arrive at a slow rate.

To evaluate EQ4, Table 5 shows the result of AED, CAR, CED, and RED measures when computing the distance with 1WD, instead of EMD. The results follow the same distribution in all cases, except in the CED measure on the BPIC17 dataset, and the RED measure on the CALL dataset. In both cases, the slight differences shown by EMD are reduced to a similar value by both techniques. For arrivals, as explained in Sect. 3, computing the distance with EMD and 1WD provide the same result, as the number of observations in both samples is the same (i.e., the number of cases). In conclusion, computing the distance using 1WD leads to similar conclusions at a lower computational cost. Thus, we recommend using 1WD when the masses of both time series are close to each other, and when the number of observations (amount of mass) is large.

Threats to Validity

The evaluation reported above is potentially affected by the following threats to the validity. First, regarding *internal validity*, the experiments rely only on 8 BPS models of one synthetic process, and 8 automatically discovered BPS models from 4 real-life processes. The results could be different for other datasets. Second, regarding *external validity*, the evaluation was assessed with real-life event logs from processes of different domains. However, the results could not be generalized for processes of domains presenting specific unseen characteristics. Third, regarding *construct validity*, we proposed a set of measures of goodness based on discretized distributions and time series. The results could be different for other measures. Finally, regarding *ecological validity*, the evaluation compares the BPS models against the original log. While this allows us to measure how well the simulation models replicate the as-is process, it does not allow us to assess the goodness of the simulation models in a what-if setting, e.g., predicting the performance of the process after a change.

6 Conclusion

We proposed a multi-perspective approach to measure the ability of a BPS model to replicate the behavior recorded in an event log. The approach decomposes simulation models into three perspectives: control-flow, temporal, and congestion. We defined measures for each of these perspectives. We evaluated the adequacy of the proposed measures by analyzing their ability to discern the impact of modifications to a BPS model. The results showed that the measures are able to detect the alterations in their corresponding perspectives. Furthermore, we analyzed the usefulness of the metrics in real-life scenarios w.r.t. their ability to uncover the relative strengths and weaknesses of two approaches for the automated discovery of BPS models. The findings showed that beyond capturing the quality of BPS model and identifying the sources of discrepancies, the measures can also assist in eliciting areas for improvement in these techniques. Finally, as some of the proposed measures present higher computational cost, we evaluated more efficient measures, finding that they perform similarly to computationally-heavy methods.

In future work, we will explore the applicability of the proposed measures to other process mining problems, e.g., concept drift detection and variant analysis. Studying how to assess the quality of BPS models in the context of object-centric event logs is another future work avenue. Lastly, we aim to study other quality measures for BPS models adapted from the field of generative machine learning, for example, by using a discriminative model that attempts to distinguish between data generated by the BPS model and real data.

Reproducibility. The scripts to reproduce the experiments, the datasets, and the results are publicly available at: https://doi.org/10.5281/zenodo.7761252. The measures have been implemented as a Python package (log-distance-measures) installable from pip, and the code is publicly available at: https://github.com/AutomatedProcessImprovement/log-distance-measures.

Acknowledgements. This work has been funded by the European Research Council (PIX Project) and the National Science and Engineering Research Council (NSERC) grants held by Opher Baron, Dmitry Krass, and Arik Senderovich.

References

1. Abel, M.: Lightning fast business process simulator. Master's thesis. Institute of Computer Science, University of Tartu (2011)
2. Augusto, A., Armas-Cervantes, A., Conforti, R., Dumas, M., Rosa, M.L.: Measuring fitness and precision of automatically discovered process models: a principled and scalable approach. IEEE Trans. Knowl. Data Eng. **34**(4), 1870–1888 (2022)
3. Augusto, A., Conforti, R., Dumas, M., Rosa, M.L., Polyvyanyy, A.: Split miner: automated discovery of accurate and simple business process models from event logs. Knowl. Inf. Syst. **59**(2), 251–284 (2019)
4. Brockwell, P.J., Davis, R.A.: Introduction to Time Series and Forecasting. Springer, New York (2002). https://doi.org/10.1007/b97391

5. Burke, A., Leemans, S.J.J., Wynn, M.T., van der Aalst, W.M.P., ter Hofstede, A.H.M.: Stochastic process model-log quality dimensions: an experimental study. In: ICPM 2022, pp. 80–87. IEEE (2022)
6. Camargo, M., Dumas, M., González, O.: Automated discovery of business process simulation models from event logs. Decis. Support Syst. **134**, 113284 (2020)
7. Camargo, M., Dumas, M., Rojas, O.G.: Discovering generative models from event logs: data-driven simulation vs deep learning. PeerJ Comput. Sci. **7**, e577 (2021)
8. Chapela-Campa, D., Dumas, M.: Modeling extraneous activity delays in business process simulation. In: ICPM 2022, pp. 72–79. IEEE (2022)
9. Dumas, M., Rosa, M.L., Mendling, J., Reijers, H.A.: Fundamentals of Business Process Management, 2nd edn. Springer, Heidelberg (2018). https://doi.org/10.1007/978-3-662-56509-4
10. Fracca, C., de Leoni, M., Asnicar, F., Turco, A.: Estimating activity start timestamps in the presence of waiting times via process simulation. In: Franch, X., Poels, G., Gailly, F., Snoeck, M. (eds.) CAiSE 2022. LNCS, vol. 13295, pp. 287–303. Springer, Cham (2022). https://doi.org/10.1007/978-3-031-07472-1_17
11. Kalbfleisch, J.D., Prentice, R.L.: The Statistical Analysis of Failure Time Data. Wiley, Hoboken (2011)
12. Leemans, S.J.J., Polyvyanyy, A.: Stochastic-aware conformance checking: an entropy-based approach. In: Dustdar, S., Yu, E., Salinesi, C., Rieu, D., Pant, V. (eds.) CAiSE 2020. LNCS, vol. 12127, pp. 217–233. Springer, Cham (2020). https://doi.org/10.1007/978-3-030-49435-3_14
13. Leemans, S.J.J., Syring, A.F., van der Aalst, W.M.P.: Earth movers' stochastic conformance checking. In: Hildebrandt, T., van Dongen, B.F., Röglinger, M., Mendling, J. (eds.) BPM 2019. LNBIP, vol. 360, pp. 127–143. Springer, Cham (2019). https://doi.org/10.1007/978-3-030-26643-1_8
14. Levina, E., Bickel, P.J.: The earth mover's distance is the mallows distance: some insights from statistics. In: ICCV 2001, vol. 2, pp. 251–256. IEEE Computer Society (2001)
15. Liao, T.W.: Clustering of time series data - a survey. Pattern Recognit. **38**(11), 1857–1874 (2005)
16. López-Pintado, O., Dumas, M.: Business process simulation with differentiated resources: does it make a difference? In: Di Ciccio, C., Dijkman, R., del Río Ortega, A., Rinderle-Ma, S. (eds.) BPM 2022. LNCS, vol. 13420, pp. 361–378. Springer, Cham (2022). https://doi.org/10.1007/978-3-031-16103-2_24
17. Martin, N., Depaire, B., Caris, A.: The use of process mining in business process simulation model construction - structuring the field. Bus. Inf. Syst. Eng. **58**(1), 73–87 (2016)
18. Martin, N., Pufahl, L., Mannhardt, F.: Detection of batch activities from event logs. Inf. Syst. **95**, 101642 (2021)
19. Muskulus, M., Verduyn-Lunel, S.: Wasserstein distances in the analysis of time series and dynamical systems. Physica D **240**(1), 45–58 (2011)
20. Pourbafrani, M., van der Aalst, W.M.P.: Discovering system dynamics simulation models using process mining. IEEE Access **10**, 78527–78547 (2022)
21. Rosenthal, K., Ternes, B., Strecker, S.: Business process simulation on procedural graphical process models. Bus. Inf. Syst. Eng. **63**(5), 569–602 (2021)
22. Rozinat, A., Mans, R.S., Song, M., van der Aalst, W.M.P.: Discovering simulation models. Inf. Syst. **34**(3), 305–327 (2009)
23. Senderovich, A., Beck, J.C., Gal, A., Weidlich, M.: Congestion graphs for automated time predictions. In: AAAI 2019, pp. 4854–4861. AAAI Press (2019)

24. Senderovich, A., et al.: Conformance checking and performance improvement in scheduled processes: a queueing-network perspective. Inf. Syst. **62**, 185–206 (2016)
25. Thomas, M.U.: Queueing systems. volume 1: theory (leonard kleinrock). SIAM Rev. **18**(3), 512–514 (1976)
26. Zhao, C., Sahni, S.: String correction using the damerau-levenshtein distance. BMC Bioinform. **20-S**(11), 277:1–277:28 (2019)

Open Access This chapter is licensed under the terms of the Creative Commons Attribution 4.0 International License (http://creativecommons.org/licenses/by/4.0/), which permits use, sharing, adaptation, distribution and reproduction in any medium or format, as long as you give appropriate credit to the original author(s) and the source, provide a link to the Creative Commons license and indicate if changes were made.

The images or other third party material in this chapter are included in the chapter's Creative Commons license, unless indicated otherwise in a credit line to the material. If material is not included in the chapter's Creative Commons license and your intended use is not permitted by statutory regulation or exceeds the permitted use, you will need to obtain permission directly from the copyright holder.

Event Abstraction for Partial Order Patterns

Chiao-Yun Li[1,2(✉)], Sebastiaan J. van Zelst[1,2], and Wil M. P. van der Aalst[1,2]

[1] RWTH Aachen University, Aachen, Germany
{chiaoyun.li,wvdaalst}@pads.rwth-aachen.de
[2] Fraunhofer FIT, Birlinghoven Castle, Sankt Augustin, Germany
sebastiaan.van.zelst@fit.fraunhofer.de

Abstract. *Process mining* endeavors to extract fact-based insights into processes based on *event data* stored in information systems. Due to the variety of processes in different fields and organizations, there does not exist a universal technique to allow for putting the process mining outcome directly into action. Various techniques have been developed to support human analysis. Meanwhile, as raw event data are often provided at the system level, the *abstraction principle* is applied to "lift" the data to a higher level for human interpretation, which is called *event abstraction*. Owing to the limitation of the information systems deployed in practice, most abstraction techniques are developed based on the assumption that all process activities are performed sequentially, ignoring the fact that there may be activities performed concurrently or the relation of the activity executions could not be clearly defined. In this paper, we propose an event abstraction framework based on partial order patterns. We extract the candidate pattern instances and abstract event data based on the pattern instances identified. Moreover, we instantiate the framework and optimize the implementation. The framework is evaluated with synthetic event data, and a case study based on a real-life process is performed, demonstrating the applicability of the framework.

Keywords: Process mining · Event abstraction · Partial orders

1 Introduction

Modern organizations rely on business processes executed with the support of information systems, which generate *event data* recorded during process execution. *Process mining* aims to extract valuable insights from such event data [2]. Numerous process mining techniques were developed to gain insights into various aspects; *process discovery* reveals the actual behavior of a process, which is often represented with a *process model* [4]; *conformance checking* identifies deviations in a process [6]; *performance analysis* detects inefficiencies and bottlenecks in a process [7]; *process enhancement* attempts to enhance a process model based on factual insights discovered [20].

© The Author(s), under exclusive license to Springer Nature Switzerland AG 2023
C. Di Francescomarino et al. (Eds.): BPM 2023, LNCS 14159, pp. 38–54, 2023.
https://doi.org/10.1007/978-3-031-41620-0_3

Fig. 1. A process model discovered based on real-life event data [19], which abstracts the behavior of 36 activities executed in 4,366 ways in a process instance.

To turn process mining results into actionable insights, the outcomes must be interpretable for humans, which is often achieved through the use of a process model annotated with relevant information for a limited number of activities. Typically, process mining techniques are directly applied to raw event data, i.e., event data as recorded in information systems, resulting in outcomes that may be too detailed or complex for human analysts as shown in Fig. 1, which is impossible for humans to derive valuable insights without going into detail. Due to the highly flexible and complex nature of real-life processes, the field of *event abstraction* emerged to abstract event data to a higher level based on the predefined or identified regularity of the execution of *activities*, i.e., well-defined process steps, for human interpretability. By abstracting event data to a higher level, the complexity is simplified, allowing stakeholders to understand and interpret the results.

As most information systems deployed in practice support sequence data, event data are often structured as a total order of the execution of activities. Consequently, most event abstraction techniques are developed based on the assumption that activities are executed sequentially. In practice, activities can be executed concurrently, e.g., when a person multitasks, and/or the order of executions cannot be clearly defined, e.g., when the executions are recorded at the granularity of *days*, which leads to *partially ordered* event data.

We propose a *framework to extract patterns from partially ordered event data*. By leveraging a *pattern class* defined by domain experts as an expected relation of the execution of concepts, e.g., activities, in a process, the framework identifies the corresponding *pattern instances*, i.e., the executions of the pattern class. The abstraction is achieved by aggregating pattern instances; thereby, the framework can be iteratively applied to construct a hierarchy of abstractions. We initiate and implement the framework with a generic approach for identifying pattern instances. Furthermore, we optimize the framework for extracting *candidate pattern instances*, i.e., potential sets of event data that *may* be pattern instances. We apply the framework to synthetic event data and experiment with the effect

Table 1. A running example of partially ordered event data. Every row is a record representing an activity instance characterized by its identifier (`AID`), the identifier of the process instance it belongs to (`cid`), the activity name (`Activity`), and the duration of the execution (`Start` and `Complete Timestamp`).

cid	Activity (Abbre.)	AID	Start Timestamp	Complete Timestamp
1	Get Appointment (A)	1	2021-03-26 10:36:09	2021-03-26 10:36:09
1	Consult (C)	2	2021-03-26 11:07:53	2021-03-26 11:20:23
1	Review History (R)	3	2021-03-26 11:07:07	2021-03-26 11:22:10
1	Phlebotomize (P)	4	2021-03-26 13:36:16	2021-03-26 13:39:27
1	Conduct Lab Test (L)	5	2021-03-29 00:00:00	2021-04-04 00:00:00
1	Conduct Lab Test (L)	6	2021-03-29 00:00:00	2021-04-04 00:00:00
1	Diagnose (D)	7	2021-04-09 15:32:02	2021-04-09 15:47:20
1	Provide Treatment (T)	8	2021-04-15 00:00:00	2022-05-22 00:00:00
1	Provide Treatment (T)	9	2021-05-04 00:00:00	2022-05-26 00:00:00
1	Provide Treatment (T)	10	2021-05-22 00:00:00	2022-09-03 00:00:00
1	Phlebotomize (P)	11	2022-09-08 20:09:40	2022-09-08 20:11:51
1	Conduct Lab Test (L)	12	2022-09-11 00:00:00	2022-09-17 00:00:00
1	Conduct Lab Test (L)	13	2022-09-13 00:00:00	2022-09-18 00:00:00
1	Evaluate (E)	14	2022-09-21 05:04:36	2022-09-21 05:32:15

of noises. To demonstrate the applicability, we conduct a case study based on real-life event data based on the abstraction obtained with the framework.

The paper is structured as follows. A running example is presented in Sect. 2. Section 3 introduces the mathematical concepts, which are applied to define the framework in Sect. 4. We introduce the implementation in Sect. 5 and show the experiments in Sect. 6. Finally, we review related work in Sect. 7 and discuss future directions in Sect. 8.

2   Running Example - A Treatment Procedure

Table 1 presents an excerpt of synthetic event data, which serves as a running example that we use throughout the paper. Every row represents an activity instance. The table records the activities executed in a treatment procedure of a patient. After the patient got an appointment at 10:36:09, he/she consulted the general practitioner and the practitioner reviewed the medical history of the patient at the same time. Then, a nurse phlebotomized the patient and sent the blood samples to two laboratories for different hematological tests. The reports were then sent back to the general practitioner for a diagnosis. Based on the outcome, the patient was sent to three specialists for further treatment. After roughly 1 year of treatment, the same blood tests are conducted again and the outcome of treatment is evaluated.

In Table 1, `Get Appointment` is executed in a time moment, as is often assumed in classical event data; `Conduct Lab Test` and `Provide Treatment` are recorded at the granularity of *days*, which causes unclear ordering, e.g., the two lab tests conducted at the first time; other activities are executed and recorded in time duration. Due to the time interleaving and different granularity recorded, such data form *partially ordered* event data[1], requiring a different abstraction mechanism compared to sequentially ordered event data.

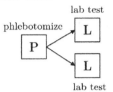

Figure 2 visualizes a pattern class, assumed to be provided by domain experts, which specifies that two lab tests must be performed concurrently *after* phlebotomization. By representing an activity instance with its abbreviated activity and its identifier as a subscript, we can extract three sets of activity instances in Table 1 given the pattern class: $\{P_4, L_5, L_6\}$, $\{P_{11}, L_{12}, L_{13}\}$, and $\{P_4, L_{12}, L_{13}\}$. Every set of activity instances forms a *pattern instance*. As the blood sample phlebotomized with P_4 is used for L_5 and L_6, and the blood sample collected with P_{11} is used for L_{12} and L_{13}, we characterize the former two pattern instances as *local pattern instances*.

Fig. 2. Visualization of a pattern class.

3   Preliminaries

Let X be an arbitrary set. $\mathcal{P}(X) = \{X' \mid X' \subseteq X\}$ denotes the powerset of X and $|X|$ denotes the number of elements in X. A sequence over X is a function $\sigma\colon \{1, 2, ..., n\} \to X$, where σ is written as $\langle x_1, x_2, \ldots, x_n \rangle$. A strict partial order is a binary relation $\prec$ on X, written as $(X, \prec)$, which is irreflexive ($\forall x \in X, x \not\prec x$), asymmetric ($\forall x, y \in X, x \prec y \implies x \neq y$), and transitive ($\forall x, y, z \in X, x \prec y \land y \prec z \implies x \prec z$). $(X, \prec)$ denotes the covering relation of $(X, \prec)$ such that $\forall x_1, x_2 \in X (x_1 \prec x_2)$, we write $x_1 \lessdot x_2$ if and only if $\nexists x' \in X (x_1 \prec x' \land x' \prec x_2)$. For simplicity, we write $(X, \prec) = X$ as the shorthand for the elements in $(X, \prec)$ and refer to a strict partial order as a partial order.

Given an arbitrary set X, l is a function of X to a set of labels; a partial order on X with such a function is called a *labeled partial order* and written as $(X, \prec, l)$. Let X and Y be two arbitrary sets. Given $(X, \prec, l_X)$ and $(Y, \prec, l_Y)$, $(X, \prec, l_X)$ and $(Y, \prec, l_Y)$ are label-preserving isomorphic, denoted as $(X, \prec, l_X) \simeq (Y, \prec, l_Y)$, iff there exists a bijective relation $b\colon X \to Y$ s.t. $\forall x_1, x_2 \in X, x_1 \prec x_2 \iff b(x_1) \prec b(x_2)$, and $\forall x \in X, l_X(x) = l_Y(b(x))$.

Let $(X, \prec, l)$ be a labeled partial order on an arbitrary set X. Let z be an arbitrary element, where $z \notin X$, and l_z is a labeling function of z. The function $\mathsf{ADD}((X, \prec, l), z, l_z)$ adds z into $(X, \prec, l)$ s.t. $\mathsf{ADD}((X, \prec, l), z, l_z) = (X', \prec', l')$ where $X' = X \cup \{z\}$, $\prec' = \prec \cup (X \times \{z\})$.

Definition 1 (Event Data). *A case is a process instance. $\mathcal{U}_{con}$ is the universe of concepts defined in a process, e.g., an activity; $\mathcal{U}_{inst}$ is the universe of the*

---

[1] A collection of time intervals must be a partial order; nevertheless, the proposed framework is based on partial orders, which is more generically applicable. The example is provided with timestamps as a motivating example.

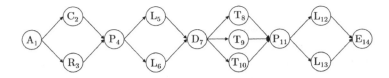

Fig. 3. Visualization of the covering relation of the activity instances in Table 1.

instances, e.g., an activity instance; $\mathcal{U}_{cid}$ *is the universe of case identifiers. A log,* $L = (\mathsf{CI}, \prec, \pi_{con}, \pi_{cid})$*, where*

- $\mathsf{CI} \subseteq \mathcal{U}_{inst}$ *is a set of instances;*
- $\prec\ =\ \mathsf{CI} \times \mathsf{CI}$ *is a partial order on* CI*;*
- $\pi_{con}: \mathsf{CI} \to \mathcal{U}_{con}$*, where* $\pi_{con}(\mathsf{ci})$ *is the concept of an instance* $\mathsf{ci} \in \mathsf{CI}$*;*
- $\pi_{cid}: \mathsf{CI} \to \mathcal{U}_{cid}$*, where* $\pi_{cid}(\mathsf{ci})$ *is the identifier of the case that an instance* $\mathsf{ci} \in \mathsf{CI}$ *belongs to.*

We let $\mathsf{CID}(L) = \{\pi_{cid}(\mathsf{ci}) \mid \mathsf{ci} \in \mathsf{CI}\}$ *denote the case identifiers in* L*. Given* $c \in \mathsf{CID}(L)$*,* $\mathsf{CI}_c = \{\mathsf{ci} \in \mathsf{CI} \mid \pi_{cid}(\mathsf{ci}) = c\}$ *and* $\mathsf{c}_L = (\mathsf{CI}_c, \prec_c)$*, where* $\prec_c\ =\ \prec \cap\, (\mathsf{CI}_c \times \mathsf{CI}_c)$*.*

Figure 3 visualizes the covering relation of the instances in the case in Table 1. Every node represents an instance, which is labeled with its identifier as a subscript and the corresponding (abbreviated) concept, i.e., the activity. The arrows indicate the covering relation among the instances.

4   Framework

We introduce and define the framework in this section. First, we outline the mechanism of the proposed framework in Sect. 4.1. Based on the mathematical notations introduced, we define a pattern class and the corresponding pattern instances in Sect. 4.2. The extraction of candidate pattern instances is introduced in Sect. 4.3. Finally, we detail the abstraction with the identification of pattern instances in Sect. 4.4.

4.1   Overview

Figure 4 presents a schematic overview of the proposed framework. We assume that a log with partially ordered event data is provided. A pattern class can be defined by a domain expert or with the knowledge obtained during the exploration of the log. Given a pattern class, the framework exhaustively extracts all the candidate pattern instances in the log. Next, one identifies partial orders of pattern instances from the candidate pattern instances where the relationship between the identified pattern instances is defined in a flexible manner. Finally, an abstracted log based on the pattern class is constructed. Since the pattern class and the extraction and the identification of pattern instances support partially ordered event data, the proposed framework can be iteratively applied to the abstracted log based on another pattern class.

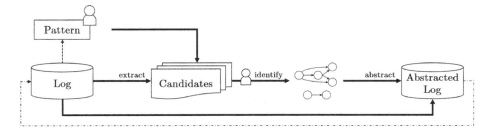

Fig. 4. A schematic overview of the framework. There are three key steps in the framework, the extraction of candidate pattern instances, the identification of partial orders of pattern instances, and the abstraction based on the pattern class.

4.2   Pattern Class and Pattern Instance

A *pattern class* is an expected relation of the execution of concepts; a *pattern instance* is, intuitively, a set of instances that adhere to the expectation. We formally define a pattern class as follows.

Definition 2 (Pattern Class). *A pattern class* $PC = (X, \prec, l)$, *where* $l\colon X \to \mathcal{U}_{con}$, *is a labeled partial order where* $|X| \geq 2$. *We assume that a pattern class is a concept defined in a process such that* $PC \in \mathcal{U}_{con}$. *We say that* $\{l(x) \mid x \in X\}$ *are the underlying concepts of a pattern class.*

A pattern instance is a labeled partial order of instances that is label-preserved isomorphic to a pattern class as defined below.

Definition 3 (Pattern Instance). *Let* L *be a log. Given* $c \in CID(L)$, *let* $c_L = (Cl_c, \prec_c)$. *Given* $Cl \subseteq \underline{c_L}$ *and a pattern class* $PC \in \mathcal{U}_{con}$, *a pattern instance* $pi = (Cl, \prec, \pi_{con})$, *where* $\prec = \prec_c \cap (Cl \times Cl)$, *is a labeled partial order over* Cl *where* $pi \simeq PC$. *We assume that a pattern instance* pi *is an instance s.t.* $pi \in \mathcal{U}_{inst}$ *and* $\pi^p_{cid}\colon \mathcal{U}_{inst} \to \mathcal{U}_{cid}$ *and* $\pi^p_{con}\colon \mathcal{U}_{inst} \to \mathcal{U}_{con}$, *where* $\pi^p_{cid}(pi) = c$ *and* $\pi^p_{con}(pi) = PC$. *We say that* $\underline{pi}$ *are the underlying instances of* pi.

We define a pattern class with at least two elements since it is trivial with a single element as it simply implies the projection of the label of the element on instances. Meanwhile, with the constraint imposed on a pattern class, a pattern instance consists of at least two instances. Note that a pattern instance is defined in the context of a case and the definition above allows for the extraction of a pattern instance where the relations of the underlying instances are undefined. Figure 5 presents the pattern instances of the pattern class PC visualized in Fig. 2. To differentiate with the visualization of event data, an element in a pattern class is visualized with a square labeled with the corresponding concept; the arrows indicate the partial order relation of the elements.

Meanwhile, a pattern instance may consist of instances that are hardly related in practice. For example, the pattern instance pi_3 visualized in Fig. 5 suggests that the lab tests are conducted based on the blood sampled one and a half

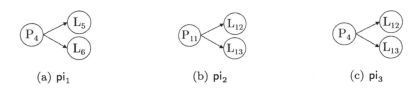

Fig. 5. Visualization of the pattern instances of the pattern class PC defined in Fig. 2 for the case in Table 1, where $\forall 1 \leq i \leq 3$, $\pi^{\mathsf{P}}_{con}(\mathsf{pi}_i) = \mathsf{PC}$ and $\pi^{\mathsf{P}}_{cid}(\mathsf{pi}_i) = 1$. Note that pi_1 and pi_2 are local pattern instances.

years ago. Compared to pi_3, the other pattern instances, pi_1 and pi_2 in Fig. 5, are more likely to be the pattern instances that one has in mind. Hence, to further identify a pattern instance with closely related instances, we characterize a pattern instance as a *local pattern instance* if the underlying instances are *related* based on the covering relation.

Definition 4 (Local Pattern Instance). *Let* $\mathsf{pi} \in \mathcal{U}_{inst}$ *be a pattern instance. We characterize* pi *as a local pattern instance iff* $\forall \mathsf{ci}_1, \mathsf{ci}_n \in \mathsf{pi}(\mathsf{ci}_1 \neq \mathsf{ci}_n)$, $\exists \langle \mathsf{ci}_1, \mathsf{ci}_2, \ldots, \mathsf{ci}_n \rangle$, *where* $\forall 1 \leq i \leq n, \mathsf{ci}_i \in \mathsf{pi}$ *and* $\forall 1 \leq i < n, \mathsf{ci}_i \prec \mathsf{ci}_{i+1} \vee \mathsf{ci}_{i+1} \prec \mathsf{ci}_i$.

By iteratively applying the framework, we identify pattern instances in every iteration. Hence, a pattern instance consists of a set of instances that may be activity instances and/or pattern instances identified in the previous iteration.

4.3   Candidate Pattern Instances

Given a pattern class, first, we search for all the possible pattern instances of a pattern class in a log, which we name as *candidate pattern instances*. Since an instance is only related to one case, the search is performed for every case and the collection of the candidate pattern instances identified for every case in a log is the candidate pattern instances in the log.

Definition 5 (Candidate Pattern Instance Extraction). *Let* L *be a log and* PC *be a pattern class. Let* $\mathsf{c}_L = (\mathsf{CI}_c, \prec_c)$, *where* $c \in \mathsf{CID}(L)$. *We define function* $\mathsf{EXT}_L \colon \mathcal{U}_{cid} \times \mathcal{U}_{con} \rightarrow \mathcal{P}(\mathcal{P}(\mathcal{U}_{inst}) \times \mathcal{P}(\mathcal{U}_{inst} \times \mathcal{U}_{inst}))$, *where* $\mathsf{EXT}_L(c, \mathsf{PC}) = \{\mathsf{can} = (\mathsf{CI}, \prec, \pi_{con}) \mid \mathsf{CI} \subseteq \mathsf{CI}_c, \prec = \prec_c \cap (\mathsf{CI} \times \mathsf{CI}), \mathsf{can} \simeq \mathsf{PC}\}$ *denotes the candidate pattern instances of* PC *in* c.

The candidate pattern instances in a case are a set of partial orders of instances that are isomorphic to the pattern class. With the same example, the partial orders of activity instances visualized in Fig. 5 are extracted as the candidate pattern instances.

4.4   Abstraction Based on Patterns

We aggregate the underlying instances of a pattern instance to construct an abstracted log. The pattern instances are identified for every case in a log and the relation between an instance in a case and a pattern instance is inferred from the underlying instances of the pattern instance. First, we generalize the identification of pattern instances as follows.

Definition 6 (Pattern Instance Identification). *Let L be a log and CAN be the candidate pattern instances in $c \in \mathsf{CID}(L)$. We define the function* $\mathsf{IDEN}_c\colon \mathcal{P}(\mathcal{P}(\mathcal{U}_{inst}) \times \mathcal{P}(\mathcal{U}_{inst} \times \mathcal{U}_{inst})) \rightarrow \mathcal{P}(\mathcal{U}_{inst}) \times \mathcal{P}(\mathcal{U}_{inst} \times \mathcal{U}_{inst})$, where $\mathsf{IDEN}_c(\mathsf{CAN}) = (\mathsf{PI}, \prec)$ *is a partial order of pattern instances* $\mathsf{PI} \subseteq \mathsf{CAN}$ *in* c, *where* $\forall \mathsf{pi}_1, \mathsf{pi}_2 \in \mathsf{PI}(\mathsf{pi}_1 \neq \mathsf{pi}_2), \mathsf{pi}_1 \prec \mathsf{pi}_2 \implies \exists \mathsf{ci} \in \underline{\mathsf{pi}_1} \forall \mathsf{CI}' \in \underline{\mathsf{pi}_2}(\mathsf{ci} \prec \mathsf{ci}')$.

The pattern instances in a case are a subset of the candidate pattern instances in the case. The identification can be initiated in a flexible manner while the partial order relation of the pattern instances identified must not violate the minimum requirement specified in Definition 6. Note that two different pattern instances may share some underlying instances.

Figure 6 motivates the necessity of the flexible instantiation of the identification of pattern instances. Suppose that we extract two local pattern instances pi_1 and pi_2 as specified in Fig. 6. By simply inferring the relation of the pattern instances based on the underlying instances, $\mathsf{pi}_1 \not\prec \mathsf{pi}_2$ since they share a_4. However, assume that the a_4 represents a milestone achieved in a waterfall process; it is more reasonable to define the relation as $\mathsf{pi}_1 \prec \mathsf{pi}_2$. Hence, we generalize the identification of pattern instances and allow one to impose the constraints that are applicable to the organization; nevertheless, a generic instantiation is also implemented and introduced in the next section.

Finally, the abstraction is realized by *aggregating* the pattern instance identified and defining the relation between the pattern instances and other instances.

Definition 7 (Abstraction). *Let* $L = (\mathsf{CI}, \prec, \pi_{con}, \pi_{cid})$ *be a log. Let* $\mathsf{PI}_L = \{\mathsf{pi} \mid c \in \mathsf{CID}(L) \colon \pi_{cid}^\mathsf{P}(\mathsf{pi}) = c\}$ *be the pattern instances in* L; $\prec_p = \mathsf{PI}_L \times \mathsf{PI}_L$ *denotes the partial order on* PI_L. *Given* $\mathsf{CI}_{rst} = \mathsf{CI} \setminus \bigcup_{\mathsf{pi} \in \mathsf{PI}_L} \underline{\mathsf{pi}}$ *and* $\prec_{rst} = \prec \cap (\mathsf{CI}_{rst} \times \mathsf{CI}_{rst})$, *i.e., the partial order of instances that are not in any pattern instances. An abstracted log,* $L' = (\mathsf{CI}', \prec', \pi'_{con}, \pi'_{cid})$, *is a log derived from* L *where*

- $\mathsf{CI}' = \mathsf{CI}_{rst} \cup \mathsf{PI}_L$ *is a set of instances;*

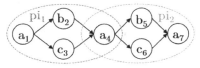 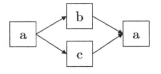

(a) Visualization of the covering relation of a case and the pattern instances identified.

(b) A pattern class.

Fig. 6. A motivating example of pattern instance identification, where the identified pattern instances, pi_1 and pi_2, are annotated and $\mathsf{pi}_1 \prec \mathsf{pi}_2$.

- $\prec' = \prec_{\mathsf{rst}} \cup \prec_{\mathsf{p}} \cup (\mathsf{Cl}_{\mathsf{rst}} \times \mathsf{PI}_L)$ *is a partial order on* Cl', *where* $\forall \mathsf{ci} \in \mathsf{Cl}_{\mathsf{rst}} \forall \mathsf{pi} \in \mathsf{PI}_L(\pi_{cid}(\mathsf{ci}) = \pi_{cid}^{\mathsf{p}}(\mathsf{pi}))$, $\mathsf{ci} \prec \mathsf{pi} \iff \forall \mathsf{ci}' \in \underline{\mathsf{pi}}(\mathsf{ci} \prec \mathsf{ci}')$ *and* $\mathsf{pi} \prec \mathsf{ci} \iff \forall \mathsf{ci}' \in \underline{\mathsf{pi}}(\mathsf{ci}' \prec \mathsf{ci})$;
- $\pi'_{con}\colon \overline{\mathsf{Cl}'} \to \mathcal{U}_{con}$, *where* $\forall \mathsf{ci} \in \mathsf{Cl}', \mathsf{ci} \in \mathsf{Cl}_{\mathsf{rst}} \iff \pi'_{con}(\mathsf{ci}) = \pi_{con}(\mathsf{ci}) \wedge \mathsf{ci} \in \mathsf{PI}_L \iff \pi'_{con}(\mathsf{ci}) = \pi_{con}^{\mathsf{p}}(\mathsf{ci})$;
- $\pi'_{cid}\colon \mathsf{Cl}' \to \mathcal{U}_{cid}$, *where* $\forall \mathsf{ci} \in \mathsf{Cl}', \mathsf{ci} \in \mathsf{Cl}_{\mathsf{rst}} \iff \pi'_{cid}(\mathsf{ci}) = \pi_{cid}(\mathsf{ci}) \wedge \mathsf{ci} \in \mathsf{PI}_L \iff \pi'_{cid}(\mathsf{ci}) = \pi_{cid}^{\mathsf{p}}(\mathsf{ci})$.

We define the key artifacts, i.e., a pattern class and a pattern instance, in this section. We further impose constraints on the relation among the underlying instances of a pattern instance to identify pattern instances that could be more suitable under certain circumstances in practice. The proposed framework is introduced and illustrated with the running example described in Sect. 2.

5   Implementation

In this section, we present the implementation of the framework. We explain the implementation of candidate pattern instances extraction in Sect. 5.1, followed by the extraction of local pattern instances in Sect. 5.2. A generic method to identify pattern instances is introduced in Sect. 5.3.

5.1   Extracting Candidate Pattern Instances

To extract the candidate pattern instances in a case, we incrementally add instances to a partial order of instances in the case and check for isomorphism between the instances selected and a pattern class. Let $\mathsf{PC} = (X, \prec, l)$ be a pattern class and L be a log. Given a case $c \in \mathsf{CID}(L)$, $\mathsf{c}_L = (\mathsf{Cl}_c, \prec_c)$. Given $\mathsf{Cl} \subseteq \mathsf{Cl}_c$ and $\prec_{\mathsf{po}} = \prec_c \cap (\mathsf{Cl} \times \mathsf{Cl})$, we let $\mathsf{po} = (\mathsf{Cl}, \prec_{\mathsf{po}}, \pi_{con})$. We define the following functions:

- $\mathsf{INIT}\colon \mathcal{P}(\mathcal{U}_{inst}) \times \mathcal{P}(\mathcal{U}_{inst} \times \mathcal{U}_{inst}) \to \mathcal{P}(\mathcal{P}(\mathcal{U}_{inst}) \times \mathcal{P}(\mathcal{U}_{inst} \times \mathcal{U}_{inst}))$, where $\mathsf{INIT}(\mathsf{c}_L)$ initiates a set of partial orders of instances to be extended where $\forall (\mathsf{Cl}', \prec') \in \mathsf{INIT}(\mathsf{c}_L)$, $\mathsf{Cl}' \subseteq \mathsf{Cl}_c$ and $\prec' = \prec_c \cap (\mathsf{Cl}' \times \mathsf{Cl}')$.
- $\mathsf{SEL}\colon \mathcal{P}(\mathcal{U}_{inst}) \times \mathcal{P}(\mathcal{U}_{inst} \times \mathcal{U}_{inst}) \times \mathcal{P}(\mathcal{U}_{inst}) \times \mathcal{P}(\mathcal{U}_{inst} \times \mathcal{U}_{inst}) \to \mathcal{P}(\mathcal{U}_{inst})$, where $\mathsf{SEL}(\mathsf{po}, \mathsf{c}_L) \subseteq \mathsf{Cl}_c$ selects a set of instances to be added into po where $\forall \mathsf{ci} \in \mathsf{SEL}(\mathsf{po}, \mathsf{c}_L), \mathsf{ci} \notin \mathsf{po}$.
- $\mathsf{CHECK}\colon \mathcal{P}(\mathcal{U}_{inst}) \times \mathcal{P}(\overline{\mathcal{U}_{inst} \times \mathcal{U}_{inst}}) \times \mathcal{U}_{inst} \times \mathcal{U}_{con} \to \{true, false\}$, where $\mathsf{CHECK}(\mathsf{po}, \mathsf{ci}, \mathsf{PC})$ checks if adding $\mathsf{ci} \in \mathsf{Cl}_c$ into po *may* form a candidate pattern instance of PC, i.e., let $\mathsf{po}' = \mathsf{ADD}(\mathsf{po}, \mathsf{ci}, \pi_{con})$, $\mathsf{CHECK}(\mathsf{po}, \mathsf{ci}, \mathsf{PC}) = true$ iff $\exists X' \subseteq \underline{\mathsf{PC}}(|X'| \geq 1 \wedge \mathsf{po}' \simeq (X', \prec_{X'}, l))$, where $\prec_{X'} = \prec \cap (X' \times X')$; if $X' = \underline{\mathsf{PC}}$, po' forms a candidate pattern instance.

Algorithm 1 illustrates the implementation of $\mathsf{EXT}_L(c, \mathsf{PC})$ in Definition 5 with the three key functions defined. The algorithm initiates a set of partial orders of instances and incrementally adds other instances. If there are no instances to be added such that it *may* form a candidate pattern instance of the input pattern class in the later iteration, we discard the partial order of instances. Otherwise, we check if the instances form a candidate pattern instance. If so, the instances form a candidate pattern instance, or the partial order of instances is added back to the open items to be checked in the next iteration.

A (labeled) partial order can be easily converted into a (labeled) directed acyclic graph (DAG). By representing a partial order of instances and a pattern class as labeled DAGs, we apply *graph edit distance* [17], i.e., a measure of similarity between two graphs that searches for the minimal cost of graph operations to make one graph isomorphic to the other, to check for the isomorphism between two partial orders.

Algorithm 1. Candidate Pattern Instance Extraction

Input: case $c \in \text{CID}(L)$, a pattern class $\text{PC} \in \mathcal{U}_{con}$
Output: a collection of candidate pattern instances of PC in c, i.e., *candidates*

 1: $c_L = (\text{CI}_c, \prec_c)$
 2: $open \leftarrow \text{INIT}(c_L)$ ▷ a set of partial orders of instances
 3: $candidates \leftarrow \{\}$ ▷ an empty set to collect candidate pattern instances
 4: **while** $open$ **do**
 5:     $\text{po} \leftarrow open.pop()$
 6:     $\text{CI}' \leftarrow \text{SEL}(\text{po}, c_L)$ ▷ instances to add incrementally
 7:     **for** $\text{ci} \in \text{CI}'$ **do**
 8:         **if** $\text{CHECK}(\text{po}, \text{ci}, \text{PC})$ **then**   ▷ if po can be a potential candidate
                                              ▷ pattern instance of PC by adding ci
 9:            $\text{po}' \leftarrow \text{ADD}(\text{po}, \text{ci}, \pi_{con})$
10:            **if** $\text{po}' \simeq \text{PC}$ **then**
11:               $candidates.add(\text{po}')$
12:            **else**
13:               $open.add(\text{po}')$ ▷ add po$'$ back to *open* for the next iteration
14:            **end if**
15:         **end if**
16:     **end for**
17: **end while**

Optimization. We optimize the implementation by reducing the search space while ensuring that the relation among the instances is not altered. The optimization of the algorithm can be easily achieved by directly projecting the relevant instances in a case, i.e., the instance of the underlying concepts of a pattern class, since the relation among the instances remains after the projection.

5.2   Extracting Local Pattern Instances

The local pattern instances may be extracted by selecting from the pattern instances identified. Alternatively, we search for candidate pattern instances with the property of local pattern instances. The search of local pattern instances may be implemented in a similar way as described in Algorithm 1, however, with the isomorphism check based on the covering relation of partial orders. The check is realized by representing the covering relation of a partial order with a DAG, which is the *transitive reduction* of the DAG representing the partial order.

Nevertheless, the optimization of the search cannot be performed in the same way as the projection of relevant instances may result in missing relations in the graph representing the covering relation of a partial order. Hence, we must ensure the connectivity of the covering relation of a partial order of instances. Let $\mathsf{PC} = (X, \prec, l)$ be a pattern class. Given start activities $\mathsf{SA} = \{l(x) \mid x \in X, \nexists x' \in X(x' \prec x)\}$, we add an artificial start node and the relation from the start node to every node labeled with activity in SA to the graph representing a case. We remove a node if there does not exist an *undirected* path from the start node to the node. Then, the projection of relevant instances may be performed and we optimize the search with the bread-first search strategy.

5.3   Identification of Pattern Instances

One may impose semantic constraints on IDEN_c to identify pattern instances. Alternatively, for a generic application of the framework, we implement IDEN_c by introducing an overlapping threshold $t \in \mathbb{R}$, where $0 \leq t < 1$. Let L be a log and CAN denote the candidate pattern instances in a case $c \in \mathsf{CID}(L)$. For any $\mathsf{pi}_1, \mathsf{pi}_1 \in \mathsf{IDEN}_c(\mathsf{CAN})$, $|\mathsf{pi}_1 \cap \mathsf{pi}_2|/|\mathsf{pi}_1 \cup \mathsf{pi}_2| \leq t$; if $t = 0$, a set of disjoint candidate pattern instances are identified as pattern instances. We determine the relation between pi_1 and pi_2 based on the majority relation of the non-shared underlying instances, i.e., given $\mathsf{Cl}_1 = \mathsf{pi}_1 \setminus \mathsf{pi}_2$ and $\mathsf{Cl}_2 = \mathsf{pi}_2 \setminus \mathsf{pi}_1$, $\mathsf{pi}_1 \prec \mathsf{pi}_2 \iff |\{\mathsf{ci}_1 \in \mathsf{Cl}_1 \mid \mathsf{ci}_2 \in \mathsf{Cl}_2(\mathsf{ci}_1 \prec \mathsf{ci}_2)\}|/|\mathsf{Cl}_1| > 0.5$; note that we assume that $\pi^{\mathsf{p}}_{con}(\mathsf{pi}_1) = \pi^{\mathsf{p}}_{con}(\mathsf{pi}_2)$ s.t. $|\mathsf{Cl}_1| = |\mathsf{Cl}_2|$. The instantiation is non-deterministic.

The abstraction, based on the given pattern instances and their relation, is straightforward to implement, following the guidelines outlined in Definition 7. As a result, an abstracted log is computed based on the pattern class, which serves as the input for the subsequent iteration of another pattern class.

6   Experiments

In this section, we present the application of the proposed framework with a synthetic log and conduct a case study based on a real-life log [19].

6.1   Evaluation

We construct a partially ordered log containing $5,000$ cases based on the process in Fig. 7, from which Table 1 is extracted. We evaluate the proposed framework in three aspects: the number of pattern instances selected, the performance with optimization applied, and the quality metrics of the process models discovered.

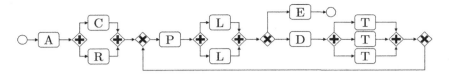

Fig. 7. A process model used for generating the synthetic log. The model is represented with BPMN [1]. The labels correspond to the abbreviation of activity labels in Table 1.

Defining Pattern Classes. We define four pattern classes. Let the pattern class visualized in Fig. 2 be pattern class PC2; other pattern classes are visualized in Fig. 8. We identify local pattern instances for PC1 (Fig. 8a) and PC2 (Fig. 2). The pattern class PC3 defined in Fig. 8b consists of three concurrent Provide Treatment. The pattern class PC4 is defined by other pattern classes and activity Diagnose (D). As the framework abstracts a log based on one pattern class at a time, the numbers indicate the corresponding iteration.

Experimental Setup. Inspired by [12], we evaluate the impact of noises on the proposed framework. We inject $n\%$, where $n \in \{10, 20, 30\}$, of noises to the log by injecting $n\%$ of noises to every case as illustrated in Algorithm 2. We randomly select $n\%$ of instances in a case and swap the relation with one of the directly succeeding instance for every instance selected, i.e., given an instance c_i, we select an instance $c_i'(c_i \neq c_i')$, where $c_i \prec c_i'$, and swap the relation. For discovering the process models, we applied Inductive Miner - Infrequent with a noise threshold of 0.2 [8].

Figure 9 shows the number of pattern instances identified for every pattern class. Since the first three pattern classes do not share common activity labels, the number of pattern instances extracted is independent; however, the number of pattern instances of Treatment Process (PC4) dramatically decreases since the pattern class is defined based on other pattern classes; hence, the number of pattern instances abstracted is also limited to the number of pattern instances of other pattern classes. The number of pattern instances identified for PC2 is much higher than other pattern classes since the pattern class can be conducted several times in a case. In addition, with the percentage of noise injected increasing, the

(a) Basic Consultation (PC1)   (b) Treatment (PC3)   (c) Treatment Process (PC4)

Fig. 8. Pattern classes defined for synthetic log. Abbreviated labels are provided for simplicity. Note that PC4 is defined based on PC2, PC3, and Diagnose.

Algorithm 2. Noise Injection of Case

Input: case $c \in \mathsf{CID}(L)$, a noise percentage $n \in \{10\%, 20\%, 30\%\}$
Output: case with noise $\mathsf{c}' = (\mathsf{CI}', \prec')$
1: $\mathsf{c}' = (\mathsf{CI}_c, \prec_c)$ ▷ initiate a partial order of instances in case c
2: $cnt \leftarrow floor(|\mathsf{CI}_c| \times n)$ ▷ number of pairs of instances to swap
3: $\mathsf{CI} \leftarrow$ randomly select cnt instances   ▷ $\mathsf{CI} \subseteq \mathsf{CI}_c$, where $|\mathsf{CI}| = cnt$
4: **for** $\mathsf{ci} \in \mathsf{CI}$ **do**
5:    $\mathsf{ci}' \leftarrow$ SelectFollowingNeighbor($\mathsf{ci}, \mathsf{c}_L$)   ▷ $\mathsf{ci} \prec \mathsf{ci}'$
6:    $\mathsf{c}' \leftarrow$ Swap($\mathsf{ci}, \mathsf{ci}'$)
7: **end for**

---

number of pattern instances identified decreases because the noise alters the relation of the instances in the log.

Figure 10 presents the average time required to abstract a case based on the pattern classes defined. Regardless of the optimization, abstracting a case based on a pattern class requires less than 1 second. With optimization, the runtime is further reduced to 4 times faster; for the pattern class of **Basic Consultation** (PC1), the optimization even results in 16 times faster. Meanwhile, we observe that the abstraction based on the pattern class **Treatment** (PC3) is much faster than other abstractions since the checking for isomorphism is much faster as the graph representing the pattern class contains only isolated nodes labeled with **Provide Treatment** and no relation, represented with edges in a graph, needs to be examined.

Figure 11 reports the quality metrics of the process models discovered based on the abstracted logs. We see that the noise has a negative impact on the fitness as shown in Fig. 11a. Except for the abstraction based on PC4 without noise injection, the fitness is correlated with the number of pattern instances abstracted as shown in Fig. 9. Compared to fitness, the noise has less impact on the precision as shown in Fig. 11b. Interestingly, with more noise injected, which causes fewer pattern instances abstracted, the precision increases. By analyzing the conformance details, we infer that it is due to the unmatched instances that are labeled with the underlying concepts of a pattern class, which impacts the number of instances in a case and further influences the metrics. The harmonic

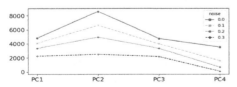

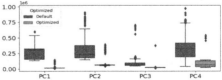

Fig. 9. Number of pattern instances abstracted per pattern class with different noise injection.

Fig. 10. Average abstraction time in microsecond per case based on the pattern classes defined.

mean of the fitness and the precision, F-measure, as shown in Fig. 11c enhances due to the increase in fitness.

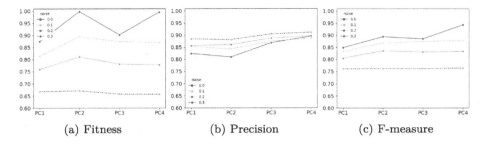

(a) Fitness   (b) Precision   (c) F-measure

Fig. 11. Quality metrics of process models discovered using abstracted logs. The ranges of the y-axis of the plots are uniformly set from 0.6 to 1.0 for comparison.

6.2   Case Study

To demonstrate the proposed framework in practice, we apply the framework to a real-life log [19]. We preprocess the log to construct a partially ordered log based on timestamps. We pair every start record with exactly one complete record based on their order to construct an activity instance. The pattern classes are defined by sets of activities indicating a successful operation. Figure 12 shows one of the pattern classes, indicating a successful application.[2]

Figure 14 presents an excerpt of the analysis. The figure shows the behavior of the process with the relative frequency projected on the visualization. The analysis shows that, since the pattern classes are defined by sets of successful

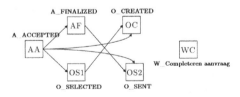

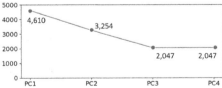

Fig. 12. Successful Application (PC1), where WC is expected to be executed concurrently in time with activities.

Fig. 13. Number of pattern instances abstracted per pattern class defined for case study.

---

[2] We define PC2 as concurrent {O_SENT_BACK, W_Nabellen offertes}, PC3 as concurrent {O_ACCEPTED, A_APPROVED, A_REGISTERED, A_ACTIVATED, W_Valideren aanvraag}, and PC4 as a sequence of PC1, PC2, and PC3 (considering that a total order is also a partial order).

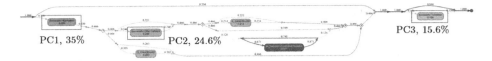

Fig. 14. An excerpt of the analysis based on an abstracted log. The visualization is based on IMflc [9] and the shade of color shows the relative importance based on the number of instances identified on the path. We highlight the defined pattern classes.

operation occurring sequentially in a process, as the process continues, the percentage of successful operations also decreases. The percentage of successful end-to-end operation, i.e., PC4, depends on the last successful operations, i.e., PC3. The analysis is further supported by the number of pattern instances abstracted based on the pattern classes defined in Fig. 13. Meanwhile, with abstraction, we can easier identify and further investigate the unsuccessful executions, e.g., other activities shown in the excerpt. In addition, we apply the abstraction with the technique proposed in [14], where we model the undefined relations in the pattern classes as parallelism in the representation of the regularity defined in the work. With the same sets of activities in the regularity defined as the pattern classes, the technique does not identify any instances.

7   Related Work

This section discusses the research in event abstraction in the field of process mining. Numerous event abstraction techniques focus on identifying regularity in event data to group activities or records of activities. The authors in [18] apply a statistical model to *predict* the class of a record at a higher level. Nguyen et al. decompose a process into sets of activities by exploiting the modularity metrics based on a graph constructed from event data [15]. In [11], the authors apply clustering based on the features encoded from fragments of a sequence of event data, which are seen as an instance at the higher level and are provided with domain knowledge. The work focuses on identifying the regularity from sequential event data with strong assumptions on a process, e.g., sensor data and milestone existence; the identification of the instances at the higher level is rather straightforward or ignored due to the assumption of classical event data.

To facilitate analyzing event data at a higher level, it is important that, not only concepts at the higher level, an instance of a concept at the higher level is also identified such that one may apply abstraction iteratively based on their needs. Some work applies the hierarchy of concepts to construct a hierarchy of abstractions [10,13]. In Lu et al. [13], an instance at the higher level is extracted by projecting the relevant records. Leemans et al. apply the discovering techniques to discover groups of process models at different levels and compose them to construct a complete model at the specified level [10]; the extraction of an instance is achieved with alignment [3]. The alignment is also exploited to identify instances of the regularity identified in [14]. In Bose and van der Aalst [5], the authors discover frequent local execution regularity and abstract accordingly. The techniques discussed consider the iterative applicability; however, except for simply projecting the records in [13], the extraction of the instances at the higher level tends to be limited to local regularity.

This paper explicitly considers partially ordered event data and focuses on the extraction of the pattern instances. For discovering the regularity, since a pattern class and a case can be represented as labeled DAGs, we argue that existing techniques for frequent graph pattern mining may be exploited [16]. With a pattern class defined based on partial order relation, the extraction achieves beyond local regularity and the reliability of the abstraction can be enhanced with the support of human analysts by providing a more comprehensible representation, i.e., the regularity that can be directly mapped to the behavior of the execution of concepts, e.g., activities, observed in real-life.

8   Conclusion

Motivated by the applicability of event abstraction in practice, we present and define a framework for identifying pattern instances based on partially ordered event data, which reflect the behavior of the activities performed in real-life. The framework abstracts a log based on a pattern class by extracting candidate pattern instances and identifying pattern instances of the pattern class. We implemented the framework and conducted experiments by constructing a hierarchy of abstractions based on a synthetic and a real-life log. The experiments demonstrate the impact of noises and how one can obtain insights from the analysis based on abstracted logs. The framework is also applicable to classical event data since a total order is also a partial order. For future work, our objectives are two-fold. First, we aim to further support in defining the order of pattern classes as specified by domain experts. Second, we plan to extend the extraction to identify unexpected behavior in partially ordered event data, while explicitly considering the causal relation of the instances.

References

1. Business process model and notation (BPMN) version 2.0. Object Management Group (2011)
2. van der Aalst, W.M.P.: Process Mining - Data Science in Action, 2nd edn. Springer, Heidelberg (2016). https://doi.org/10.1007/978-3-662-49851-4
3. van der Aalst, W.M.P., Adriansyah, A., van Dongen, B.F.: Replaying history on process models for conformance checking and performance analysis. WIREs Data Mining Knowl. Discov. **2**(2), 182–192 (2012)
4. Augusto, A., et al.: Automated discovery of process models from event logs: review and benchmark. IEEE Trans. Knowl. Data Eng. **31**(4), 686–705 (2019)
5. Jagadeesh Chandra Bose, R.P., van der Aalst, W.M.P.: Abstractions in process mining: a taxonomy of patterns. In: Dayal, U., Eder, J., Koehler, J., Reijers, H.A. (eds.) BPM 2009. LNCS, vol. 5701, pp. 159–175. Springer, Heidelberg (2009). https://doi.org/10.1007/978-3-642-03848-8_12

6. Carmona, J., van Dongen, B.F., Solti, A., Weidlich, M.: Conformance Checking - Relating Processes and Models. Springer, Cham (2018). https://doi.org/10.1007/978-3-319-99414-7
7. Hornix, P.T.G.: Performance analysis of business processes through process mining. Master's Thesis, Eindhoven University of Technology (2007)
8. Leemans, S.J.J., Fahland, D., van der Aalst, W.M.P.: Discovering block-structured process models from event logs containing infrequent behaviour. In: Lohmann, N., Song, M., Wohed, P. (eds.) BPM 2013. LNBIP, vol. 171, pp. 66–78. Springer, Cham (2014). https://doi.org/10.1007/978-3-319-06257-0_6
9. Leemans, S.J.J., Fahland, D., van der Aalst, W.M.P.: Using life cycle information in process discovery. In: Reichert, M., Reijers, H.A. (eds.) BPM 2015. LNBIP, vol. 256, pp. 204–217. Springer, Cham (2016). https://doi.org/10.1007/978-3-319-42887-1_17
10. Leemans, S.J.J., Goel, K., van Zelst, S.J.: Using multi-level information in hierarchical process mining: Balancing behavioural quality and model complexity. In: van Dongen, B.F., Montali, M., Wynn, M.T. (eds.) 2nd International Conference on Process Mining, ICPM 2020, Padua, Italy, 4–9 October 2020, pp. 137–144. IEEE (2020)
11. de Leoni, M., Dündar, S.: Event-log abstraction using batch session identification and clustering. In: Hung, C., Cerný, T., Shin, D., Bechini, A. (eds.) SAC 2020: The 35th ACM/SIGAPP Symposium on Applied Computing, online event, [Brno, Czech Republic], 30 March–3 April 2020, pp. 36–44. ACM (2020)
12. de Leoni, M., Marrella, A.: Aligning real process executions and prescriptive process models through automated planning. Expert Syst. Appl. **82**, 162–183 (2017)
13. Lu, X., Gal, A., Reijers, H.A.: Discovering hierarchical processes using flexible activity trees for event abstraction. In: van Dongen, B.F., Montali, M., Wynn, M.T. (eds.) 2nd International Conference on Process Mining, ICPM 2020, Padua, Italy, 4–9 October 2020, pp. 145–152. IEEE (2020)
14. Mannhardt, F., de Leoni, M., Reijers, H.A., van der Aalst, W.M.P., Toussaint, P.J.: Guided process discovery - a pattern-based approach. Inf. Syst. **76**, 1–18 (2018)
15. Nguyen, H., Dumas, M., ter Hofstede, A.H.M., Rosa, M.L., Maggi, F.M.: Stage-based discovery of business process models from event logs. Inf. Syst. **84**, 214–237 (2019)
16. Nguyen, L.B.Q., Zelinka, I., Snásel, V., Nguyen, L.T.T., Vo, B.: Subgraph mining in a large graph: a review. WIREs Data Mining Knowl. Discov. **12**(4), e1454 (2022)
17. Sanfeliu, A., Fu, K.: A distance measure between attributed relational graphs for pattern recognition. IEEE Trans. Syst. Man Cybern. **13**(3), 353–362 (1983)
18. Tax, N., Sidorova, N., Haakma, R., van der Aalst, W.M.P.: Event abstraction for process mining using supervised learning techniques. In: Bi, Y., Kapoor, S., Bhatia, R. (eds.) IntelliSys 2016. LNNS, vol. 15, pp. 251–269. Springer, Cham (2018). https://doi.org/10.1007/978-3-319-56994-9_18
19. van Dongen, B.F.: BPI challenge 2012 (2012). https://data.4tu.nl/articles/dataset/BPI_Challenge_2012/12689204
20. Yasmin, F.A., Bukhsh, F.A., de Alencar Silva, P.: Process enhancement in process mining: a literature review. In: Ceravolo, P., López, M.T.G., van Keulen, M. (eds.) Proceedings of the 8th International Symposium on Data-driven Process Discovery and Analysis (SIMPDA 2018), Seville, Spain, 13–14 December 2018. CEUR Workshop Proceedings, vol. 2270, pp. 65–72. CEUR-WS.org (2018)

Incremental Discovery of Process Models Using Trace Fragments

Daniel Schuster[1,2]([envelope]) [iD], Niklas Föcking[1] [iD], Sebastiaan J. van Zelst[1,2] [iD],
and Wil M. P. van der Aalst[1,2] [iD]

[1] Fraunhofer Institute for Applied Information Technology FIT,
Sankt Augustin, Germany
{daniel.schuster,niklas.foecking,sebastiaan.van.zelst}@fit.fraunhofer.de
[2] RWTH Aachen University, Aachen, Germany
wvdaalst@pads.rwth-aachen.de

Abstract. Process discovery learns process models from event data and is a crucial discipline within process mining. Most existing approaches are fully automated, i.e., event data is provided, and a process model is returned. Thus, process analysts cannot interact and intervene besides parameter settings. In contrast, Incremental Process Discovery (IPD) enables users to actively participate in the discovery phase by gradually selecting process behavior to be incorporated into a process model. Further, most discovery approaches assume process executions, also termed traces, recorded in event data to be complete—complete traces span the actual process from start to completion. Incomplete traces are usually removed in the event data preparation as most discovery algorithms cannot handle them respectively treat them simply as full traces. This paper presents a novel IPD approach that can incorporate process behavior recorded in trace fragments, thus supporting incomplete data. Our experiments show promising results indicating that using trace fragments within IPD leads to high-quality process models.

Keywords: Process mining · Process discovery · Alignments

1 Introduction

Process discovery, i.e., learning process models from event data, is a critical discipline within *process mining*. Discovered models are vital artifacts as they capture the actual execution of processes. Further, many subsequently applied process mining techniques require models as input, for example, generating temporal performance diagnostics and conformance checking statistics [9]. Moreover, these models are used for specifying process-aware information systems [15].

Most process discovery approaches are fully automated [3,22,23]. They take event data as input and return a process model. Apart from parameter settings, users cannot interact with these algorithms despite choosing the input event data and post-processing the discovered model. Moreover, most process discovery

© The Author(s), under exclusive license to Springer Nature Switzerland AG 2023
C. Di Francescomarino et al. (Eds.): BPM 2023, LNCS 14159, pp. 55–73, 2023.
https://doi.org/10.1007/978-3-031-41620-0_4

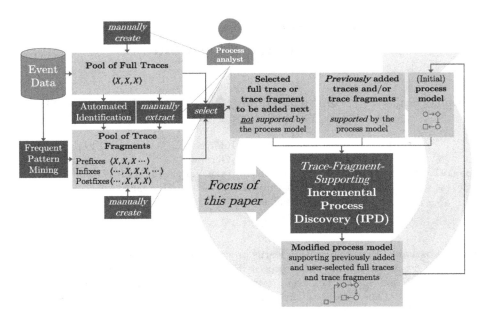

Fig. 1. Overview of Incremental Process Discovery (IPD) with trace fragments and potential origins of trace fragments. Process analysts gradually select process behavior, i.e., full traces and trace fragments, that is added to the process model

algorithms consider the process executions, i.e., *traces*, recorded in the event data to be *complete*—traces are assumed to span the process from start to completion. *Trace fragments* are usually filtered during event data preparation [6,7].

In contrast to conventional process discovery, taking event data as input and returning a process model, *domain-knowledge-utilizing process discovery* utilizes additional information besides event data, e.g., user feedback in an interactive discovery phase [10] or explicitly-specified knowledge like precedence constraint among activities [12]. We provide a review of such approaches in [19].

This paper focuses on *Incremental Process Discovery (IPD)* [17], which discovers process models from event data by gradually extending a model by new process behavior, cf. Fig. 1. The central research question addressed is: *How can trace fragments, i.e., trace prefixes/infixes/postfixes, be (incrementally) added to a process model?* We answer this question by proposing a novel IPD approach that allows gradually discovering models from trace fragments and full traces. Thus, the proposed approach utilizes incomplete process behavior, i.e., trace fragments. We evaluate the proposed approach on real-life event data. The results indicate that incorporating trace fragments is beneficial and yields high quality process models. The results further show that a distinction between trace fragments and full traces can lead to better models compared to approaches that do not support fragments respectively consider all traces as full. Finally, we implemented the approach in the open-source process mining tool *Cortado* [20].[1]

Consider Fig. 1; trace fragments may originate from different sources. First, event data itself may contain incomplete respectively partial traces that often occur when event data of a specific time range is extracted from information systems. Since trace fragments are usually not labeled as such and are considered complete by state-of-the-art process discovery approaches, event data preparation techniques must be used, e.g., [6], to identify trace fragments that can then be added to the fragment pool, cf. Fig. 1. Second, users can manually extract relevant fragments from full traces. Reasons to proceed in this way are manifold. For instance, an analyst does not want certain variations from full traces that cover specific process stages in the discovered process model; instead, the analyst is only interested in specific fragments. Finally, users can manually create trace fragments during IPD if particular process behavior is not present in the data but should be reflected by the discovered model.

This paper addresses current challenges within business process management (BPM) and process mining. Central research challenges of the BPM discipline are identified in [4]. One challenge is the augmentation "of process mining with common sense and domain knowledge" [4, p. 3]. Domain knowledge about the process under study is often available besides event data; however, process mining techniques often do not utilize such domain knowledge. IPD itself and the proposed trace-fragment-supporting IPD approach allow such exploitation of domain knowledge because 1) users gradually select process behavior (full traces and fragments) to be incorporated into the process model under discovery and 2) can specify how traces from event data are interpreted, i.e., either as full or prefix/infix/postfix traces. By manually creating trace fragments (cf. Fig. 1), another means to incorporate domain knowledge exists. The authors [4] further argue that domain knowledge utilization is beneficial to overcome event data quality issues.

The remainder of this paper is structured as follows. Section 2 presents related work, while Sect. 3 introduces preliminaries. We present the proposed trace-fragment-supporting IPD approach in Sect. 4. An initial evaluation of the proposed approach is presented in Sect. 5. Finally, Sect. 6 concludes this paper.

2   Related Work

Plenty of conventional, automated process discovery algorithms exist; we refer to [3,22,23] for overviews. These mainly differ in the model formalism used, guarantees of the discovered model concerning model properties (e.g., soundness), concerning the event data provided (e.g., replay fitness), and computational complexity. From an input/output perspective, however, these algorithms work similarly, i.e., they take event data as input and automatically learn a model. Domain-knowledge-utilizing process discovery approaches are significantly less

---

[1] https://cortado.fit.fraunhofer.de (from version 1.10.0).

common than conventional process discovery. We provide a systematic literature review in [19]. As considered in this work, IPD has been initially proposed in [17]. In [21], the authors also use the term incremental process discovery. However, they create multiple transition systems describing process behavior and then incrementally compose them into a single transition system from which a process model is eventually discovered. Thus, their definition of incremental discovery is unrelated to ours, as illustrated in Fig. 1.

Process model repair [2,11,14] is related to IPD, as elaborated in [19]. Model repair techniques extend process models by non-fitting process behavior. Although repair techniques are not intended to be used incrementally, they can be used similar as illustrated in Fig. 1. However, they create a process model as close as possible to the input process model since the focus is on repairing and not on discovering. This objective is an essential difference from IPD, where this objective does not necessarily exist and is even disadvantageous because the model in the discovery process is constantly being developed and changed.

To the best of our knowledge, no discovery approach, neither conventional nor domain-knowledge-utilizing, and no model repair approach addresses trace fragments explicitly. Thus, even if there is domain knowledge that allows traces to be identified as fragments and labeled as such, no existing approach supports it. However, note that discovery approaches utilizing region theory [5] can handle prefixes. On the contrary, trace fragments—most approaches consider all traces complete—are typically filtered from event logs to obtain better models [6,7]. Thus, supporting trace fragments within IPD is a novelty.

3   Preliminaries

Let X be a set. We denote the universe of multi-sets over X by $\mathcal{B}(X)$ and the set of all sequences over X as X^*, e.g., $[a^3, c] \in \mathcal{B}(\{a, b, c\})$ and $\langle a, b, b \rangle \in \{a, b, c\}^*$. Given two multi-sets $M, M' \in \mathcal{B}(X)$, we denote their union by $M \uplus M'$. We denote the length of a sequence σ by $|\sigma|$. For $1 \leq i \leq |\sigma|$, $\sigma(i)$ represents the i-th element of σ. Given sequences σ and σ', we denote their concatenation by $\sigma \cdot \sigma'$, e.g., $\langle a \rangle \cdot \langle b, c \rangle = \langle a, b, c \rangle$. We extend the $\cdot$ operator to sets of sequences, i.e., let $S_1, S_2 \subseteq X^*$ then $S_1 \cdot S_2 = \{\sigma_1 \cdot \sigma_2 \mid \sigma_1 \in S_1 \wedge \sigma_2 \in S_2\}$. For sequences σ, σ', the set of all interleaved sequences is denoted by $\sigma \diamond \sigma'$, e.g., $\langle a, b \rangle \diamond \langle c \rangle = \{\langle a, b, c \rangle, \langle a, c, b \rangle, \langle c, a, b \rangle\}$. We extend the $\diamond$ operator to sets of sequences. Let $S_1, S_2 \subseteq X^*$, $S_1 \diamond S_2$ denotes the set of interleaved sequences, i.e., $S_1 \diamond S_2 = \bigcup_{\sigma_1 \in S_1, \sigma_2 \in S_2} \sigma_1 \diamond \sigma_2$.

For $\sigma \in X^*$ and $X' \subseteq X$, we define the projection function $\sigma_{\downarrow_{X'}} : X^* \to (X')^*$ with: $\langle \rangle_{\downarrow_{X'}} = \langle \rangle$, $(\langle x \rangle \cdot \sigma)_{\downarrow_{X'}} = \langle x \rangle \cdot \sigma_{\downarrow_{X'}}$ if $x \in X'$, and $(\langle x \rangle \cdot \sigma)_{\downarrow_{X'}} = \sigma_{\downarrow_{X'}}$ otherwise.

Let $t = (x_1, \ldots, x_n) \in X_1 \times \ldots \times X_n$ be an n-tuple over n sets. We define projection functions that extract a specific element of t, i.e., $\pi_1(t) = x_1, \ldots, \pi_n(t) = x_n$, e.g., $\pi_2((a, b, c)) = b$. For a sequence $\sigma = \langle (x_1^1, \ldots, x_n^1), \ldots, (x_1^m, \ldots, x_n^m) \rangle \in (X_1 \times \ldots \times X_n)^*$ containing n-tuples, we define projection functions $\pi_1^*(\sigma) = \langle x_1^1, \ldots, x_1^m \rangle, \ldots, \pi_n^*(\sigma) = \langle x_n^1, \ldots, x_n^m \rangle$; e.g., $\pi_2^*(\langle (a, b), (c, d), (c, b) \rangle) = \langle b, d, b \rangle$.

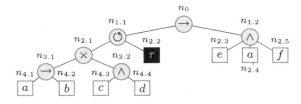

Fig. 2. Exemplary process tree $T_0 = (V_0, E_0, \lambda_0, n_0)$ with $V_0 = \{n_o, \ldots, n_{4.4}\}$, $E_0 = \{(n_0, n_{1.1}), \ldots, (n_{3.2}, n_{4.4})\}$, and $\lambda(n_0) = \to, \ldots, \lambda(n_{4.4}) = d$

3.1 Event Data

Event logs are a collection of event data describing the execution of a process. Individual process executions, referred to as *traces*, are considered sequences of executed activities. For instance, $\sigma = \langle a_1, a_2, \ldots, a_n \rangle$ is a trace consisting of n activities, and $L = \left[\langle c, b, d \rangle^3, \langle a, e, d \rangle^2\right]$ is an event log that consists of 5 traces.

Definition 1 (Trace & Event Log). *Let $\mathcal{A}$ be the universe of activities. A trace σ is a sequence of activities, i.e., $\sigma \in \mathcal{A}^*$. An event log L is a multi-set of traces, i.e., $L \subseteq \mathcal{B}(\mathcal{A}^*)$.*

3.2 Process Models

We use process trees that represent a subclass of *sound workflow nets* and are an important model formalism used by many discovery approaches [8,13,17]. Process trees are rooted, labeled, ordered trees where leaf nodes represent activities and inner nodes control-flow operators that specify the execution of its subtrees. We distinguish four operators: sequence ($\to$), parallel ($\wedge$), loop ($\circlearrowleft$), and exclusive choice ($\times$). Figure 2 depicts an example process tree T_0.

Definition 2 (Process Tree). *Let $\mathcal{A}$ be the universe of activities with $\tau \notin \mathcal{A}$ and $\bigoplus = \{\to, \times, \wedge, \circlearrowleft\}$ be the process tree operators. We define a process tree as a labeled, rooted tree $T = (V, E, \lambda, r)$ consisting of a totally ordered set of nodes V, edges $E \subseteq V \times V$, a labeling function $\lambda : V \to \mathcal{A} \cup \{\tau\} \cup \bigoplus$, and a root $r \in V$.*

- *$(\{n\}, \emptyset, \lambda, n)$ with $\lambda(n) \in \mathcal{A} \cup \{\tau\}$ is a process tree.*
- *Given $k > 1$ trees $T_1 = (V_1, E_1, \lambda_1, r_1), \ldots, T_k = (V_k, E_k, \lambda_k, r_k)$, node $r \notin V_1 \cup \ldots \cup V_k$, and $\forall 1 \leq i < j \leq k (V_i \cap V_j = \emptyset)$ then $T = (V, E, \lambda, r)$ is a process tree where:*
  - *$V = V_1 \cup \cdots \cup V_k \cup \{r\}$,*
  - *$E = E_1 \cup \cdots \cup E_k \cup \{(r, r_1), \ldots, (r, r_k)\}$,*
  - *$\lambda(x) = \lambda_j(x)$ for all $j \in \{1, \ldots, k\}, x \in V_j$, and*
  - *$\lambda(r) \in \bigoplus$ and $\lambda(r) = \circlearrowleft \Rightarrow k = 2$.*

We denote the universe of process trees by $\mathcal{T}$.

For an arbitrary $T=(V, E, \lambda, r) \in \mathcal{T}$, we define function $c^T : V \rightarrow V^*$ that returns the child nodes of a given node sorted accordingly; for instance, $c^{T_0}(n_{1.2}) = \langle n_{2.3}, n_{2.4}, n_{2.5}\rangle$. We refer to the *Lowest Common Ancestor* (LCA) of two nodes $n, n' \in V$ as $lca^T(n, n') \in V$; e.g., $lca^{T_0}(n_{4.4}, n_{2.2}) = n_{1.1}$. The function $\Delta^T(n) \rightarrow \mathcal{T}$ returns the *subtree* rooted at $n \in V$ from T. We write $T' \sqsubseteq T$ to denote that T' is a subtree of T.

We define the semantics of process trees via *running sequences* consisting of 2-tuples where the first entry is a node n and the second entry is either the label if n is a leaf node or a label indicating the opening or closing of the subtree rooted at n. Figure 3 shows one running sequence ρ of process tree T_0. Note that ρ corresponds to the trace $\left(\pi_2^*(\rho)\right)_{\downarrow_\mathcal{A}} = \langle d, c, e, f, a\rangle$.

$$
\rho = \langle (n_0, open),
$$
$$
(n_{1.1}, open),
$$
$$
(n_{2.1}, open),
$$
$$
(n_{3.2}, open),
$$
$$
(n_{4.4}, d), (n_{4.3}, c),
$$
$$
(n_{3.2}, close),
$$
$$
(n_{2.1}, close),
$$
$$
(n_{1.1}, close),
$$
$$
(n_{1.2}, open),
$$
$$
(n_{2.3}, e), (n_{2.5}, f), (n_{2.4}, a),
$$
$$
(n_{1.2}, close),
$$
$$
(n_0, close)\rangle
$$

Fig. 3. Running sequence $\rho \in \mathcal{RS}(T_0)$

Definition 3 (Running Sequence). *Let $\mathcal{A}$ be the universe of activities with $\tau, open, close \notin \mathcal{A}$. Let $S = V \times (\mathcal{A} \cup \{\tau, open, close\})$ be the set of steps. For $T=(V, E, \lambda, r) \in \mathcal{T}$, we recursively define its running sequences $\mathcal{RS}(T) \subseteq S^*$.*

- *if $\lambda(r) \in \mathcal{A} \cup \{\tau\}$ (T is a leaf node): $\mathcal{RS}(T) = \{\langle (r, \lambda(r))\rangle\}$*
- *if $\lambda(r) = \rightarrow$ with child nodes $c^T(r) = \langle v_1, \ldots, v_k\rangle$ for $k \geq 1$:*
  $\mathcal{RS}(T) = \{\langle (r, open)\rangle\} \cdot \mathcal{RS}(\Delta^T(v_1)) \cdot \ldots \cdot \mathcal{RS}(\Delta^T(v_k)) \cdot \{\langle (r, close)\rangle\}$
- *if $\lambda(r) = \times$ with child nodes $c^T(r) = \langle v_1, \ldots, v_k\rangle$ for $k \geq 1$:*
  $\mathcal{RS}(T) = \{\langle (r, open)\rangle\} \cdot \mathcal{RS}(\Delta^T(v_1)) \cup \ldots \cup \mathcal{RS}(\Delta^T(v_k)) \cdot \{\langle (r, close)\rangle\}$
- *if $\lambda(r) = \wedge$ with child nodes $c^T(r) = \langle v_1, \ldots, v_k\rangle$ for $k \geq 1$:*
  $\mathcal{RS}(T) = \{\langle (r, open)\rangle\} \cdot \mathcal{RS}(\Delta^T(v_1)) \diamond \ldots \diamond \mathcal{RS}(\Delta^T(v_k)) \cdot \{\langle (r, close)\rangle\}$
- *if $\lambda(r) = \circlearrowright$ with child nodes $c^T(r) = \langle v_1, v_2\rangle$:*
  $\mathcal{RS}(T) = \{\langle (r, open)\rangle \cdot \sigma_1 \cdot \sigma_1' \cdot \sigma_2 \cdot \sigma_2' \cdot \ldots \cdot \sigma_m \cdot \langle (r, close)\rangle \mid m \geq 1 \wedge \forall 1 \leq i \leq m (\sigma_i \in \mathcal{RS}(\Delta^T(v_1))) \wedge \forall 1 \leq i < m (\sigma_i' \in \mathcal{RS}(\Delta^T(v_2)))\}$

The *language* of a tree $T \in \mathcal{T}$ is a set of supported traces, i.e., $\mathcal{L}(T) = \{\pi_2^*(\sigma)_{\downarrow_\mathcal{A}} \mid \sigma \in \mathcal{RS}(T)\} \subseteq \mathcal{A}^*$. Further, we define its prefix/infix/postfix language.

- $\mathcal{L}_{prefix}(T) = \{\sigma_1 \mid \sigma_1, \sigma_2 \in \mathcal{A}^* \wedge \sigma_1 \cdot \sigma_2 \in \mathcal{L}(T)\}$
- $\mathcal{L}_{infix}(T) = \{\sigma_2 \mid \sigma_1, \sigma_2, \sigma_3 \in \mathcal{A}^* \wedge \sigma_1 \cdot \sigma_2 \cdot \sigma_3 \in \mathcal{L}(T)\}$
- $\mathcal{L}_{postfix}(T) = \{\sigma_2 \mid \sigma_1, \sigma_2 \in \mathcal{A}^* \wedge \sigma_1 \cdot \sigma_2 \in \mathcal{L}(T)\}$

Finally, we formally introduce fitness-preserving process discovery.

Definition 4 (Fitness-Preserving Process Discovery). *Let $L \subseteq \mathcal{B}(\mathcal{A}^*)$ be an event log. We define a fitness-preserving, automated process discovery algorithm as a function $disc : \mathcal{B}(\mathcal{A}^*) \rightarrow \mathcal{T}$ such that $L \subseteq \mathcal{L}(disc(L))$.*

≫	≫	≫	≫	c	d	≫	≫	≫	≫	≫	a	f	≫	≫
$(n_0,$ open$)$	$(n_{1.1},$ open$)$	$(n_{2.1},$ open$)$	$(n_{3.2},$ open$)$	$(n_{4.3},$ c$)$	$(n_{4.4},$ d$)$	$(n_{3.2},$ close$)$	$(n_{2.1},$ close$)$	$(n_{1.1},$ close$)$	$(n_{1.2},$ close$)$	$(n_{2.3},$ open $e)$	$(n_{2.4},$ a$)$	$(n_{2.5},$ f$)$	$(n_{1.2},$ close$)$	$(n_0,$ close$)$

(a) Optimal *full alignment* γ_1 for the full trace $\langle c,d,a,f \rangle$ and T_0

≫	≫	≫	≫	a	b	≫	≫	≫	≫	≫	d
$(n_0,$ open$)$	$(n_{1.1},$ open$)$	$(n_{2.1},$ open$)$	$(n_{3.1},$ open$)$	$(n_{4.1},$ a$)$	$(n_{4.2},$ b$)$	$(n_{3.1},$ close$)$	$(n_{2.1},$ close$)$	$(n_{2.2},$ $\tau)$	$(n_{2.1},$ open$)$	$(n_{3.2},$ open$)$	$(n_{4.4},$ d$)$

(b) Optimal *prefix alignment* γ_2 for the trace prefix $\langle a,b,d \rangle$ and T_0

d	f	≫	≫	≫	≫	≫	d
$(n_{4.4},$ d$)$	≫	$(n_{3.2},$ close$)$	$(n_{2.1},$ close$)$	$(n_{2.2},$ $\tau)$	$(n_{2.1},$ open$)$	$(n_{3.2},$ open$)$	$(n_{4.4},$ d$)$

(c) Optimal *infix alignment* γ_3 for the trace infix $\langle d,f,d \rangle$ and T_0

e	e	f	≫	≫
≫	$(n_{2.3},$ e$)$	$(n_{2.5},$ f$)$	$(n_{1.2},$ close$)$	$(n_0,$ close$)$

(d) Optimal *postfix alignment* γ_4 for the trace postfix $\langle e,e,f \rangle$ and T_0

Fig. 4. Examples of optimal alignments for process tree T_0 (cf. Fig. 2)

3.3 Alignments

Alignments [1] are a state-of-the-art conformance-checking technique [9] to compare process models with traces. Full alignments, often referred to as alignments, and prefix alignments have been introduced in [1]. In [16], we define infix and postfix alignments and describe their computation. Fig. 4 shows an exemplary full, prefix, infix, and postfix alignment for different trace (fragments) and tree T_0. In general, the first row, i.e., the *trace part*, of any alignment corresponds to the trace (fragment) when ignoring the skip symbol ≫. Analogously, the second row, i.e., the *model part*, corresponds to a running sequence (fragment) of the tree when ignoring ≫. Each column represents an alignment move; we generally distinguish five types: synchronous moves , log moves , visible model moves , invisible model moves , and opening & closing model moves .

Definition 5 (Full/Prefix/Infix/Postfix Alignment). *Let $\mathcal{A}$ be the universe of activity labels, $\gg\,\notin\mathcal{A}\cup\{\tau\}$, $\sigma \in \mathcal{A}^*$ be a trace (fragment), $T=(V,E,\lambda,r) \in \mathcal{T}$, and $S = V \times (\mathcal{A}\cup\{\tau, open, close\})$ be the set of running sequence steps. Sequence $\gamma \in \big((\mathcal{A}\cup\{\gg\})\times(S\cup\{\gg\})\big)^*$ is a full/prefix/infix/postfix alignment if:*

1. *$\sigma = \pi_1^*(\gamma)_{\downarrow\mathcal{A}}$,*
2. - *Full alignment:*   *$\pi_2^*(\gamma)_{\downarrow s} \in \mathcal{RS}(T)$,*
   - *Prefix alignment:*   *$\exists\rho \in S^*\big(\pi_2^*(\gamma)_{\downarrow s} \cdot \rho \in \mathcal{RS}(T)\big)$,*
   - *Infix alignment:*   *$\exists\rho_1,\rho_2 \in S^*\big(\rho_1 \cdot \pi_2^*(\gamma)_{\downarrow s} \cdot \rho_2 \in \mathcal{RS}(T)\big)$,*
   - *Postfix alignment:* *$\exists\rho \in S^*\big(\rho \cdot \pi_2^*(\gamma)_{\downarrow s} \in \mathcal{RS}(T)\big)$,*
3. *$(\gg,\gg) \notin \gamma$,*
4. *$\forall 1\leq i\leq|\gamma|\ \big(\pi_1(\gamma(i))\in\mathcal{A} \wedge \pi_2(\pi_2(\gamma(i)))\in\mathcal{A} \Rightarrow \pi_1(\gamma(i))=\pi_2(\pi_2(\gamma(i)))\big)$, and*
5. *$\forall 1\leq i\leq|\gamma|\ \big(\pi_2(\pi_2(\gamma(i)))\in\{open, close\} \Rightarrow \pi_1(\gamma(i))=\gg\big)$.*

We denote the universe of full/prefix/infix/postfix alignments for T and σ by $\Gamma_{full}(T,\sigma)$, $\Gamma_{prefix}(T,\sigma)$, $\Gamma_{infix}(T,\sigma)$, and $\Gamma_{postfix}(T,\sigma)$. We denote the universe of alignments as $\Gamma(T,\sigma) = \Gamma_{full}(T,\sigma)\cup\Gamma_{prefix}(T,\sigma)\cup\Gamma_{infix}(T,\sigma)\cup\Gamma_{postfix}(T,\sigma)$.

Consider Fig. 4d, $\gamma_4 = \langle(e, \gg), (e, (n_{2.3}, e)), (f, (n_{2.5}, f)), (\gg, (n_{1.2}, close)),$ $(\gg, (n_0, close)))\rangle$ is a postfix alignment. Let $\gamma \in \Gamma(T, \sigma)$ and $\gamma(i)$ for $1 \leq i \leq |\gamma|$ be an alignment move. We introduce abbreviations for ease of reading.

- $traceLabel(\gamma(i)) = \pi_1(\gamma(i))$
- $modelNode(\gamma(i)) = \begin{cases} \pi_1(\pi_2(\gamma(i))) & \text{if } \pi_2(\gamma(i)) \in S \\ \gg & \text{otherwise} \end{cases}$
- $modelLabel(\gamma(i)) = \begin{cases} \pi_2(\pi_2(\gamma(i))) & \text{if } \pi_2(\gamma(i)) \in S \\ \gg & \text{otherwise} \end{cases}$

For example, consider postfix alignment γ_4 shown in Fig. 4d: $traceLabel(\gamma_4(2)) = e$, $modelNode(\gamma_4(2)) = n_{2.3}$, and $modelLabel(\gamma_4(2)) = e$.

For an alignment $\gamma \in \Gamma(T, \sigma)$, we say that the alignment move $\gamma(i)$ for $1 \leq i \leq |\gamma|$ *indicates a deviation* if it is a log move, i.e., $traceLabel(\gamma(i)) = \gg$, or a visible model move, i.e., $traceLabel(\gamma(i)) = \gg \wedge modelLabel(\gamma(i)) \in \mathcal{A}$.

Let $\square \in \{full, prefix, infix, postfix\}$, $T \in \mathcal{T}$, and $\sigma \in \mathcal{A}^*$. Since many alignments exist for a given trace (fragment) and a process tree, the concept of *optimality* exists. An alignment is optimal if the number of visible model moves and log moves is minimal. We define four functions $align_{\square} : \mathcal{T} \times \mathcal{A}^* \to \Gamma_{\square}(T, \sigma)$ that return a $\square$ alignment for $\sigma \in \mathcal{A}^*$ and $T \in \mathcal{T}$. We write $align_{\square}^{opt}$ to indicate that we compute an *optimal* $\square$ alignment.

4   Trace-Fragment-Supporting IPD

This section describes the proposed trace-fragment-supporting IPD approach (cf. Fig. 1) that builds on the IPD approach presented in [17], which only features full traces. The basic idea, however, remains. When a full trace/trace fragment is added that is not contained in the language of the current process tree, the proposed approach determines relevant subtrees in the given process tree causing the deviation and rediscovers these deviating subtrees such that previous added traces/trace fragments and additionally the given trace (fragment) are supported. The remainder of this section presents the algorithm in detail. Section 4.1 presents the core part of the algorithm and introduces a running example. Next, Sect. 4.2 describes how deviating subtrees are identified, and Sect. 4.3 introduces the corresponding sub-log calculation needed for rediscovery.

4.1   Overview

This section provides an overview of the proposed approach. Further, we introduce a running example demonstrating various critical steps of the approach, cf. Fig. 5. Below, we list the required four *inputs*.

1. trace (fragment) $\sigma_{next} \in \mathcal{A}^*$ to be added next to the current process tree
2. interpretation $\square \in \{full, prefix, infix, postfix\}$ of trace (fragment) σ_{next}
3. *previously added* traces and trace fragments, divided into full traces $L_{full} \subseteq \mathcal{A}^*$, prefixes $L_{prefix} \subseteq \mathcal{A}^*$, infixes $L_{infix} \subseteq \mathcal{A}^*$, and postfixes $L_{postfix} \subseteq \mathcal{A}^*$

4. process tree $T \in \mathcal{T}$, supporting previously added traces/trace fragments, i.e., $L_{full} \subseteq \mathcal{L}(T)$, $L_{prefix} \subseteq \mathcal{L}_{prefix}(T)$, $L_{infix} \subseteq \mathcal{L}_{infix}(T)$, and $L_{postfix} \subseteq \mathcal{L}_{postfix}(T)$

Note that the proposed approach requires an initial process tree as input to start the incremental process discovery in the very first iteration. For example, this initial process tree can consist of only a single (invisible) activity. Alternatively, users can manually model an initial tree or use a conventional process discovery algorithm to discover one.

The *output* of the trace-fragment-supporting IPD algorithm is a process tree T that, in addition to the previously added traces and trace fragments, contains σ_{next} in its language—depending on σ_{next}'s interpretation, $\sigma_{next} \in \mathcal{L}_{\square}(T)$. Subsequently, we introduce the overall algorithm, presented in Algorithm 1, and exemplify critical steps with the running example shown in Fig. 5.

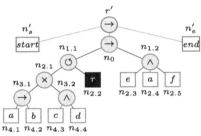

(a) Input process tree T, i.e., tree T_0 extended by a *start* and *end* activity

$\sigma_{next} = \langle a, a, f, end \rangle$

$\square = postfix$

$L_{full} = [\sigma_1 = \langle start, a, b, a, b, f, e, a, end \rangle]$

$L_{prefix} = [\sigma_2 = \langle start, d, c, e, a \rangle]$

$L_{infix} = [\sigma_3 = \langle c, f, a \rangle, \sigma_4 = \langle b, d, c, e, a \rangle]$

$L_{postfix} = [\sigma_5 = \langle f, a, end \rangle]$

(b) Input trace (fragment) σ_{next} to be added to T, its interpretation $\square$, previously added full traces, and trace fragments

1	2	3	4	5	6	7
a	a	f	$\gg$	$\gg$	end	$\gg$
$(n_{2.4},$	$\gg$	$(n_{2.5},$	$(n_{1.2},$	$(n_0,$	$(n'_e,$	$(r',$
$a)$		$f)$	$close)$	$close)$	$end)$	$close$

(c) Optimal postfix alignment for σ_{next} and T—the 2. alignment move indicates a deviation (Alg. 1 line 2)

$lca^T(n_{2.4}, n_{2.5}) = n_{1.2}$

$T_{LCA} = \triangle^{T'}(n_{1.2})$

$n_{1.2}$

(d) Problematic subtree $T_{LCA} \sqsubseteq T$ (Alg. 1 line 3)

$L_{LCA} = [\quad \langle e, a, a, f \rangle \qquad$ *derived from σ_{next}*
$\qquad\quad \langle f, e, a \rangle \qquad$ *derived from σ_1*
$\qquad\quad \langle e, a, f \rangle^2 \qquad$ *derived from σ_2, σ_4*
$\qquad\quad \langle f, a, e \rangle \qquad$ *derived from σ_3*
$\qquad\quad \langle e, f, a \rangle] \qquad$ *derived from σ_5*

(e) Sub-log L_{LCA} for T_{LCA} (Alg. 1 line 4) that represents trace fragments T_{LCA} must support to avoid the deviation indicated in the postfix alignment shown in Fig. 5c

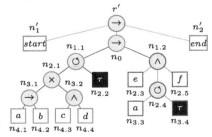

(f) Replacing $T_{LCA} \sqsubseteq T$ by $disc(L_{LCA}) \in \mathcal{T}$ (Alg. 1 line 5)

Fig. 5. Running example of a full execution of the proposed IPD approach

Input Preparation. Algorithm 1 requires that the input artifacts, the tree T and the traces/trace fragments as described above, are extended by an artificial start and end activity. These artificial activities are needed to ensure the correct integration of trace infixes and postfixes into T.

1. Process tree T is extended by *start*, *end* $\notin \mathcal{A}$ activities, e.g., cf. Fig. 5a.
2. Trace (fragment) to be added next σ_{next} and previously added traces/trace fragments are correspondingly extended by *start* and *end* activities to match the extended process tree T, for example, consider Fig. 5b.

Extending Process Tree T. This section describes Algorithm 1. First, we calculate an optimal full/prefix/infix/postfix alignment γ according to σ_{next}'s interpretation. In case γ does not indicate a deviation, we know that $\sigma_{next} \in \mathcal{L}_\square(T)$ and return. Otherwise, we call Algorithm 2 in line 3 that determines the subtree $T_{LCA} \sqsubseteq T$ that causes the first cohesive block of deviations, as indicated in γ. Hereinafter, assume that T_{LCA} exists. Next, we calculate sub-log L_{LCA} for the determined subtree T_{LCA}, cf. line 4. The sub-log corresponds to all sub-traces that T_{LCA} must be able to replay, i.e., $L_{LCA} \not\subseteq \mathcal{L}(T_{LCA})$. Sub-log L_{LCA} is therefore calculated based on previous added full/prefix/infix/postfix traces and the trace (fragment) to be added next, i.e., σ_{next}. Next, we replace $T_{LCA} \sqsubseteq T$ by a new subtree $disc(L_{LCA}) \in \mathcal{T}$ that fully supports the computed sub-log, i.e., $L_{LCA} \subseteq \mathcal{L}(disc(L_{LCA}))$ (line 5). Again, we compute alignment γ for the modified tree T and σ_{next} (line 2). If γ still indicates deviations, we repeat the procedure described above until all deviations are resolved. Note that the termination of Algorithm 1 is guaranteed since in each iteration of the while block (line 2–6), the first contiguous block of deviations is resolved. Thus, eventually $\sigma_{next} \in \mathcal{L}_\square(T)$.

Algorithm 1: *TraceFragmentSupportingIPD*

---

input : $T = (V, E, \lambda, r) \in \mathcal{T}$, // process tree to be extended
$\square \in \{full, prefix, infix, postfix\}, \sigma_{next} \in \mathcal{A}^*$, // $\square$ trace σ_{next} to be added to $\mathcal{L}_\square(T)$
$L_{full}, L_{prefix}, L_{infix}, L_{postfix} \subseteq \mathcal{B}(\mathcal{A}^*)$, // previously added full traces/trace fragments
output: $T \in \mathcal{T}$ // $\sigma_{next} \in \mathcal{L}_\square(T), L_{full} \subseteq \mathcal{L}(T), L_{prefix} \subseteq \mathcal{L}_{prefix}(T)$,
$L_{infix} \subseteq \mathcal{L}_{infix}(T), L_{postfix} \subseteq \mathcal{L}_{postfix}(T)$
begin

1    $L_\square \leftarrow L_\square \uplus [\sigma_{next}]$;   // add σ_{next} to the corresponding log L_{next}

2   **while** $\gamma \leftarrow align_\square^{opt}(T, \sigma_{next})$ *indicates a deviation* **do**   // $\sigma_{next} \notin \mathcal{L}_\square(T)$

3      $T_{LCA} \leftarrow DetermineSubtree(T, \gamma)$;   // Alg. 2

     if T_{LCA} **then**

4        $L_{LCA} \leftarrow SubLog(T, T_{LCA}, L_{full}, L_{prefix}, L_{infix}, L_{postfix})$;   // Alg. 3

5        $T \leftarrow$ replace $T_{LCA} \sqsubseteq T$ by $disc(L_{LCA}) \in \mathcal{T}$;

     else   // no subtree causing the deviation could be determined

6        $T \leftarrow$ extend T according to Fig. 6;

7   **return** T;

Consider the running example, cf. Fig. 5. The postfix alignment (cf. Fig. 5c) indicates a deviation—the second move is a log move, i.e., activity a cannot be replayed twice in the model. Next, we compute the subtree T_{LCA} that causes the deviation, cf. Fig. 5d. T_{LCA} supports the postfix $\langle a, f \rangle$ but not the postfix $\langle a, a, f \rangle$. Since we want to replace respectively rediscover T_{LCA}, we calculate a corresponding sub-log L_{LCA}, cf. Fig. 5e. The actual computation is explained in a subsequent section. However, note that the calculated sub-log L_{LCA} contains $\langle e, a, a, f \rangle$. Thus, when discovering a tree from L_{LCA} using a fitness-preserving discovery algorithm (cf. Definition 4), the discovered tree supports the execution of two subsequent a activities.

So far, we assumed that T_{LCA} causing the deviation(s) as indicated in γ could be determined. However, one case exists in which T_{LCA} cannot be determined, i.e., σ_{next} is an infix, and T does not contain any of its activities. Hence, γ includes only log moves. Thus, we do not have any reference point in the tree

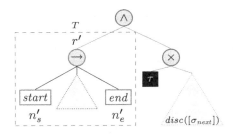

Fig. 6. Extending tree T by an optional parallel subtree supporting σ_{next}

Algorithm 2: *DetermineSubtree* (called in Alg. 1 line 3)

---

input : $T \in \mathcal{T}$, $\gamma \in \Gamma(T, \sigma_{next})$ // alignment for trace (fragment) σ_{next} and T

output: $T_{LCA} \sqsubseteq T$ // subtree that is responsible for the first deviation (block)

begin

1   **forall** $1 \leq i \leq |\gamma|$ **do**

2   **if** $\gamma(i)$ *indicates a deviation* **then**

3        $i_{before} \leftarrow$ *closest move* $\gamma(i_{before})$ *before* $\gamma(i)$ *that is a synchronous move or an invisible model move* (if possible, otherwise *null*);

4        $i_{after} \leftarrow$ *closest move* $\gamma(i_{after})$ *after* $\gamma(i)$ *that is a synchronous move or an invisible model move* (if possible, otherwise *null*);

5   **if** $i_{before} \wedge i_{after}$ **then**   // both corresponding moves exist

6   **return** $\Delta^T \left(lca^T \left(modelNode(\gamma(i_{before})), modelNode(\gamma(i_{after})) \right) \right)$;

7   **else if** i_{before} **then**   // only a corresponding move before exists

8   **return** $\Delta^T \left(modelNode(i_{before}) \right)$;

9   **else if** i_{after} **then**   // only a corresponding move after exists

10   **return** $\Delta^T \left(modelNode(i_{after}) \right)$;

11   **else**

12   **return** *null*;

where the infix should happen. In this case, we extend tree T as depicted in Fig. 6, i.e., we discover a subtree $disc([\sigma_{next}])$, make it optional, and add it in parallel to T (cf. line 6). This extension guarantees that $\sigma_{next} \in \mathcal{L}_\square(T)$. This described procedure is however usually very The next sub-sections introduce the algorithms *DetermineSubtree* and *SubLog* called in Algorithm 1 (line 3 and 4).

4.2 Subtree Detection

Algorithm 2 describes the subtree detection *DetermineSubtree* of T_{LCA}. As input, Algorithm 1 provides tree T and alignment $\gamma \in \Gamma_\square(T, \sigma_{next})$ indicating a deviation. The central idea is to find the first deviation (block) in γ and the closest alignment moves that: surround the found deviation (block), do not indicate a deviation, and correspond to an executed leaf node of T, cf. line 3 and 4. If such two moves can be found, we compute an LCA from the corresponding leaf nodes of these moves. We know that the subtree rooted at the computed LCA is causing the deviation (block), and hence, we return it (line 6). If we can only find one of the two moves, we return the subtree rooted at the corresponding node—this subtree consists of only a leaf node and indicates that around this leaf node, a deviation occurs regarding σ_{next}. In the particular case that no surrounding move can be found, we return nothing, cf. line 12. Note that this case can only happen if we have an infix alignment containing only log moves—all

---

Algorithm 3: *SubLog* (called in Alg. 1 line 4)

input : $T \in \mathcal{T}$, $T_{LCA} \sqsubseteq T$, $L_{full}, L_{prefix}, L_{infix}, L_{postfix} \subseteq \mathcal{B}(\mathcal{A}^*)$
output: $L_{LCA} \subseteq \mathcal{B}(\mathcal{A}^*)$ // sub-log for T_{LCA}
begin

1     $L_{LCA} \leftarrow []$;   // initialize sub-log for T_{LCA}
2   forall $\sigma \in L_{full}$ do
3       $\gamma \leftarrow align_{full}^{opt}(T, \sigma)$;
4       $L_{LCA} \leftarrow L_{LCA} \uplus ExtractSubTraces(T_{LCA}, \gamma, \{1, \ldots, |\gamma|\})$;   // Alg. 4
5   forall $\sigma \in L_{prefix}$ do
6       $\gamma \leftarrow align_{prefix}^{opt}(T, \sigma) \cdot align_{postfix}(T, \langle\rangle)$ such that $\gamma \in \Gamma_{full}(T, \sigma)$;
7       $I \leftarrow \{1, \ldots, i\}$ such that $\langle\gamma(1), \ldots, \gamma(i)\rangle = align_{prefix}^{opt}(T, \sigma)$;
8       $L_{LCA} \leftarrow L_{LCA} \uplus ExtractSubTraces(T_{LCA}, \gamma, I)$;   // Alg. 4
9   forall $\sigma \in L_{infix}$ do
10      $\gamma \leftarrow align_{prefix}(T, \langle\rangle) \cdot align_{infix}^{opt}(T, \sigma) \cdot align_{postfix}(T, \langle\rangle)$ such that $\gamma \in \Gamma_{full}(T, \sigma)$;
11      $I \leftarrow \{i, \ldots, i+n\}$ such that $\langle\gamma(i), \ldots, \gamma(i+n)\rangle = align_{infix}^{opt}(T, \sigma)$;
12      $L_{LCA} \leftarrow L_{LCA} \uplus ExtractSubTraces(T_{LCA}, \gamma, I)$;   // Alg. 4
13   forall $\sigma \in L_{postfix}$ do
14      $\gamma \leftarrow align_{prefix}(T, \langle\rangle) \cdot align_{postfix}^{opt}(T, \sigma)$ such that $\gamma \in \Gamma_{full}(T, \sigma)$;
15      $I \leftarrow \{i, \ldots, |\gamma|\}$ such that $\langle\gamma(i), \ldots, \gamma(|\gamma|)\rangle = align_{postfix}^{opt}(T, \sigma)$;
16      $L_{LCA} \leftarrow L_{LCA} \uplus ExtractSubTraces(T_{LCA}, \gamma, I)$;   // Alg. 4
17   return L_{LCA};

| full alignment | | | | | | | | | | | | |
| prefix alignment $align_{prefix}(T, \langle\rangle)$ | | | | | postfix alignment $align_{postfix}^{opt}(T, \sigma_{next})$ (cf. Fig. 5c) | | | | | | | |
1	2	...	11	12	13	14	15	16	17	18	19	20
≫	≫	...	≫	≫	≫	a	a	f	≫	≫	end	≫
$(r',$ open)	$(n_s,$ start)	...	$(n_{1.1},$ close)	$(n_{1.2},$ open)	$(n_{2.3},$ e)	$(n_{2.4},$ a)	≫	$(n_{2.5},$ f)	$(n_{1.2},$ close)	$(n_o,$ close)	$(n_e',$ end)	$(r',$ close)

T_{LCA} opens ⟶ T_{LCA} closes

Fig. 7. Extending the postfix alignment from the running example (cf. Fig. 5c)

other alignments have at least a synchronous move on the initially added *start* or *end* activity (cf. Fig. 5a).

Consider alignment γ from the running example (cf. Fig. 5c). Its first and only deviation is at the second move, surrounded by two synchronous moves representing the execution of node $n_{2.4}$ and $n_{2.5}$. Thus, we compute $LCA^T(n_{2.4}, n_{2.5}) = n_{1.2}$, and return subtree T_{LCA} rooted at $n_{1.2}$ (cf. Fig. 5d) because this subtree does not support executing two a activities, as indicated by γ.

4.3 Sub-log Calculation for Detected Subtree

Algorithm 3 describes the sub-log calculation *SubLog* called in Algorithm 1 line 4 for the determined subtree T_{LCA}. The output of the sub-log calculation is an event log L_{LCA} that the determined subtree T_{LCA} must support. To this end, all traces and trace fragments including σ_{next} are aligned with $T \sqsupseteq T_{LCA}$ to identify the corresponding sub-traces that T_{LCA} must support.

For example, consider postfix alignment γ (cf. Fig. 5c) and T_{LCA} with root node $n_{1.2}$ (cf. Fig. 5d). Adding $\sigma_{next} = \langle a, a, f \rangle$ to sub-log L_{LCA} would result in an unnecessary imprecise subtree because when replacing T_{LCA} by a rediscovered tree from L_{LCA}, activity e would be optional. However, no previously added trace (fragment) nor σ_{next} requires activity e being optional. Thus, we extend postfix alignment γ to a full one such that T_{LCA} is fully executed within the model part. Figure 7 exemplifies such an extension of γ. For each full execution of T_{LCA}, we generate a sub-trace. T_{LCA} is opened in move 12, and closed in move 17. All moves in between that represent the execution of a leaf node (i.e., move 13, 14, and 16) are contained in T_{LCA}. Thus, we add the sub-trace $\langle e, a, a, f \rangle$ to L_{LCA}. We proceed similarly for previously added traces and trace fragments.

Algorithm 4: *ExtractSubTraces* (called in Alg. 3)

input : $T_{LCA}=(V_{LCA}, E_{LCA}, \lambda_{LCA}, r_{LCA}) \sqsubseteq T$, $\gamma \in \Gamma_{full}(T, \sigma)$, $I \subseteq \{1, \ldots, |\gamma|\}$
output: $L \subseteq \mathcal{B}(\mathcal{A}^*)$ // sub-log for T_{LCA}
begin

1    $L \leftarrow [\,]$;   // initialize sub-log for T_{LCA}
     forall $1 \le i \le |\gamma|$ **do**   // iterate over alignment moves

2       $\sigma' \leftarrow \langle\rangle$;

3   **if** $V_{LCA} = \{r_{LCA}\}$ **then**   // T_{LCA} is leaf node

4   **while** $modelNode(\gamma(i)) \ne r_{LCA}$ **do**

5   **if** $\gamma(i)$ *is log move* **then**

6                $\sigma' \leftarrow \sigma' \cdot \langle traceLabel(\gamma(i))\rangle$;   // add log moves

7             $i \leftarrow i+1$;

         if $modelNode(\gamma(i)) = r_{LCA}$ **then**   // r_{LCA} is executed

8             $\sigma' \leftarrow \sigma' \cdot \langle modelLabel(\gamma(i))\rangle$;   // $modelLabel(\gamma(i)) = \lambda_{LCA}(r_{LCA})$

9   **if**

               $\forall i<j\le|\gamma| \big(\gamma(j)$ *is neither a sync. nor an invisible model move* $\big)$
               then

10                 $\sigma' \leftarrow \sigma' \cdot \langle traceLabel(\gamma(j)), \ldots, traceLabel(\gamma(|\gamma|))\rangle_{\downarrow_{\mathcal{A}}}$;

11          $L \leftarrow L \uplus [\sigma']$;

12   **else**   // T_{LCA} is a subtree with more than one node

13   **if** $modelNode(\gamma(i)) = r_{LCA} \;\wedge\; modelLabel(\gamma(i)) = open$ **then**

14   **while** $modelNode(\gamma(i)) \ne r_{LCA} \;\vee\; modelLabel(\gamma(i)) \ne close$ **do**
              // consider all subsequent moves until r_{LCA} is closed

15   **if** $modelNode(\gamma(i)) \in V_{LCA} \;\wedge\; \big[\gamma(i)$ *is synchronous move* $\vee$
                 $\big(\gamma(i)$ *is visible model move* $\wedge\, i \notin I\big)\big]$ **then**

16                    $\sigma' \leftarrow \sigma' \cdot \langle modelLabel(\gamma(i))\rangle$;

               else if $traceLabel(\gamma(i)) \in \mathcal{A}$ **then**

17                    $\sigma' \leftarrow \sigma' \cdot \langle traceLabel(\gamma(i))\rangle$;

18                $i \leftarrow i+1$;

19             $L \leftarrow L \uplus [\sigma']$;

20   **return** L;

Algorithm 3 provides the sub-log calculation. For full traces, we calculate a full alignment (line 4) and extract the corresponding sub-traces. For trace fragments, we compute a corresponding prefix/infix/postfix alignment and expand this into a full alignment, as exemplified in Fig. 7. Extending to full alignments is required as T_{LCA} might span larger parts of the process and hence might be only partially executed within the prefix/infix/postfix alignment. For instance, consider Fig. 7. The depicted postfix alignment does not contain a full execution of T_{LCA}.

Algorithm 4 defines the extraction of sub-trace(s) for T_{LCA} from full alignments. If T_{LCA} is a leaf node (line 3), we add all log moves until T_{LCA} is executed.

If afterwards T_{LCA} is never executed again, potential log moves after the last execution of T_{LCA} are also added to the sub-trace σ'. Thus, per execution of T_{LCA} one sub-trace is added to L. Note that log moves only occur for the trace to be added, for all other previously added traces/trace fragments log moves do not occur in γ. If T_{LCA} is a leaf node (line 12), we search for the opening of T_{LCA} (line 13). All activities from visible model/synchronous moves that belong to T_{LCA} (line 16) and log moves (line 17) are added to σ' until T_{LCA} is closed.

5 Evaluation

We present an initial evaluation of the proposed trace-fragment-supporting IPD approach. The central goal of the evaluation is to showcase that distinguishing trace fragments from full traces within IPD leads to comparable or even better process models than classic IPD [17], considering all traces as full ones.

5.1 Experimental Setup

We compare trace-fragment-supporting IPD (TFS-IPD) with IPD [17] and automated conventional process discovery algorithms: Inductive Miner (IM) [13], IM infrequent (IMf) [13], and evolutionary tree miner (ETM) [8]. All listed approaches discover process trees. We use publicly available real-life event logs.[2] Note that event logs generally consider all traces recorded as full traces. Thus, to obtain trace fragments, we proceed as follows.

1. Removing cases containing events in the first or last 20% of the period covered by the event log (objective: filtering incomplete traces)
2. Iterating over remaining traces. With probability $\frac{1}{2}$ we alter a full trace. If so, we apply with probability $\frac{1}{3}$ one of the following options (for $x = max\{1, 20\%$ avg. trace length$\}$).
   (a) we remove the first x activities (results in a trace prefix)
   (b) we remove the last x activities (results in a trace postfix)
   (c) we remove the first x and last x activities (results in a trace infix)

If the above procedure yields empty traces or empty trace fragments, we ignore them. We calculate fitness and precision using the log after the first step, as described above. For (TFS-)IPD, we discover an initial model from the 1% most frequent full trace variants using IM. The source code of our experiments, of (TFS-)IPD, and further results are available online.[3]

---

[2] BPI Ch. 2020–Request for Payment (DOI: 10.4121/uuid:52fb97d4-4588-43c9-9d04-3604d4613b51)

Road Traffic Fine Management (DOI: 10.4121/uuid:270fd440-1057-4fb9-89a9-b699b4 7990f)

Receipt log (DOI: 10.4121/uuid:a07386a5-7be3-4367-9535-70bc9e77dbe6).

[3] https://github.com/fit-daniel-schuster/trace-fragment-supporting-incremental-process-discovery.

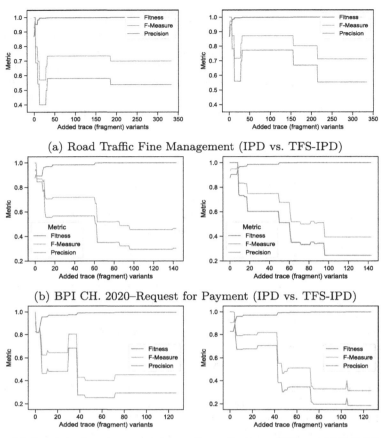

(a) Road Traffic Fine Management (IPD vs. TFS-IPD)

(b) BPI CH. 2020–Request for Payment (IPD vs. TFS-IPD)

(c) Receipt of environmental permit applications (IPD vs. TFS-IPD)

Fig. 8. Comparing IPD (left) and TFS-IPD (right)

5.2 Results

Figure 8 compares IPD with TFS-IPD for three different event logs. Both approaches start from the identical initial model, and we add the same trace (fragment) variants in the same order (starting from the most frequent one). Across all logs, TFS-IPD significantly outperforms IPD in most cases. Especially in the beginning, we quickly obtain models with fitness around >90% where TFS-IPD outperforms IPD regarding precision. Note that the goal of IPD and TFS-IPD is not to necessarily incorporate all behavior because real-life event logs often contain noise and are, therefore, typically filtered.

Table 1 lists the results for the different discovery approaches. Note that only TFS-IPD distinguishes between full traces and trace fragments; other approaches treat trace fragments as full traces. We observe that TFS-IPD often discovers

Table 1. Model quality metrics rounded to two decimal points for the different approaches and different percentage values of added trace (fragment) variants

Approach	% of added trace (fragment) variants	Event log								
		Road Traffic Fine Management			BPI Ch. 2020–Request for Payment			Receipt of environmental permit applications		
		F-measure	fitness	precision	F-measure	fitness	precision	F-measure	fitness	precision
Trace-fragment-supporting IPD	20	0.87	1.00	0.77	0.75	0.99	0.60	0.82	0.97	0.70
	40	0.87	1.00	0.77	0.68	1.00	0.51	0.48	0.99	0.31
	60	0.80	1.00	0.67	0.51	1.00	0.34	0.33	1.00	0.20
	80	0.71	1.00	0.55	0.39	1.00	0.25	0.33	1.00	0.20
	100	0.71	1.00	0.55	0.39	1.00	0.25	0.31	1.00	0.19
IPD [17]	20	0.73	1.00	0.58	0.72	0.98	0.57	0.64	0.97	0.48
	40	0.73	1.00	0.58	0.72	0.98	0.57	0.40	0.99	0.25
	60	0.70	1.00	0.54	0.49	1.00	0.32	0.45	1.00	0.29
	80	0.70	1.00	0.54	0.45	1.00	0.29	0.45	1.00	0.29
	100	0.70	1.00	0.54	0.46	1.00	0.30	0.45	1.00	0.29
IM [13]	20	0.76	1.00	0.61	0.63	1.00	0.46	0.76	0.97	0.62
	40	0.72	1.00	0.57	0.68	1.00	0.24	0.42	0.84	0.28
	60	0.56	1.00	0.39	0.44	1.00	0.28	0.25	1.00	0.15
	80	0.65	1.00	0.48	0.39	1.00	0.24	0.28	1.00	0.17
	100	0.67	1.00	0.50	0.37	1.00	0.23	0.33	1.00	0.20
IMf (0.9) [13]	20	0.81	0.78	0.84	0.52	0.54	0.50	0.76	0.97	0.62
	40	0.81	0.78	0.84	0.26	0.64	0.17	0.42	0.84	0.28
	60	0.75	0.66	0.86	0.17	0.65	0.10	0.25	1.00	0.15
	80	0.71	0.66	0.77	0.43	0.86	0.29	0.28	1.00	0.17
	100	0.71	0.66	0.77	0.17	0.64	0.10	0.33	1.00	0.20
ETM (default settings, 60s timeout) [8]	20	0.51	0.99	0.34	0.68	0.92	0.54	0.75	0.90	0.64
	40	0.51	1.00	0.34	0.69	0.97	0.53	0.71	0.86	0.60
	60	0.82	0.77	0.89	0.63	0.96	0.47	0.60	0.87	0.46
	80	0.54	0.98	0.38	0.69	0.95	0.54	0.62	0.87	0.48
	100	0.52	0.98	0.35	0.64	0.96	0.48	0.67	0.87	0.54

process models of similar or even higher quality regarding the metrics shown. As expected, the more trace fragments are added by an approach, the higher the fitness, but the precision decreases. In short, TFS-IPD often learns more precise process models for comparable fitness values than other approaches.

6   Conclusion

We presented an IPD approach supporting trace fragments—prefix, infix, and postfix traces. Supporting trace fragments and thus incomplete data within process discovery is a novelty, as the general practice regarding trace fragments usually focuses on filtering or considering fragments as full traces. We have implemented the proposed approach, including functionalities for handling trace fragments, in the open-source process mining tool *Cortado* [20]. Our experimental results indicate distinguishing trace fragments from full traces leads to high-quality models. While this paper focused on the foundational algorithmic aspects of supporting trace fragments in IPD, we plan to conduct a case study to investigate how process analysts can utilize trace fragments in real-world settings.

Further, we plan to extend the sub-model freezing functionality for IPD [18] to support trace fragments as well.

References

1. Adriansyah, A.: Aligning observed and modeled behavior. Ph.D. thesis, Eindhoven University of Technology (2014). https://doi.org/10.6100/IR770080
2. Armas Cervantes, A., van Beest, N.R.T.P., La Rosa, M., Dumas, M., García-Bañuelos, L.: Interactive and incremental business model repair. In: Panetto, H., et al. (eds.) OTM 2017. LNCS, vol. 10573, pp. 53–74. Springer, Cham (2017). https://doi.org/10.1007/978-3-319-69462-7_5
3. Augusto, A., et al.: Automated discovery of process models from event logs: review and benchmark. IEEE TKDE **31**(4), 686–705 (2019). https://doi.org/10.1109/TKDE.2018.2841877
4. Beerepoot, I., et al.: The biggest business process management problems to solve before we die. Comput. Ind. **146**, 103837 (2023). https://doi.org/10.1016/j.compind.2022.103837
5. Bergenthum, R., Desel, J., Lorenz, R., Mauser, S.: Process mining based on regions of languages. In: Alonso, G., Dadam, P., Rosemann, M. (eds.) BPM 2007. LNCS, vol. 4714, pp. 375–383. Springer, Heidelberg (2007). https://doi.org/10.1007/978-3-540-75183-0_27
6. Bernard, G., Andritsos, P.: Truncated trace classifier. Removal of incomplete traces from event logs. In: Nurcan, S., Reinhartz-Berger, I., Soffer, P., Zdravkovic, J. (eds.) BPMDS/EMMSAD -2020. LNBIP, vol. 387, pp. 150–165. Springer, Cham (2020). https://doi.org/10.1007/978-3-030-49418-6_10
7. Bezerra, F., Wainer, J., van der Aalst, W.M.P.: Anomaly detection using process mining. In: Halpin, T., et al. (eds.) BPMDS/EMMSAD -2009. LNBIP, vol. 29, pp. 149–161. Springer, Heidelberg (2009). https://doi.org/10.1007/978-3-642-01862-6_13
8. Buijs, J., van Dongen, B.F., van der Aalst, W.M.P.: A genetic algorithm for discovering process trees. In: Congress on Evolutionary Computation. IEEE (2012)
9. Carmona, J., van Dongen, B.F., Solti, A., Weidlich, M.: Conformance Checking. Springer, Cham (2018). https://doi.org/10.1007/978-3-319-99414-7
10. Dixit, P.M., Buijs, J.C.A.M., van der Aalst, W.M.P.: Prodigy: human-in-the-loop process discovery. In: 12th International Conference on Research Challenges in Information Science (RCIS). IEEE (2018). https://doi.org/10.1109/RCIS.2018.8406657
11. Fahland, D., van der Aalst, W.M.P.: Repairing process models to reflect reality. In: Barros, A., Gal, A., Kindler, E. (eds.) BPM 2012. LNCS, vol. 7481, pp. 229–245. Springer, Heidelberg (2012). https://doi.org/10.1007/978-3-642-32885-5_19
12. Greco, G., Guzzo, A., Lupia, F., Pontieri, L.: Process discovery under precedence constraints. ACM Trans. Knowl. Discov. Data **9**(4), 1–39 (2015)
13. Leemans, S.J.J.: Robust Process Mining with Guarantees. Springer, Cham (2022). https://doi.org/10.1007/978-3-030-96655-3
14. Polyvyanyy, A., van der Aalst, W.M.P., ter Hofstede, A.H.M., Wynn, M.T.: Impact-driven process model repair. ACM Trans. Softw. Eng. Methodol. **25**(4), 1–60 (2017). https://doi.org/10.1145/2980764
15. Reichert, M., Weber, B.: Enabling Flexibility in Process-Aware Information Systems. Springer, Heidelberg (2012). https://doi.org/10.1007/978-3-642-30409-5

16. Schuster, D., Föcking, N., van Zelst, S.J., van der Aalst, W.M.P.: Conformance checking for trace fragments using infix and postfix alignments. In: Sellami, M., Ceravolo, P., Reijers, H.A., Gaaloul, W., Panetto, H. (eds.) CoopIS 2022. LNCS, vol. 13591, pp. 299–310. Springer, Cham (2022). https://doi.org/10.1007/978-3-031-17834-4_18

17. Schuster, D., van Zelst, S.J., van der Aalst, W.M.P.: Incremental discovery of hierarchical process models. In: Dalpiaz, F., Zdravkovic, J., Loucopoulos, P. (eds.) RCIS 2020. LNBIP, vol. 385, pp. 417–433. Springer, Cham (2020). https://doi.org/10.1007/978-3-030-50316-1_25

18. Schuster, D., van Zelst, S.J., van der Aalst, W.M.P.: Freezing sub-models during incremental process discovery. In: Ghose, A., Horkoff, J., Silva Souza, V.E., Parsons, J., Evermann, J. (eds.) ER 2021. LNCS, vol. 13011, pp. 14–24. Springer, Cham (2021). https://doi.org/10.1007/978-3-030-89022-3_2

19. Schuster, D., van Zelst, S.J., van der Aalst, W.M.P.: Utilizing domain knowledge in data-driven process discovery: a literature review. Comput. Ind. **137**, 103612 (2022). https://doi.org/10.1016/j.compind.2022.103612

20. Schuster, D., van Zelst, S.J., van der Aalst, W.M.P.: Cortado: a dedicated process mining tool for interactive process discovery. SoftwareX **22**, 101373 (2023). https://doi.org/10.1016/j.softx.2023.101373

21. Solé, M., Carmona, J.: Incremental process discovery. In: Jensen, K., Donatelli, S., Kleijn, J. (eds.) Transactions on Petri Nets and Other Models of Concurrency V. LNCS, vol. 6900, pp. 221–242. Springer, Heidelberg (2012). https://doi.org/10.1007/978-3-642-29072-5_10

22. van Dongen, B.F., Alves de Medeiros, A.K., Wen, L.: Process mining: overview and outlook of petri net discovery algorithms. In: Jensen, K., van der Aalst, W.M.P. (eds.) Transactions on Petri Nets and Other Models of Concurrency II. LNCS, vol. 5460, pp. 225–242. Springer, Heidelberg (2009). https://doi.org/10.1007/978-3-642-00899-3_13

23. de Weerdt, J., de Backer, M., Vanthienen, J., Baesens, B.: A multi-dimensional quality assessment of state-of-the-art process discovery algorithms using real-life event logs. Inf. Syst. **37**(7), 654–676 (2012). https://doi.org/10.1016/j.is.2012.02.004

Approximating Multi-perspective Trace Alignment Using Trace Encodings

Alessandro Gianola[(✉)], Jonghyeon Ko, Fabrizio Maria Maggi, Marco Montali, and Sarah Winkler

Free University of Bozen-Bolzano, Bolzano, Italy
{gianola,maggi,montali,winkler}@inf.unibz.it, jongko@unibz.it

Abstract. Alignments provide sophisticated diagnostics that pinpoint deviations in a trace with respect to a process model. One crucial aspect is to consider, in the alignment task, not only the control flow perspective but also other sources of information available in event logs like data payloads. However, the combination of these dimensions makes the problem of multi-perspective trace alignment highly challenging since the number of traces accepted by the model is typically infinite. In this paper, we address this problem by proposing an approximate approach to alignment computation: instead of computing the optimal alignments based on the complete knowledge about a process trace available in the log, we perform approximate alignments based on lossy trace encodings that only consider certain information about the trace. The advantage of this approach is twofold. First, the trace alignment task is much faster. Second, the analyst can choose what type of information is relevant for computing the alignments by selecting the encodings that represent a trace based on that information. Our experiments show that the approximate approach is faster than the optimal one and, for encodings sufficiently rich, able to provide accurate results.

Keywords: Conformance checking · Trace Encoding · Multi-Perspective process mining · SMT

1 Introduction

Conformance checking is one of the central tasks of process mining [2]. Its main goal is to compare a reference process model with an event log containing actual process executions to understand whether such concrete executions deviate from the model. Within the family of conformance checking techniques, a prominent approach is to measure and explain deviations through *alignments*.

An alignment is intuitively a sequence of pairs, called moves, consisting of an event from the log and a transition in the process model. Given a suitable

This research has been partially supported by the Italian Ministry of University and Research (MUR) under the PRIN project PINPOINT Prot. 2020FNEB27, and by the Free University of Bozen-Bolzano with the ADAPTERS and CAT projects.

© The Author(s), under exclusive license to Springer Nature Switzerland AG 2023
C. Di Francescomarino et al. (Eds.): BPM 2023, LNCS 14159, pp. 74–91, 2023.
https://doi.org/10.1007/978-3-031-41620-0_5

function that assigns cost to moves, an optimal alignment is an alignment whose overall cost is minimal. This is notoriously challenging to compute, as it requires to solve an optimization problem over a finite portion of the space of model traces, where the portion to be considered depends on the length of the trace under scrutiny, and comes with the additional computational burden of computing trace distances. A plethora of techniques have been therefore defined to tackle the problem in an optimal [2] or approximate [1] way.

In the alignment task, however, not only the control flow perspective is crucial, but also other sources of information from event logs like data payloads. This has led to a recent series of approaches to tackle data-aware conformance checking [3,4,16,17]. There, Data Petri Nets (DPNs) [10,16] are the reference model to represent a process that accounts for control-flow and data, with process variables that can carry data values of different types.

The standard way for measuring the distance between a log trace and a DPN is to compute *optimal* alignments, based on a notion of distance that tackles at once the events, their orderings, and their data payloads. However, in the presence of models with rich data and control flow perspectives, computing optimal alignments can be extremely costly in terms of performance. This is also due to the fact that even by bounding the maximum length of model traces, the number of them is usually infinite, because of data. This calls for sophisticated techniques to handle the data component.

In this paper, we propose an alternative approach for data-aware conformance checking, which *approximates* optimal alignments based on machine learning techniques, in particular lossy trace encodings [12,14]. To this end, we do not work directly on models, but on sets of abstract traces. Roughly, our approach proceeds in three stages:

(1) We build a set $\mathcal{T}$ of *abstract traces*, i.e., classes of traces representative for all possible behaviors of the process. For this set, we propose two possibilities: (1a) For the class of DPNs whose transition guards are only variable-to-constant comparisons, we show how all possible behaviors up to a bounded length can be succinctly represented by a *finite* set of abstract traces. (1b) For DPNs with numeric variables but more complex guards, such a representation is in general not possible. In this case, our approach can be applied by taking as $\mathcal{T}$ simply a set of "happy paths", i.e., traces that are considered representative of the process behavior (e.g., obtained by collecting sufficiently many cases).

(2) We use machine learning techniques known as *encodings* to represent $\mathcal{T}$ as a set of vectors in a vector space. Here, different encodings can be employed to preserve from the abstract traces the information that is deemed most relevant for the conformance checking task. The result, called *behavior encoding space* is a compact numeric representation of all relevant behaviors.

(3) In order to check the conformance of a concrete trace, we apply the encoding from the previous stage to it, obtaining a vector $\mathbf{X}$, and subsequently compute the k vectors from the behavior encoding space that are closest to

X, using a kNN-based method. From these vectors, we can then get back the abstract traces that are considered closest to the input trace.

Note that the class of DPNs in (1a) has been found expressive and useful in practice, and is amenable to automatic discovery techniques [8,11]. Moreover, it is known that the process run in an optimal alignment can be upper-bounded in length in terms of the given trace [3], and $\mathcal{T}$ is a complete set of representatives. Therefore, the conformance checking task can be reformulated as the task to select a suitable abstract trace from $\mathcal{T}$, without loss of precision, which justifies the subsequent approximation approach in stages (2) and (3).

We experimentally validate our approach for both settings (1a) and (1b), comparing the results with the conformance checker CoCoMoT [3]. These experiments show that abstract traces (1a) together with smart trace encodings and vector space distance measures allow for a good approximation of the optimal alignments, in terms of precision and similarity. Moreover, we show that even when using a plain trace set as a representation of the process behaviors (1b), the encoding-based approach approximates the optimal one with high precision.

The remainder of this paper is structured as follows: We first recall background about DPNs and alignments (Sect. 2). Then, we present our notions of trace-based distance function and abstract traces (Sect. 3), and subsequently, trace encodings (Sect. 4). We evaluate our approach in Sect. 5. Finally, we discuss related work (Sect. 6) and conclude (Sect. 7).

2 Background and Preliminaries

We use a restricted but significant class of Data Petri nets (DPNs) for modeling multi-perspective processes, adopting the same formalization as in [3,16].

Let V be a set of *process variables*, each with a type and an associated domain: integers (`int`), or rationals (`rat`).[1] We consider two disjoint sets of annotated variables $V^r = \{v^r \mid v \in V\}$ and $V^w = \{v^w \mid v \in V\}$ to be read and written by process activities, as explained below. Based on these, we define constraints according to the grammar for c:

$$c ::= v_z \odot z \mid v_r \odot q \mid c \wedge c$$

where $v_z \in V_{\text{int}}$, $z \in \mathbb{Z}$, $v_q \in V_{\text{rat}}$, and $q \in \mathbb{Q}$, and $\odot$ is in $\{\geq, \leq, >, <, =\}$. Our constraints are thus more restrictive than in other sources [3], permitting only variable-to-constant comparisons, but this will allow us to define precise abstract traces. The set of constraints over variables V is denoted $\mathcal{C}(V)$; they are used for read and write operations in process activities.

Definition 1 (DPN). *A tuple* $\mathcal{N} = (P, T, F, \ell, A, V, guard)$ *is a* Petri net with data *(DPN), where:*

- *(P, T, F, ℓ) is a Petri net with two non-empty disjoint sets of places P and transitions T, a flow relation $F : (P \times T) \cup (T \times P) \rightarrow \mathbb{N}$ and a labeling function $\ell : T \rightarrow A \cup \{\tau\}$, where A is a finite set of activity labels and τ is a special symbol denoting silent transitions;*

---

[1] Booleans and strings can be encoded as integers, as commonly done [3,17].

- V *is a set of typed process variables; and*
- *guard*: $T \rightarrow \mathcal{C}(V^r \cup V^w)$ *is a guard assignment; for* $t \in T$ *with* $\ell(t) = \tau$ *we assume that guard*(t) *does not use variables in* V^w.

Transition guards serve to simultaneously read and write variables. For instance, a transition with guard $(x^r > 3)$ can only be taken if the current value of variable x is greater than 3 (the superscript r indicates that the guard is on the current, or *read*, variable). On the other hand, a guard $(x^w > 1) \wedge (x^r < 4)$ requires that the current value of x is smaller than 4 and, at the same time, it non-deterministically *writes* to x a new value that is greater than 1 (superscripts w refer to *written* values). Note that transition guards with disjunctions can be simulated by employing multiple transitions between the same places.

As customary, given $x \in P \cup T$, we use $^\bullet x := \{y \mid F(y,x) > 0\}$ to denote the *preset* of x and $x^\bullet := \{y \mid F(x,y) > 0\}$ to denote the *postset* of x. In order to refer to the variables read and written by a transition t, we use the notations $read(t) = \{v \mid v^r \in \mathcal{V}ar(guard(t))\}$ and $write(t) = \{v \mid v^w \in \mathcal{V}ar(guard(t))\}$.

To represent the current values of variables, we consider a *state variable assignment*, i.e., a (possibly partial) function α that assigns a value (of the right type) to each variable in V. We denote by DOM(α) the domain of α. A *state* in a DPN $\mathcal{N}$ is a pair (M, α) constituted by a marking $M: P \rightarrow \mathbb{N}$ for the underlying Petri net (P, T, F, ℓ), plus a total state variable assignment α. Therefore, a state simultaneously accounts for the control flow progress and for the current values of all variables in V, as specified by α.

We fix one state (M_I, α_0) as *initial*, where M_I is the initial marking of the underlying Petri net and α_0 specifies the initial value of all variables in V. Similarly, we denote the final marking as M_F, and call *final* any state of the form (M_F, α_F) for some α_F.

A *transition variable assignment* is a partial function β with DOM(β) $\subseteq$ $V^r \cup V^w$ that assigns a value to annotated variables, namely $\beta(x) \in \mathcal{D}(type(x))$, with $x \in V^r \cup V^w$. Transition variable assignments are used to specify how variables change as the result of activity executions (cf. Definition 2).

We now define when a Petri net transition may fire from a given state.

Definition 2 (Transition firing). *A transition* $t \in T$ *is* enabled *in state* (M, α) *if there exists a transition variable assignment* β *such that:*

- DOM(β) $= \mathcal{V}ar(guard(t))$: β *is defined for the variables in the guard;*
- $\beta(v^r) = \alpha(v)$ *for every* $v \in read(t)$, *i.e.,* β *is as* α *for read variables;*
- $\beta \models guard(t)$, *i.e.,* β *satisfies the guard; and*
- $M(p) \geq F(p, t)$ *for every* $p \in {}^\bullet t$.

An enabled transition may fire, producing a new state (M', α'), *s.t.* $M'(p) = M(p) - F(p,t) + F(t,p)$ *for every* $p \in P$, *and* $\alpha'(v) = \beta(v^w)$ *for every* $v \in write(t)$, *and* $\alpha'(v) = \alpha(v)$ *for every* $v \notin write(t)$. *A pair* (t, β) *as above is called (valid)* transition firing, *and we denote its firing by* $(M, \alpha) \xrightarrow{(t,\beta)} (M', \alpha')$.

Informally, a transition firing between the current state (M, α) and the next state (M', α') is a couple (t, β) where: *i)* $t \in T$ is a transition that is enabled in the 'token game' sense of standard Petri nets; *ii)* β is a function connecting the values of the read variables (matching the values assigned by α in the current state) to the values of the write variables (matching the values assigned by α' in the next state); *iii)* β satisfies the guard associated to t.

Based on this single-step transition firing, we say that a state (M', α') is *reachable* in a DPN with initial state (M_I, α_0) iff there exists a sequence of valid transition firings of the form $\mathbf{f} = \langle (t_1, \beta_1), \ldots, (t_n, \beta_n) \rangle$ such that $(M_I, \alpha_0) \xrightarrow{(t_1, \beta_1)} \ldots \xrightarrow{(t_n, \beta_n)} (M', \alpha')$. Moreover, such a sequence $\mathbf{f}$ is called a *process run* of $\mathcal{N}$ if $(M_I, \alpha_0) \xrightarrow{\mathbf{f}} (M_F, \alpha_F)$ for some α_F, i.e., if the run leads to a final state. As in [3,17], we restrict to DPNs where a final state is reachable. We denote the set of transition firings of $\mathcal{N}$ by $\mathcal{F}(\mathcal{N})$, and the set of process runs by $Runs(\mathcal{N})$.

Example 1. Let $\mathcal{N}$ be as shown below (with initial marking $[p_0]$ and final marking $[p_3]$). $Runs(\mathcal{N})$ contains, e.g., $\langle (\mathsf{a}, \{x^w \mapsto 12\}), (\mathsf{b}, \{y^w \mapsto 1\}), (\mathsf{c}, \{x^r \mapsto 12\}) \rangle$ and $\langle (\mathsf{a}, \{x^w \mapsto 1\}), (\mathsf{b}, \{y^w \mapsto 1\}), (\mathsf{d}, \{x^r \mapsto 1\}) \rangle$, for $\alpha_0 = \{x, y \mapsto 0\}$.

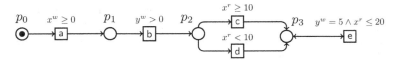

Given a set S, we denote S^* as the set of sequences of elements from S, and $\mathcal{M}(S)$ as the set of multisets over S. For a set A of activity labels, an *event* is a pair (b, α) for $b \in A$ and α a (typically partial) state variable assignment, associating values to variables in V.

Definition 3 (Log trace, event log). *Given a set $\mathcal{E}$ of events, a* log trace $\mathbf{e} \in \mathcal{E}^*$ *is a sequence of events in $\mathcal{E}$ and an* event log $L \in \mathcal{M}(\mathcal{E}^*)$ *is a multiset of log traces from $\mathcal{E}$.*

Conformance checking aims at constructing an *alignment* of a given log trace $\mathbf{e}$ wrt the DPN $\mathcal{N}$, by matching events in the log trace against transitions firings in a process run. Since not every event can typically be put in correspondence with a transition firing, a "skip" symbol $\gg$ is used. Let $\mathcal{E}^{\gg} = \mathcal{E} \cup \{\gg\}$ and, given $\mathcal{N}$, the extended set of transition firings $\mathcal{F}^{\gg} = \mathcal{F}(\mathcal{N}) \cup \{\gg\}$.

Given a DPN $\mathcal{N}$ and a set $\mathcal{E}$ of events as above, a pair $(e, f) \in \mathcal{E}^{\gg} \times \mathcal{F}^{\gg} \setminus \{(\gg, \gg)\}$ is called *move*. A move (e, f) is a *log move* if $e \in \mathcal{E}$ and $f = \gg$; a *model move* if $e = \gg$ and $f \in \mathcal{F}(\mathcal{N})$; and *synchronous move* if $(e, f) \in \mathcal{E} \times \mathcal{F}(\mathcal{N})$.

For a sequence of moves $\gamma = (e_1, f_1), \ldots, (e_n, f_n)$, the *log projection* $\gamma|_L$ of γ is the maximal subsequence of $e_1, \ldots, e_n$ in $\mathcal{E}^*$, and the *model projection* $\gamma|_M$ of γ is the maximal subsequence of $f_1, \ldots, f_n$ in $\mathcal{F}(\mathcal{N})^*$ (i.e., without $\gg$ symbols).

Definition 4 (Alignment). *Given $\mathcal{N}$, a sequence of legal moves γ is an* alignment *of a log trace $\mathbf{e}$ if $\gamma|_L = \mathbf{e}$, and it is* complete *if $\gamma|_M \in Runs(\mathcal{N})$.*

Example 2. The sequences γ_1 and γ_2 below are possible complete alignments of the log trace $\mathbf{e} = \langle (\mathsf{a}, \{x \mapsto 2\}), (\mathsf{b}, \{y \mapsto 1\}), (\mathsf{d}, \emptyset) \rangle$ wrt the DPN from Example 1:

$$
\gamma_1 = \begin{array}{|c c|c c|c|} \hline \mathsf{a} & x \mapsto 2 & \mathsf{b} & y \mapsto 1 & \mathsf{d} \\ \hline \mathsf{a} & x^w \mapsto 2 & \mathsf{b} & y^w \mapsto 1 & \mathsf{d} \\ \hline \end{array}
\qquad
\gamma_2 = \begin{array}{|c c|c c|c|c|} \hline \mathsf{a} & x \mapsto 2 & \mathsf{b} & y \mapsto 1 & \mathsf{d} & \gg \\ \hline \mathsf{a} & x^w \mapsto 12 & \mathsf{b} & y^w \mapsto 1 & \gg & \mathsf{c} \\ \hline \end{array}
$$

We denote by $Align(\mathcal{N}, \mathbf{e})$ the set of complete alignments for a log trace $\mathbf{e}$ wrt $\mathcal{N}$. A *cost function* is a mapping $\kappa\colon Moves_\mathcal{N} \to \mathbb{R}^+$ that assigns a cost to every move. It is naturally extended to alignments as follows.

Definition 5 (Cost). *Given $\mathcal{N}$, $\mathbf{e}$ and $\gamma = (e_1, f_1), \ldots, (e_n, f_n) \in Align(\mathcal{N}, \mathbf{e})$, the* cost *of γ is obtained by summing up the costs of its moves, that is, $\kappa(\gamma) = \sum_{i=1}^n \kappa(e_i, f_i)$. Moreover, γ is* optimal *for $\mathbf{e}$ if $\kappa(\gamma)$ is minimal among all complete alignments for $\mathbf{e}$, namely there is no $\gamma' \in Align(\mathcal{N}, \mathbf{e})$ with $\kappa(\gamma') < \kappa(\gamma)$.*

For instance, using the standard cost function from [3, Def. 6] and the alignments in Example 2, we would have $\kappa(\gamma_1) = 0$ and $\kappa(\gamma_2) = 3$. We denote the cost of an optimal alignment for $\mathbf{e}$ wrt $\mathcal{N}$ by $\kappa_\mathcal{N}^{opt}(\mathbf{e})$.

3   Trace-Based Conformance Checking

In this section, we develop notions to perform (approximated) conformance checking on the basis of trace classes rather than the model itself.

Abstract Trace. In order to simulate the conformance checking procedure, we first extract a set of abstract traces that are representative for the given DPN. To that end, we use the following definitions, for a DPN with data variables V. A *variable range assignment* ι is a (possibly partial) function from the set of data variables V to intervals, such that for all $v \in V$, $\iota(v)$ is of the form $[l, u]$, $]l, u]$, $[l, u[$, or $]l, u[$, for l, u numeric values in DOM(v). Then, given the set T of transitions, an *abstract event* is a pair (t, ι), where $t \in T$ and ι is a variable range assignment, and an *abstract trace* is a sequence of abstract events.

A trace $\mathbf{e} = \langle e_1, \ldots, e_n \rangle$ *matches* an abstract trace $\mathbf{f} = \langle f_1, \ldots, f_m \rangle$ if $m = n$ (same length); and for all $1 \le i \le n$, if $e_i = (l, \alpha)$ with corresponding $f_i = (t, \iota)$, it holds that $l = \ell(t)$, i.e., they have the same label; and DOM(α) = DOM(ι), and for all $v \in$ DOM(α), the value $\alpha(v)$ is in the interval $\iota(v)$. Finally, a finite set of abstract traces $\mathcal{T}$ is *representative* for a DPN $\mathcal{N}$ up to *length* k if for every trace $\mathbf{e}$ with $|\mathbf{e}| \le k$ and $\kappa_\mathcal{N}^{opt}(\mathbf{e}) = 0$, the trace $\mathbf{e}$ matches some $\mathbf{e}_a \in \mathcal{T}$.

Our approach exploits that for a given log trace and DPN, the length of a process run in an optimal alignment can be bounded upfront. More precisely:

Lemma 1 ([5, Lem. 2]). *Given a log trace $\mathbf{e}$ of length n and a DPN $\mathcal{N}$, there is a computable function $maxlen(\mathcal{N}, n)$ s.t. $\mathbf{e}$ has an optimal alignment γ wrt the standard cost function s.t. $\gamma|_M$ has length at most $maxlen(\mathcal{N}, n)$.*

Let a trace **e** *correspond* to a process run **f** if, for $\langle f_1, \ldots, f_n \rangle$ the subsequence of non-silent transitions in **f**, $\mathbf{e} = \langle e_1, \ldots, e_n \rangle$, and for all i, $1 \leq i \leq n$, if $f_i = (t, \beta)$ then $e_i = (l, \alpha)$ such that $\ell(t) = l$ and $\alpha(v) = \beta(v^w)$ for all $v \in \text{DOM}(\beta)$.

Using this notion, we get the following useful corollary of Lemma 1:

Theorem 1. *Let a set of abstract traces T be representative for a DPN $\mathcal{N}$ up to $maxlen(\mathcal{N}, n)$. Then, for every trace **e** with $|\mathbf{e}| \leq n$, there is an optimal alignment γ such that T has an abstract trace $\mathbf{e}_a$ that corresponds to $\gamma|_M$.*

This means that, in order to find the process run associated with an optimal alignment for a given log trace, it suffices to consider abstract traces in a set of representative abstract traces T.

Computing Representative Sets of Abstract Traces. We now show one concrete method to compute a representative set T for a DPN $\mathcal{N}$.

1. For a given k, we enumerate all transition sequences of $\mathcal{N}$ from the initial to a final marking that have length at most k, and select from these the subset T' of sequences which correspond to actual process runs. This filtering can be done, e.g., by checking with an SMT encoding (as done in CoCoMoT) whether the sequence of transitions is satisfiable.
2. For every sequence $\langle t_1, \ldots, t_n \rangle$ in T' and $1 \leq i \leq n$, we define a trace range substitution ι_i as follows. First, a variable $v \in V$ is in $dom(\iota_i)$ iff $v \in write(t_i)$. For such v, let j (s.t. $i < j \leq n$) be the smallest number such that either $j = n$, or $v \in write(t_{j+1})$. Thus, the value of v written in t_i persists until instant j. All guards in $t_i, \ldots, t_j$ are, by construction, conjunctions of variable-to-constant comparisons. Let L be the greatest lower bound set for v, and U the smallest upper bound set for v in $t_i, \ldots, t_j$; if no respective bound occurs, $L = -\infty$ or $U = \infty$. We fix $\iota_i(v)$ to either $[L, U]$, $[L, U[$, $]L, U]$ or $]L, U[$, depending on whether L and U are included or not. Finally, T consists of all $\langle (t_1, \iota_1) \ldots, (t_n, \iota_n) \rangle$ such that $\langle t_1, \ldots, t_n \rangle$ is in T'.

It can be checked that the set T constructed in this way is indeed a representative set of abstract traces.

Example 3. For $\mathcal{N}$ as in Example 1, a representative set of abstract traces up to length 4 consists of $\langle (\mathsf{a}, x \mapsto [0, 10[), (\mathsf{b}, y \mapsto]0, \infty[), (\mathsf{d}, \emptyset), (\mathsf{e}, y \mapsto [5, 5]) \rangle$, $\langle (\mathsf{a}, x \mapsto [10, \infty[), (\mathsf{b}, y \mapsto]0, \infty[), (\mathsf{c}, \emptyset) \rangle$, $\langle (\mathsf{a}, x \mapsto [0, 10[), (\mathsf{b}, y \mapsto]0, \infty[), (\mathsf{d}, \emptyset) \rangle$, and $\langle (\mathsf{a}, x \mapsto [10, 20]), (\mathsf{b}, y \mapsto]0, \infty[), (\mathsf{c}, \emptyset), (\mathsf{e}, y \mapsto [5, 5]) \rangle$.

Measuring the Distance Between Two Traces. In conformance checking, one usually measures the distance between a trace and a model run. Here, we approximate such a cost by taking the distance between two traces:

Definition 6. *For log traces $\mathbf{e} = \langle e_1, \ldots, e_m \rangle$ and $\mathbf{e}' = \langle e'_1, \ldots, e'_n \rangle$, the trace distance $\delta(\mathbf{e}|_i, \mathbf{e}'|_j)$ is recursively defined for all $0 \leq i \leq m$ and $0 \leq j \leq n$:*

$$\delta(\epsilon, \epsilon) = 0 \quad \delta(\mathbf{e}|_{i+1}, \epsilon) = Q_L(e_{i+1}) + \delta(\mathbf{e}|_i, \epsilon) \quad \delta(\epsilon, \mathbf{e}'|_{j+1}) = Q_L(e'_{j+1}) + \delta(\epsilon, \mathbf{e}'|_j)$$

$$\delta(\mathbf{e}|_{i+1}, \mathbf{e}'|_{j+1}) = \min \begin{cases} Q_=(e_{i+1}, e'_{j+1}) + \delta(\mathbf{e}|_i, \mathbf{e}'|_j) \\ Q_L(e_{i+1}) + \delta(\mathbf{e}|_i, \mathbf{e}'|_{j+1}) \\ Q_L(e'_{j+1}) + \delta(\mathbf{e}|_{i+1}, \mathbf{e}'|_j) \end{cases}$$

Here $Q_=$ and Q_L are two *penalty functions*, the former for synchronous moves and the latter for asynchronous moves in one of the logs. These penalties can be instantiated in different ways. We adapt the *standard cost function* [3,17] to two traces and set

$$Q_L(b,\alpha) = 1 \quad Q_=((b,\alpha),(b',\alpha')) = \begin{cases} |\{v \in \mathrm{DOM}(\alpha) \mid \alpha(v) \neq \alpha'(v)\}| & \text{if } b = b' \\ \infty, & \text{otherwise} \end{cases}$$

For instance, for the log trace $\mathbf{e} = \langle (\mathsf{a}, \{x \mapsto 2\}), (\mathsf{b}, \{y \mapsto 1\}), (\mathsf{d}, \emptyset) \rangle$ from Example 2 and $\mathbf{e}' = \langle (\mathsf{a}, \{x \mapsto 12\}), (\mathsf{b}, \{y \mapsto 1\}), (\mathsf{c}, \emptyset) \rangle$ (matching the process run of γ_2), we have $\delta(\mathbf{e}, \mathbf{e}) = 0$ and $\delta(\mathbf{e}, \mathbf{e}') = 3$.

4   Approximating Alignments with Trace Encodings

In this section, we introduce an encoding approach for abstract traces (Sect. 4.1) and then a kNN-based method to obtain an approximate solution of the trace alignment problem (Sect. 4.2).

4.1   Encodings for Abstract Traces

To have a lossy representation of abstract traces, we use an encoding $\mathcal{E} : \mathcal{T} \to \mathbb{R}^n$ with $n \in \mathbb{N}$ that transforms each abstract trace into a vector of the n-dimensional Euclidean space $\mathbb{R}^n$. We call the resulting set of vectors $\mathcal{E}(\mathcal{T})$ *behavior encoding space*. The literature provides encoding functions to represent strings [6], which we can directly employ for representing the control-flow dimension of the abstract traces. For example, the *boolean* encoding represents a trace through a vector of boolean values each indicating if a specific activity label is present or not in the trace. The *frequency-based* encoding, instead of boolean values, represents the control flow in a trace with the frequency of each activity label in the trace. Another way of encoding a trace is by taking into account also information about the order in which events occur in it, as in the *simple index* encoding. Here, each dimension corresponds to a position in the trace and its value is a numeric code representing the activity label occurring in that position.

A more complex control-flow encoding is obtained by associating each dimension in $\mathbb{R}^n$ to a different sub-trace of size p (i.e., p-grams). Each feature of this encoding represents how frequently and "compactly" a sub-trace appears in the trace of interest. For simplicity, we consider 2-grams, but the following can be easily generalized to p-grams. Given an abstract trace $\mathbf{e}_a$, we employ a simplified version of the encoding from [14] to transform $\mathbf{e}_a$ into a vector in $\mathbb{R}^n$ in two steps. First, we identify all 2-grams occurring in all the abstract traces in $\mathcal{T}$. Then, we construct a vector in $\mathbb{R}^n$ where each dimension of the vector is a real number representing the frequency and the compactness of a specific 2-gram. E.g., for the 2-gram ab, this value is given by $\mathcal{E}_{ab}(\mathbf{e}_a) = \sum_{1 \leq i \leq |\mathbf{e}_a|-1} \lambda^i [\Lambda^i]_{ab}$, where $[\Lambda^i]_{ab} \to 0, 1$ indicates the occurrences of ab at distance i in $\mathbf{e}_a$, and $\lambda \in]0, 1]$ is a parameter that represents the penalty provided for less compact 2-grams. Lower values of λ correspond to a higher distance between the numeric

Table 1. Encoding of traces *caba*, *caa* and *cb*.

	aa	ab	ac	ba	bb	bc	ca	cb	cc
caba	λ^2	λ	0	λ	0	0	$\lambda + \lambda^3$	λ^2	0
caa	λ	0	0	0	0	0	$\lambda + \lambda^2$	0	0
cb	0	0	0	0	0	0	0	λ	0

representation of more compact 2-grams wrt less compact ones (if λ is equal to 1 the compactness has no influence on the feature values of this encoding).

Example 4. Table 1 shows the 2-gram encodings of some traces over the activity labels $A = \{a, b, c\}$. Trace *cb* has only one non-zero dimension $\mathcal{E}_{cb}(cb) = \lambda$; trace *caa* has two non-zero dimensions: $\mathcal{E}_{ca}(caa) = \lambda + \lambda^2$ (*ca* occurs once with *c* and *a* at distance 1, i.e., *c̲a̲a*, and once with *c* and *a* at distance 2, i.e., *c̲aa̲*), and $\mathcal{E}_{aa}(caa) = \lambda$ (*a* is repeated after a single step in the trace only once, i.e., *ca̲a̲*).

In addition to control-flow, abstract traces include variable range assignments from the set of data variables V linked to each activity. For instance, if variable *Amount* $\in [10, 20[$ triggers activity *b* after *a* in e_a, then, the abstract event corresponding to *b* contains interval $[10, 20[$ for variable *Amount*. All the possible intervals for a variable (derived from all the abstract traces for a given DPN) have to be transformed into specific values to apply existing methods for trace encoding like the ones in [12]. To do so, we encode each variable v of each abstract event e_i using a feature space of *interval features* that are boolean features composed of v and a possible interval I. In this way, for each abstract event e_i, we have a set of interval features with values $\mathbb{D}_{e_i,v,I}$ given by:

$$\mathbb{D}_{e_i,v,I} = \begin{cases} 1 & \text{if } \iota(v) \subseteq I, \\ 0 & \text{otherwise} \end{cases} \tag{1}$$

where $\iota(v)$ is the variable range assignment in e_i. As an example, given a set of possible intervals for variable *Amount* $\{[0, 10[, [10, 20[, [10, 30[\}$, an abstract event $e_i = (l, Amount = [16, 24])$ is encoded by three boolean features: $\mathbb{D}_{e_i,Amount:[0,10[} = 0$, $\mathbb{D}_{e_i,Amount:[10,20[} = 0$, and $\mathbb{D}_{e_i,Amount:[10,30[} = 1$.

We use these boolean features as "event attributes" of each abstract event in an abstract trace. In this way, we can directly apply existing trace encodings [12] to abstract traces. These encodings can include control flow features (that can range from a simple boolean encoding to more complex encodings like the one based on *p*-grams), and data-flow features (derived from the interval features introduced above). We point out here that the choice of the encoding is part of the analysis. The analyst can select the information that is more relevant for computing the alignments depending on the specific process and the specific context in which the alignments are computed. The encoded abstract traces in the behavior encoding space $\mathcal{E}(\mathcal{T})$ are used to identify the top-k alignments of a (concrete) log trace $\mathbf{e}$ as described in the next section.

4.2 Approximating Alignment Computation Using kNN

In the alignment problem, we assume to have a set of (abstract) traces $\mathbf{e}_a \in \mathcal{T}$ and a set of non-conforming (concrete) log traces $\mathbf{e}$. The trace alignment task consists in searching the model trace $\mathbf{e}_a^* \in \mathcal{T}$ that is the closest to each log trace according to a given distance/cost function $\delta_\mathcal{E} : \mathbb{R}^n \times \mathbb{R}^n \to \mathbb{R}$.

When in the alignment task, together with the control-flow also other perspectives available in event logs like timestamps and data payloads are taken into consideration, this multi-perspective trace alignment becomes a challenging problem. However, if we use encodings to represent model and log traces, we can select the most relevant information needed to compute the alignments and, at the same time, reduce the time needed to perform the task. The log traces and the data space including the possible model traces (i.e., the possible alignments) can be explored, once a log trace to be aligned is given, by using k-nearest neighbors (kNN) algorithms. By using trace encoding, we can compute the encodings of the log trace $\mathcal{E}(\mathbf{e})$ and of all the possible model traces $\mathcal{E}(\mathbf{e}_a)$, and compute their distance using a distance function $\delta_\mathcal{E}(\mathbf{e}_a, \mathbf{e}) := \langle \mathcal{E}(\mathbf{e}_a), \mathcal{E}(\mathbf{e}) \rangle$. Then, the trace alignment problem is solved by computing the approximate alignment for an observed log trace $\mathbf{e}$ as $\min \arg_{\mathbf{e}_a} \delta_\mathcal{E}(\mathbf{e}_a, \mathbf{e})$. In particular, the approximate alignment(s) can be computed using kNN algorithms that find the k nearest data points to a *query* x from a set $\mathcal{X}$ of *data points* according to a distance function. By casting the trace alignment problem to a kNN problem, we can find the best k alignments of a log trace $\mathbf{e}$ in the space of the model traces. This can be done by using ad-hoc data structures like Ball-Tree and KD-Tree [13] to retrieve the k-neighborhood of $\mathbf{e}$ by pre-ordering (*indexing*) the space of the (embedded) model traces wrt a distance function $\delta_\mathcal{E}$.

kNN algorithms, being unsupervised, give to all features in $\mathcal{E}(\mathbf{e}_a)$ the same weight. However, since control-flow is, in general, represented by a lower number of features wrt the data flow, an equal distribution of the weights would penalize the control-flow, which is, instead, crucial for the alignment task. Moreover, since each variable is divided into interval features to represent the data-flow of the abstract trace, based on the possible intervals for that variable, a variable having a higher number of possible intervals would have a higher weight.

To overcome this problem, it is possible to use weighted kNN algorithms and force a different distribution of weights. This can be done in two steps. First, we separate $\mathcal{E}(\mathbf{e}_a)$ into $\mathcal{E}_A(\mathbf{e}_a)$ and $\mathcal{E}_V(\mathbf{e}_a)$ containing the control-flow and the data-flow features for trace $\mathbf{e}_a$ respectively. Then, we fix a parameter $s \in [0, 1]$ and assign weights s and $1 - s$, to $\mathcal{E}_A(\mathbf{e}_a)$ and $\mathcal{E}_V(\mathbf{e}_a)$. In this way, if, for instance, $s = 0.4$, we can assign weight 0.4 to the entire set of control flow features and weight 0.6 to the entire set of data features. Secondly, to avoid that a data variable having more possible intervals in $\mathcal{E}_V(\mathbf{e}_a)$ gets a higher weight, we uniformly distribute the weight $1 - s$ over the data variables and not over the interval features. For instance, if we have two data variables *Amount* and *Point* with possible intervals $Amount \in \{[0, 10[, [10, 20[,$ $[10, 30[\}$ and $Point \in \{[0, 5[, [5, 10[\}$, respectively, weight $1 - s = 0.6$ is uniformly distributed over the two data variables and, then, the resulting weights

(0.3 for each data variable) are distributed over the corresponding interval features, leading the weight distribution $(0.1, 0.1, 0.1, 0.15, 0.15)$ over the interval features ($Amount : [0, 10[$, $Amount : [10, 20[$, $Amount : [10, 30[$, $Point : [0, 5[$, and $Point : [5, 10[$). Therefore, each interval feature for a data variable v_1 having more possible intervals gets a lower weight, but the sum of the weights of all the interval features of each data variable is the same. In this way, considering all the features $l_1, ..., l_p$ from $\mathcal{E}_A(\mathbf{e}_a)$ and $v_1, ..., v_q$ from $\mathcal{E}_V(\mathbf{e}_a)$, we define a variable weight $w = (w_{l_1}, ..., w_{l_p}, w_{v_1}, ..., w_{v_q})$, where $\sum w = 1$, $\sum(w_{l_1}, ..., w_{l_p}) = s$, and $\sum(w_{v_1}, ..., w_{v_q}) = 1 - s$. In the kNN algorithm, we multiply the features in $\mathcal{E}(\mathbf{e}_a)$ by w to normalize them.

The computation of approximate alignments is more efficient than the computation of optimal alignments. However, this computational gain comes with a loss in precision. It is well-known that the generation of precise encodings for graph data with loops is NP-complete [6]. To keep the information preserved at most, we investigate different encodings recently provided by the process mining community, as well as the proposed simplified p-grams encoding, in next section.

5   Experimentation

In this section, we report on experiments that contrast our approximate, encoding-based approach with precise conformance checking techniques. As outlined in the introduction, we performed two experiments to that end. (a) First, we compare our encoding-based approach with the state-of-the-art data-aware conformance checker CoCoMoT [3]: here, for a given trace, we compare the best-matching model run as computed by CoCoMoT, with the best-matching abstract trace as estimated by our approximate approach (recall that every model run corresponds to a unique abstract trace). (b) Second, we consider the situation where the process behaviors are specified by a set of traces that plays the role of a reference log. Here, for a given trace, we compare the best-matching trace from the reference log according to the distance function from [17], with the best-matching trace as estimated by our approximate approach. This second setting can be of interest if no DPN is available; but the experiment also helps to study specifically how well encodings can emulate the distance function.

Consequently, for stage (1) of our approach, process behaviors were represented as follows: (1a) We represented all behaviors of a DPN with variable-to-constant comparisons by a complete set $\mathcal{T}$ of abstract traces, as described in Sect. 3. (1b) We took as $\mathcal{T}$ a plain set of traces. Then, we (2) applied the trace encodings discussed in Sect. 4.1 to obtain the behavior encoding space, and (3) compared the returned approximate alignments with the optimal ones. The implementation we used for the experiments is publicly available at https://github.com/jonghyeonk/Multi-Trace-Alignment.

Datasets. For (1a), we used a DPN modeling a road fine management process [10, Fig. 13] that was mined automatically and where all guards are variable-to-constant comparisons. With the proposed abstract trace, we generated a rep-

Table 2. Descriptive statistics of the trace representations $\mathcal{T}$ for the datasets.

Road Fines				Sepsis			
# of abstract traces	639	# of events	3,422	# of traces	1,079	# of events	15,214
# of activities	9	trace length	1~6	# of activities	16	trace length	3~185
# of data variables	5			# of data variables	2		

resentative set $\mathcal{T}$ with 639 abstract traces, as well as a test set L of 1,885 log traces with random values which comply with the DPN to a varying degree.

For (1b), we considered the *Sepsis* [15] event log, which represents the pathway of patients with symptoms of sepsis in a Dutch hospital. Here, we took $\mathcal{T}$ as all traces in the above dataset. As test set, we generated a log L consisting of 30 non-compliant log traces by modifying 10 traces in $\mathcal{T}$ based on three types of deviations: (i) modification of an activity label, (ii) modification of a categorical feature ('Diagnosis'), and (iii) modification of a numerical feature ('CRP') obtained by multiplying the original value by 10. Statistics of the trace representations $\mathcal{T}$ are summarized in Table 2.

Experiment Setup. For (1a) the *Road Fines* experiment, for each trace **e** in the test set L, we computed the optimal alignment γ and the associated process run $\gamma|_M$ using CoCoMoT, as well as its (unique) matching abstract trace $\mathbf{e}_a$. Then, we compared $\mathbf{e}_a$ with the result of the encoding-based approach. To compute $\gamma|_M$, we used the default settings of CoCoMoT, including its heuristic to determine $maxlen(\mathcal{N}, |\mathbf{e}|)$, which was obtained as $|\mathbf{e}| + m$, where m is the length of the shortest trace accepted by $\mathcal{N}$.

For (1b) the *Sepsis* experiment, for each trace **e** in the test set L, we computed the trace $\mathbf{e}'$ in $\mathcal{T}$ such that $\delta(\mathbf{e}, \mathbf{e}')$ is minimal according to the trace distance (Definition 6), and compared it with the result of the encoding-based approach.

Experimental Settings. For both experiments, we used the same settings: we set the split parameter s for feature weights to 0.5, selected the top k alignments with $k = \{k_{(10\%)}, k_{(20\%)}, k_{(30\%)}\}$ (k is the percentage of abstract traces returned) using a kNN algorithm,[2] and used three standard distance metrics (*Cosine, Manhattan, Euclidean*) with the following five encodings [12,14]:

- *aggregate*: the control-flow is represented using numerical features indicating the frequency of each activity label. Similarly, for categorical data variables, we use numerical features indicating the frequency of each possible categorical value. For numerical data variables, instead, we use the average, standard deviation, max, min, and sum of the values;
- *boolean*: the numerical data variables are represented as in the aggregate encoding. The control-flow and the categorical data variables are, instead,

---

[2] We used a function provided by the *sklearn* Python library, using the parameter *auto* for the selection of the algorithm, which makes the function able to select the most appropriate algorithm based on the input.

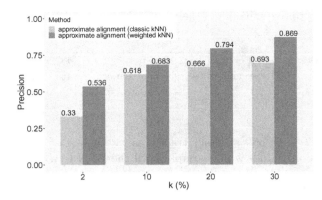

Fig. 1. Performance improvement on the *Road Fines* experiment after integrating the variable weight function (for *complexindex* encoding and *Euclidean* distance).

represented through boolean features (true/false) indicating whether a certain activity label or a certain categorical value is present in the trace;

- *complexindex*: this is the complex-index encoding introduced in [12]. The control flow is represented using the simple-index encoding, i.e., each control-flow feature corresponds to a position in the trace and the value of the feature is the activity occurring in that position of the trace. Similarly, for each data variable, we have different features representing that data variable in different positions of the trace and the value of the feature is the value of the data variable if the variable is numeric and, if the variable is instead categorical, a code representing its categorical value;
- *laststate*: this encoding represents the control-flow with the simple-index encoding and the data variables using the latest payload of a trace, i.e., data variables are treated as static features without taking into consideration their evolution over time;
- *p-gram+aggregate*: we used the encoding based on p-grams introduced in Sect. 4.1 with $p = 2$ and $\lambda = 0.7$. We integrated in this encoding the data perspective in the same way as done in the aggregate encoding.

In the evaluation, we measure (i) whether among the top-k alignments returned by the kNN algorithm, there is the optimal alignment returned by CoCoMoT (*precision*), (ii) how similar the top-k alignments returned by the approximate method are wrt the optimal alignment returned by CoCoMoT (*similarity*), and (iii) the execution times (*time*). For computing precision, we count the number of true positives TP, i.e., how many times the top-k alignments include the optimal alignment returned by CoCoMoT, and the number of false positives FP, i.e., how many times the top-k alignments do not include the optimal alignment returned by CoCoMoT. Then, the precision is computed as: $Precision = TP/(TP + FP)$. For similarity, we calculate the average Euclidean distance $dist$ between the top-k alignments and the optimal alignment returned by CoCoMoT and we compute the similarity score as: $similarity = 1 - dist$.

Table 3. Precision of approximate trace alignment with different parameters in the *Road Fines* experiment. For each k, the best precision is highlighted in bold.

Encoding method	Distance metric	k (%)	Precision (ref = CoCoMot)	Time (sec)
aggregate	Cosine	(10%/20%/30%)	(0.613/0.740/0.803)	(0.08/0.09/0.10)
aggregate	Euclidean	(10%/20%/30%)	(0.195/0.501/0.551)	(0.19/0.13/0.17)
aggregate	Manhattan	(10%/20%/30%)	(0.195/0.518/0.555)	(0.14/0.21/0.16)
boolean	Cosine	(10%/20%/30%)	(0.580/0.712//0.756)	(0.08/0.17/0.10)
boolean	Euclidean	(10%/20%/30%)	(0.597/0.725/0.808)	(0.13/0.19/0.13)
boolean	Manhattan	(10%/20%/30%)	(0.596/0.729/0.828)	(0.16/0.15/0.16)
complexindex	Cosine	(10%/20%/30%)	(0.683/0.775/0.838)	(0.28/0.24/0.27)
complexindex	Euclidean	(10%/20%/30%)	(0.683/0.794/0.869)	(0.36/0.33/0.37)
complexindex	Manhattan	(10%/20%/30%)	(0.481/0.790/0.862)	(0.70/0.73/0.67)
laststate	Cosine	(10%/20%/30%)	(0.420/0.688/0.800)	(0.08/0.08/0.17)
laststate	Euclidean	(10%/20%/30%)	(0.494/0.712/0.845)	(0.12/0.13/0.14)
laststate	Manhattan	(10%/20%/30%)	(0.510/0.734/0.882)	(0.16/0.18/0.16)
p-gram+aggregate	Cosine	(10%/20%/30%)	(0.705/0.776/0.817)	(0.07/0.08/0.08)
p-gram+aggregate	Euclidean	(10%/20%/30%)	(0.715/0.776/0.853)	(0.11/0.12/0.12)
p-gram+aggregate	Manhattan	(10%/20%/30%)	(**0.719/0.798/0.898**)	(0.12/0.14/0.18)

Table 4. Precision of approximate trace alignment with different parameters in *Sepsis* experiment. For each k, the best precision is highlighted in bold.

Encoding method	Distance metric	k (%)	Precision (ref = CoCoMot)	Time (sec)
aggregate	Cosine	(10%/20%/30%)	(0.813/0.833/0.841)	(0.01/0.01/0.01)
aggregate	Euclidean	(10%/20%/30%)	(0.888/0.898/0.906)	(0.02/0.02/0.02)
aggregate	Manhattan	(10%/20%/30%)	(0.888/0.898/0.906)	(0.02/0.03/0.02)
boolean	Cosine	(10%/20%/30%)	(0.776/0.800/0.808)	(0.01/0.01/0.02)
boolean	Euclidean	(10%/20%/30%)	(0.854/0.864/0.873)	(0.02/0.02/0.02)
boolean	Manhattan	(10% / 20%/30%)	(0.852/0.864/0.873)	(0.02/0.02/0.03)
complexindex	Cosine	(10%/20%/30%)	(0.816/0.816/0.888)	(0.04/0.04/0.04)
complexindex	Euclidean	(10%/20%/30%)	(0.891/0.931/0.949)	(0.05/0.04/0.05)
complexindex	Manhattan	(10%/20%/30%)	(**0.924/0.938/0.951**)	(0.08/0.07/0.07)
laststate	Cosine	(10%/20%/30%)	(0.822/0.822/0.822)	(0.01/0.01/0.01)
laststate	Euclidean	(10%/20%/30%)	(0.857/0.923/0.924)	(0.01/0.01/0.02)
laststate	Manhattan	(10%/20%/30%)	(0.855/0.923/0.924)	(0.02/0.02/0.02)
p-gram+aggregate	Cosine	(10%/20%/30%)	(0.864/0.881/0.891)	(0.01/0.01/0.01)
p-gram+aggregate	Euclidean	(10%/20%/30%)	(0.914/0.926/0.939)	(0.02/0.02/0.02)
p-gram+aggregate	Manhattan	(10%/20%/30%)	(0.914/0.926/0.939)	(0.04/0.03/0.03)

We computed the execution times (in seconds) by running the alignment tools on an Intel Core i9-12900H CPU with 2.5 GHz, 40 GB RAM, with MS Windows 11, and measuring the total time needed to align all log traces.

Experimental Results. Tables 3 and 4 show the precision of the approximate method achieved by varying three parameters (encoding method, distance metric, and k) for the *Road Fines* experiment and the *Sepsis* experiment. The results

Table 5. Similarity between the approximate trace alignment and the CoCoMoT alignment with different parameters in the *Road Fines* experiment.

Encoding method	Distance metric	k	Similarity (ref = CoCoMot)	Time (sec)
aggregate	Cosine	(1/3/5/10)	(0.971/0.971/0.971/0.973)	(0.07/0.07/0.07/0.07)
aggregate	Euclidean	(1/3/5/10)	(0.963/0.963/0.963/0.960)	(0.11/0.12/0.12/0.12)
aggregate	Manhattan	(1/3/5/10)	(0.933/0.938/0.942/0.948)	(0.18/0.13/0.16/0.14)
boolean	Cosine	(1/3/5/10)	(0.978/0.978/0.978/0.977)	(0.06/0.07/0.08/0.08)
boolean	Euclidean	(1/3/5/10)	(0.978/0.978/0.978/0.977)	(0.16/0.12/0.12/0.17)
boolean	Manhattan	(1/3/5/10)	(0.978/0.978/0.978/0.977)	(0.13/0.13/0.14/0.14)
complexindex	Cosine	(1/3/5/10)	(0.991/0.991/0.991/0.990)	(0.22/0.25/0.25/0.25)
complexindex	Euclidean	(1/3/5/10)	(0.991/0.991/0.991/0.990)	(0.37/0.31/0.35/0.35)
complexindex	Manhattan	(1/3/5/10)	(0.978/0.979/0.980/0.981)	(0.68/0.68/0.70/0.64)
laststate	Cosine	(1/3/5/10)	(0.985/0.985/0.984/0.984)	(0.07/0.07/0.07/0.07)
laststate	Euclidean	(1/3/5/10)	(0.985/0.985/0.984/0.984)	(0.17/0.12/0.12/0.17)
laststate	Manhattan	(1/3/5/10)	(0.985/0.985/0.984/0.984)	(0.13/0.18/0.13/0.14)
p-gram+aggregate	Cosine	(1/3/5/10)	**(0.996/0.996/0.996/0.995)**	(0.06/0.07/0.06/0.07)
p-gram+aggregate	Euclidean	(1/3/5/10)	**(0.996/0.996/0.996/0.995)**	(0.10/0.10/0.15/0.11)
p-gram+aggregate	Manhattan	(1/3/5/10)	**(0.996/0.996/0.996/0.995)**	(0.15/0.12/0.13/0.12)

Table 6. Similarity of approximate trace alignment in respect to the CoCoMoT alignment with different parameters on the *Sepsis* event log.

Encoding method	Distance metric	k	Similarity (ref = CoCoMot)	Time (sec)
aggregate	Cosine	(1/3/5/10)	(0.609/0.610/0.609/0.608)	(0.01/0.01/0.01/0.01)
aggregate	Euclidean	(1/3/5/10)	(0.608/0.608/0.608/0.608)	(0.02/0.02/0.02/0.02)
aggregate	Manhattan	(1/3/5/10)	(0.608/0.608/0.608/0.608)	(0.02/0.02/0.02/0.02)
boolean	Cosine	(1/3/5/10)	(0.606/0.606/0.607/0.606)	(0.01/0.01/0.01/0.01)
boolean	Euclidean	(1/3/5/10)	(0.606/0.606/0.606/0.607)	(0.02/0.02/0.01/0.02)
boolean	Manhattan	(1/3/5/10)	(0.606/0.606/0.606/0.606)	(0.02/0.02/0.02/0.02)
complexindex	Cosine	(1/3/5/10)	**(0.620/0.620/0.620/0.620)**	(0.04/0.04/0.04/0.04)
complexindex	Euclidean	(1/3/5/10)	(0.619/**0.620/0.620**/0.620)	(0.04/0.04/0.04/0.04)
complexindex	Manhattan	(1/3/5/10)	(0.619/**0.620/0.620**/0.621)	(0.05/0.05/0.05/0.05)
laststate	Cosine	(1/3/5/10)	(0.603/0.602/0.601/0.602)	(0.01/0.01/0.01/0.01)
laststate	Euclidean	(1/3/5/10)	(0.603/0.601/0.600/0.600)	(0.02/0.02/0.02/0.02)
laststate	Manhattan	(1/3/5/10)	(0.603/0.601/0.600/0.601)	(0.01/0.02/0.02/0.02)
p-gram+aggregate	Cosine	(1/3/5/10)	(0.604/0.607/0.607/0.607)	(0.01/0.01/0.01/0.01)
p-gram+aggregate	Euclidean	(1/3/5/10)	(0.605/0.607/0.607/0.607)	(0.03/0.02/0.02/0.02)
p-gram+aggregate	Manhattan	(1/3/5/10)	(0.605/0.607/0.606/0.607)	(0.02/0.02/0.02/0.02)

show that, as expected, when k increases, the top-k alignments are more likely to include the optimal trace or abstract trace. In the cases in which the precision is higher for lower values of k, the approximate approach results to be more effective. Regarding the encoding methods, *complexindex* and *p-grams+aggregate* have a higher precision overall wrt the other encodings. This was expected since the *aggregate* and *boolean* encodings are less rich in the representation of the control flow information and the *laststate* reflects less accurately the data flow.

Figure 1 highlights the effectiveness of the variable weight function we have introduced to guide the kNN algorithms in finding the top k alignments.

Concerning the similarity measure, Tables 5 and 6 report the average similarity between the top k alignments returned by the approximate approach with $k \in \{1,3,5,10\}$ and the optimal abstract trace returned by CoCoMoT (for *Road Fines*), or the closest trace (for *Sepsis*). As the number of traces is much lower wrt the number of traces used in Tables 3 and 4 ($k_{(10\%)} = 64$ and $k_{(10\%)} = 108$ for *Road Fines* and *Sepsis*, respectively), the precision becomes much lower. However, even if the optimal alignment could not be easily identified, the high similarity values shown in Table 5 indicate that the approximate approach returns alignments that are very close to the optimal one. For *Sepsis*, where we do not perform trace alignment wrt a DPN, the similarity is lower. This is due to the fact that, in this case, the alignment task is more challenging since the returned alignments are concrete traces whereas, in the *Road Fines* case, the alignment returned is an abstract trace, i.e., a class of traces. However, as shown in Table 4, when a larger number of possible alignments are returned, this issue does not affect the precision of the approximate approach.

For what concerns the execution times, the approximate approach is, on average, 100 times faster than CoCoMoT while producing alignments that are very similar to the optimal one. For instance, the alignment task for the *Road Fines* experiment has been completed in 22.02 s using CoCoMot, but it has been completed in 0.18 s using the approximate approach with the *p-grams + aggregate* encoding, *Manhattan* distance and $k(\%) = 30\%$. We also highlight that, although the precision is high only when k is sufficiently large, the fact that the approximate approach can return a fraction of the input log that likely contains the optimal alignment can render this approach useful as a sampling method, in combination with an optimal approach: the approximation technique can be applied as a preprocessor to find a set of candidates C for optimal alignments. Afterwards, less performant but optimal tools like CoCoMoT [3] can be applied only to this limited set of candidates (either using trace distances or incorporating the information from C into the DPN).

6   Related Work

A few papers provide extensions over the computation of control-flow alignments so that also the data dimension is considered. The approach in [9] first takes into account the control-flow and, after it has been aligned, aligns case attributes. The one in [8] is similar, but makes use of DPNs as process models, where the data-perspective is taken into consideration by augmenting the computed alignment with write operations over process variables via MILP solving. An improvement of this is obtained in [16,17], where in a faster A*-based technique the process and data dimensions are considered at the same time. Another recent notable approach is the one we use as baseline in this paper, i.e., the one from [3], where a very general multi-perspective conformance checking problem based on an abstract notion of cost function is solved via state-of-the-art SMT solving. Differently from our approach, all the above techniques compute optimal alignments.

As the current approaches for alignment computation have the main problem of the complexity both in space and time, various approximate solutions for control-flow alignment have been proposed. In [1], the authors provide a statistical approach to conformance checking that employs trace sampling from the event log and result approximation in order to derive conformance results in an efficient manner. This solution is orthogonal to our technique since, instead of sampling the log, we take abstractions from the model. Moreover, general approximation schemes for alignment have been proposed in [19]. The work casts a recursive strategy to solve the alignment problem by splitting ILP models into small pieces. The same authors present in [20] a technique to decrease both in execution time and memory the computation of alignments via the reduction of the given process model and the event log. A decomposition-based method is proposed in [7] for an approximation of the alignments with good precision and low execution time. Recently, an approximate alignment approach based on process trees has been proposed [18]. The approach splits the problem of alignments into smaller sub-problems along the tree hierarchy and solves them individually and in parallel. The approximate approaches presented so far consider the alignment problem at the control-flow level, while there is no existing work for handling the multi-perspective approximate alignment problem.

7   Conclusion

In this paper, we showed how trace encoding methods can be used to compute multi-perspective trace alignments. By opportunely selecting the encoding methods, the analyst can choose the relevant information for computing the alignments. This also makes the alignment task faster. Our experiments show that the approximate approach is 100 times faster than CoCoMoT. The results are accurate in terms of precision (identification of the optimal alignment) and similarity between the approximate and the optimal alignments. In future work, we want to extend this approach to probabilistic trace alignment both in the standard conformance checking scenario, in which a log trace is aligned with a DPN, and in the case in which a trace is aligned wrt a log of "happy paths". In the latter, the probability of a path can be computed using clustering methods, more precisely by taking the density of the cluster to which the path belongs.

References

1. Bauer, M., van der Aa, H., Weidlich, M.: Sampling and approximation techniques for efficient process conformance checking. Inf. Syst. **104**, 101666 (2022)
2. Carmona, J., van Dongen, B.F., Solti, A., Weidlich, M.: Conformance Checking - Relating Processes and Models. Springer, Cham (2018). https://doi.org/10.1007/978-3-319-99414-7

3. Felli, P., Gianola, A., Montali, M., Rivkin, A., Winkler, S.: CoCoMoT: conformance checking of multi-perspective processes via SMT. In: Proceedings of BPM 2021 (2021)

4. Felli, P., Gianola, A., Montali, M., Rivkin, A., Winkler, S.: Conformance checking with uncertainty via SMT. In: Proceedings of BPM 2022 (2022)

5. Felli, P., Gianola, A., Montali, M., Rivkin, A., Winkler, S.: Data-aware conformance checking with SMT. Inf. Syst. **117** (2023). https://doi.org/10.1016/j.is.2023.102230

6. Gärtner, T., Flach, P., Wrobel, S.: On graph kernels: hardness results and efficient alternatives. In: Proceedings of COLT 2003 (2003)

7. Lee, W.L.J., Verbeek, H., Munoz-Gama, J., van der Aalst, W.M., Sepúlveda, M.: Recomposing conformance: closing the circle on decomposed alignment-based conformance checking in process mining. Inf. Sci. **466**, 55–91 (2018)

8. de Leoni, M., van der Aalst, W.M.P.: Data-aware process mining: discovering decisions in processes using alignments. In: Proceedings of 13th SAC. ACM (2013)

9. de Leoni, M., van der Aalst, W.M.P., van Dongen, B.F.: Data- and resource-aware conformance checking of business processes. In: Proceedings of BIS 2012 (2012)

10. de Leoni, M., Felli, P., Montali, M.: A holistic approach for soundness verification of decision-aware process models. In: Proceedings of 37th ER (2018)

11. de Leoni, M., Felli, P., Montali, M.: Integrating BPMN and DMN: modeling and analysis. J. Data Semant. **10**(1), 165–188 (2021)

12. Leontjeva, A., Conforti, R., Di Francescomarino, C., Dumas, M., Maggi, F.M.: Complex symbolic sequence encodings for predictive monitoring of business processes. In: Proceedings of BPM 2015 (2015)

13. Liu, T., Moore, A.W., Gray, A., Cardie, C.: New algorithms for efficient high-dimensional nonparametric classification. J. Mach. Learn. Res. **7**(6) (2006)

14. Lodhi, H., Saunders, C., Shawe-Taylor, J., Cristianini, N., Watkins, C.J.C.H.: Text classification using string kernels. J. Mach. Learn. Res. **2**, 419–444 (2002)

15. Mannhardt, F.: Sepsis cases - event log (2016). https://data.4tu.nl/articles/dataset/Sepsis_Cases_-_Event_Log/12707639/1

16. Mannhardt, F.: Multi-perspective process mining. Ph.D. thesis, Technical University of Eindhoven (2018)

17. Mannhardt, F., de Leoni, M., Reijers, H., van der Aalst, W.: Balanced multi-perspective checking of process conformance. Computing **98**(4), 407–437 (2016)

18. Schuster, D., van Zelst, S., van der Aalst, W.M.P.: Alignment approximation for process trees. In: Leemans, S., Leopold, H. (eds.) ICPM 2020. LNBIP, vol. 406, pp. 247–259. Springer, Cham (2021). https://doi.org/10.1007/978-3-030-72693-5_19

19. Taymouri, F., Carmona, J.: A recursive paradigm for aligning observed behavior of large structured process models. In: Proceedings of BPM 2016 (2016)

20. Taymouri, F., Carmona, J.: Model and event log reductions to boost the computation of alignments. In: Ceravolo, P., Guetl, C., Rinderle-Ma, S. (eds.) SIMPDA 2016. LNBIP, vol. 307, pp. 1–21. Springer, Cham (2018). https://doi.org/10.1007/978-3-319-74161-1_1

POWL: Partially Ordered Workflow Language

Humam Kourani[(✉)] and Sebastiaan J. van Zelst

Fraunhofer FIT - Data Science and Artificial Intelligence, Sankt Augustin, Germany
{humam.kourani,sebastiaan.van.zelst}@fit.fraunhofer.de

Abstract. Process models are used to represent processes in order to support communication and allow for the simulation and analysis of the processes. Many real-life processes naturally define partial orders over the activities they are composed of. Partial orders can be used as a graph-like representation of process behavior. On the one hand, partially ordered graph representations allow us to easily model concurrent and sequential behavior between activities while ensuring simplicity and scalability. On the other hand, partial orders lack the support for typical process constructs such as choice and loop structures. Therefore, in this paper, we present a novel process modeling notation, i.e., the Partially Ordered Workflow Language (POWL). A POWL model is a partially ordered graph extended with control-flow operators for modeling choice and loop structures. A POWL model has a hierarchical structure; i.e., POWL models can be combined into a new model either using a control-flow operator or as a partial order. We propose an initial approach to demonstrate the feasibility of using POWL models for process discovery, and we evaluate our approach based on real-life data.

Keywords: POWL · process modeling · partial order · process tree

1 Introduction

A process model provides an illustration of a process that supports communication and allows for the simulation and analysis of the process. Process models can either be created by hand or discovered using process discovery techniques [2]. Organizations use information systems to track and record data about the execution of their processes, and this data is used for the discovery of process models. Process models might provide insights for organizations and allow them to analyze their processes in order to detect problems and bottlenecks. This can help to automate processes and to make better decisions.

Different modeling notations are used to model processes. *Petri nets* are a powerful modeling notation widely used to formally describe the behavior of processes. A sub-class of Petri nets, called *Workflow nets (WF-nets)*, is usually used to model *business processes*. WF-nets adhere to structural quality requirements; e.g., they define a clear notion for marking the start and end of processes. However, WF-nets may still suffer from behavioral quality issues. For instance, it is

© The Author(s), under exclusive license to Springer Nature Switzerland AG 2023
C. Di Francescomarino et al. (Eds.): BPM 2023, LNCS 14159, pp. 92–108, 2023.
https://doi.org/10.1007/978-3-031-41620-0_6

possible to construct a Workflow net with dead parts that can never be reached. WF-nets that do not suffer from such quality anomalies are called *sound*.

A *process tree* [16] is a hierarchical modeling notation, i.e., a mathematical tree, in which the leaves represent activities and the internal vertices represent control-flow operators for modeling behavioral dependencies between their children. Process trees represent a strict subset of WF-nets; i.e., any process tree can be transformed into a WF-net modeling the same behavior, but not all WF-nets can be modeled as process trees. Process trees are guaranteed to be sound by construction as they are limited to modeling hierarchical structures.

Partial orders are used as a representation of the execution order of activities for many real-life processes. In a partial order, some activities may have a strict order with respect to each other (e.g., activity "a" must happen before activity "b"), while other activities are concurrent (e.g. activities "b" and "c" may happen in any order). This reflects the reality of many business processes, where there may be multiple ways to accomplish a goal. Several partial-order-based modeling notations have been introduced, e.g., prime event structures [25] and conditional partial order graphs [22]. These notations allow us to model concurrency and sequential dependencies in an efficient and compact manner; however, none of them properly support cyclic process behavior, which is very common in practice. Moreover, in a partial order over activities, we assume all activities to be executed, and thus, modeling a choice is not supported.

On the one hand, process trees fail to model non-hierarchical dependencies that can be easily described by a partial order. On the other hand, we cannot model loop or choice structures in a partially ordered graph. We propose a new modeling notation that combines hierarchical modeling notations with partial orders. We call our modeling language *Partially Ordered Workflow Language (POWL)*. A POWL model is a partially ordered graph extended with control-flow operators for modeling choice and loop structures; i.e., a POWL model is a hierarchical model that allows for defining partial orders over sub-models.

The remainder of the paper is structured as follows. We start with a motivating example in Sect. 2. We discuss related work in Sect. 3, and we briefly present preliminaries in Sect. 4. We define POWL models in Sect. 5, and we introduce an initial approach for the discovery of POWL models in Sect. 6. We evaluate our approach using real-life data in Sect. 7. Finally, we provide a brief summary of the paper and propose ideas for future work in Sect. 8.

2 Motivation

In this section, we motivate our contribution based on a simple example.

We consider a process for purchasing items from an online shop. The user starts an order by logging in to their account (a). Then, the user simultaneously selects the items to purchase (b) and sets a payment method (c). Afterward, the user either pays (d) or completes an installment agreement (e). After selecting the items, the user chooses between multiple options for a free reward (f). Since the reward value depends on the purchase value, this step is done after selecting the items, but it is independent of the payment activities. Finally, the items are

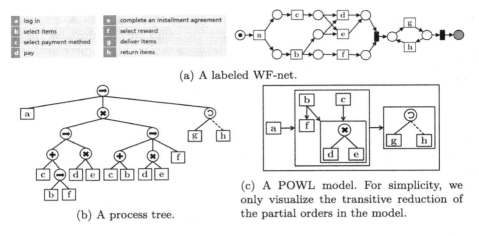

(a) A labeled WF-net.

(b) A process tree.

(c) A POWL model. For simplicity, we only visualize the transitive reduction of the partial orders in the model.

Fig. 1. Process models.

delivered to the user (g). The user may exchange received items. The user can return some items (h), and each time items are returned, a new delivery is made afterward. The WF-net shown in Fig. 1a precisely models this process.

A process tree models hierarchical behavioral structures using the control-flow operators →, ×, ↺, and +. The → operator models a sequential execution of blocks; × models an exclusive choice; + models concurrency; ↺ models a do-redo loop between two blocks (i.e., the first block is executed once first, and every time the second block is executed it is followed by another execution of the first block). Figure 1b shows a process tree modeling the behavior of our example process. This tree contains a choice of two sub-tree over the same set of activities (b, c, d, e, and f). Process trees are limited to modeling hierarchical structures; i.e., without duplicating activities, a process tree cannot precisely model the dependencies between the activities b, c, d, e, and f.

Figure 1c shows a POWL model precisely modeling the behavior of our example process. The outer layer of the hierarchy is a partial order modeling a sequence between the activity sets $\{a\}$, $\{b, c, d, e, f\}$, and $\{g, h\}$. Another partial order is used to model the non-hierarchical dependencies between the activity sets $\{b\}$, $\{c\}$, $\{d, e\}$, and $\{f\}$. The process tree operators × and ↺ are used to model the choice between d and e and the loop between g and h respectively.

Compared to the process tree and the WF-net, the POWL model has a simpler structure with fewer nodes and edges (i.e., no places and no duplication of activities). Moreover, the POWL model shows non-hierarchical dependencies without duplicating activities, while sub-models can still be easily identified in the hierarchy and the soundness guarantee is preserved.

3   Related Work

Different modeling notations are used among process mining tools and techniques. We refer to [3] for an overview of process modeling notations in process mining. In [16], Leemans introduce the inductive mining framework and multiple process discovery approaches implementing it. The approach we propose for the discovery of POWL models extends the inductive mining framework.

Many ideas for combining different modeling notations have been proposed. In [4], a hybrid Petri net is defined as a Petri net extended with informal arcs connecting transitions. In [27], another type of hybrid process models is defined by combining imperative and declarative modeling languages.

Partial orders are used for data representation and process modeling. An overview of the use of partial orders in process mining is provided in [17]. In [21], Mannila et al. propose an approach for the discovery of frequent episodes, where an episode is defined as a partially ordered set of events. This approach is adapted in [15] to discover partially ordered sets of activities in event logs. In [14], the authors create partially ordered representations of activities and combine them into a workflow graph. In [13], the authors suggest an approach for generating prime event structures from event logs. A prime event structure [25] is a partially ordered graph enriched with a conflict relation. This approach is able to model choice due to the conflict relation; however, loops remain a major challenge for prime event structures. In [22], the authors present a method for deriving conditional partial order graphs from event logs. A conditional partial order graph [23] is a compact representation of a family of partial orders that is able to model choice structures, but it fails to capture cyclic behavior.

In [26], the authors introduce a flow language (BPEL) that allows for combining web service primitives using advanced control-flow constructs (including event handlers). BPEL additionally allows for imposing an execution order over primitives executed in parallel using control links. BPEL is a powerful language for implementing web services; however, BPEL is very complex for end users and can be viewed as a programming language rather than a modeling language [5].

4   Preliminaries

In this section, we present basic preliminaries that ease this paper's readability.

$\mathbb{N}=\{1,2,3,...\}$ denotes the set of natural numbers. We use $\mathbb{N}_{odd}=\{1,3,5,...\}$ to denote the set of odd numbers, and we use $\mathbb{N}_{even}=\{2,4,6,...\}$ to denote the set of even numbers.

$\mathcal{P}(X)=\{X'\subseteq X\}$ denotes the powerset of a set X. For n sets $X_1,...,X_n$, we define the n-ary Cartesian product as the set $X_1\times...\times X_n=\{(x_1,...,x_n) \mid x_i\in X_i$ for $1\leq i\leq n\}$. An n-ary relation over $X_1,...,X_n$ is a subset of the n-ary Cartesian product $X_1\times...\times X_n$.

Let X and Y be two sets, and $f\colon X{\rightarrow}Y$ be a function. f is injective if $\underset{x,x'\in X}{\forall} f(x){=}f(x'){\Rightarrow}x{=}x'$. f is surjective if $\underset{y\in Y}{\forall}\ \underset{x\in X}{\exists}\ f(x){=}y$. f is bijective if it is injective and surjective. We use $\mathcal{B}(X,Y)$ to denote the set of all bijective functions from X to Y. A multi-set generalizes the notion of a set and allows for multiple occurrences of the same element. The order of occurrences of the elements in a multi-set is irrelevant. We define a multi-set M over a set X as a function $M\colon X{\rightarrow}\mathbb{N}\cup\{0\}$. We write a multi-set as $M=[x_1{}^{c_1},...,x_n{}^{c_n}]$ where

$M(x_i)=c_i$ for $1\leq i\leq n$ (for $x\in X$ with $M(x)=1$, we omit the superscript; in case $M(x)=0$, we omit x). We use $\mathcal{M}(X)$ to denote the set of all multi-sets over X.

A *sequence* is an ordered collection of elements. We define a sequence over a set X as a function $\sigma\colon \{1,\ldots,n\}\rightarrow X$, and we write $\sigma=\langle\sigma(1),\ldots,\sigma(n)\rangle$. We use $|\sigma|=n$ to denote the length of σ and X^* to denote the set of all sequences over X. We use $\sigma_1\cdot\sigma_2$ to denote the *concatenation* of two sequences σ_1 and σ_2, e.g., $\langle x_1\rangle\cdot\langle x_2,x_1\rangle=\langle x_1,x_2,x_1\rangle$. We overload notation and, for two sets of sequences L_1 and L_2, we write $L_1\cdot L_2=\{\sigma_1\cdot\sigma_2 \mid \sigma_1\in L_1\wedge\sigma_2\in L_2\}$. We use $\sigma{\uparrow}_Y$ to denote the *projection* of a sequence σ on a set Y. For example, $\langle x_1,x_2,x_1\rangle{\uparrow}_{\{x_1,x_3\}}=\langle x_1,x_1\rangle$.

Let $\prec\subseteq X\times X$ be a 2-ary relation over a set X. For $(x_1,x_2)\in X\times X$, we write $x_1\prec x_2$ to denote that $(x_1,x_2)\in\prec$, and we write $x_1\not\prec x_2$ to denote that $(x_1,x_2)\notin\prec$. $\prec$ is a *strict partial order* if it is *irreflexive* (i.e., $x\not\prec x$ for all $x\in X$) and *transitive* (i.e., if $x_1\prec x_2$ and $x_2\prec x_3$, then $x_1\prec x_3$)[1]. In the remainder of the paper, we use the term *partial order* to refer to a strict partial order. We refer to $\rho=(X,\prec)$ as a *partially ordered set (poset)*. The language of ρ is defined as the set of sequences $\mathcal{L}(\rho)=\{\sigma\in X^* \mid \underset{f\in\mathcal{B}(\{1,\ldots,|\sigma|\},X)}{\exists}\ \underset{1\leq i\leq|\sigma|}{\forall}\ \sigma(i)=f(i)\wedge \underset{1\leq j\leq|\sigma|}{\forall}\ f(i)\prec f(j)\Rightarrow i<j\}$. The *transitive reduction* of $\prec$ is defined as $\prec^-=\{(x_1,x_3)\in\prec \mid \underset{x_2\in X}{\nexists}\ x_1\prec x_2\wedge x_2\prec x_3\}$. We use $\Pi(X)$ to denote the set of all posets over X. Let X and Y be two sets, $\rho=(X,\prec)$ be a poset, and $\gamma\colon X\rightarrow Y$ be a *labeling* function. The triple $\rho'=(X,\prec,\gamma)$ is called a *labeled partial order* over X and Y. The language of ρ' is defined as the set of sequences $\mathcal{L}(\rho')=\{\langle\gamma(\sigma(1)),\ldots,\gamma(\sigma(|\sigma|))\rangle \mid \sigma\in\mathcal{L}(\rho)\}$. We use $\Pi(X,Y)$ to denote the set of all labeled partial orders over X and Y.

We use Σ to denote the universe of activities, and we use $\tau\notin\Sigma$ to denote the *silent activity* (τ is also referred to as the *unobservable activity*).

5   POWL Language

In this section, we introduce the Partially Ordered Workflow Language (POWL). We define POWL models, their semantics, and an approach for transforming POWL models into sound WF-nets.

A POWL model is a partially ordered graph representation of a process, extended with control-flow operators for modeling choice and loop structures. We define three types of POWL models. The first type is the *base case* consisting of a single activity. For the second type, we use the existing process tree operators $\times$ and $\circlearrowright$ (defined in [16]) to combine multiple POWL models into a new model. We use the operator $\times$ to model an exclusive choice of $n\geq 2$ POWL models and the operator $\circlearrowright$ to model a do-redo loop of two POWL models. The third type of POWL models is defined as a poset of $n\geq 2$ POWL models. We interpret unconnected nodes in a poset to be concurrent and connections between nodes as sequential dependencies. Figure 1c shows an example POWL model.

---

[1] Irreflexivity and transitivity imply *asymmetry*; i.e., if $x_1\prec x_2$, then $x_2\not\prec x_1$.

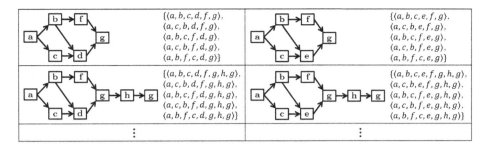

	$\{\langle a,b,c,d,f,g\rangle,$ $\langle a,c,b,d,f,g\rangle,$ $\langle a,b,c,f,d,g\rangle.$ $\langle a,c,b,f,d,g\rangle,$ $\langle a,b,f,c,d,g\rangle\}$		$\{\langle a,b,c,e,f,g\rangle,$ $\langle a,c,b,e,f,g\rangle,$ $\langle a,b,c,f,e,g\rangle,$ $\langle a,c,b,f,e,g\rangle,$ $\langle a,b,f,c,e,g\rangle\}$
	$\{\langle a,b,c,d,f,g,h,g\rangle,$ $\langle a,c,b,d,f,g,h,g\rangle,$ $\langle a,b,c,f,d,g,h,g\rangle,$ $\langle a,c,b,f,d,g,h,g\rangle,$ $\langle a,b,f,c,d,g,h,g\rangle\}$		$\{\langle a,b,c,e,f,g,h,g\rangle,$ $\langle a,c,b,e,f,g,h,g\rangle,$ $\langle a,b,c,f,e,g,h,g\rangle,$ $\langle a,c,b,f,e,g,h,g\rangle,$ $\langle a,b,f,c,e,g,h,g\rangle\}$
⋮		⋮	

Fig. 2. Translation of the POWL model shown in Fig. 1c into a set of labeled partial orders. For simplicity, we only show the labels of the transitions. We also show the language of each partial order (i.e., as a set of activity sequences).

Definition 1 (POWL Model). *A POWL model is recursively defined as follows:*

- *Any activity $a \in \Sigma \cup \{\tau\}$ is a POWL model.*
- *Let ψ_1 and ψ_2 be two POWL models. $\circlearrowleft(\psi_1, \psi_2)$ is a POWL model.*
- *Let $P = \{\psi_1, ..., \psi_n\}$ be a set of $n \geq 2$ POWL models.*
  - *$\times(\psi_1, ..., \psi_n)$ is a POWL model.*
  - *A poset $\rho = (P, \prec) \in \Pi(P)$ is a POWL model.*

We use Ψ to denote the universe of POWL models. We define the execution semantics for POWL models. Since a partially ordered set of POWL models is a POWL model, we define the semantics of POWL models in terms of partial orders as well. However, choice and loop structures cannot be described using a single partial order. Hence, we define the semantics of a POWL model by transforming it into a set of labeled partial orders over a set of newly generated nodes (we call them *transitions*), and we use activities as labels. Figure 2 shows the result of applying this transformation on the POWL model shown in Fig. 1c.

For the base case (i.e., a single activity), the POWL model is transformed into a single partial order with a transition having the corresponding activity as a label. For a silent activity, we create an empty labeled partial order. For the operator $\times$, the language is defined as the union of the languages of the sub-models. For the operator $\circlearrowleft$, we combine labeled partial orders from the languages of the sub-models such that the first order is from the do-part and each order from the redo-part is followed by an order from do-part. When combining these orders, we replace every transition from the orders of the languages of the sub-models by a new transition having the same label. For a poset of POWL models, labeled partial orders are generated by combining orders from the languages of the sub-models such that the partial order of the sub-models is preserved.

Definition 2 (Partial Order Semantics). *Let $\mathcal{T}$ be the universe of transitions. $\Gamma : \Psi \rightarrow \mathcal{P}(\Pi(\mathcal{T}, \Sigma))$ is a function recursively defined to transform a POWL model into a set of labeled partial orders as follows.*

- *For $a{\in}\Sigma$, $\Gamma(a){=}\{(\{t\},\emptyset,(t,a))\}$ where $t{\in}\mathcal{T}$ is a new transition.*
- $\Gamma(\tau){=}\{(\emptyset,\emptyset,\emptyset)\}$.
- *Let $P{=}\{\psi_1,...,\psi_n\}$ be a set of $n{\geq}2$ POWL models.*
  - $\Gamma({\times}(\psi_1,...,\psi_n)){=} \bigcup\limits_{1\leq i\leq n} \Gamma(\psi_i).$
  - $\Gamma(\circlearrowleft(\psi_1,\psi_2)){=}$
    $$\bigcup\limits_{n\in\mathbb{N}_{odd}} \{(T,\prec,\gamma) \mid \underset{(T_1,\prec_1,\gamma_1)\in\Gamma(\psi_{\tilde{1}}),...,(T_n,\prec_n,\gamma_n)\in\Gamma(\psi_{\tilde{n}}),f\in\mathcal{B}(\bigcup\limits_{1\leq i\leq n}T_i,T)}{\exists}$$
    $$\underset{1\leq i\leq n,t_i\in T_i}{\forall} \Big(\gamma_{\tilde{i}}(t_i){=}\gamma(f(t_i))$$
    $$\wedge \underset{1\leq j\leq n,t_j\in T_j}{\forall} f(t_i){\prec}f(t_j) \Leftrightarrow ((i{=}j \wedge t_i{\prec}_it_j) \vee i{<}j)\big)\}$$
    where $T{\subseteq}\mathcal{T}$ refers to a set of new transitions and $\tilde{i}$ refers to the transformation of an index $i \in \mathbb{N}$ defined as: $\tilde{i} = \begin{cases} 1 & \textit{if } i \in \mathbb{N}_{odd}, \\ 2 & \textit{if } i \in \mathbb{N}_{even}. \end{cases}$
  - *For a poset $\rho{=}(P,\prec)$, $\Gamma(\rho){=}\{(T,\prec',\gamma) \mid \underset{(T_1,\prec_1,\gamma_1)\in\Gamma(\psi_1),...,(T_n,\prec_n,\gamma_n)\in\Gamma(\psi_n)}{\exists}$*
    $$T{=} \bigcup\limits_{1\leq i\leq n} T_i$$
    $$\wedge \underset{1\leq i\leq n,t_i\in T_i}{\forall} \Big(\gamma(t_i){=}\gamma_i(t_i)$$
    $$\wedge \underset{1\leq j\leq n,t_j\in T_j}{\forall} t_i{\prec}'t_j \Leftrightarrow ((i{=}j \wedge t_i{\prec}_it_j) \vee \psi_i{\prec}\psi_j)\big)\}.$$

After transforming a POWL model into a set of labeled partial orders, we can derive the set of activity sequences that can be generated by the model. We overload notation by defining the language of a POWL model $\psi{\in}\Psi$ as the set of activity sequences $\mathcal{L}(\psi){=}\{\sigma{\in}\Sigma^* \mid \underset{\rho=(T,\prec,\gamma)\in\Gamma(\psi)}{\exists} \sigma{\in}\mathcal{L}(\rho)\}$.

Similar to process trees, POWL models can be recursively transformed into WF-nets. The transformation approach is schematically presented in Fig. 3. The generated workflow net is guaranteed to be sound; the soundness can be proven by the composition theorem ([1, Theorem 3]).

6   Discovery of POWL Models

In this section, we demonstrate the feasibility of using POWL models in process discovery by extending the base inductive miner [16] to mine for POWL models.

6.1   Event Log

Organizations use information systems to track and record information about the execution of their processes. Data can be stored in different forms. In process discovery, we assume data to be provided in the form of an *event log*. We define an *event log* $L{\in}\mathcal{M}(\Sigma^*)$ as a multi-set of activity sequences. A *trace* $\sigma{\in}L$ is a sequence of activities that represents the execution of a single process instance.

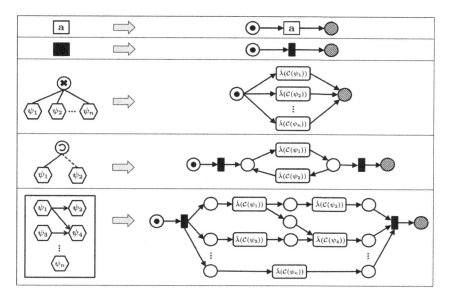

Fig. 3. POWL to WF-net converter $\mathcal{C}$. For a WF-net W, we use $\hat{\lambda}(N)$ to denote the Petri net that results after removing the source and sink places of W.

Let $L \in \mathcal{M}(\Sigma^*)$ be an event log. $\Sigma_L = \{a \in \Sigma \mid \underset{\sigma \in L, 1 \leq i \leq |\sigma|}{\exists} \sigma(i) = a\}$ denotes the set of activities that occur in L. We use $L_\triangleright = \{a \in \Sigma_L \mid \underset{\sigma \in L}{\exists} \sigma(1) = a\}$ to denote the set of *start activities* and $L_\square = \{a \in \Sigma_L \mid \underset{\sigma \in L}{\exists} \sigma(|\sigma|) = a\}$ to denote the set of *end activities*. The *directly-follows graph (DFG)* is a 2-ary relation $\mapsto_L \subseteq \Sigma_L \times \Sigma_L$ that captures direct successions between activities; i.e., $a \mapsto_L b$ iff $\underset{\sigma \in L, 1 \leq i < |\sigma|}{\exists} \sigma(i) = a \wedge \sigma(i+1) = b$. The *eventually-follows graph (EFG)* $\leadsto_L \subseteq \Sigma_L \times \Sigma_L$ captures direct and indirect successions between activities; i.e., $a \leadsto_L b$ iff $\underset{\sigma \in L, 1 \leq i < j \leq |\sigma|}{\exists} \sigma(i) = a \wedge \sigma(j) = b$.

$L_1 = [\langle a, b, c \rangle^3, \langle a, b, d \rangle^2]$ is an example event log. This event log consists of five traces with $\Sigma_{L_1} = \{a, b, c, d\}$, $L_{1\triangleright} = \{a\}$, $L_{1\square} = \{c, d\}$, $\mapsto_{L_1} = \{(a, b), (b, c), (b, d)\}$, and $\leadsto_{L_1} = \{(a, b), (a, c), (a, d), (b, c), (b, d)\}$.

6.2 Inductive Miner

The inductive miner [16] is one of the leading approaches in process discovery. It provides formal guarantees such as soundness, perfect fitness (i.e., it discovers a model that covers all behavior recorded in the log), and rediscoverability of certain process structures. There are several variants of the inductive miner (e.g., for handling incompleteness or infrequent behavior). In this paper, we extend the base variant of the inductive miner that assumes a noise-free event log and returns a model that perfectly fits the input event log.

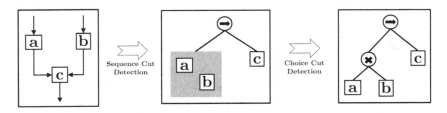

Fig. 4. Two steps of process tree cut detection: → and × cuts.

The inductive miner is a recursive top-down approach. The algorithm tries to detect a *cut*, i.e., it tries to detect a behavioral pattern in the directly-follows graph and a partitioning of the activities according to this pattern. The inductive miner supports four cuts corresponding to the four operators of process trees, and it recursively generates a process tree based on the detected cuts.

Definition 3 (Process Tree Cut). *Let $L \in \mathcal{M}(\Sigma^*)$ be an event log. A process tree cut $(\oplus, A_1, ..., A_n)$ of L is tuple of a control-flow operator $\oplus \in \{\to, \times, +, \circlearrowleft\}$ and a partitioning of the activities into $n \geq 2$ subsets; i.e., $\Sigma_L = A_1 \cup ... \cup A_n$ and $A_i \cap A_j = \emptyset$ for $1 \leq i < j \leq n$.*

After detecting a cut, the event log is projected into the different groups of the partitioning, creating several sub-logs. The same approach is then recursively applied to all sub-logs until a *base case* of the recursion is reached. A base case is defined as an event log whose activity set consists of a single activity. A base case can be easily transformed into a process tree: either into a single node or using the operators × or $\circlearrowleft$ to model an optional activity or a self-loop.

For the formal description of the different steps of the inductive miner, we refer to [16]. Figure 4 shows an example directly-follows graph and two steps of process tree cut detection based on it. First, a sequence cut $(\to, \{a, b\}, \{c\})$ is detected in the initial directly-follows graph; i.e., a sequential dependency between these groups of activities is discovered. The event log is then projected into the two groups of activities, creating sub-logs. The second sub-log is a base case. For the first sub-log, a choice cut $(\times, \{a\}, \{b\})$ is detected, and again, two sub-logs are generated. Both sub-logs are base cases, and the algorithm terminates returning the process tree $\to(\times(a, b), c)$.

If neither a base case nor a cut is detected, then the inductive miner invokes a *fall-through function*. This function always returns a cut that might correspond to an under-fitting model (i.e., a model that does not precisely capture the behavior recorded in the log), but it allows for continuing the recursion. For example, the fall-through function might return a concurrency cut between an activity that occurs exactly once in every trace and the rest of the activities. All steps of the algorithms are fitness-preserving [16]; i.e., all traces in the input event log are guaranteed to be included in the language of the generated model.

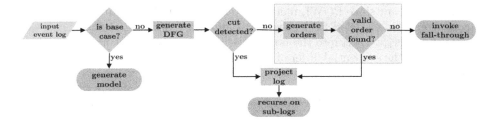

Fig. 5. Approach for the discovery of POWL models.

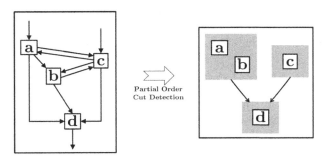

Fig. 6. Partial order cut detection.

6.3 Partial Order Cut

We adapt the inductive miner to discover POWL models instead of process trees. If the algorithm fails to detect a base case or a process tree cut, we mine for partial orders before invoking the fall-through function; we generate partial orders over all possible partitionings of activities, and we validate these orders using certain rules. If a *valid* order is found, then the event log is projected on the partitioning of activities and the recursion continues on the sub-logs; otherwise, the fall-through function is invoked. An overview of the different steps of our approach for the discovery of POWL models is shown in Fig. 5.

We define a *partial order cut* as a partitioning of the activities and a partial order over the partitioning. Since a partial order is transitive, we use the eventually-follows graph instead of the directly-follows graph for the detection of partial order cuts (we discuss the detection step in Sect. 6.4). Figure 6 shows an example eventually-follows graph with a partial order cut detected based on it. This cut consist of a partitioning of activities $P=\{\{a,b\},\{c\},\{d\}\}$ and a partial order $\prec$ defined by two ordering relations $\{a,b\}\prec\{d\}$ and $\{c\}\prec\{d\}$.

Definition 4 (Partial Order Cut). *Let* $L\in\mathcal{M}(\Sigma^*)$ *be an event log. A partial order cut of L is a poset* $\rho=(\{A_1,...,A_n\},\prec)$ *over a partitioning of the activities into* $n\geq2$ *subsets; i.e.,* $\Sigma_L=A_1\cup...\cup A_n$ *and* $A_i\cap A_j=\emptyset$ *for* $1\leq i<j\leq n$.

Our approach tries to detect a process tree cut before mining for a partial order. In case a sequence or a concurrency cut is detected, we transform it into a partial order cut since POWL models do not support the operators $\rightarrow$ and $+$. A concurrency is modeled as a poset with an empty ordering relation; i.e., we transform $(+, A_1, ..., A_n)$ into the poset $(\{A_1, ..., A_n\}, \emptyset)$. A sequence is modeled as poset using the sequential order of the nodes; i.e., we transform $(\rightarrow, A_1, ..., A_n)$ into the poset $(\{A_1, ..., A_n\}, \prec)$ where $A_i \prec A_j$ iff $1 \leq i < j \leq n$.

The base inductive miner only detects cuts that preserve perfect fitness [16]; i.e., all traces observed in the event log are guaranteed to be included in the language of the generated model. Similarly, we ensure perfect fitness for our approach.

Definition 5 (Fitness-Preserving Partial Order Cut). *Let $L \in \mathcal{M}(\Sigma^*)$ be an event log and $\rho = (P, \prec)$ be a partial order cut of L. ρ is fitness-preserving iff for any $A_1, A_2 \in P$: $A_1 \prec A_2 \Rightarrow \{\sigma \restriction_{A_1 \cup A_2} \mid \sigma \in L\} \subseteq A_1^* \cdot A_2^*$.*

6.4   Detection of Partial Order Cut

Our approach mines for a partial order cut before invoking the fall-through function. We use a brute-force approach that generates all possible partitionings of activities, and for each partitioning, we mine for a *valid* partial order. We define a valid partial order cut as a behavioral pattern in the eventually-follows graphs that corresponds to a partial order over the partitioning of activities.

A valid order contains an ordering edge between two groups of activities if and only if all activities of the first group are eventually followed by all activities of the second group and none of the activities of the second group is eventually followed by an activity of the first group. Moreover, two groups are not connected through any ordering edges if and only if they are concurrent. We define two groups to be concurrent if every activity of each group is eventually following all activities of the other group. Finally, we ensure that groups with no preceding groups with respect to the order contain start activities and groups with no succeeding groups with respect to the order contain end activities.

Definition 6 (Valid Partial Order Cut). *Let $L \in \mathcal{M}(\Sigma^*)$ be an event log and $\rho = (P, \prec)$ be a partial order cut of L. ρ is valid if the following conditions hold for all $A_i, A_j \in P$; $A_i \neq A_j$:*

1. $(A_i \prec A_j \wedge A_j \not\prec A_i)$ iff $\underset{a_i \in A_i, a_j \in A_j}{\forall} a_i \rightsquigarrow_L a_j \wedge a_j \not\rightsquigarrow_L a_i$.
2. $(A_i \not\prec A_j \wedge A_j \not\prec A_i)$ iff $\underset{a_i \in A_i, a_j \in A_j}{\forall} (a_i \rightsquigarrow_L a_j \wedge a_j \rightsquigarrow_L a_i)$.
3. *if* $\underset{A_k \in P}{\nexists} A_k \prec A_i$, *then* $A_i \cap L_\triangleright \neq \emptyset$.
4. *if* $\underset{A_k \in P}{\nexists} A_i \prec A_k$, *then* $A_i \cap L_\square \neq \emptyset$.

The partial order cut shown in Fig. 6 is valid. Note that if a valid partial order cut over a partitioning of activities exists, then it is unique (the first condition of Definition 6 uniquely defines a relation). Moreover, a valid partial order cut is fitness-preserving; i.e., the partial order cut detection step is fitness-preserving.

Theorem 1. *Let $L \in \mathcal{M}(\Sigma^*)$ be an event log and $\rho = (P, \prec)$ be a valid partial order cut of L. ρ is fitness-preserving.*

Proof. Let $A_i, A_j \in P$ with $A_i \prec A_j$. Then, $A_j \not\prec A_i$ since $\prec$ is asymmetric.

$$\Rightarrow \quad \underset{a_i \in A_i, a_j \in A_j}{\forall} \quad a_i \leadsto_L a_j \wedge a_j \not\leadsto_L a_i.$$

$$\Rightarrow \quad \underset{\sigma \in L, 1 \leq k < l \leq |\sigma|}{\nexists} \quad \sigma(k) \in A_j \wedge \sigma(l) \in A_i.$$

$$\Rightarrow \quad \{\sigma\!\uparrow_{A_i \cup A_j} \mid \sigma \in L\} \subseteq \{\sigma_i \cdot \sigma_j \mid \sigma_i \in A_i{}^* \wedge \sigma_j \in A_j{}^*\} = A_i{}^* \cdot A_j{}^*. \qquad \square$$

6.5   Discussion: Scalability, Fitness, Maximality

Our approach serves as a proof of concept to demonstrate the feasibility of using POWL models for process discovery. The step of partial order cut detection needs to be improved in terms of efficiency. We use a brute force approach for the step of partial order cut detection. Our approach generates all possible partitionings of activities until a valid order over one of these partitionings is found. For a large number of activities, this step becomes very time-consuming unless a partial order cut is detected in an early stage. A possible improvement for future work is to exploit the eventually-follows graph to dynamically prune the search space instead of generating all partitionings of activities.

The inductive mining framework guarantees perfect fitness for the generated models if all steps of the discovery are fitness-preserving [16, Corollary 4.2]. Our approach extends the base inductive miner (IM) by adding the step of partial order cut detection. All steps of IM are fitness-preserving [16], and we mine for valid partial order cuts, which are also fitness-preserving (Theorem 1). Therefore, our approach is guaranteed to discover fitting models.

Precision is another criterion used to assess the quality of process discovery approaches. A precise process model is a model that does not allow for behavior not observed in the log. Process discovery approaches aim at creating a balance between fitness and precision. As our approach guarantees perfect fitness, our goal is to maximize precision by discovering a model that allows for less behavior as possible. In Definition 6, we defined valid partial order cuts by exploiting the eventually-follows graph. This definition ensures the uniqueness of a valid cut for a given partitioning of activities. However, a general notion of maximality among different partitionings is missing. Currently, we only maximize the size of the partitioning; we generate the partitioning of maximal size first and try to detect a valid partial order cut over it, then we decrease the size of the partitioning gradually. For future work, we would like to have a stronger notion of maximality for valid partial order cuts over different partitionings.

7   Evaluation

We implemented our approach for the discovery of POWL models in PM4Py (http://pm4py.org/), and we evaluate it using real-life event logs. We compare our approach (IM$_P$) with the base inductive miner (IM) and a more advanced

Table 1. Evaluation results. We highlighted differences in precision for models discovered by IM$_P$ compared to IM: increases in green and decreases in red.

Event log	#Act.	Time (sec)				Precision				Fitness				simplicity			
		IM	IM$_P$	IM$_C$	SM	IM	IM$_P$	IM$_C$	SM	IM	IM$_P$	IM$_C$	SM	IM	IM$_P$	IM$_C$	SM
BPI 2017	8	1.54	1.22	6.03	8.85	0.37	0.65	0.27	0.56	1	1	1	1	0.67	0.74	0.62	0.65
BPI 2017	12	18.02	5.66	13.79	16.86	0.23	0.23	0.22	0.34	1	1	1	1	0.65	0.68	0.64	0.57
BPI 2018	8	25.79	16.31	11.48	45.14	0.35	0.32	0.29	0.29	1	1	0.99	1	0.65	0.63	0.64	0.54
BPI 2018	12	59.75	112.29	17.25	55.36	0.2	0.21	0.27	0.19	1	1	0.98	1	0.61	0.63	0.63	0.48
BPI 2019	8	2.93	2.95	8.5	13.48	0.62	0.78	0.78	0.73	1	1	1	1	0.64	0.67	0.67	0.54
BPI 2019	12	5.56	3.3	13.15	13.46	0.55	0.7	0.7	0.7	1	1	1	1	0.64	0.64	0.65	0.49
Dom. Decl.	8	0.08	0.26	0.5	0.57	0.4	0.4	0.39	0.9	1	1	1	1	0.65	0.66	0.65	0.57
Dom. Decl.	12	0.13	38.16	0.98	0.61	0.5	0.51	0.37	0.84	1	1	1	1	0.61	0.67	0.62	0.59
Int. Decl.	8	0.05	0.08	0.4	0.54	0.5	0.54	0.59	0.66	1	1	1	1	0.67	0.71	0.69	0.54
Int. Decl.	12	0.12	0.31	1.6	0.68	0.47	0.51	0.53	0.56	1	1	0.98	0.89	0.65	0.69	0.69	0.38
Travel Permit	8	0.17	0.2	1.05	1.58	0.51	0.51	0.68	0.56	1	1	0.96	0.92	0.67	0.67	0.71	0.44
Travel Permit	12	0.45	280.02	0.79	0.91	0.33	0.33	0.6	0.41	1	1	0.95	0.99	0.65	0.67	0.68	0.49
Travel Costs	8	0.04	0.19	0.12	0.19	0.43	0.39	0.35	0.55	1	1	1	0.99	0.63	0.68	0.67	0.56
Travel Costs	12	0.15	276.87	0.22	0.28	0.23	0.23	0.24	0.54	1	1	1	0.83	0.58	0.68	0.63	0.38
Pay. Request	8	0.05	0.19	0.27	0.43	0.75	0.75	0.29	0.91	1	1	0.9	1	0.63	0.68	0.66	0.55
Pay. Request	12	0.09	41.44	0.5	0.5	0.49	0.49	0.38	0.82	1	1	1	1	0.6	0.67	0.64	0.55
Sepsis	8	0.08	0.1	0.07	0.21	0.51	0.51	0.51	0.42	1	1	1	1	0.64	0.65	0.64	0.5
Sepsis	12	0.2	0.14	0.16	0.3	0.34	0.35	0.4	0.31	1	1	1	1	0.64	0.65	0.63	0.47
Fine	8	0.49	0.71	9.21	5.35	0.76	0.76	0.76	0.91	1	1	1	1	0.66	0.67	0.68	0.56
Fine	11	0.69	7.13	17.52	5.49	0.58	0.58	0.78	0.92	1	1	1	1	0.62	0.63	0.64	0.51
Hosp. Billing	8	0.59	0.69	9.09	10.67	0.78	0.78	0.6	0.9	1	1	1	1	0.67	0.67	0.64	0.56
Hosp. Billing	12	0.99	1.55	14.17	5.61	0.6	0.6	0.46	0.86	1	1	1	1	0.65	0.66	0.62	0.53

variant of the inductive miner that handles incompleteness (IM$_C$) [16]. We additionally apply another state-of-the-art discovery approach: the split miner (SM) [6]. Since both IM and IM$_P$ guarantee perfect fitness, we set the filtering threshold of the split miner to 0; for the other parameters, we use the default values.

We transform the discovered models into WF-nets, and we assess their quality using three conformance-checking metrics implemented in PM4Py: fitness [7], precision [24], and simplicity [8]. Fitness quantifies how well the discovered model reproduces the behavior recorded in the event log. Precision quantifies the degree to which the model is restricted to the behavior recorded in the event log. The simplicity metric implemented in PM4Py evaluates a model as simple if it has a low average degree of arcs (i.e., a low number of arcs per place or transition).

We use multiple real-life event logs for the evaluation. We use an event log that records sepsis cases from a hospital [19], an event log of a system managing road traffic fines [18], an event log for a hospital billing system [20], BPI Challenge 2017 [9], BPI Challenge 2018 [12], BPI Challenge 2019 [10], and the five event logs of the BPI Challenge 2020 [11]: Request For Payment, Prepaid Travel Costs, Travel Permit Data, International Declarations, and Domestic Declarations. We filter the event logs to only keep the most frequent activities using two values for this filter: 8 and (at most) 12 activities.

The results of the evaluation are shown in Table 1. We report the time required for discovering each model and the obtained conformance-checking scores.

On the one hand, IM_P led to better time performance than the other approaches in some cases (e.g., BPI Challenge 2017). On the other hand, IM_P was more time-consuming in other cases. Compared to IM, the time increased from 0.17 s to 0.2 s for the travel permit log with 8 activities and from 0.45 s to 280.02 s for the log with 12 activities. This shows how increasing the number of activities can dramatically worsen the time performance. These results were expected as discussed in Sect. 6.5. IM_P serves as proof of concept, and it needs to be improved in terms of scalability in future work.

As expected, both IM and IM_P led to perfectly fitting models, while for IM_C and SM, we observe lower fitness values in some cases. As discussed in Sect. 6.5, IM_P preserves the fitness guarantee of IM as the step of partial order cut detection is fitness-preserving.

In general, the three variants of the inductive miner led to simpler models than the split miner. Our approach achieved the highest simplicity score on average (0.67), while SM achieved the lowest score on average (0.52). Note that these scores only evaluate the simplicity of the discovered models after transforming them into WF-nets; i.e., we are not evaluating the simplicity of the three different types of models (IM and IM_C produce process trees [16], IM_P produces POWL models, and SM produces BPMN models [6]).

We observe that precision varies among the different event logs. On the one hand, SM led to significantly higher precision values and lower simplicity values compared to the inductive miner in many cases (e.g., the fine management logs). On the other hand, we observe cases where the inductive miner performed better in terms of both precision and simplicity (e.g., the sepsis cases log). By comparing the three variants of the inductive miner with each other, we observe that IM_P led to the highest precision on average. In Table 1, we highlighted differences in precision between the models discovered by IM and IM_P as IM_P extends IM aiming at improving precision.

In general, our approach led to more precise models than IM. For some of these cases, this is due to the handling of incompleteness. Our approach uses the eventually-follows graph instead of the directly-follows graph for detecting partial order cuts. This design decision helps to handle incompleteness in the event log since the directly-follows graph is a subset of the eventually-follows graph; i.e., if the event log is incomplete and some connections are missing in the directly-follows graph, these connections might still be present in the eventually-follows graph. For instance, the precision of the BPI Challenge 2019 log with 12 activities increased from 0.55 to 0.7. However, the POWL model discovered by IM_P does not contain any structures that cannot be captured by a process tree. It models the same behavior of the process tree discovered by IM_C.

Figure 7 shows the POWL model discovered by IM_P for the BPI Challenge 2017 event log with 8 activities. The POWL model achieved a precision of 0.68 compared to 0.37 achieved by the model discovered by the base inductive miner. IM_P discovers local dependencies between activities IM fails to discover. For example, IM_P discovered a simple sequential relation between the activities "A_Concept" and "A_Accepted". This simple sequence cannot be discovered

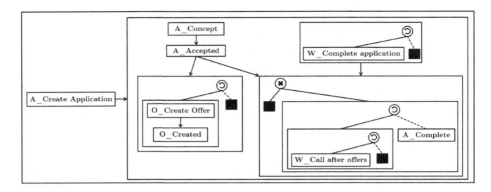

Fig. 7. POWL model discovered for the BPI Challenge 2017 event log.

by IM as it only represents a local dependency; i.e., it does not correspond to a global process tree cut covering all activities. The POWL model shows a non-hierarchical structure that cannot be captured using a process tree (i.e., the partial order over {A_Concept}, {A_Accepted}, {O_Create Offer, O_Created}, {W_Complete application}, and {W_Call after offers, A_Complete}).

Although IM_P led to more precise models compared to IM in most cases, we observe some exceptions. For instance, for the BPI Challenge 2018 event log with 8 activities, the precision decreased from 0.35 to 0.32. This is an example where invoking the fall-through function led to better results than detecting a partial order. In order to continue the recursion, the fall-through function returned a concurrency cut between an activity that occurs in every trace at most once and the rest of the activities.

To sum up, our evaluation shows that our approach discovers structures that cannot be captured by a process tree, and it leads to high precision and simplicity in general. However, the approach needs to be improved in terms of scalability.

8   Conclusion

Different modeling notations are used to model processes. Partial orders provide a compact representation of concurrent systems, but they are not able to represent cyclic or choice behavior. Process trees use control-flow operators for modeling processes as mathematical trees, but they are limited to hierarchical structures. A POWL model is a partially ordered graph representation, extended with control-flow operators for modeling choice and loop. POWL models can be converted into sound Workflow nets. We proposed an approach to demonstrate the feasibility of using POWL models in process discovery. We evaluated our approach using real-life event logs, and the evaluation showed that our approach is able to discover dependencies that cannot be captured by a process tree.

We propose multiple ideas for future work. First, our approach needs to be improved in terms of scalability. Moreover, it is possible to develop other types of approaches for the discovery of POWL models. Our approach is based on the base inductive miner; i.e., it is a top-down recursive approach. Developing a bottom-up approach and comparing it with the top-down approach is an interesting idea for future work. Moreover, we can develop a discovery approach for POWL models that exploits life-cycle information in event logs where each event has a duration (i.e., each event has a start timestamp and an end timestamp). We usually assume event logs to be totally ordered; i.e., we define a trace as a sequence of activities. However, event logs might also be partially ordered. We suggest developing an approach for the discovery of POWL models from partially ordered event logs. Finally, the idea of combining different modeling notations to create new types of process models is not restricted to POWL models. This idea can be applied to combine other types of process models.

References

1. Aalst, W.M.P.: Workflow verification: finding control-flow errors using petri-net-based techniques. In: van der Aalst, W., Desel, J., Oberweis, A. (eds.) Business Process Management. LNCS, vol. 1806, pp. 161–183. Springer, Heidelberg (2000). https://doi.org/10.1007/3-540-45594-9_11
2. Aalst, W.M.P.: Business process simulation revisited. In: Barjis, J. (ed.) EOMAS 2010. LNBIP, vol. 63, pp. 1–14. Springer, Heidelberg (2010). https://doi.org/10.1007/978-3-642-15723-3_1
3. van der Aalst, W.: Process Mining - Data Science in Action. Springer, Heidelberg (2016). https://doi.org/10.1007/978-3-662-49851-4
4. van der Aalst, W.M.P., De Masellis, R., Di Francescomarino, C., Ghidini, C., Kourani, H.: Discovering hybrid process models with bounds on time and complexity: when to be formal and when not? Inf. Syst. **116**, 102214 (2023)
5. van der Aalst, W.M.P., Dumas, M., ter Hofstede, A.H.M., Russell, N., Verbeek, H.M.W., Wohed, P.: Life after BPEL? In: Bravetti, M., Kloul, L., Zavattaro, G. (eds.) EPEW/WS-FM -2005. LNCS, vol. 3670, pp. 35–50. Springer, Heidelberg (2005). https://doi.org/10.1007/11549970_4
6. Augusto, A., Conforti, R., Dumas, M., Rosa, M.L., Polyvyanyy, A.: Split miner: automated discovery of accurate and simple business process models from event logs. Knowl. Inf. Syst. **59**(2), 251–284 (2019)
7. Berti, A., van der Aalst, W.M.P.: Reviving token-based replay: increasing speed while improving diagnostics. In: van der Aalst, W.M.P., Bergenthum, R., Carmona, J. (eds.) Proceedings of the International Workshop on Algorithms and Theories for the Analysis of Event Data, Satellite event of Petri Nets 2019 and ACSD 2019. CEUR Workshop Proceedings, vol. 2371, pp. 87–103. CEUR-WS.org (2019)
8. Blum, F.R.: Metrics in process discovery. Technical report. TR/DCC-2015-6, Computer Science Department, Universidad de Chile, Chile (2015)
9. van Dongen, B.: BPI Challenge 2017 (2017)
10. van Dongen, B.: BPI Challenge 2019 (2019)
11. van Dongen, B.: BPI Challenge 2020 (2020)
12. van Dongen, B., Borchert, F.: BPI Challenge 2018 (2018)
13. Dumas, M., García-Bañuelos, L.: Process mining reloaded: event structures as a unified representation of process models and event logs. In: Devillers, R., Valmari, A. (eds.) PETRI NETS 2015. LNCS, vol. 9115, pp. 33–48. Springer, Cham (2015). https://doi.org/10.1007/978-3-319-19488-2_2

14. Golani, M., Pinter, S.S.: Generating a process model from a process audit log. In: van der Aalst, W.M.P., Weske, M. (eds.) BPM 2003. LNCS, vol. 2678, pp. 136–151. Springer, Heidelberg (2003). https://doi.org/10.1007/3-540-44895-0_10

15. Leemans, M., van der Aalst, W.M.P.: Discovery of frequent episodes in event logs. In: Ceravolo, P., Russo, B., Accorsi, R. (eds.) SIMPDA 2014. LNBIP, vol. 237, pp. 1–31. Springer, Cham (2015). https://doi.org/10.1007/978-3-319-27243-6_1

16. Leemans, S.J.J.: Robust Process Mining with Guarantees - Process Discovery, Conformance Checking and Enhancement, LNBIP, vol. 440. Springer, Cham (2022). https://doi.org/10.1007/978-3-030-96655-3

17. Leemans, S.J., van Zelst, S.J., Lu, X.: Partial-order-based process mining: a survey and outlook. Knowl. Inf. Syst. **65**, 1–29 (2022)

18. de Leoni, M.M., Mannhardt, F.: Road Traffic Fine Management Process (2015)

19. Mannhardt, F.: Sepsis Cases - Event Log (2016)

20. Mannhardt, F.: Hospital Billing - Event Log (2017)

21. Mannila, H., Toivonen, H., Verkamo, A.I.: Discovery of frequent episodes in event sequences. Data Min. Knowl. Discov. **1**(3), 259–289 (1997)

22. Mokhov, A., Carmona, J.: Event log visualisation with conditional partial order graphs: from control flow to data. In: Proceedings of the International Workshop on Algorithms & Theories for the Analysis of Event Data, Satellite event of Petri Nets 2015 and ACSD 2015. CEUR Workshop Proceedings, vol. 1371, pp. 16–30. CEUR-WS.org (2015)

23. Mokhov, A., Yakovlev, A.: Conditional partial order graphs: model, synthesis, and application. IEEE Trans. Comput. **59**(11), 1480–1493 (2010)

24. Muñoz-Gama, J., Carmona, J.: A fresh look at precision in process conformance. In: Hull, R., Mendling, J., Tai, S. (eds.) BPM 2010. LNCS, vol. 6336, pp. 211–226. Springer, Heidelberg (2010). https://doi.org/10.1007/978-3-642-15618-2_16

25. Nielsen, M., Plotkin, G.D., Winskel, G.: Petri nets, event structures and domains, part I. Theor. Comput. Sci. **13**, 85–108 (1981)

26. Ouyang, C., Verbeek, E., van der Aalst, W.M.P., Breutel, S., Dumas, M., ter Hofstede, A.H.M.: Formal semantics and analysis of control flow in WS-BPEL. Sci. Comput. Program. **67**(2–3), 162–198 (2007)

27. Slaats, T., Schunselaar, D.M.M., Maggi, F.M., Reijers, H.A.: The semantics of hybrid process models. In: Debruyne, C., Panetto, H., Meersman, R., Dillon, T., Kühn, O'Sullivan, D., Ardagna, C.A. (eds.) OTM 2016. LNCS, vol. 10033, pp. 531–551. Springer, Cham (2016). https://doi.org/10.1007/978-3-319-48472-3_32

Polynomial-Time Conformance Checking for Process Trees

Eduardo Goulart Rocha[1,2]([✉]) [iD] and Wil M. P. van der Aalst[1,2][iD]

[1] Celonis Labs GmbH, Munich, Germany
[2] Process and Data Science (PADS) Chair, RWTH Aachen University,
Aachen, Germany
e.goulartrocha@celonis.com, wvdaalst@pads.rwth-aachen.de

Abstract. Conformance-checking is the field of process mining relating modeled and observed behavior. State-of-the-art conformance checking techniques do not scale for large process models and event logs, which hampers its broader adoption.

In this paper, we present a polynomial-time method to compute the markovian-based fitness and precision metrics for process trees. For that, we first show that this is equivalent to the problem of computing the set of substrings of length at most k of the model's language. Then, we show how to exploit the tree structure to compute this set in a compositional way. The experimental evaluation shows that the proposed method outperforms state-of-the-art conformance-checking techniques by orders of magnitude, while still providing quality guarantees.

Keywords: Process mining · Conformance Checking · Process Trees

1 Introduction

Conformance checking is the field of process mining relating desired and observed behavior. Given an event log and a process model, conformance checking aims at identifying and quantifying differences between the event log and the process model. An important use-case for conformance checking is to assess the quality of automatically discovered process models in the form of a single number evaluation metric. For that, multiple conformance metrics with distinct runtime and quality characteristics have been proposed in the literature [1,8,11]. Unfortunately, most state-of-the-art methods still require a runtime that is exponential on the number of activities or do not satisfy all the desired axioms for a conformance metric [13].

A notable exception are the Projected Conformance Checking (PCC) fitness and precision metrics [7], which provide strong runtime and quality guarantees for certain classes of models (process trees with unique activities and no invisible labels). Nevertheless, the PCC metrics require multiple passes over the event log, which makes them expensive to compute for large datasets.

In this work, we focus on the problem of efficiently computing conformance metrics for process trees. Process trees are a well-established modeling formalism

© The Author(s), under exclusive license to Springer Nature Switzerland AG 2023
C. Di Francescomarino et al. (Eds.): BPM 2023, LNCS 14159, pp. 109–125, 2023.
https://doi.org/10.1007/978-3-031-41620-0_7

in process mining because of its soundness guarantees and simple structure. For instance, many state-of-the-art process discovery algorithms return process trees. We provide two important contributions: First, we present a simplified, yet more expressive, definition of the k-th order markovian abstraction first presented in [2]. Next, we show how to compute the k-th order markovian abstraction of a process tree in polynomial time by exploiting the tree structure. The method achieves an improvement of orders of magnitude in computation time for models with a high degree of parallelism. Furthermore, the method scales linearly with the size of the event log, making it suitable for very large event-logs.

The remainder of the paper is organized as follows: Sect. 2 presents basic notations and concepts from automata theory, which are the backbone of the presented technique, Sect. 3 presents the general framework for computing the k-th order markovian abstraction of a process tree, Sect. 4 compares the approach to other state-of-the-art methods, Sect. 5 presents related work in the field. Finally, Sect. 6 concludes the paper with directions for future work.

2   Preliminaries

This section presents the basic concepts upon which the method is based. For a given finite alphabet Σ, Σ^k is the set of all finite words of length k formed with this alphabet and $\Sigma^* = \bigcup_{k \geq 0} \Sigma^k$. The projection of a word $w \in \Sigma^*$ in a set of symbols $S \subseteq \Sigma$ is written w_S. The concatenation of two words u, v is written uv. Similarly, the concatenation of two languages $U, V \subseteq \Sigma^*$ is written as $UV = \bigcup_{u \in U, v \in V} uv$. Given a word $w = w_1 w_2 \cdots w_n$ and $1 \leq i \leq j \leq n$, $w^{i \to j} = w_i w_{i+1} \cdots w_j$ denotes a substring of w (written $\gamma \sqsubseteq w$). We further write $pref^k(w)$, $suff^k(w)$, and $sub^k(w)$ to denote the set of non-empty prefixes, suffixes, and substrings of w with length less than or equal to k. The definitions of $pref^k$, $suff^k$, and sub^k are extended to languages too. Finally, the paper assumes familiarity with basic algorithms of automata theory [5]. We provide common notations for finite automata below:

Definition 1 (Labeled Directed Graph). *A Labeled Directed Graph is a triple $G = (V, \Sigma, E)$ where V is the set of vertices, Σ is the set of labels and $E \subseteq V \times \Sigma \times V$ is the set of edges. Given an edge $e = (v, l, v\prime)$, functions $\pi_{src}(e) = v$, $\pi_{tgt}(e) = v\prime$, and $\pi_l(e) = l$ return its source and target vertices and its label respectively*

For this paper, all considered graphs are labeled directed graphs. A *path* p in the a graph is a sequence of edges $p = e_1 e_2 \cdots e_n$ such that $\pi_{tgt}(e_i) = \pi_{src}(e_{i+1}) \; \forall 1 \leq i < n$. We define $\pi_l(p) = \pi_l(e_1)\pi_l(e_2) \cdots \pi_l(e_n)$ as the path's *labeling* $(= \epsilon$ if the path is empty). And $\pi_v(p, i) = \begin{cases} \pi_{src}(e_1) & i = 0 \\ \pi_{tgt}(e_i) & i > 0 \end{cases}$ as the i-th vertex visited by the path, where $\pi_v(p, 0)$ is the path's *start vertex*.

Definition 2 (Nondeterministic Finite Automaton). *Let ϵ be the empty string. A Nondeterministic Finite Automaton (NFA) is a 5-tuple $N =$*

$(Q, \Sigma, \delta, q_0, F)$, where Q is the set of states, Σ is the alphabet, $\delta : Q \times (\Sigma \cup \{\epsilon\}) \rightarrow \mathcal{P}(Q)$ is the transition function, q_0 is the initial state and $F \subseteq Q$ is the set of final states.

Any NFA $N = (Q, \Sigma, \delta, q_0, F)$ is associated to a graph $G = (Q, \Sigma \cup \{\epsilon\}, E)$ where $E = \{(q, l, q\prime) \in Q \times (\Sigma \cup \{\epsilon\}) \times Q \mid q\prime \in \delta(q, l)\}$ (called the *NFA's graph*). Given $q, q\prime \in Q$, we define $q[w\rangle q\prime = \{p \in E^* \mid \pi_l(p) = w \wedge \pi_v(p, 0) = q \wedge \pi_v(p, |p|) = q\prime\}$ as all the paths from q to $q\prime$ labeled by w. If $q = q_0$, $q\prime \in F$, and $q[w\rangle q\prime \neq \varnothing$, then N *accepts* w and p is an *accepting path*. The accepted language of N is defined as $\mathcal{L}(N) = \{w \in \Sigma^* \mid \exists f{\in}F \text{ s.t. } q_0[w\rangle f{\neq}\varnothing\}$. Similarly, we denote $q[w_1\rangle q\prime[w_2\rangle q\prime\prime = \{p_1 p_2 \mid p_1 \in q[w_1\rangle q\prime \wedge p_2 \in q\prime[w_2\rangle q\prime\prime\}$. Last, we define $q[w\rangle = \{q\prime \in Q \mid q[w\rangle q\prime \neq \varnothing\}$ as the set of states reachable from q by replaying w.

In an NFA N, a state is *dead* if it is not reachable from the start state and it is a *trap* if there is no path q leading from the state to a final state. We say that an NFA is *trimmed* if it has no dead or trap states. For any trimmed NFA, any path p in its graph is such that $\pi_l(p)$ is a substring of $\mathcal{L}(N)$.

Definition 3 *(Deterministic Finite Automaton (DFA))*. *A Deterministic Finite Automaton (DFA) is an NFA where* $\delta(q, \epsilon) = \varnothing \; \forall q \in Q$ *and* $|\delta(q, l)| \leq 1 \; \forall q \in Q, l \in \Sigma$.

Every DFA has the property that two paths in its graph starting from the same node are equal if and only if their labelings are the same, i.e. $|q[w\rangle| \leq 1$. We will abuse notation and write $q[w\rangle q\prime$ to refer to the single element of this set (when it exists). Given an NFA N, there exists an unique (up to isomorphism) DFA D with a minimal number of states such that $\mathcal{L}(N) = \mathcal{L}(D)$ that can be obtained via *determinization*. This can be achieved via the *powerset construction* followed by a *minimization* step [5]. In the worst case, D has exponentially many more states than N. If the DFA's graph is acyclic, we call it a *Deterministic Acyclic Finite State Automaton* (DAFSA). Given a finite language $L \subseteq \Sigma^*$, it is possible to construct a minimal DAFSA accepting L in linearithmic time [4].

While finite automata can be used to represent any regular language, process analysts need a compact and understandable modeling formalism. Among which, process trees [3] stand out for their soundness guarantees and block structure. Process trees are graphs with a tree structure. In a process tree, the leaf nodes represent activities in Σ or skips (τ) and the internal nodes represent one of four possible operators: *exclusive* ($\times$), *sequence* ($\rightarrow$), *loop* ($\circlearrowleft$), and *parallel* ($\wedge$). The tree's accepted language is defined recursively as follows:

Definition 4 *(Process Trees Semantics)*. *Let* $\sqcup\!\sqcup$ *be the shuffle product of two words, defined as:*
$$\begin{cases} w \sqcup\!\sqcup \epsilon = \epsilon \sqcup\!\sqcup w = \{w\} & w \in \Sigma^* \\ xu \sqcup\!\sqcup yv = \{x\}(u \sqcup\!\sqcup yv) \cup \{y\}(xu \sqcup\!\sqcup v) & x, y \in \Sigma \wedge u, v \in \Sigma^* \end{cases}$$

For languages A, B, *define* $A \sqcup\!\sqcup B = \bigcup_{w_a \in A, w_b \in B} w_a \sqcup\!\sqcup w_b$. *Then, the accepted language of a process tree is recursively defined as:*

- $\mathcal{L}(\tau) = \{\epsilon\}$
- $\mathcal{L}(a) = \{a\}$
- $\mathcal{L}(\times(T_1, \cdots, T_n)) = \bigcup_{i=1}^{n} \mathcal{L}(T_i)$
- $\mathcal{L}(\rightarrow(T_1, \cdots, T_n)) = \mathcal{L}(T_1)\mathcal{L}(T_2)\cdots\mathcal{L}(T_n)$
- $\mathcal{L}(\circlearrowright(T_1, \cdots, T_n)) = \mathcal{L}(T_1)(\mathcal{L}(\times(T_2, \cdots, T_n))\mathcal{L}(T_1))^*$
- $\mathcal{L}(\wedge(T_1, \cdots, T_n)) = \mathcal{L}(T_1) \sqcup\!\sqcup \mathcal{L}(T_2) \sqcup\!\sqcup \cdots \sqcup\!\sqcup \mathcal{L}(T_n)$

Given a process tree T, there exists a unique minimal DFA D such that $\mathcal{L}(T) = \mathcal{L}(D)$ [3]. However, the size of D might be exponential with the size of T. This exponential blow-up is the bottleneck for most conformance checking techniques, including the metrics based on the k-th order markovian abstraction presented in [2]. This paper focuses on improving the runtime for computing the k-th order markovian abstraction. For this, we use a slightly different definition than the one originally introduced in [2], based on the set of k-trimmed substrings of a language:

Definition 5 (K-Trimmed Substrings). *Let Σ be an alphabet and $k \geq 2$. Given a word $w \in \Sigma^*$, the set of k-trimmed substrings $s^k(w)$ is defined as:*

$$s^k(w) = \begin{cases} \{w\} & \text{if } |w| \leq k \\ \{w^{i \to i+k-1} \mid 1 \leq i \leq |w| - k + 1\} & \text{otherwise} \end{cases}$$

We extend the definition of s^k to languages as $s^k(L) = \bigcup_{w \in L} s^k(w)$. Consider languages $X = \{abc\}$ and $Y = \{i, ijk\}$ (these will serve as a running example for the remainder of the paper), then $s^1(X) = \{a, b, c\}$, $s^2(X) = \{ab, bc\}$, and $s^k(X) = \{abc\}$ for $k \geq 3$ and $s^1(Y) = \{i, j, k\}$, $s^2(Y) = \{i, ij, jk\}$, and $s^k(Y) = \{i, ijk\}$ for $k \geq 3$. The k-th order markovian abstraction (defined below) is similar to the set of k-trimmed substrings, but with special start/end markers $(+/-)$ to track the language's prefixes/suffixes.

Definition 6 (The Modified k-th Order Markovian Abstraction). *Let Σ be an alphabet, $+/- \notin \Sigma$ be special start/end markers, and $k \geq 2$. Given a word $w \in \Sigma^*$, the k-order markovian abstraction of w is defined as follows:*

$$m^k(w) = s^k(+w-)$$

Similarly, m^k of a language $L \subseteq \Sigma^*$ is defined as $\bigcup_{w \in L} m^k(w)$. In principle, m^k is defined for arbitrary languages, but throughout the rest of this work we focus on computing m^k for regular languages. We will always assume that $+, - \notin \Sigma$ and write $\Sigma^{\pm} = \Sigma \cup \{+, -\}$ and $+L- = \{+\}L\{-\}$. For any language $L \subseteq \Sigma^*$, $m^k(L)$ represents a finite set of finite words and thus can be associated to a unique minimal DAFSA (written M_L^k). Figure 1a shows the minimal DAFSAs M_X^3 and M_Y^3 accepting $m^3(X) = \{+ab, abc, bc-\}$ and $m^3(Y) = \{+i-, +ij, ijk, jk-\}$ respectively. In general, M_L^k has a very specific structure, detailed below:

Proposition 1 (Basic Properties of M_L^k). *Let Σ be an alphabet, $L \subseteq \Sigma^*$ be an arbitrary language, $k \geq 2$, and $M_L^k = (Q, \Sigma, \delta, q_0, F)$ the minimal DAFSA accepting $m^k(L)$, then:*

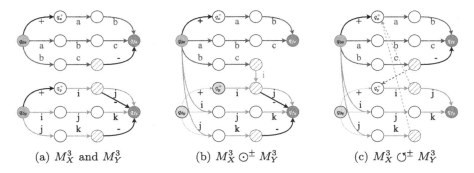

(a) M_X^3 and M_Y^3 (b) $M_X^3 \odot^\pm M_Y^3$ (c) $M_X^3 \circlearrowleft^\pm M_Y^3$

Fig. 1. M_X^3 and M_Y^3 for $X = \{abc\}$ and $Y = \{ijk, i\}$ and their sequence and loop concatenations. Start states are colored orange, final states are colored green, dead states are gray, states in Q^- are hatched, and ϵ-transitions are represented as dashed lines.

1. $m^k(L) \subseteq m^k(\Sigma^*) = \left(\bigcup_{0 \leq i < k-1} + \Sigma^i - \right) \cup (+\Sigma^{k-1}) \cup (\Sigma^{k-1}-) \cup \Sigma^k$
2. M_L^k has only one final state, i.e. $F = \{q_f\}$
3. M_L^k has exactly one edge labeled $+$. This edge has q_0 as its source and we write q^+ to represent its target, i.e. $\delta(q_0, +) = \{q^+\}$
4. All $-$-labeled edges in M_L^k lead to its unique final state. We define $Q^- = \{q \in Q \mid q_f \in \delta(q, -)\}$
5. For every $\gamma \in sub^k(+L-)$, there exists a path p in M_L^k such that $\pi_l(p) = \gamma$, and for every path p in M_L^k, there exists $\gamma \in sub^k(+L-)$ such that $\gamma = \pi_l(p)$.
6. For every path p in M_L^k, $+ \in \pi_l(b) \iff \pi_l(p) \in pref^k(+L-)$. Similarly, $- \in \pi_l(p) \iff \pi_l(p) \in suff^k(+L-)$
7. $|Q| \leq |\Sigma|^{k-1} + 2$

Finally, Definition 7 presents the (modified) markovian-based fitness and precision metrics. The metrics return almost the same (but not the same) values as the ones presented in [2], because the original definition counts words of length smaller than k twice. However, monotonicity still holds for our definition of m^k, i.e. if $A \subseteq B \Rightarrow m^k(A) \subseteq m^k(B)$, such that the proofs of the axioms presented in [2] are still valid.

Definition 7 (Markovian-Based Fitness and Precision with the Binary Cost Function). *Let $\mathbb{L}$ be an event log with language $L \subseteq \Sigma^*$, $\mathbb{P}$ be a process model with language $P \subseteq \Sigma^*$, $k \geq 2$, and $\#_\mathbb{L}(\gamma)$ the number of occurrences of substring γ in $\mathbb{L}$. Then $MAF^k(L, P) = 1 - \frac{\sum_{\gamma \in (m^k(L) \setminus m^k(P))} \#_\mathbb{L}(\gamma)}{\sum_{\gamma \in m^k(L)} \#_\mathbb{L}(\gamma)}$ and $MAP^k(L, P) = 1 - \frac{|m^k(P) \setminus m^k(L)|}{|m^k(P)|}$ are the markovian-based fitness and precision metrics respectively.*

The metrics are the set difference of the languages' substrings. The fitness metric is normalized by the substring frequency. Since that the current setting does not consider any notion of trace frequency for process models, the precision metric is normalized by $1/|m^k(P)|$. It is possible to obtain a variation of the metric by changing the cost function (see [2]). The markovian-based fitness and precision metrics were empirically shown to agree with other state-of-the-art conformance metrics such as escaping edges and PCC. However, the original method for computing m^k requires the computation of the process model's DFA and thus does not scale for larger models. In the next section, we show how to compute m^k for process trees without computing its state space, hence improving scalability.

3   General Framework

This section shows how to efficiently compute m^k for arbitrary process trees. First, we present a method to compute s^k for arbitrary regular languages. Next, we show a compositional approach to compute m^k for binary and uniquely-labeled process trees which works by recursively computing m^k for each tree node from the m^k of its child nodes. Last, we show how to generalize it for arbitrary process trees.

3.1   Computing s^k of a Regular Language

This section presents a method to compute $s^k(L)$ from a DFA accepting L.

Definition 8 (All-Substrings NFA). *Given a trimmed DFA $D = (Q, \Sigma, \delta, q_0, F)$, its all-substrings NFA is defined as $Sub_D = (Q \cup \{\hat{q}\}, \Sigma, \hat{\delta}, \hat{q}, Q \cup \{\hat{q}\})$, where $\hat{\delta}$ is defined as follows:*

$$\hat{\delta}(q, l) = \begin{cases} \delta(q, l) & \text{if } l \in \Sigma \land q \in Q \\ Q & \text{if } l = \epsilon \land q = \hat{q} \\ \varnothing & \text{otherwise} \end{cases}$$

Notice that Sub_D accepts all substrings of DFA D. This can be used to efficiently compute s^k as follows:

Lemma 1 (Computing s^k). *Let $L \subseteq \Sigma^*$ be a regular language and $D = (Q, \Sigma, \delta, q_0, F)$ a DFA accepting L. Then $s^k(L)$ can be computed in $\mathcal{O}(|Q||\Sigma|^k)$.*

Proof. We assume D to be trimmed, otherwise D can be trimmed in $\mathcal{O}(|Q|)$. We define $s^{=k}(L) = \{w \in s^k(L) \mid |w| = k\}$ and $s^{<k}(L) = s^k(L) \setminus s^{=k}(L)$. Notice that $s^{<k}(L)$ can be computed in $\mathcal{O}(|\Sigma|^{k-1})$ time by running BFS from the start node with a maximum depth of $k-1$. We prove that $\mathcal{L}(Sub_D) \cap \Sigma^k = s^{=k}(L)$:

($\subseteq$) Any $w \in \mathcal{L}(Sub_D) \cap \Sigma^k$ is such that there exists an accepting path $p \in \hat{q}[\epsilon\rangle q_i[w\rangle q_{i+k}$ in Sub_D. And since D is trimmed, there exists $u, v \subseteq \Sigma^*$ such that $q_0[u\rangle q_i \neq \varnothing$ and $q_{i+k}[v\rangle f \neq \varnothing$, $f \in F$. Hence, $q_0[u\rangle q_i[w\rangle q_{i+k}[v\rangle f \neq \varnothing \Rightarrow uwv \in L \Rightarrow w \in s^{=k}(L)$.

($\supseteq$) For any $w \in s^k(L)$, $\exists t \in L \mid t = uwv$, so $p = q_0[u\rangle q_i[w\rangle q_{i+k}[v\rangle q_n$ is a path in D. But this directly implies that there exists a unique path $p \in \hat{q}[\epsilon\rangle q_i[w\rangle q_{i+k}$ and that p is an accepting path of Sub_D. And since that $w \in \Sigma^k$, then $w \in (\mathcal{L}(Sub_D) \cap \Sigma^k)$.

The runtime bound is achieved by computing $\mathcal{L}(Sub_D) \cap \Sigma^k$ without determinizing Sub_D. The product construction builds a DAFSA. It expands at most $|\Sigma|^k$ nodes, where each node expansion has cost bound by $|Q|$. $\qquad\square$

3.2   Leaf and Exclusive Nodes

This section shows how to compute m^k for leaf and exclusive nodes, which do not require any special constructs:

Lemma 2 (Leaf Nodes). For $a \in \Sigma \cup \{\epsilon\}$, $m^k(a) = \begin{cases} \{+-\} & k \geq 2, \ a = \epsilon \\ \{+a, a-\} & k = 2, \ a \in \Sigma \\ \{+a-\} & k > 2, \ a \in \Sigma \end{cases}$

Proof. Follows directly from Definition 6. $\qquad\square$

Lemma 3 (Exclusive Node). Let $A, B \subseteq \Sigma^*$ be arbitrary languages. Then:

$$m^k(A \cup B) = m^k(A) \cup m^k(B)$$

Proof. From Definition 6, $m^k(A \cup B) = s^k(+(A \cup B)-) = s^k(+A - \cup + B-) = s^k(+A-) \cup s^k(+B-) = m^k(A) \cup m^k(B)$. $\qquad\square$

3.3   Sequence Node

For the sequence node, m^k is computed based on automata operations. For that, we first define the markovian sequence concatenation $\odot^\pm$ as follows:

Definition 9 (Markovian Sequence Concatenation). Let A, B be arbitrary languages with disjoint alphabets Σ_A, Σ_B, and $M_A^k = (Q_A, \Sigma_A^\pm, \delta_A, q_{0a}, \{q_{fa}\})$ and $M_B^k = (Q_B, \Sigma_B^\pm, \delta_B, q_{0b}, \{q_{fb}\})$ be the minimal DAFSAs accepting $m^k(A)$ and $m^k(B)$ respectively. The markovian sequence concatenation $M_A^k \odot^\pm M_B^k$ builds the DFA $(Q_A \cup Q_B, (\Sigma_A \cup \Sigma_B)^\pm, \hat{\delta}, q_{0a}, \{q_{fa}, q_{fb}\})$ where $\hat{\delta}$ is defined as follows:

$$\hat{\delta}(q, l) = \begin{cases} \delta_A(q, l) & q \in Q_A, l \in \Sigma_A \cup \{+\} \\ \delta_B(q_{0b}, l) & q = q_{0a}, l \in \Sigma_B \\ \delta_B(q_b^+, l) & q \in Q_A^-, l \in \Sigma_B \cup \{-\} \\ \delta_B(q, l) & q \in Q_B, l \in \Sigma_B^\pm \\ \varnothing & otherwise \end{cases}$$

Figure 1b shows $M_X^3 \odot^\pm M_Y^3$, which accepts $\{+ab, abc, bcij, bci-, ijk, jk-\}$. Intuitively, the markovian sequence concatenation is merging the transition function of state q_b^+ into the transition functions of states in Q_A^-. Notice that $s^3(\mathcal{L}(M_X^3 \odot^\pm M_Y^3)) = \{+ab, abc, bci, cij, ci-, ijk, jk-\} = m^3(XY)$. Lemma 4 below formalizes this fact, which can be used to compute m^k for the sequence node:

Lemma 4 (m^k *of Language Concatenation*). *Let A, B be arbitrary languages with disjoint alphabets Σ_A, Σ_B, and $M_A^k = (Q_A, \Sigma_A^\pm, \delta_A, q_{0a}, \{q_{fa}\})$ and $M_B^k = (Q_B, \Sigma_B^\pm, \delta_B, q_{0b}, \{q_{fb}\})$ the minimal DAFSAs accepting $m^k(A)$ and $m^k(B)$ respectively. Then:*

$$m^k(AB) = s^k(\mathcal{L}(M_A^k \odot^\pm M_B^k))$$

Proof. ($\subseteq$) For any $w \in +AB-$, there exists $\hat{w}_a \in A, \hat{w}_b \in B$ such that $w = +\hat{w}_a\hat{w}_b-$. Then for any $\gamma \in s^k(+\hat{w}_a\hat{w}_b-)$ one of the following holds:

$$\gamma = \begin{cases} \gamma_a & \gamma_a \sqsubseteq +\hat{w}_a \\ \gamma_b & \gamma_b \sqsubseteq \hat{w}_b- \\ \gamma_a\gamma_b & \gamma_a \in suff^k(+\hat{w}_a), \gamma_b \in pref^k(\hat{w}_b-) \end{cases}$$

For Case 1, the condition implies $|\gamma_a| = k$ and so $\gamma_a \in m^k(A)$. Let p be the path in M_A^k accepting γ_a. Since that $- \notin \gamma_a$, and that only $-$-labeled edges were removed from M_A^k, then p is also an accepting path in $M_A^k \odot^\pm M_B^k$.

Similarly for Case 2, $|\gamma_b| = k$ and $\gamma_b \in m^k(B)$. Let $p = q_{0b}[\gamma_b\rangle q_{fb}$ be the path in M_B^k accepting γ_b. Then $\hat{p} = q_{0a}[\gamma_b\rangle q_{fb}$ is an accepting path in $M_A^k \odot^\pm M_B^k$.

For Case 3, the condition implies $|\gamma_a|, |\gamma_b| < k$, $\gamma_a- \sqsubseteq +\hat{w}_a-$ and $+\gamma_b \sqsubseteq +\hat{w}_b-$. And since $|\gamma_a - |, | + \gamma_b| \le k$ then there exists α_a, β_b such that $\alpha_a\gamma_a- \in s^k(+\hat{w}_a-) \subseteq m^k(A)$ and $+\gamma_b\beta_b \in s^k(+\hat{w}_b-) \subseteq m^k(B)$ (Proposition 1-5), and so $p_a = q_{0a}[\alpha_a\rangle\hat{q}_a[\gamma_a\rangle q_a^-[-\rangle q_{fa}$ and $p_b = q_{0b}[+\rangle q_b^+[\gamma_b\rangle\hat{q}_b[\beta_b\rangle q_{fb}$ are accepting paths in M_A^k and M_B^k. Then $p = q_{0a}[\alpha_a\rangle\hat{q}_a[\gamma_a\rangle q_a^-[\gamma_b\rangle\hat{q}_b[\beta_b\rangle$ is a path in $M_A^k \odot^\pm M_B^k$ such that $\pi_l(p) = \alpha_a\gamma\beta_b$, and since $|\gamma| = k$, then $\gamma \in s^k(\alpha_a\gamma\beta_b) \subseteq s^k(\mathcal{L}(M_A^k \odot^\pm M_B^k))$.

($\supseteq$) Notice that $M_A^k \odot^\pm M_B^k$ is acyclic. Therefore, for every accepting path p in $M_A^k \odot^\pm M_B^k$ accepting $\pi_l(p) = w \in \mathcal{L}(M_A^k \odot^\pm M_B^k)$, there exists $0 \le j \le n$ such that $\pi_s(p, i) \in Q_A \,\forall i \le j$ and $\pi_s(p, i) \in Q_B \,\forall i > j$. If $j = 0$, then $w \in s^k(+B-)$ and $+ \notin w \Rightarrow w \in sub^k(B-)$. If $j = n$, then $w \in s^k(+A-)$ and $- \notin w \Rightarrow w \in sub^k(+A)$. In both cases, $|w| = k$, thus $w \in s^k(+AB-)$.

If $0 < j < n$, it holds that $e_j = (q_{j-1}, w_j, q_j)$ where $q_{j-1} \in Q_A^-$, $w_j \in \Sigma_B$, and $q_j \in \delta_B(q_b^+, w_j)$. So $w^{1 \to j-1} - \in s^k(+A-)$ and $+w^{j \to n} \in s^k(+B-)$ (Proposition 1-6). Which implies that w is a substring of $+AB-$. Now if $|w| \le k$, then $+, - \in w$, which implies that $w \in +AB-$ and thus $\{w\} = s^k(w) \subseteq s^k(+AB-)$. If $|w| > k$, then it follows directly that $s^k(w) \subseteq s^k(+AB-)$. $\qquad\square$

$M_A^k \odot^\pm M_B^k$ is a DAFSA with $|Q_A| + |Q_B| \in \mathcal{O}(|\Sigma_A \cup \Sigma_B|^k)$ states. From Lemma 1, it follows that $m^k(AB)$ can be computed in $\mathcal{O}(|\Sigma_A \cup \Sigma_B|^{2k})$.

3.4 Loop Node

In Sect. 3.3, we have seen how to concatenate two DAFSAs to compute m^k for language concatenation. Similarly, we define the loop concatenation as a construct to compute m^k for the loop node's language. We first define an NFA constructed from both markovians' DAFSAs and show how this relates to the markovian of the loop node. Then, we show that determinizing this construct is polynomial-time due to its specific structure, thus still being efficient.

Definition 10 (Markovian Loop Concatenation). *Let A, B be arbitrary languages with disjoint alphabets Σ_A and Σ_B and $M_A^k = (Q_A, \Sigma_A^\pm, \delta_A, q_{0a}, \{q_{fa}\})$ and $M_B^k = (Q_B, \Sigma_B^\pm, \delta_B, q_{0b}, \{q_{fb}\})$ the minimal DAFSAs accepting $m^k(A)$ and $m^k(B)$ respectively. The markovian loop concatentation $M_A^k \circlearrowleft^\pm M_B^k$ builds the NFA $N = (Q_A \cup Q_B, (\Sigma_A \cup \Sigma_B)^\pm, \hat{\delta}, q_{0a}, \{q_{fa}, q_{fb}\})$ where $\hat{\delta}$ is defined as follows:*

$$\hat{\delta}(q, l) = \begin{cases} \delta_A(q, l) & q \in Q_A, \ l \in \Sigma_A^\pm \\ \delta_B(q_{0b}, +) & q \in Q_A^-, \ l = \epsilon \\ \delta_A(q_{0a}, +) & q \in Q_B^-, \ l = \epsilon \\ \delta_B(q, l) & q \in Q_B, \ l \in \Sigma_B \cup \{+\} \\ \delta_B(q_{0b}, l) & q = q_{0a}, \ l \in \Sigma_B \\ \varnothing & otherwise \end{cases}$$

Figure 1c shows $M_X^3 \circlearrowleft^\pm M_Y^3$ accepting $\{+ab, abc, bc-, bcij, bciab, ijk, jkab\}$. Notice that $s^3(\mathcal{L}(M_X^3 \circlearrowleft^\pm M_Y^3)) = \{+ab, abc, bc-, bci, cij, cia, iab, ijk, jka, kab\} = m^3(A(BA)^*)$. Lemma 5 below formalizes this fact, which can be used to compute m^k for loop nodes:

Lemma 5 (m^k of the Loop Node). *Let A, B be arbitrary languages with disjoint alphabets Σ_A and Σ_B, and $M_A^k = (Q_A, \Sigma_A^\pm, \delta_A, q_{0a}, \{q_{fa}\})$ and $M_B^k = (Q_B, \Sigma_B^\pm, \delta_B, q_{0b}, \{q_{fb}\})$ the minimal DAFSAs accepting $m^k(A)$ and $m^k(B)$ respectively. Then:*

$$m^k(A(BA)^*) = s^k(\mathcal{L}(M_A^k \circlearrowleft^\pm M_B^k))$$

Proof. We define sets $\hat{A} = \{w_a \in A \mid |w_a| \le k - 2\}$ and $\hat{B} = \{w_b \in B \mid |w_b| \le k - 2\}$. The graph of $M_A^k \circlearrowleft^\pm M_B^k$ is such that, for every $\hat{w}_a \in \hat{A}$, there exists $q_a^- \in Q_A^-$ such that $q_a^-[\hat{w}_a\rangle q_a^- \ne \varnothing$ (analogous for $\hat{B}$).

($\subseteq$) For every $w \in +A(BA)^*-$, then $w = +w_{a,1}w_{b,1}w_{a,2}\cdots w_{a,n}-$ s.t. $w_{a,i} \in \hat{A}$, $\forall 1 \le i \le n$ and $w_{b,i} \in \hat{B}$, $\forall 1 \le i < n$. We distinguish between two cases:

Case 1: ($|w| \le k$) Then $|+w_{a,i}-| \le k$ for every $i \le n$, which implies that $+w_{a,i}- \in \mathcal{L}(M_A^k)$. Therefore, $p_{a,i} = q_{0a}[+\rangle q_a^+[w_{a,i}\rangle q_{a,i}^-[-\rangle q_{fa}$ in M_A^k is such that $q_{a,i}^- \in Q_A^-$. Similarly for B, for every $1 \le i \le n-1$, $p_{b,i} = q_{0b}[+\rangle q_b^+[w_{b,i}\rangle q_{b,i}^-[-\rangle q_{fb}$ is such that $q_{b,i}^- \in Q_B^-$. Thus, the set $q_{0a}[+\rangle q_a^+[w_{a,i}\rangle q_{a,i}^-(\epsilon) q_b^+[w_{b,i}\rangle q_{b,i}^-[-\rangle q_{fb}$ is not empty and contains paths $\hat{p}$ in $M_A^k \circlearrowleft^\pm M_B^k$ such that $\pi_l(\hat{p}) = +w_{a,i}w_{b,i}-$. This can be continued to find a path p in $M_A^k \circlearrowleft^\pm M_B^k$ such that $\pi_l(p) = w$. And since $|w| \le k$, then $w \in s^k(\mathcal{L}(M_A^k \circlearrowleft^\pm M_B^k))$.

Case 2: ($|w| > k$) Then for every $\gamma \in s^k(w)$, it holds:

$$\gamma \in \begin{cases} suff^k(+A)\hat{B}(\hat{A}\hat{B})^* pref^k(A-) \\ suff^k(+A)(\hat{B}\hat{A})^* pref^k(B) \\ suff^k(B)(\hat{A}\hat{B})^* pref^k(A-) \\ suff^k(B)\hat{A}(\hat{B}\hat{A})^* pref^k(B) \\ s^k(+A-) \\ s^k(B) \end{cases} \qquad (1)$$

For the first 4 cases, it is possible to apply an argument similar to Case 1, observing that prefixes/suffixes of A and B lead to q_a^+/Q_A^- and q_b^+/Q_B^- (Proposition 1-6). For the fifth case, since that M_A^k is fully contained in $M_A^k \circlearrowleft^\pm M_B^k$, then $\gamma \in m^k(A) \subseteq m^k(\mathcal{L}(M_A^k \circlearrowleft^\pm M_B^k))$. For the sixth case, $+, - \notin \gamma$ which implies that there exist path $q_{0b}[\gamma^{1\to1}\rangle\hat{q}_1[\gamma^{2\to k}\rangle q_{fb}$ in M_B^k. And so $q_{0a}[\gamma^{1\to1}\rangle\hat{q}_1[\gamma^{2\to k}\rangle q_{fb} \neq \varnothing$ in the graph of $M_A^k \circlearrowleft^\pm M_B^k$.

($\supseteq$) We first show that every $w = w_1 w_2 \cdots w_n \in \mathcal{L}(M_A^k \circlearrowleft^\pm M_B^k)$ is a substring of $+A(BA)^*-$. For that, consider all accepting paths p in $M_A^k \circlearrowleft^\pm M_B^k$. If p only passes through edges in M_A^k, then $\pi_l(p) \in m^k(A) \subseteq m^k(A(BA)^*)$. Else, if $w_1 \in \Sigma_B$ and p does not pass through an ϵ edge, then $p \in q_{0a}[w_1\rangle\hat{q}[w^{2\to n}\rangle q_{fb}$ in $M_A^k \circlearrowleft^\pm M_B^k$ and so $q_{0b}[w_1\rangle\hat{q}[w^{2\to n}\rangle q_{fb}$ is a path in $M_B^k \Rightarrow w \in m^k(B)$. And since that p does not pass through an $+/-$-labeled edge (they were removed from M_B^k), then $|w| = k$, which implies $w \in m^k(A(BA)*)$. Else, if p passes through an ϵ edge, then one of the following holds:

$$p \in \begin{cases} q_{0a}[\gamma_{a_1}\rangle q_a^-[\epsilon\rangle q_b^+[\gamma_{b_1}\rangle q_b^-[\epsilon\rangle q_a^+[\gamma_{a_2}\rangle \cdots q_a^+[\gamma_{a_i}\rangle q_{fa} \\ q_{0a}[\gamma_{a_1}\rangle q_a^-[\epsilon\rangle q_b^+[\gamma_{b_1}\rangle q_b^-[\epsilon\rangle q_a^+[\gamma_{a_2}\rangle \cdots q_b^+[\gamma_{b_i}\rangle q_{fb} \\ q_{0a}[\gamma_{b_1}\rangle q_b^-[\epsilon\rangle q_a^+[\gamma_{a_1}\rangle q_a^-[\epsilon\rangle q_b^+[\gamma_{b_2}\rangle \cdots q_a^+[\gamma_{a_i}\rangle q_{fa} \\ q_{0a}[\gamma_{b_1}\rangle q_b^-[\epsilon\rangle q_a^+[\gamma_{a_1}\rangle q_a^-[\epsilon\rangle q_b^+[\gamma_{b_2}\rangle \cdots q_b^+[\gamma_{b_i}\rangle q_{fb} \end{cases} \quad (2)$$

We only prove the first case (the other cases are analogous). For this case, it holds that $\gamma_{a_1} \in suff^k(+\hat{A})$, $\gamma_{a_i} \in pref^k(\hat{A}-)$ and $\gamma_{a_j} \in \hat{A} \; \forall 1 < j < i$ and $\gamma_{b_j} \in \hat{B} \; \forall 1 \leq j < i$. This all implies that $w \in suff^k(+A)\hat{B}(\hat{A}\hat{B})^* pref^k(A-)$ (notice the correspondence to the first case of (1)) and that w is a substring of $+\hat{A}\hat{B}(\hat{A}\hat{B})^*\hat{A}- \subseteq +A(BA)^*-$. Now if $|w| = k$, then $w \in m^k(+A(BA)^*-)$. Else if $|w| < k$, then $|\gamma_{a_1}| < k - 1 \Rightarrow |\gamma_{a_1} -| < k$ and since that $\gamma_{a_1 -} \in m^k(A)$, then $+ \in \gamma_{a_1}$. Similarly, we derive that $- \in \gamma_{a_i}$ and thus $w \in +\hat{A}(\hat{B}\hat{A})^*- \Rightarrow w \in m^k(+A(BA)^*-)$.   □

Lemma 5 shows that $s^k(\mathcal{L}(M_A^k \circlearrowleft^\pm M_B^k) = m^k(A(BA)*)$. But $M_A^k \odot^\pm M_B^k$ is an NFA and the algorithm from Lemma 1 requires a DFA as input. NFA determinization is worst-case exponential in its size. The following lemma shows that this does not happen for $M_A^k \circlearrowleft^\pm M_B^k$ due to its specific structure. The basic idea is that the ϵ transitions are the only source of non-determinism and that tokens of non-determinism "die" after at most k steps.

Lemma 6 (Determinizing the Markovian Loop Concatenation Does Not Explode). *Let A, B be arbitrary languages with disjoint alphabets Σ_A, Σ_B, and $M_A^k = (Q_A, \Sigma_A^\pm, \delta_A, q_{0a}, \{q_{fa}\})$ and $M_B^k = (Q_B, \Sigma_B^\pm, \delta_B, q_{0b}, \{q_{fb}\})$ the minimal DAFSAs accepting $m^k(A)$ and $m^k(B)$ respectively, $Q_{AB} = Q_A \cup Q_B$ and $\Sigma_{AB} = \Sigma_A \cup \Sigma_B$. Then runtime to determinize $M_A^k \circlearrowleft^\pm M_B^k$ is in $\mathcal{O}(k|Q_{AB}|^k)$.*

Proof. Start by noticing that all ϵ-edges in $M_A^k \circlearrowleft^\pm M_B^k$ lead to either q_a^+ or q_b^+. That means, that the ϵ-closure $\hat{S}$ of any state $S \subseteq Q_{AB}$ reached during the powerset construction is such that $\hat{S} \subseteq S \cup \{q_a^+, q_b^+\}$. Furthermore, the construction is such that for any reachable state $q \in Q_{AB}$, and

$l_a \in \Sigma_A^{\pm}$, if $|\hat{\delta}(q, l_a)| \neq \varnothing$, then $\hat{\delta}(q, l_a) \subseteq Q_A$ and similarly for every state $q \in Q_{AB}$ and $l_b \in \Sigma_B$, if $|\hat{\delta}(q, l_b)| \neq \varnothing$, then $\hat{\delta}(q, l_b) \subseteq Q_B$. This implies that every reachable state S in the powerset construction is such that either $S \subseteq Q_A$ or $S \subseteq Q_B$. Therefore, any path in the powerset construction is such that $\pi_s(p) = (S_{a_{1,1}} S_{a_{1,2}} \cdots S_{a_{1,n_1}})(S_{b_{1,1}} S_{b_{1,2}} \cdots S_{b_{1,m_1}})(S_{a_{2,1}} S_{a_{2,2}} \cdots S_{a_{2,n_2}}) \cdots$, where $S_{a_{i,j}} \subseteq \mathcal{P}(Q_A)$, $S_{b_{i,j}} \subseteq \mathcal{P}(Q_B)$, and $|S_{a_{i,1}}| = |S_{b_{i,1}}| = 1$.

Let $q_{a_{i,1}}$ and $q_{b_{i,1}}$ be the single elements of $S_{a_{i,1}}$ and $S_{b_{i,1}}$. Now consider the path $S_{a_{i,1}}[w_a] S_{a_{i,n_i}}$, $w_a \in (\Sigma_A^{\pm})^*$ in the powerset construction. Then $S_{a_{i,n_i}} \subseteq q_{a_{i,1}}[w_a\rangle \cup \bigcup_{1 < i \leq n} q_a^+[w_a^{i \to n}\rangle$ (in M_A^k). But since that M_A^k is a DAFSA with maximum word length k, then $q_a^+[w_a^{i \to n}\rangle = \varnothing$ if $|w_a^{i \to n}| \geq k$, which implies that $|S_{a,n}| \leq k$. A similar argument applies for $w_b \in \Sigma_B^{\pm}$. Therefore, the powerset construction expands at most $|Q_{AB}|^k$ nodes, with each node expansion costing at at most k, where $|Q_{AB}| \leq |\Sigma_A|^{k-1} + |\Sigma_A|^{k-1} + 2$ (Proposition 1-7).   □

From Lemma 1, it follows that m^k of the loop node can be computed in $\mathcal{O}(k|\Sigma|^{k^2})$. This exponent seems very high at first, but in practice it does not happen. This is related to the fact that if one of the subtrees does not accept the empty word, then there is no real non-determinism in $M_A^k \circlearrowleft^{\pm} M_B^k$.

3.5   Parallel Node

Finally, we consider the parallel node. Parallel nodes largely contribute to the original method's inefficiency because they inevitably lead to an explosion in the state space's size. Before presenting the construction for the parallel node, we must define the parallel composition of two languages [5]:

Definition 11 *(Parallel Composition).* *Given languages $A \subseteq \Sigma_A^*, B \subseteq \Sigma_B^*$, the parallel composition $A \parallel B \subseteq (\Sigma_A \cup \Sigma_B)^*$ is such that:*

$$w \in A \parallel B \iff w_{\Sigma_A} \in A \wedge w_{\Sigma_B} \in B$$

The parallel composition is closely related to the shuffle product. In fact, if $\Sigma_A \cap \Sigma_B = \varnothing$, then $A \parallel B = A \sqcup B$. Lemma 7 shows how to exploit this relation to compute m^k for parallel nodes:

Lemma 7 (m^k of the Shuffle Product). *Let A, B be arbitrary languages such that $\Sigma_A \cap \Sigma_B = \varnothing$, and $\Sigma_{AB} = \Sigma_A \cup \Sigma_B$. Then:*

$$m^k(A \sqcup B) = sub^k(m^k(A)) \parallel sub^k(m^k(B)) \parallel m^k(\Sigma_{AB})$$

which can be computed in $\mathcal{O}(|\Sigma_{AB}|^{2k})$.

Proof. Observe that $\Sigma_A \cap \Sigma_B = \varnothing \Rightarrow m^k(A \sqcup B) = s^k(+(A \sqcup B)-) = s^k(+A- \parallel +B-)$ and that $sub^k(m^k(A)) = sub^k(+A-)$ (Proposition 1-5).

($\subseteq$) Consider $w \in +A- \parallel +B-$. Then for every $\gamma \in s^k(w)$, it holds that $\gamma_{\Sigma_A^{\pm}} \sqsubseteq w_{\Sigma_A^{\pm}}$. And since $w_{\Sigma_A^{\pm}} \in +A-$ and $|\gamma| \leq k$, then $\gamma_{\Sigma_A^{\pm}} \in sub^k(+A-)$.

Analogously, $\gamma_{\Sigma_B^\pm} \in sub^k(+B-)$. Finally, since $\gamma = \gamma_{\Sigma_{AB}^\pm}$ and $\gamma \in s^k(+(A \sqcup B)-) \subseteq m^k(\Sigma_{AB}^*)$, then $\gamma \in sub^k(+A-) \parallel sub^k(+B-) \parallel m^k(\Sigma_{AB}^*)$.

($\supseteq$) For every $\gamma \in sub^k(+A-) \parallel sub^k(+B-) \parallel m^k(\Sigma_{AB}^*)$, there exists $w_a = \alpha_a \gamma_{\Sigma_A^\pm} \beta_a \in +A-$ and $w_b = \alpha_b \gamma_{\Sigma_B^\pm} \beta_b \in +B-$. Notice that $+$ is only present at most once in γ and $\gamma \in \gamma_{\Sigma_A^\pm} \parallel \gamma_{\Sigma_B^\pm}$, therefore $+ \in \alpha_a \iff + \in \alpha_b$. Similarly, $- \in \beta_a \iff - \in \beta_b$. Also notice that $+ \notin \alpha_a, \alpha_b$ implies $\alpha_a = \alpha_b = \epsilon$ and $- \notin \beta_a, \beta_b$ implies $\beta_a = \beta_b = \epsilon$.

If $+ \in \alpha_a$, then $+ \in \alpha_b \Rightarrow \alpha_a = +\hat{\alpha}_a$, $\alpha_b = +\hat{\alpha}_b$ and we define $\alpha = +\hat{\alpha}_a \hat{\alpha}_b$. Else, if $+ \notin \alpha_a$ then $\alpha = \epsilon$. Similarly, we define $\beta = \hat{\beta}_a \hat{\beta}_b-$ or $\beta = \epsilon$. In all cases, $\alpha \gamma \beta \in +A- \parallel +B- \Rightarrow s^k(\alpha \gamma \beta) \subseteq s^k(+A- \parallel +B-)$. Notice that if $|\gamma| = k \Rightarrow \gamma \in s^k(\alpha \beta \gamma)$. And that if $|\gamma| < k$, then $\alpha = \beta = \epsilon$ (since that $\gamma \in m^k(\Sigma_{AB}^*)$), $\Rightarrow \gamma \in s^k(\alpha \gamma \beta)$. Putting it together, $\gamma \in s^k(\alpha \gamma \beta) \subseteq s^k(+A- \parallel +B-)$.

For the runtime bound, notice that $sub^k(m^k(A))$ and $sub^k(m^k(B))$ can be computed in $\mathcal{O}(|\Sigma_{AB}|^{2k})$ and that the computation of the network automaton [5] associated to $sub^k(m^k(A)) \parallel sub^k(m^k(B)) \parallel m^k(\Sigma_{AB}^*)$, expands at most $|\Sigma_{AB}|^k$ states.   □

Total Runtime Boundary As shown above, computing m^k for each tree node is in $\mathcal{O}(k|\Sigma|^{k^2})$. For a process tree T containing n operator nodes, the runtime to compute $m^k(T)$ is in $\mathcal{O}(kn|\Sigma|^{k^2})$. Oftentimes, n is linear with $|\Sigma|$.

3.6   Handling Arbitrary Process Trees

The previous sections have shown how to compute m^k for binary process trees with unique visible label nodes. Notice that any process tree can be transformed into a binary tree accepting the same language (hence having the same m^k). For trees with repeated labels, the results below show that it suffices to first map each visible label node in the tree T to a unique label, and then map $m^k(\mathcal{L}(T))$ back to the original labels.

Lemma 8 (m^k of a Remapped Language). *Let $A \subseteq \Sigma_A^*$ and $B \subseteq \Sigma_B^*$ be arbitrary languages and $\lambda : \Sigma_B \to \Sigma_A$ s.t. $A = \lambda(B)$, then $s^k(A) = \lambda(s^k(B))$.*

Proof. Notice that for all w_a, w_b such that $w_a = \lambda(w_b)$, then $\forall_{i \le j}\ w_a^{i \to j} \in s^k(w_a) \iff w_b^{i \to j} \in s^k(w_b)$.

($\subseteq$) For any $w_a \in A$, there exists $w_b \in B$ s.t. $w_a = \lambda(w_b)$. For all $\gamma_a \in s^k(w_a)$, $\gamma_a = w^{i \to j}$ for some $i \le j$. Thus, $\gamma_a = \lambda(w_b^{i \to j})$ and since $|w_a| = |w_b|$, then $w_b^{i \to j} \in s^k(w_b) \Rightarrow s^k(w_a) \subseteq \lambda(s^k(w_b)) \subseteq \lambda(s^k(B))$.

($\supseteq$) For any $w_b \in B$, there exists $w_a \in A$ s.t. $w_a = \lambda(w_b)$. For all $\gamma_b \in s^k(w_b)$, it holds that $\gamma_b = w_b^{i \to j}$ for some $i \le j$. Thus, $\lambda(\gamma_b) = w_a^{i \to j}$ and since $|w_a| = |w_b|$, then $w_a^{i \to j} \in s^k(w_a) \Rightarrow s^k(A) \supseteq s^k(w_a) \supseteq \lambda(s^k(w_b))$.   □

The result above can be extended to m^k by defining $\lambda^\pm : \Sigma_B^\pm \to \Sigma_A^\pm$ such that

$$\lambda^\pm(l) = \begin{cases} \lambda(l) & l \in \Sigma_B \\ l & l \in \{+, -\} \end{cases}.$$ Then $+A- = \lambda^\pm(+B-) \Rightarrow m^k(A) = \lambda^\pm(m^k(B))$.

This mapping function can always be constructed for a process tree as follows:

Lemma 9 (m^k of Arbitrary Process Trees). *For an arbitrary process tree T_A with alphabet Σ_A and visible label nodes $N = \{n_1, \cdots n_i\}$. Given an alphabet Σ_B such that $|\Sigma_B| = k$ and a bijective mapping $r : N \to \Sigma_B$ defining a map $\lambda : \Sigma_B \to \Sigma_A$ as $\lambda(b) = \mathcal{L}(r^{-1}(b))$, then the process tree $\hat{T}$ obtained by relabeling each visible node n of T with $r(n)$ is such that $\mathcal{L}(T) = \lambda(\mathcal{L}(\hat{T}))$.*

Proof. It follows directly from Definition 4 by noticing that $\lambda(AB) = \lambda(A)\lambda(B)$ and $\lambda(A \sqcup B) = \lambda(A) \sqcup \lambda(B)$. $\square$

The results from this Section show that it is possible to compute m^k for arbitrary process trees in polynomial time. It is also possible to compute m^k for event logs in linear time. Thus, the markovian conformance metrics (Definition 7) can also be computed in polynomial time.

4   Experimental Evaluation

This section compares the proposed method with the previous approach described in [2] and state-of-the-art techniques in terms of runtime and the induced metrics. For a fair comparison, all techniques are implemented in pure Python[1].

4.1   Effect of Parallelism

The first experiment measures the influence of parallelism in the runtime. For that, we generate artificial process trees with a fixed number of activities (30) and varying degrees of parallelism (0.2 to 0.5). For each configuration, 50 process trees are generated. For each tree, an event log consisting of 2000 distinct variants is sampled and a small amount of noise is injected into the logs by adding, removing, and swapping activities.

We compare the runtime to compute three types of conformance artifacts: trace alignment (align), the model and log projections required by the PCC framework (PCC), and the markovian abstraction. For the latter two metrics, we vary their k parameter, indicating the projection size and substring size respectively, from 2 to 4 and break down the runtime for each method by the time taken to process the log and the model. For the markovian abstraction, we compare the method originally presented in [2] (m^k-orig) and the proposed method (m^k-opt). For each experiment run, we set a timeout of 20 min.

The results are summarized in Table 1. Trace alignment is by far the slowest method, with an average execution time of over five minutes and multiple time-outs. For comparison, none of the other methods timed out. PCC is arguably the second-slowest method, being the slowest in all but two scenarios. m^k-opt is the fastest method in all scenarios.

---

[1] The datasets and experiment results can be found at: https://github.com/EduardoGoulart1/efficient-mk/.

Table 1. The effect of parallelism on the runtime required to compute conformance artifacts, broken down by log and model processing times (if applicable). The number of timeouts (if any) is indicated in parenthesis.

k	Method	Time(s) [Log | Model]							
		par=0.2		par=0.3		par=0.4		par=0.5	
	align	343.4 (14)		368.4 (10)		454.2 (10)		439.9 (18)	
2	PCC	0.262	0.054	0.228	0.050	0.310	0.051	0.258	0.048
	m^k-orig	0.009	0.108	0.009	0.320	0.010	0.564	0.010	1.465
	m^k-opt		0.029		0.034		0.047		0.051
3	PCC	4.376	0.507	3.424	0.476	5.442	0.487	3.976	0.445
	m^k-orig	0.010	0.358	0.009	0.895	0.011	1.727	0.010	4.127
	m^k-opt		0.143		0.241		0.354		0.438
4	PCC	41.704	3.388	36.459	3.133	36.795	3.207	37.661	2.909
	m^k-orig	0.011	2.005	0.010	4.327	0.012	8.751	0.011	20.408
	m^k-opt		0.925		1.503		2.680		3.663

For models with little parallelism, m^k-orig and m^k-opt perform similarly well, with m^k-opt being slightly faster. This is explained by the fact that these models have a small and linear state-space. However, increasing the amount of parallelism from 0.2 to 0.5 causes a tenfold increase in the runtime for m^k-orig, while for m^k-opt it increases by a factor of at most 4, to which we conclude that m^k-opt can better handle large models. In comparison, PCC is unaffected by the degree of parallelism. Instead, its runtime is dominated by the event log projections.

In general, the experiment shows the shortcomings of trace alignment and the PCC framework in terms of runtime, especially considering large event logs. It also shows that m^k-orig struggles to process large models. m^k-opt emerged as the clear winner in terms of performance. For event logs, m^k-opt can be up to 400 times faster than PCC. At the same time, computing m^k for process models takes roughly the same time as computing the tree projections.

4.2   Real Datasets

Next, we evaluate the markovian-based conformance metrics on two real-world datasets: the Italian Road Fines event log, and the BPI Challenge 2015 event log (BPIC-15). We filter the BPIC-15 log for the municipality 1, subprocess 8, and remove repeated activities. This preprocessing is needed as otherwise the used process discovery methods would only return flower constructs. For each event log, we mine four process trees with the Inductive Miner infrequent variant with noise thresholds of 02 and 05 (IMf02 and IMf05 respectively), the Inductive Miner incomplete variant (IMc) and the flower miner. We use alignment-based trace fitness [1] (AL) as the ground-truth fitness measure and escaping edges precision [9] (ETC) as the ground truth precision measure. We vary the respective k parameter of PCC, MAF, and MAP from 2 to 4. The results are summarized in Table 2.

The first thing to notice is that the basic property that language inclusion implies fitness of 1.0 is fulfilled by metrics for the IMc and Flower models for both datasets. Next, for both datasets, PCC and MAF generate the same fitness rankings as the ground truth alignment-based fitness measure (AL) for all k-s. As k increases, the difference in fitness between models IMf02 and IMf05 increases.

For the Road Fines datasets, all metrics induce different precision rankings. ETC is assigning a higher precision to the flower model than to the IMc model. For $k = 2, 3$, PCC and MAP agree on their rankings, but assign IMc as being more precise than IMf02. This is counter-intuitive since that the IMc model has a lot more parallelism and self-loops. For $k = 4$, PCC even assigns IMc as the most precise model. For the BPIC 2015 datasets, all metrics agree on the model rankings. However, the PCC metric will assign a relatively high precision for models such as IMc and the Flower model.

In summary, the experiment shows that MAF and MAP induce similar fitness and precision rankings as other state-of-the-art techniques. Notice that for both datasets, as k increases, MAP tends towards zero. This is expected from the definition of MAP, which does not consider any notion of substring frequency.

Table 2. Quality evaluation of fitness and precision metrics.

Miner	Road Fines				BPIC 2015			
	IMf02	IMf05	IMc	Flower	IMf02	IMf05	IMc	Flower
AL	0.982	0.784	1.0	1.0	0.899	0.773	1.0	1.0
Fitness PCC2	0.986	0.857	1.0	1.0	0.985	0.962	1.0	1.0
PCC3	0.976	0.755	1.0	1.0	0.977	0.937	1.0	1.0
PCC4	0.967	0.664	1.0	1.0	0.968	0.912	1.0	1.0
MAF2	0.965	0.953	1.0	1.0	0.881	0.737	1.0	1.0
MAF3	0.936	0.826	1.0	1.0	0.833	0.475	1.0	1.0
MAF4	0.899	0.745	1.0	1.0	0.805	0.422	1.0	1.0
ETC	0.895	0.653	0.318	0.325	0.497	0.817	0.261	0.127
Precision PCC2	0.946	0.949	0.931	0.593	0.735	0.924	0.660	0.635
PCC3	0.814	0.838	0.831	0.497	0.624	0.824	0.551	0.534
PCC4	0.658	0.703	0.718	0.423	0.529	0.722	0.462	0.451
MAP2	0.735	0.949	0.830	0.542	0.549	0.729	0.316	0.225
MAP3	0.277	0.389	0.353	0.134	0.115	0.411	0.047	0.020
MAP4	0.082	0.122	0.106	0.020	0.015	0.199	0.005	0.001

5   Related Work

Conformance checking is the field of process mining focused on comparing a process' desired to its observed behavior. The process model describes the desired behavior. It is often encoded as a Petri net or any equivalent model with execution semantics (YAWL, Process Trees, etc.).

Conformance-checking is especially challenging because computing the process model's behavior has often worst-case exponential time due to the state explosion problem. Hence, most state-of-the-art methods such as token-based replay [11], alignments [1], entropia [10], or Earth mover's distance [8] have worst-case exponential time. This also includes the original method for computing markovian-based conformance metrics presented in [2].

A notable exception is the *Projected Conformance Checking* framework (PCC framework) [7] which uses projections on subsets of activities to significantly alleviate the state explosion problem. In fact, for certain classes of process trees the runtime is polynomial. Nevertheless, PCC requires multiple passes over the event log, which is impractical for large production datasets, as shown in Sect. 4.

The idea of exploiting the tree structure to speed up computations is not new. In [12] a method is presented to approximate alignments by constructing an equivalent optimization problem from the tree structure. In [14], a method is presented to repair alignments for iterative scenarios, for the use-case where alignments need to be computed for similar process trees. Our work differs from them in which we provide a speed up in computation time without the need to approximate. Finally, in [15] a method is presented to compute trace probabilities by transforming the tree into a probabilistic context-free grammar, this transformation is only possible because of the process tree' structure.

Last, sampling techniques [6] can be orthogonally applied to any conformance method, including our technique. However, sampling only provides a linear speed-up and previously exponential techniques will remain exponential. In production settings, where controllable runtime is important, exponential factors are rarely a good idea.

6   Conclusion

This paper provides two important contributions. First, we presented an alternative definition of the markovian abstraction that can be more easily manipulated using techniques from automata theory. Next, we showed how to exploit the tree-structure of process trees to perform polynomial-time conformance checking with guarantees. The experimental evaluation shows an improvement of multiple orders of magnitude in the runtime compared to the original approach presented in [2] and other state-of-the-art conformance checking techniques, while at the same time still generating similar fitness and precision rankings. Most importantly, the runtime of the approach is bounded by a polynomial, making it more controllable.

As future work, we plan to apply the proposed technique to optimization-based discovery techniques such as the evolutionary tree miner [3], which requires repetitive computation of conformance metrics. We also plan to explore the stochastic perspective, by computing the probability of each substring in the process tree's language, to address the problem of vanishing precision values for MAP.

Acknowledgments. We thank the Alexander von Humboldt (AvH) Stiftung for supporting our research.

References

1. Adriansyah, A., van Dongen, B., van der Aalst, W.: Conformance checking using cost-based fitness analysis. In: 2011 IEEE 15th International Enterprise Distributed Object Computing Conference, pp. 55–64 (2011)
2. Augusto, A., Armas-Cervantes, A., Conforti, R., Dumas, M., Rosa, M.L.: Measuring fitness and precision of automatically discovered process models: a principled and scalable approach. IEEE Trans. Knowl. Data Eng. **34**(4), 1870–1888 (2022)
3. Buijs, J.: Flexible evolutionary algorithms for mining structured process models. Technische Universiteit Eindhoven. **57** (2014)
4. Daciuk, J., Mihov, S., Watson, B.W., Watson, R.E.: Incremental construction of minimal acyclic finite-state automata. Comput. Linguist. **26**(1), 3–16 (2000)
5. Esparza, J., Blondin, M.: Automata Theory: An Algorithmic Approach. MIT Press, Cambridge (2023)
6. Bauer, M., van der Aa, H., Weidlich, M.: Estimating process conformance by trace sampling and result approximation. In: Hildebrandt, T., van Dongen, B.F., Röglinger, M., Mendling, J. (eds.) BPM 2019. LNCS, vol. 11675, pp. 179–197. Springer, Cham (2019). https://doi.org/10.1007/978-3-030-26619-6_13
7. Leemans, S.J.J., Fahland, D., van der Aalst, W.: Scalable process discovery and conformance checking. Softw. Syst. Model. **17**, 599–631 (2016)
8. Leemans, S.J., van der Aalst, W.W., Brockhoff, T., Polyvyanyy, A.: Stochastic process mining: earth movers' stochastic conformance. Inf. Syst. **102**, 101724 (2021)
9. Muñoz-Gama, J., Carmona, J.: A fresh look at precision in process conformance. In: Hull, R., Mendling, J., Tai, S. (eds.) BPM 2010. LNCS, vol. 6336, pp. 211–226. Springer, Heidelberg (2010). https://doi.org/10.1007/978-3-642-15618-2_16
10. Polyvyanyy, A., Solti, A., Weidlich, M., Ciccio, C.D., Mendling, J.: Monotone precision and recall measures for comparing executions and specifications of dynamic systems. ACM Trans. Softw. Eng. Methodol. **29**(3), 1–41 (2020)
11. Rozinat, A., van der Aalst, W.: Conformance checking of processes based on monitoring real behavior. Inf. Syst. **33**(1), 64–95 (2008)
12. Schuster, D., van Zelst, S., van der Aalst, W.M.P.: Alignment approximation for process trees. In: Leemans, S., Leopold, H. (eds.) ICPM 2020. LNBIP, vol. 406, pp. 247–259. Springer, Cham (2021). https://doi.org/10.1007/978-3-030-72693-5_19
13. Syring, A.F., Tax, N., van der Aalst, W.M.P.: Evaluating conformance measures in process mining using conformance propositions. In: Koutny, M., Pomello, L., Kristensen, L.M. (eds.) Transactions on Petri Nets and Other Models of Concurrency XIV. LNCS, vol. 11790, pp. 192–221. Springer, Heidelberg (2019). https://doi.org/10.1007/978-3-662-60651-3_8
14. Vázquez-Barreiros, B., van Zelst, S.J., Buijs, J.C.A.M., Lama, M., Mucientes, M.: Repairing alignments: striking the right nerve. In: Schmidt, R., Guédria, W., Bider, I., Guerreiro, S. (eds.) BPMDS/EMMSAD -2016. LNBIP, vol. 248, pp. 266–281. Springer, Cham (2016). https://doi.org/10.1007/978-3-319-39429-9_17
15. Watanabe, A., Takahashi, Y., Ikeuchi, H., Matsuda, K.: Grammar-based process model representation for probabilistic conformance checking. In: 2022 4th International Conference on Process Mining (ICPM), pp. 88–95. IEEE (2022)

Engineering

Investigating the Influence of Data-Aware Process States on Activity Probabilities in Simulation Models: Does Accuracy Improve?

Massimiliano de Leoni[1], Francesco Vinci[1(✉)], Sander J. J. Leemans[2], and Felix Mannhardt[3]

[1] University of Padua, Padua, Italy
`massimiliano.deleoni@unipd.it, francesco.vinci.1@phd.unipd.it`
[2] RWTH Aachen, Aachen, Germany
`s.leemans@bpm.rwth-aachen.de`
[3] Eindhoven University of Technology, Eindhoven, The Netherlands
`f.mannhardt@tue.nl`

Abstract. Business process simulation enables analysts to run a process in different scenarios, compare its performances and consequently provide indications on how to improve a business process. Process simulation requires one to provide a simulation model, which should accurately reflect reality to ensure the reliability of the simulation findings. An accurate simulation model passes through a correct stochastic modelling of the activity firings: activities are associated with the probability of each to fire. Literature determines these probabilities by looking at the frequency of the activity occurrences when they are enabled. This is a coarse determination, because this way does not consider the actual process state, which might influence the probabilities themselves (e.g., a thorough loan assessment is more likely for larger loan requests). The process state is as a faithful abstraction of the process instance execution so far, including the process-variable values, the activity firing history, etc. This paper aims to investigate how process states can be leveraged to improve activity firing probabilities. A technique has been put forward and compared with the baseline where basic branching probabilities are employed. Experimental results show that, indeed, business simulation models are more accurate to replicate the real process' behavior.

Keywords: Process Simulation · Stochastic Models · Branching Probabilities · Process Mining

1 Introduction

Business process simulation refers to techniques for the simulation of business process behavior on the basis of a process simulation model, a process model extended with additional information for a probabilistic characterization of the different run-time aspects (case arrival rate, activity durations and probabilities,

© The Author(s), under exclusive license to Springer Nature Switzerland AG 2023
C. Di Francescomarino et al. (Eds.): BPM 2023, LNCS 14159, pp. 129–145, 2023.
https://doi.org/10.1007/978-3-031-41620-0_8

roles, etc.). Simulation provides a flexible approach to analyse and improve business processes. Through simulation experiments, various 'what if' questions can be answered, and redesigning alternatives can be compared with respect to some key performance indicators. The main idea of business process simulation is to carry out a significantly large number of runs, in accordance with a simulation model. Statistics over these runs are collected to gain insight into the processes, and to determine the possible issues (bottlenecks, wastes, costs, etc.). By applying different changes to the simulation model, one can assess the consequences of these changes without putting them in production, and consequently can explore dimensions to possible process improvements.

A successful application of business process simulation for process improvement relies on a simulation model that reflects the real process behavior; conclusions drawn on an unrealistic simulation model lead to process redesigns that may not yield improvements, or even may worsen the performances. A large body of research work has already focused on accurately creating process models and several of their run-time aspects (cf. Sect. 5), including case arrival rate, activity durations, and roles. However, no recent work has focused on accurately estimating, given a set of enabled activities at run time, the probability of each to occur. Currently, this determination is solely based on the branching probabilities. Of course, this is a course determination, which does not consider that probabilities of activities to occur may vary as function of the characteristics of the process instances, the activities that previously occurred in the same process instance, time-related information, etc. Consider, for example, a loan application process: the probability of executing an activity about a thorough assessment grows, e.g., with the amount of the requested loan, and decreases with annual salary of the applicant. Also, the probability of rejecting an application may grow with the number of requests to the applicant of providing further documents.

This paper introduces the concept of process state, which is a faithful abstraction of a case, and investigates the research question how a proper choice of this abstraction allows a more accurate estimation of the activity firing probabilities of simulation models, with respect to the simple branching probabilities. More accurate firing probabilities lead to more accurate simulation models.

In order to answer this research question, the paper builds upon Petri nets, and discovers a so-called weight function for each transition, which is defined over a process state, which, e.g., can include process variables, and the transition-firing history. Then, for the cases that are simulated, the current process state is computed and used to evaluate the weight function of the transitions that are enabled at that point. The probability to fire a transition is thus obtained as the ratio of its weight and the sum of the weights of all enabled transitions. It follows that the higher is the weight of a transition, the higher is the probability of that transition to fire. Since the case characteristics that may influence the probability may depend on the specific process that is being simulated, we propose a framework where the process-state abstraction can be customized to include or exclude certain characteristics. In Sect. 2, some examples for process states are provided. In this paper, we report on the use of logistic regression to learn the weight function, but other approaches could be alternatively employed, such as regression trees.

The research question is finally answered by applying the aforementioned framework to two real-life processes, each with a real-life event log. For each process, five different definitions of process states have been considered to compute weights of the Petri-net models of the two processes to be simulated. The results show that the simulation using models with transition probabilities based on process states allows obtaining simulation results that are more accurate, if compared with simulation models based on branching probabilities.

Section 2 starts introducing the preliminary concepts of event logs, and then continues (i) introducing the novel notion of the process-state abstraction and (ii) their usage with stochastic labelled Data Petri nets, a Petri-net extension to associate weights to transitions. Section 3 discusses how to compute the transition weights as function of a customizable abstraction of the process state. Section 4 illustrates how Stochastic labelled Data Petri nets can be represented via Coloured Petri nets in CPN Tools, and reports on the benefits for simulation models to use transition probabilities based on process states. Section 5 reports on the related works, and Sect. 6 concludes the paper.

2   Event Logs and Stochastic Data Petri Nets

The determination of the weights is obtained from an analysis of an event log of the process that aims to be simulated:

Definition 1 (Events). *Let $\mathcal{A}$ be the set of process' activities. Let $\mathcal{V}$ the set of process attributes. Let $\mathcal{W}_\mathcal{V}$ be a function that assigns a domain $\mathcal{W}_\mathcal{V}(x)$ to each process attribute $x \in \mathcal{V}$. Let $\overline{\mathcal{W}} = \cup_{x \in \mathcal{V}}\mathcal{W}_\mathcal{V}(x)$. An event is a tuple $(a, v) \in \mathcal{A} \times (\mathcal{V} \nrightarrow \overline{\mathcal{W}})$ where a is the event activity, v is a partial function assigning values to process attributes with $v(x) \in \mathcal{W}_\mathcal{V}(x)$.*

A trace is a sequence of events, the same event can potentially occur in different traces, namely attributes are given the same assignment in different traces. This means that potentially the entire same trace can appear multiple times. This motivates why an event log is to be defined as a multiset of traces:[1]

Definition 2 (Traces & Event Logs). *Let $\mathcal{E} = \mathcal{A} \times (\mathcal{V} \nrightarrow \overline{\mathcal{W}})$ be the universe of events. A trace σ is a sequence of events, i.e. $\sigma \in \mathcal{E}^*$. An event-log $\mathcal{L}$ is a multiset of traces, i.e. $\mathcal{L} \subset \mathbb{B}(\mathcal{E}^*)$.*

In this paper, simulation models are provided in form of so-called Stochastic Labelled Data Petri Nets (SLDPNs). While SLDPNs are not able to represent every aspect relevant for simulation models, they are simple, yet sufficient to discuss and formalize the concepts behind activity probabilities. In SLDPNs, a transition firing consists in executing a transition and assigning values to some process attributes. The sequence of transition firings determines the process state:

---

[1] $\mathbb{B}(X)$ indicates the set of all multisets with the elements in set X.

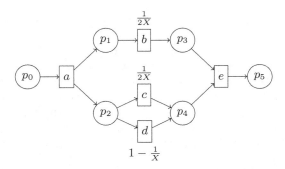

Fig. 1. Example of a Stochastic Data Petri Net. Transitions are annotated with the weights, when the latter are present.

Definition 3 (Process State). *Let T be a set of transitions. Let $\mathcal{V}$ the set of process attributes. Let $\mathcal{W}_\mathcal{V}$ be a function that assigns a domain $\mathcal{W}_\mathcal{V}(x)$ to each process attribute $x \in \mathcal{V}$. Let $\overline{\mathcal{W}} = \cup_{x \in \mathcal{V}} \mathcal{W}_\mathcal{V}(x)$. Let Δ be the set of process states. A process-state function maps a sequence of transition firings to a process state:* $\mathcal{S}_\Delta : (T \times (\mathcal{V} \not\to \overline{\mathcal{W}}))^* \to \Delta.$

Note that the marking is not part of the process state (see below). A process-state function can be customized, according to the specific domain. For instance, if one wants to account for the process attributes in the set $\mathcal{V}$, the set $\overline{\Delta}$ of possible process states consists of tuples $(x_1, \ldots, x_n)$ where x_i is the value assigned to variable $v_i \in \mathcal{V}$, after defining an ordering of the attributes in $\mathcal{V}$. In particular, for a sequence $\sigma = \langle (t_1, f_1), \ldots, (t_m, f_m) \rangle$ of transition firings, $S_{\overline{\Delta}}$ returns a tuple $(x_1, \ldots, x_n)$ in which x_i is the latest value assigned to v_i in σ, namely there is a transition firing $(t_j, f_j) \in \sigma$ such that $f_j(v_i) = x_i$ and, for all $j < k \leq m$, v_i is not in the domain of f_k.

In SLDPNs, each transition is associated to a weight function that is dependent on a process state.

Definition 4 (Stochastic Labelled Data Petri Net - syntax). *Let A be a set of activities, and Δ the set of possible process states. A stochastic labelled data Petri net (SLDPN) is a tuple $(P, T, F, \lambda, \Delta, M_0, \mathfrak{w})$, such that (P, T, F) is a Petri Net, $\lambda : T \not\to A$ be a labelling function, M_0 is the initial marking, and $\mathfrak{w} : T \times \Delta \to \mathbb{R}^+$ is a weight function.*

Example. Figure 1 shows an example of an SLDPN. The control flow of this SLDPN consists of an AND split followed by the parallel executions of b and a choice between c and d. The transitions are annotated with weight functions depending on the continuous variable X.

The state of an SLDPN is the combination of a marking and an process state $d \in \Delta$. Hereafter, when clear from the context, the process state is simply referred to as state. The marking determines which transitions are enabled, while the process state influences the probabilities of transitions:

Definition 5 (Stochastic Labelled Data Petri Net - semantics). *Let* $N = (P, T, F, \lambda, \Delta, M_0, \mathfrak{w})$ *be an SLDPN. Let* $\sigma \in (T \times (\mathcal{V} \nrightarrow \overline{\mathcal{W}}))^*$ *be a sequence of transition firings, leading to marking* M. *Let* S_Δ *be the process-state function, and* $E(M) \subseteq T$ *be the set of transitions enabled at marking* M *of Petri net* (P, T, F). *The probability to fire* t *after* σ *is:*

$$Pr_N(t, M, \sigma) = \frac{\mathfrak{w}(t, S_\Delta(\sigma))}{\sum_{t' \in E(M)} \mathfrak{w}(t', S_\Delta(\sigma))}.$$

3 Framework for Determination of Weights

In this section, we introduce a framework that, given an event log $\mathcal{L}$, a Petri net (P, T, F), a labelling function λ and an initial marking M_0, can be used to determine the weights of the transitions, thereby transform the Petri net into an SLDPN. The framework can be instantiated for a process-state function S_Δ, generalising the proposal in [10], and a parameterised weight function $\mathfrak{w}$, and consists of four steps:

Step 1 For each trace $\sigma \in \mathcal{L}$, reconstruct the corresponding path of transitions that σ took through the model. This reconstruction is performed using a sequence of *moves*: a synchronous move combines an event (a, v) in the log trace with a transition t on the model path such that $\lambda(t) = a$; a model move is a transition on the model path; while a log move is an event in the log trace. Such a sequence of moves, where the projection of the sequential synchronous and log moves yields the trace, and the projection of the sequential synchronous and model moves yields the path, is called an *alignment* [1].

For a given trace of the event log, an optimal alignment is an alignment with a minimal number of log and model moves[2], over all paths in the model. Note that this alignment does not need to take the data values or weight functions into account and can be computed solely based on the regular Petri net and each trace of the event log.

Step 2 For one optimal alignment of each trace in the event log, we use the process-state function S_Δ to reconstruct the sequence of process states Δ. By definition, any path of the model starts in the initial marking M_0 and the process state $S_\Delta(\langle\,\rangle)$. For each synchronous or model move m in the optimal alignment, we have a transition t available. As the moves are sequential, we can take the partial function assigning values to process attributes (v) from the last synchronous move in the alignment, before m. If no such last move exists, we take an empty function.

As such, we obtain a sequence of transition firings (t, v). Through S_Δ, this sequence yields a sequence of process states $\delta \in \Delta$. Similarly, the sequence of markings can be derived from the model and the sequence of transition firings.

---

[2] For the minimalisation, we do not count model moves on unlabelled transitions.

Step 3 For each transition t in the model, we gather the observations made in each trace σ of the log related to this transition: each time that t was enabled and fired in the sequence of transition firings, the associated process state before firing t is recorded as positive observation. Each time that t was enabled but another transition fired a negative observation using the process state before firing that transition is recorded. Given these collected multisets of positive process states $\Delta_{t+} \subset \mathbb{B}(\Delta)$ and negative process states $\Delta_{t-} \subset \mathbb{B}(\Delta)$ we build a training set to learn the influence of process states on the weight of transition t as:

$$\biguplus_{\delta \in \Delta_{t+}} [(\delta, 1)] \cup \biguplus_{\delta \in \Delta_{t-}} [(\delta, 0)]$$

Step 4 We leverage any suitable machine learning model that supports the process state representation chosen as input to serve as parameterised weight function $\mathfrak{w}$. Such a model should assign higher weights, e.g., the 1 in the training set, for those process states in which the transition t was observed to occur opposed to those in which another transition was observed. We can obtain the overall parameterised weight function by fitting a separate model for each transition t since the weights obtained for enabled transitions in the SLDPN are not required to sum up to 1.

As an example, we instantiate our framework with a process state function $\mathcal{S}_\Delta$ that takes into account (i) the event attribute values observed for the first event in a trace and (ii) the count of the activity occurrences in the history of the process instance:

$$\mathcal{S}_\Delta(\langle (t_1, v_1), \ldots, (t_n, v_n) \rangle) = (v_1, [t_1, \ldots t_n]).$$

Our process state is, thus, defined as $\Delta = (\mathcal{V} \nrightarrow \overline{\mathcal{W}}) \times \mathbb{B}(T)$. We use the logistic function over the same attribute values and history as parameterised weight function $\mathfrak{w}$.

Assume the example trace $\sigma = \langle a^{X=3}, d^{X=4}, d^{X=5}, b^{X=3}, e^{X=5} \rangle$ with a single process attribute X. We align σ in Step 1 to the model shown in Fig. 1. This alignment is:

Log	$a^{X=3}$	$d^{X=4}$	$d^{X=5}$	$b^{X=3}$	$e^{X=5}$
Model	a	d	-	b	e

In Step 2, we transform this alignment into a sequence of process states:

$$\langle (X = 3, [a]), (X = 3, [a, d]), (X = 3, [a, b, d]), (X = 3, [a, b, d, e]) \rangle$$

Then, Step 3 constructs the observations:

$$\Delta_{a+} = [(X = \bot, [])] \qquad \Delta_{a-} = []$$
$$\Delta_{b+} = [(X = 3, [a, d])] \qquad \Delta_{b-} = [(X = 3, [a])]$$
$$\Delta_{c+} = [] \qquad \Delta_{c-} = [(X = 3, [a])]$$
$$\Delta_{d+} = [(X = 3, [a])] \qquad \Delta_{d-} = []$$
$$\Delta_{e+} = [(X = 3, [a, b, d])] \qquad \Delta_{e-} = []$$

In the final Step 4 the weight function is then approximated using logistic regression for each of the transitions using the training sets build from positive and negative observations. We use logistic regression since it provides white-box explanations and is more usable for simulators such as CPN Tools. Moreover, white-box simulators can be used for what-if analysis, whereas deep learning models cannot [6]. Logistic regression, and many other machine learning models, requires input variables to be numeric. Thus, we need to transform the multiset of activity occurrences into several variables, one for each activity in the process model. Similarly, we could use one-hot encoding for categorical variables. Finally, we obtain the coefficients for the logistic function and obtain the final SLDPN including the learned weight function.

4 Experiments

The experiment focuses on verifying the similarity between the original event logs and those obtained from simulation. Our probability model of activity firing is only supported by CPN Tools, which includes simulation features.

Section 4.1 illustrates the case studies that are used to assess the validity of the proposed approach. Section 4.2 discusses the experimental setup, including how CPN Tools has been employed to define our probability model. Finally, Sect. 4.3 reports the results, and discusses the findings.

4.1 Case Studies

The approach has been evaluated in two different case studies for which a public event log and a reference model is available: Road-Traffic Fines and Sepsis (see [9] for details).

The Road Traffic Fines event log describes the process of managing road traffic fines by a local police force in Italy. The event log contains 150370 traces and 11 different activities, with 12 data attributes. Sepsis case study is a real-life event log obtained from an Enterprise Resource Planning (ERP) system of a regional hospital in The Netherlands. It contains 1050 traces, 16 different activities and several data attributes, most of them binary.

Figures 2 and 3 depicts the Petri Net models used for our evaluations, which are shown in Fig. 12.8 and Fig. 13.3 of [9]. The Road-Traffic Fines model allows

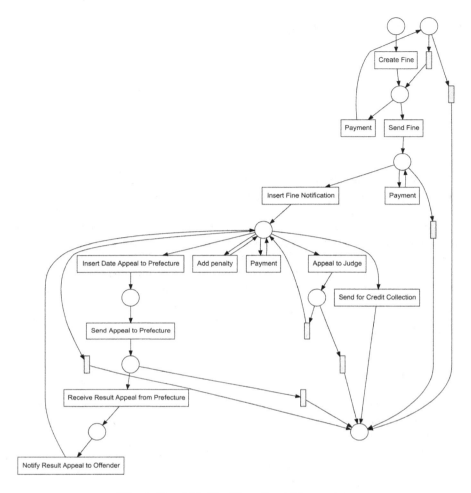

Fig. 2. Road-Traffic Fines Petri Net model.

executions with only one activity, i.e. `Create Fine`, however, the event log does not contain any trace such that. This led us to improve the model: a place and a transition after the first `Payment` are added to maintain the loop of it and remove the possibility of ending the trace after `Create Fine`.

4.2   Experimental Setup

The assessment is based on comparing the similarity of the original event logs with those obtained via simulation. The simulation models are constructed using five different characterization of the process state: with data only, namely using the values of the process attributes; with history only, namely using the number of occurrences of process activities; and with data and history, as well as

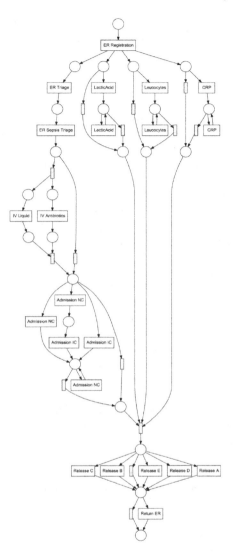

Fig. 3. Sepsis Petri Net model.

with combinations of so-called binary history. History is also delineated in the binary version, where we only consider whether or not an activity has happened, irrespectively of the number of occurrences. As baseline of comparison, we also built the simulation model using branching probabilities, which were computed through the Multi-perspective Process Explorer [11]. The comparison of the original event log and those obtained via simulation is computed through the Earth-Movers' Stochastic distance introduced in [8]. This measure considers the stochastic characteristics of the event logs: which activities are executed in which order, and how often a particular order of activities is executed.

For evaluating our approach, we divided the original logs into training and test sets using a temporal split: 70% to the training set and 30% to the test set. The training set was utilized to calculate the weights of the corresponding SLDPNs, and also the branching probabilities of the comparison baseline. By comparing the results of them, we can determine the effectiveness of our approach and assess the impact of different processing techniques on performance.

We instantiated the framework for determining the weights described in Sect. 3 in ProM by encoding the process state obtained into a set of attributes and implemented the parameterised weight function as a set of logistic regression models over that set of attributes and the binary dependent response variable. We use ridge regression as implemented in WEKA 3.8 [17] and implemented an export functionality to obtain the resulting coefficients $\beta_0, \ldots, \beta_n$ of the logistic model that are sufficient to determine the weight. More complex models may be added to the implementation in the future as long as their parameters can be used to compute weights based on process states. After exporting the logistic regression coefficients, we use them in the simulation models.

The simulation models are implemented using CPN Tools. In fact, one of the key advantages of CPN Tools is the ability to model and analyze complex systems. Additionally, the Standard ML programming language can be used to implement custom functions, making it possible to adapt the model to the specific needs of the simulation.

We illustrate how SLDPNs can be represented via CPN Tools through a simple example. In particular, we focus on the SLDPN in Fig. 1. The CPN model consists of several parts, each with a specific function. The black part represents the Petri Net underneath, while the blue part focuses on the simulation of the CPN Tools: n_sim process instances are simulated one at time until the previous instance is completed. Note that this does not affect the event log similarity since the evaluation is based on control-flow and not on time-related measurements. This is done here for the sake of simplicity: it is trivial to extend it to allow for multiple process instance executions at the same time.

The brown part is related to generating the values of the process attributes for the different simulated traces. The literature proposes to assign a suitable statistical distribution to each process attribute (cf. Sect. 5): values are then sampled from those distributions. However, this might introduce noise in the experiments, if suitable distributions are not found. This ultimately leads to an unfair comparison of the original event log and that obtained from simulation. Recall that we simulate as many process instances, i.e. traces, as those in the original event log. We leverage on this, and, for each trace of the original event log, we use the same set of process-attribute values for one process instance that is simulated.

The green part of the model checks the enabled transitions and the trace history. Finally, the purple part is responsible for computing the probabilities of

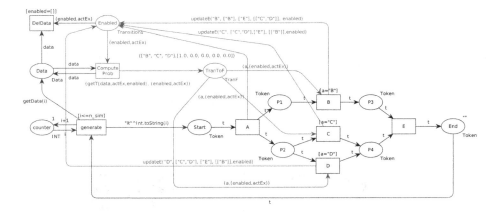

Fig. 4. Implementation of the SLDPN in Fig. 1 in CPN Tools.

firing the enabled transitions using·the weights obtained by the logistic regression models.

The arc inscriptions in the CPN Tools model example are used to implement the SLDPN models: the `getData` function returns a list of real value representing the i-th data state; the `updateE` function returns a tuple (`enabled, actEx`), where `enabled` is the list of strings of the enabled transitions after having executed a certain transition, and `actEx` is a list of real values of length number of activities in which each value represents how many time the correspondent activity has been performed; `getT` returns the transition to fire according to the probabilities computed using a function that implements the formula in Definition 5, where ɯ is the logistic regression using the coefficients exported from the Prom Plug-In (Fig. 4).

This procedure is then applied to the Road-Traffic Fine and Sepsis SLDPNs. Each simulated model has been used to generate as many traces as those in the original event log (respectively, 150370 and 1050 traces). The simulation reports generated by CPN Tools have been converted to obtain event logs in XES files using a parser which we implemented in Python.[3]

4.3   Experimental Results

For each case study, we have thus computed six simulation models, using probabilities based on the branching probabilities and on the weights with five different process-state abstractions. Each simulation model has been run ten times, to mitigate the stochasticity of simulation.

Table 1 shows the final results for each simulation model and case study. The reported measures are the average of the Earth Movers' distances between each simulated event log and the real one. The value in brackets are the maximum

---

[3] https://github.com/franvinci/InfluenceofDataawareProcessStatesonActivityProbabilities.

Table 1. Average Earth Movers' distance (with absolute error) for each configuration.

Case Study	Branching Prob.	Data	History	Data & History	Bin History	Data & Bin. History
Road-Traffic Fines	0.8793	0.8831	0.9526	0.9542	0.9352	0.9427
	(±0.001)	(±0.001)	(±0.001)	(±0.001)	(±0.001)	(±0.001)
Sepsis	0.5309	0.6245	-	-	0.5906	0.6201
	(±0.008)	(±0.009)	-	-	(±0.005)	(±0.006)

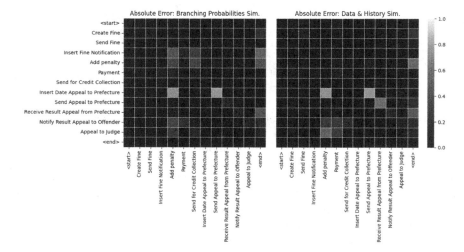

Fig. 5. Road-Traffic Fines absolute errors of the probabilities of occurrence of an activity a_j (columns) immediately after a_i (rows) between the real and the simulated event logs. Left: using branching probabilities. Right: using data and history. Darker colours indicate lower errors.

variation of similarity for each of ten simulated logs obtained for each configuration, wrt. the original event log. Note how the variation is negligible (less than 1%), thus positively hinting at the reliability of the results.

In the Road-Traffic Fines case study, it can be noticed a slight improvement in log similarity using only data features. However, by adding history, our approach outperforms the baseline of 7.5%: this suggests that, for this case, the data alone are insufficient for capturing the control flow stochasticity, while the history features provide valuable insights. Binary history also leads to good performance. However, configurations with history states perform better.

On the other hand, in the Sepsis case study, using the data features without history results in a 9.4% increase in baseline performance, proving that computing firing probabilities based on these features can lead to more accurate simulation models. We were unable to perform the experiments for the Sepsis case study when the process state abstraction includes the history. Indeed, the

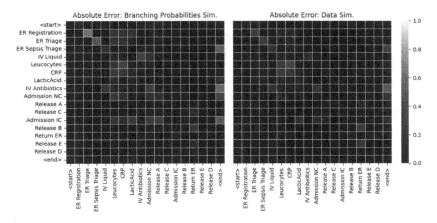

Fig. 6. Sepsis absolute errors of the probabilities of occurrence of an activity a_j (columns) immediately after a_i (rows) between the real and the simulated event logs. Left: using branching probabilities. Right: using data. Darker colours indicate lower errors.

simulation was often blocked in a livelock: the weight function associated to the transition with label `Leucocytes` in the loop was typically returning weights close to 1 in most of process states, due to the history being included in the process state. This causes a loop to be infinitely repeating. This issue is certainly due to the nature of the method to synthesize the weight function, namely the use of logistic regression: it is likely the case that, because logistic regression is only able to synthesize weight function of a certain nature, the synthesis is unable to find a suitable weight function for transition with label `Leucocytes`. Thus, we decided to employ a binary history, namely that only accounts whether a transition has fired or not in the past, independently of the number of times. The configuration using only binary history outperforms the baseline of 6%. However, the results with the configuration with both data and binary history states are similar to those using only data: this could be due to correlations between the features.

To investigate which aspects of the simulation are improved in the best configuration for each case study compared to the baseline, we computed for each activity pair (a_i, a_j), the probability of a_j executing immediately after a_i, given that a_i has been executed. This probability has been computed as the ratio between the number of times a_j immediately follows a_i and the number of times a_i is executed: $|a_i \rightarrow a_j|/|a_i|$. When a given activity $\bar{a}$ is the first of the trace, we have a fictitious previous activity $start$ to indicate this, namely a pair $(start, \bar{a})$. Similarly, we introduce a fictitious activity end. We then compared these probabilities with those from the real event log. Given any immediately follow relation $a_i \rightarrow a_j$, we computed the absolute error as $\left| p_{ij}^{real} - p_{ij}^{sim} \right|$, where p_{ij}^{real} and p_{ij}^{sim} are the probability described above for the real and simulated logs, respectively.

Figures 5 and 6 illustrate the absolute errors for the two cases. In particular, in the Road-Traffic Fines case study, our approach seems to work better when `Insert Fine Notification` and `Add penalty` are executed. Indeed, it can be noticed that the probabilities of `Add penalty`, `Send for Credit Collection` and `end` immediately occurring after `Insert Fine Notification` are more similar to the real ones using our approach, since the error is closer to zero. Similarly for `Add penalty`.

In the Sepsis case, our method is more accurate after the execution of `ER Registration` and `ER Triage`, and in the final activities: looking at the cells (`ER Registration`, `ER Triage`), (`ER Triage`, `ER Sepsis Triage`), and the cells of the `end` column, errors are smaller using data than using simply branching probabilities.

There are some cases in which the error is greater with our approach than with the basic one. This is probably due to the monotonicity of the logistic regression, which cannot control the loop parts of the model. However, the overall error decreases with our method.

5   Related Works

The determination of the activity probabilities in simulation model is related to stochastic process discovery [3,10,15]. However, they do not focus on building simulation models, nor do they verify whether or not their approaches are beneficial to increase the accuracy of business simulation models. In particular, Burke et al. use a similar concept of weights to determine probabilities [3], but, instead of using a more elaborated process state, they only consider the occurrences of previous activities, which our experiments show to not always be beneficial. Mannhardt et al. also use weights to determine probabilities [3], but they only focus on process attributes, which is similarly not always beneficial for more accurate simulation models. This paper brings together the works by Burket et al. and by Mannhardt et al. allowing both process-state abstractions, their combination, and also richer abstraction (e.g., includining resource and time information).

Related work also focuses on Markov-based stochastic models [2,16]. This class of models is, however, unable to feature concurrency, data and silent transitions, which are all crucial aspects of business process models, including when used for simulation.

On the other hand, several studies have been conducted in the simulation domain to discover and improve simulation models. However, none of these works have focused on discovering accurate activity firing probabilities. In [12], Pouarbafrani et al. proposed a framework with a simulation tool for enriched process trees, and in [14], the authors used historical execution data to provide simulation models. However, in these works, the probability of an activity occurring is not dependent on the data states. Camargo et al. proposed hybrid approaches using Deep Learning techniques to generate simulated logs in [6,7], but the control flow depends on constant branching probabilities that do not consider the data and history. In [4,5], Simod has been introduced as a framework for the automated discovery of simulation models, but the activity firing probabilities are again based on branching probabilities. GenCPN [13] was proposed as a tool that extracts process parameters, including constant branching probabilities, directly from an event log and uses Process Mining techniques to convert them into a CPN model. However, these simulators use constant branching probabilities, which are not dependent on process states.

6   Conclusions

Business process simulation is a flexible approach to analyze operational processes, and, when different sorts of issues are observed, to evaluate alternative scenarios wit the aim to overcome them. The advantage of business process simulation is that the evaluation is performed through the digital twin of the actual process, and hence these alternative scenarios are assessed without risking them in real production environments. Of course, the requisite of valuable applications of process simulations is that this digital twin is an accurate representation of the real process. The digital twin is modelled through a process simulation model, which hence needs to be accurate in order to profit from the advantage of simulation.

A process simulation model consists of the actual model of the process to be simulated that is extended with the run-time characterization of the process (case arrival rate, resources and roles, activity durations, etc.). While a large body of research has recently been carried out to create accurate simulation models, an accurate stochastic modelling of activity firings in these models has been totally overlooked (cf. Sect. 5): it is simply assumed that these probabilities can be well determined by looking at the frequency of occurrence of these activities in the log (e.g., the so-called branching probabilities). Unfortunately, this is a coarse determination because the activity firing probabilities usually depend on characteristics of the single process instances that are simulated, and these characteristics might significant change from instance to instance. For example, the chance of executing the activity to reject a loan application might depend on the amount requested and/on the number of further clarification requests.

This paper aims to assess whether our intuition that a more accurate modelling of the activity firing probabilities passes on having them depend on the process state, which models the characteristics of the cases. To answer this question, we compute the weights of activities as function of the process state, where these weights are proportional to the activity firing probability. The technique has been operationalized using stochastic labelled Data Petri nets, a simple language, yet sufficiently expressive to model weights, probabilities, and process states. Each Petri-net transition is associated to a weight function: weights are then computed by evaluate the weight functions on the current process state.

Since the process state is in fact an abstraction, we make it customizable as function of the process to simulate. Indeed, these probabilities are possibly linked to different process characteristics, depending on the process to simulate. We conducted experiments with two different process and five alternative process-state definition to answer our business questions. The results confirm that our weight-based estimation of activity-firing probabilities allowed us to build more accurate simulation models, and that the best process-state abstraction changes with the process to simulate.

Currently, weight functions are synthesized using logistic regression, which imposes a certain shape of the function. This has also caused the livelock problem in simulation discussed in Sect. 4.3, because it was not possible to synthesize a suitable weight function when the process state includes the history. As a future work, we aim to replace the synthesis based on logistic regression with alternative synthesis methods, which possibly can return better approximations of the real weight function. Also, we aim to asses how fitness and precision of the process models influence the quality of the simulations.

Acknowledgment. The research is financially supported by MUR (PNRR) and University of Padua, by the Department of Mathematics of University of Padua, through the BIRD project "Data-driven Business Process Improvement" (code **BIRD215924/21**), and by the "Smart Journey Mining project" funded by the Research Council of Norway (grant no. **312198**).

References

1. van der Aalst, W.M.P., Adriansyah, A., van Dongen, B.F.: Replaying history on process models for conformance checking and performance analysis. WIREs Data Mining Knowl. Discov. **2**(2), 182–192 (2012)
2. Baier, C., Katoen, J.: Principles of Model Checking. MIT Press, Cambridge (2008)
3. Burke, A., Leemans, S.J.J., Wynn, M.T.: Stochastic process discovery by weight estimation. In: Leemans, S., Leopold, H. (eds.) ICPM 2020. LNBIP, vol. 406, pp. 260–272. Springer, Cham (2021). https://doi.org/10.1007/978-3-030-72693-5_20
4. Camargo, M., Dumas, M., González, O.: Automated discovery of business process simulation models from event logs. Decis. Support Syst. **134**, 113284 (2020). https://doi.org/10.1016/j.dss.2020.113284
5. Camargo, M., Dumas, M., Rojas, O.G.: SIMOD: a tool for automated discovery of business process simulation models. In: Proceedings of the Dissertation Award, Doctoral Consortium, and Demonstration Track at BPM 2019 co-located with 17th International Conference on Business Process Management, BPM 2019, Vienna, Austria, 1–6 September 2019. CEUR Workshop Proceedings, vol. 2420, pp. 139–143. CEUR-WS.org (2019). https://ceur-ws.org/Vol-2420/paperDT5.pdf
6. Camargo, M., Dumas, M., Rojas, O.G.: Discovering generative models from event logs: data-driven simulation vs deep learning. PeerJ Comput. Sci. **7**, e577 (2021). https://doi.org/10.7717/peerj-cs.577
7. Camargo, M., Dumas, M., Rojas, O.G.: Learning accurate business process simulation models from event logs via automated process discovery and deep learning. In: Franch, X., Poels, G., Gailly, F., Snoeck, M. (eds.) Advanced Information Systems Engineering - 34th International Conference, CAiSE 2022. LNCS, vol. 13295, pp. 55–71. Springer, Cham (2022). https://doi.org/10.1007/978-3-031-07472-1_4, https://doi.org/10.1007/978-3-031-07472-1_4

8. Leemans, S.J.J., Syring, A.F., van der Aalst, W.M.P.: Earth movers' stochastic conformance checking. In: Hildebrandt, T., van Dongen, B., Röglinger, M., Mendling, J. (eds.) Business Process Management Forum - BPM Forum 2019. LNIBP, vol. 360, pp. 127–143. Springer, Cham (2019). https://doi.org/10.1007/978-3-030-26643-1_8
9. Mannhardt, F.: Multi-perspective process mining. Ph.D. thesis, Mathematics and Computer Science (2018). proefschrift
10. Mannhardt, F., Leemans, S.J.J., Schwanen, C.T., de Leoni, M.: Modelling data-aware stochastic processes - discovery and conformance checking. In: Gomes, L., Lorenz, R. (eds.) Application and Theory of Petri Nets and Concurrency. PETRI NETS 2023. LNCS, vol. 13929, pp. 77–98. Springer, Cham (2023). https://doi.org/10.1007/978-3-031-33620-1_5
11. Mannhardt, F., de Leoni, M., Reijers, H.A.: The multi-perspective process explorer. In: Proceedings of the BPM Demo Session 2015 Co-located with the 13th International Conference on Business Process Management (BPM 2015), Innsbruck, Austria, September 2, 2015. CEUR Workshop Proceedings, vol. 1418, pp. 130–134. CEUR-WS.org (2015). http://ceur-ws.org/Vol-1418/paper27.pdf
12. Pourbafrani, M., van der Aalst, W.M.P.: Interactive process improvement using simulation of enriched process trees. In: Hacid, H., et al. (eds.) Service-Oriented Computing - ICSOC 2021 Workshops, vol. 13236. LNCS, pp. 61–76. Springer International Publishing, Cham (2022). https://doi.org/10.1007/978-3-031-14135-5_5
13. Pourbafrani, M., Balyan, S., Ahmed, M., Chugh, S., van der Aalst, W.M.P.: GenCPN: automatic CPN model generation of processes. In: Proceedings of the ICPM Doctoral Consortium and Demo Track 2021 co-located with 3rd International Conference on Process Mining, ICPM Doctoral Consortium / Demo Track 2021, Eindhoven, The Netherlands, November 2021. CEUR Workshop Proceedings, vol. 3098, pp. 23–24 (2021). http://ceur-ws.org/Vol-3098/demo_192.pdf
14. Pourbafrani, M., van Zelst, S.J., van der Aalst, W.M.P.: Supporting automatic system dynamics model generation for simulation in the context of process mining. In: Abramowicz, W., Klein, G. (eds.) Business Information Systems - BIS 2020. LNBIP, vol. 389, pp. 249–263. Springer, Cham (2020). https://doi.org/10.1007/978-3-030-53337-3_19
15. Rogge-Solti, A., van der Aalst, W.M.P., Weske, M.: Discovering stochastic petri nets with arbitrary delay distributions from event logs. In: Lohmann, N., Song, M., Wohed, P. (eds.) BPM 2013. LNBIP, vol. 171, pp. 15–27. Springer, Cham (2014). https://doi.org/10.1007/978-3-319-06257-0_2
16. Rogge-Solti, A., Weske, M.: Prediction of business process durations using non-Markovian stochastic petri nets. Inf. Syst. **54**, 1–14 (2015). https://doi.org/10.1016/j.is.2015.04.004
17. Witten, I.H., Frank, E., Hall, M.A.: Data Mining: Practical Machine Learning Tools and Techniques, 3rd edn. Morgan Kaufmann, Elsevier (2011)

DyLoPro: Profiling the Dynamics of Event Logs

Brecht Wuyts[1]([⊠])[iD], Hans Weytjens[1][iD], Seppe vanden Broucke[1,2][iD],
and Jochen De Weerdt[1][iD]

[1] LIRIS, Faculty of Economics and Business, KU Leuven, Leuven, Belgium
brecht.wuyts@kuleuven.be
[2] Department of Business Informatics and Operations Management, Ghent
University, Ghent, Belgium

Abstract. Modern business processes are often characterized by continuous change, which can lead to bias in the results of process mining techniques that assume a static process. This bias is caused by concept drift, which can manifest in many forms and affect various process perspectives. Current research on concept drift in process mining has focused on drift detection techniques in the control-flow perspective, with limited capabilities for comprehensive dynamic profiling of event logs. To address this gap, this paper presents the *DyLoPro* framework, a generic approach that facilitates the exploration of event log dynamics over time using *visual analytics*. The framework caters to all types of event logs and allows for the exploration of event log dynamics from various process perspectives, both individually and combined with the performance perspective. Additionally, the framework is accompanied by an efficient and user-friendly Python library, rendering it a valuable instrument for both researchers and practitioners. A case study using large real-life event logs demonstrates the effectiveness of the framework.

Keywords: Process Mining · Event Logs · EDA · Concept Drift · Visual Analytics

1 Introduction

Most Process Mining (PM) techniques are premised on the assumption of a stable underlying process. However, business processes are often subject to continuous change. These changes are referred to as concept drift and can occur in many different forms (sudden, gradual, recurring, incremental) and apply to different perspectives (control-flow, resource, data, performance) [5]. Applying PM on event logs in which this stationarity assumption does not hold, i.e. in which one or more drifts occur in the underlying process, can induce a significant yet oftentimes unnoticed bias in the results, leading to incorrect insights. The impact of drifts in different perspectives on the results of the process mining techniques varies depending on the technique used. *Process discovery* techniques

© The Author(s), under exclusive license to Springer Nature Switzerland AG 2023
C. Di Francescomarino et al. (Eds.): BPM 2023, LNCS 14159, pp. 146–162, 2023.
https://doi.org/10.1007/978-3-031-41620-0_9

generalize control-flow information into one static process model. Any significant control-flow changes over time would result in the process model representing an extensive set of behavior, falsely suggesting that we are dealing with a very flexible process, while in reality merely a series of more rigid process models are contained within the data. Moreover, both *conformance checking* as well as *performance analysis* techniques produce measures (e.g. fitness scores, or waiting times) as computed over an entire event log. Hence, both sets of techniques depict a static image of measures that, in effect, vary and change over time. In this manner, any crucial patterns, trends or changes over time remain largely undetected. Furthermore, event log dynamics are also crucial for *predictive process monitoring (PPM)* techniques, as PPM algorithms typically learn based on case-specific information derived from multiple process perspectives so as to optimize the prediction of outcome, remaining time, or next event(s). Despite being often neglected, it is recommended that prior to training and evaluating a predictive model, a thorough exploratory analysis of the dynamics over time is conducted to uncover potential data leakage, perform feature selection, and identify and account for significant changes and patterns that could affect the validity or usefulness of a train-test split choice, in particular when the preferred out-of-time evaluation setup is chosen.

These examples indicate the importance of properly exploring the dynamics in event logs before applying such PM techniques. However, concept drift detection in PM is largely confined to the detection of sudden drifts in the control-flow perspective [7], and mainly evaluated on artificial data. Even for recent techniques that initiate a widening beyond control-flow, comprehensive *dynamic profiling of event logs* is well beyond their capabilities. To the best of our knowledge, no framework has been developed to comprehensively and efficiently explore the dynamics in an event log over time.

Therefore, this paper introduces *Dynamic Log Profiling (DyLoPro)*, a comprehensive *visual analytics* framework designed to explore event log dynamics over time. *DyLoPro*'s comprehensiveness is achieved through the incorporation of the main process perspectives - the control-flow, data (including resources) and performance, along two orthogonal dimensions of *log concepts* and *representation types*. It incorporates six log concepts to capture all essential information from event logs, including *variants* and *directly-follows relations* for the control-flow perspective, and *categorical and numeric case and event features* for the data perspective. These six log concepts can be represented using five representation types, including four performance-oriented ones (*throughput time, number of events per case, outcome*, and *directly-follows-relations' performance*) and one generic type. With this two-dimensional approach, end users can gain a nuanced and holistic view of event log dynamics, efficiently identifying patterns, temporary or permanent changes, and trends of interest from multiple perspectives. Upon identification, they can further analyze these patterns and trends, ultimately leading to more appropriate application of downstream process mining techniques.

Accordingly, the remainder of this paper is organized as follows. Section 2 discusses related work, followed by preliminaries in Sect. 3. The *DyLoPro* framework is formally introduced in Sect. 4. In Sect. 5, the effectiveness of the framework is assessed on large real-life event logs with the aid of its associated Python library, before concluding the paper in Sect. 6.

2   Related Work

Concept drift in PM is branded as one of the main challenges in the field [1]. Bose et al. [5] were first to propose a concept drift detection technique for event logs. Later techniques primarily concerned the detection of (sudden) drifts in the control-flow perspective, mostly evaluated using synthetic data [7]. More recently, Adams et al. [2] introduce a framework that adds a cause-effect analysis on top of concept drift detection, and thereby enables relating drifts in different perspectives to each other. However, these detection algorithms are still subject to tedious parameter tuning, time-consuming to run on large real-life event logs, and not proven to be robust against noise.

Furthermore, Exploratory Data Analysis (EDA) [10], is widely considered as the first crucial step in any data analysis project. However, partly due to the complex sequential and multi-perspective nature of event logs, EDA in PM is usually carried out in an ad-hoc way [12], if at all, with comprehensiveness difficult to attain without a considerable time and effort investment. On top of that, the EDA phase in PM is often limited to control-flow exploration using interactive process maps, thereby focusing on the most frequent paths and variants and getting a feel for the degree of structuredness. Additionally, PM practitioners often derive some summary statistics and observe the distribution of activities and data attributes. However, these summary statistics and distributions are typically aggregated over the whole log, thereby already incorporating the bias induced by potential concept drift. Visualizations in general however, unveil vital information omitted by summary statistics [6].

Avoiding biased PM results due to a failure to recognize and account for non-stationary effects in event data, can be addressed by means of visualization. One of the few techniques offering the visualization of the dynamics of an event log over time, is the dotted chart technique [8]. While flexibly configurable, the dotted chart is strongly event focused, and therefore lacks capabilities to provide more aggregated insights. Furthermore, Yeshchenko et al. [11] propose a methodology in which a drift detection technique is complemented by visualizations to further explain the detected drifts. However, this technique is restricted to detecting and visualizing drifts in the control-flow perspective. A third relevant technique, implemented in the ProM [4] framework, is the Performance Spectrum Miner (PSM) [3]. This technique specifically focuses on performance visualization of any occurrence of an individual process segment over time. PSM's notions of process segments and the corresponding performances are highly similar to our definitions of *Directly-Follows Relationships (DFRs)* and *DFR Performances (see Definitions 6 and 7)*. However, PSM only focuses on the dynamic profiling of DFR performances, whereas our framework provides four additional ways in which DFR characteristics can be visualized over time. Moreover, we do not limit our method to DFRs, but to many other concepts catering to different perspectives, as illustrated in subsequent chapters.

3   Preliminaries

In PM, *event logs* record the activity executions within a case or instance of a business process. We define an *event* as follows:

Definition 1 (Event). *Let set A denote the universe of possible activity labels. An event is a tuple $e = (a, c, t, (cf_1, cfv_1), \ldots, (cf_{m_1}, cfv_{m_1}), (ef_1, efv_1), \ldots, (ef_{m_2}, efv_{m_2}))$ with $a \in A$ the activity label, c the case ID, t the timestamp, $(cf_1, cfv_1), \ldots, (cf_{m_1}, cfv_{m_1})$ (with $m_1 \geq 0$) the potential case features with their values and $(ef_1, efv_1), \ldots, (ef_{m_2}, efv_{m_2})$ (with $m_2 \geq 0$) the potential event features and their values. All elements comprising the event tuple e can be accessed individually. E.g., $e(cf_1)$ returns the value cfv_1 of case feature cf_1, and $e(a)$ returns the activity label corresponding to that event.*

The complete sequence of events logged for one particular *case* forms a *trace*. The terms *case* and *trace* will be used interchangeably.

Definition 2 (Trace). *A trace is a non-empty sequence $\sigma = [e_1, \ldots, e_n]$ such that:*

- *All events within the same trace share the same case ID: $\forall i, j \in [1 \ldots n]$ $e_i(c) = e_j(c)$*
- *Events in a trace are ordered chronologically: $\forall e_i, e_j \in \sigma : i < j \Rightarrow e_i(t) \leq e_j(t)$*
- *All events within the same trace share the same value for all case features (if any): $\forall i, j \in [1 \ldots n]; \forall \alpha \in [1, \ldots, m_1] : e_i(cf_\alpha) = e_j(cf_\alpha)$*

Furthermore, we introduce the following trace-level projection functions:

- *$|\sigma|$ $(= n)$: the function returning the case length (in number of events).*
- *$\sigma(cf_\alpha) = cfv_\alpha$ $(\alpha = 1, \ldots, m_1)$: the function returning the constant value of case feature cf_α for a trace σ.*
- *$\sigma(ef_\beta) = [efv_{\beta,1}, \ldots, efv_{\beta,n}]$ $(\beta = 1, \ldots, m_2)$: the function returning the vector of values of event feature ef_β that occurred in the events of a trace σ. Consequently, $[efv_{\beta,1}, \ldots, efv_{\beta,n}] = [e_1(ef_\beta), \ldots, e_n(ef_\beta)]$.*

An *event log* can then be defined as a collection or set of *traces* recording executions of completed cases.

Definition 3 (Event Log). *An Event Log L is a set of traces describing completed cases. Formally: $L = \{\sigma_i | 1 \leq i \leq |L|\}$. The number of traces in L as is indicated by $|L|$, and the time interval over which the traces in the event log are recorded is denoted by $T(L) = [t^{min} : t^{max}]$, with t^{min} and t^{max} being the earliest and latest recorded timestamp over all events in the log.*

Other meaningful case features can be derived based on the information available in the event log. One such informative performance measure is the *throughput time*.

Definition 4 (Throughput Time). *The throughput time of a trace $\sigma = [e_1, \ldots, e_{|\sigma|}]$ corresponds to the total time that a case was processed. This can be formally expressed as $tt(\sigma) = e_{|\sigma|}(t) - e_1(t)$ and can be expressed in every preferred time unit.*

Furthermore, for each trace, the ordered sequence of activity labels can be derived. In most processes, often many different permutations of these sequences can be executed to complete one particular case. Each of those sequences is called a *(control-flow) variant*.

Definition 5 (Set of (Control-flow) Variants). *Given a an event log L, there exists a set $Var(L)$ of unique variants v such that:*

$$\forall\ \sigma = [e_1, \ldots, e_n] \in L : \exists!\ v = [v_1, \ldots, v_n] \in Var(L)\ s.t.\ \forall i \in [1 \ldots n] :$$
$e_i(a) = v_i$. *Furthermore, let $var(\sigma) = v\ (= [e_1(a), \ldots, e_n(a)])$ be the function that maps a trace σ to its corresponding variant v.*

Hence, a control-flow variant $v = [v_1, \ldots, v_n]$ *(with $v_i \in A, \forall i \in [0, \ldots, n]$)* is simply a sequence of activity labels.

Another concept that relates to the control-flow information, is the concept of *Directly-Follows Relations (DFRs)*. More specifically, two activities belonging to events of the same trace, are said to be in a directly-follows relation if the second activity directly succeeds the first. Formally:

Definition 6 (Directly-Follows Relation (DFR)). *Given a trace $\sigma\ (\in L)$, then two activities x and y $(x, y \in var(\sigma)$ are said to be in a directly-follows relationship $(x, y) \iff \exists e_i, e_j (\in \sigma) : e_i(a) = x, e_j(a) = y \wedge j = i + 1$.*

Furthermore, since every trace is an ordered sequence of events, the total number of DFRs present in a given trace equals $|\sigma| - 1$. Additionally, let us define:

- *The list of DFRs present in a given trace σ as: $dfr(\sigma) = [(a_1, a_2), (a_2, a_3), \ldots (a_{n-1}, a_n)]$ (with $\forall i \in [1, \ldots, n] : a_i = e_i(a)$).*
- *The number of times that a given dfr (x, y) occurs in trace σ as $n^{(x,y)}(\sigma)$. In case of rework and/or loops in the process, this could be greater than 1.*

The DFR performance of a certain DFR (x, y) in a given trace, can then be defined as the time elapsed between the completion time of activity x and the completion time of activity y.

Definition 7 (DFR Performance). *Given a DFR (x, y) of a trace $\sigma\ (\in L)$, i.e. $(x, y) \in dfr(\sigma)$ as defined in Definition 6, the DFR performance of (x, y) in that particular trace σ can then be defined as $dfr_{perf}((x, y), \sigma) = e_y(t) - e_x(t)$, with $e_x, e_y \in \sigma$ being the events corresponding to the occurrence of activity x and y respectively. Since each DFR can occur multiple times within a given trace, it is important to note that the projection function dfr_{perf} can yield more than one performance measure for the same trace. I.e. given that $(x, y) \in dfr(\sigma)$, $dfr_{perf} : ((x, y), \sigma) \mapsto \mathbb{R}^\alpha$ with $\alpha \geq 1$.*

Analyzing the performances of certain (important) directly-follows relations might for example reveal certain bottlenecks in the process.

4   The *DyLoPro* Framework

The *DyLoPro* framework consists of three main stages to construct and visualize time series that characterize event log dynamics: (1) log discretization, (2) domain definition, and (3) time series construction & visualization.

4.1   Stage 1: Log Discretization

First, an event log L should be discretized into a chronologically ordered set of sublogs $D(L) = \{L_1, \ldots, L_{|T|}\}$. Hereto, we start by splitting up the time interval $T(L)$ into smaller equal-length intervals as follows:

Definition 8 (Set of Time Intervals). *The chronologically ordered set of equal-length time intervals T is defined as:*
$$T = \{t_1^p, \ldots, t_{|T|}^p\} \; s.t. \; \forall i,j \in [1, \ldots, |T|], i < j \; : \; t_i^{p,end} < t_j^{p,start}, \; \forall i \in [1, \ldots, |T|] : t_i^p \subseteq T(L),$$

with p the *interval length*, and $|T|$ the number of time intervals created.

Secondly, based on the ordered set of intervals T, each trace has to be assigned to exactly one interval so as to create the *log discretization* $D(L)$ of ordered sublogs. However, since traces might not be fully contained within one single subinterval t^p, there are multiple options for assigning a trace to an interval, e.g. based on the timestamp of the first event, or last event, or based on the interval in which most events of a trace occur. For the sake of simplicity, we will assume that cases are assigned to the time period that contains the timestamp of a trace's first event. However, in the implementation, other assignment conditions can be chosen.

Definition 9 (Set of Sublogs). *Given event log L and its set of time intervals T, then the log discretization produces a set of chronologically ordered sublogs $D(L) = \{L_1, \ldots, L_{|T|}\}$, with each sublog corresponding to a certain time period, such that:*

- *Each sublog L_i contains all the traces that were initialized in time period t_i^p: $\forall i \in [1, \ldots, |T|] : L_i = \{\sigma \in L | e_1^\sigma(t) \in t_i^p\}$.*
- *Each trace $\sigma \in L$ is assigned to exactly one sublog L_i: $\forall \sigma \in L : \exists! \, L_i \in D(L)$ s.t. $\sigma \in L_i$.*

For example, an event log covering a one year time interval $T(L)$ and with the subinterval length p^1 set to one week, will be *discretized* in a chronologically ordered set of 52 sublogs $D(L) = \{L_1, \ldots, L_{52}\}$.

4.2   Stage 2: Domain Definition

Secondly, after log discretization, we need to define how to capture and represent event log dynamics, referred to as *domain definition*. This entails the specification of *log concepts*, i.e. the main dimensions along which we capture event log dynamics, and *representation types*, i.e. how the event log dynamics should be represented and analyzed for each log concept. Concretely, each *log concept* -

---

[1] What a good value for p constitutes, depends on the arrival frequency of cases. Each resulting sublog should be populated enough such that the derived measures are representative, but not too populated, as aggregating over a too long period and/or over too many cases could level out interesting trends and patterns.

representation type combination will translate itself into a domain-specific *mapping function*, mapping each of the sublogs onto a real-valued vector. Table 1 presents an extensive but non-exhaustive set of log concepts and representation types, along with their respective mapping functions. A mapping function can be formally defined as:

Definition 10 (Mapping Function g). *Let L_i be a sublog, let C^o be the selected log concept, let R^t be the selected representation type of C^o and let θ^{C^o} be the set of concept-specific configuration parameters. The mapping function g then maps the sublog L_i to a set of real-valued vectors M, with the exact implementation of $g\left(L_i, (C^o, R^t), \theta^{C^o}\right)$ depending on concept C^o, the concept-specific configuration parameters θ^{C^o}, and representation type R^t. The θ^{C^o} is omitted for readability. This can be formally expressed as $g\left(L_i, (C^o, R^t)\right) = M$, with $M = \{M_1, \ldots, M_\xi\}$ $(\xi \geq 1)^2$ and $\forall i \in [1, \ldots, \xi] : M_i \in \mathbb{R}^K$ (with $K \geq 1)^3$.*

For each log concept C^o, regardless of the *representation type* R^t, a specific set of one or more configuration parameters θ^{C^o} is provided. These configuration parameters denote all configuration options on top of the domain definition that provide the practitioner with additional flexibility, and include, i.a., choosing *aggregation operator* '$Agg(X)$' and K dimensions. More specifically, most functions in Table 1 include an *aggregation operator* '$Agg(X)$', which aggregates over a set of values X derived from a sublog and returns a single number. A sensible default implementation of '$Agg(X)$' is the mean, but could e.g. also be the median, minimum, maximum and so on. Additionally, also the dimensionality K as discussed in Definition 10 has to be determined. The exact realization of these dimensions depends on C^o, and will be explained when discussing each of the concepts. Now that we have introduced the dimensionality K, we can rewrite mapping function $g\left(L_i, (C^o, R^t)\right)$ as $[g_1\left(L_i, (C^o, R^t)\right), \ldots, g_K\left(L_i, (C^o, R^t)\right)]$, in which each $g_k\left(L_i, (C^o, R^t)\right)$ maps to a set of one or more scalars $M_k = \{M_{1,k}, \ldots, M_{\xi,k}\}$ $(\xi \geq 1; \forall k \in [1, \ldots, K])$. These are also the functions displayed in Table 1. To retrieve the set of real-valued vectors M as defined in Definition 10, one simply has to apply $g_k\left(L_i, (C^o, R^t)\right)$ for each of the selected dimensions k. Figure 1 illustrates how mapping function $g_k\left(L_i, (C^o, R^t)\right)$ maps the fourth sublog L_4 to two scalars $M_k = \{M_{1,k}, M_{2,k}\}$. As discussed in Sect. 4.3, applying $g_k\left(L_i, (C^o, R^t)\right)$ to each consecutive sublog $L_1, \ldots, L_{|\mathcal{T}|}$ allows us to construct and visualize the two corresponding time series.

Log Concepts

Variants. The *Variants* log concept focuses on individual variants as defined in Definition 5. Variants are one way to conceptualize the control-flow perspective.

---

[2] For our proposed set of mapping functions displayed in Table 1, ξ is either 1 or 2. However, this framework is rather meant as a starting point, and its verbosity can be extended at the discretion of the user.

[3] The dimensionality K, which is constant for every vector M_i $(i \in [1, \ldots, \xi])$, is completely determined by the concept-specific configuration parameters θ^{C^o}. We do not include this in the notation for simplicity.

Table 1. *Domain Definition and their respective mapping functions:* Overview of the proposed Log Concepts C^o, together with their Representation Types R^t. For each (C^o, R^t) combination, a dedicated function g_k $(L_i, (C^o, R^t))$ that maps a sublog L_i to a set of 1 or more real-valued numbers M_k, is proposed. For each (C^o, R^t)-combination, the resulting set of vectors M_k is provided in the cells below. Non-compatible (C^o, R^t)-combinations are indicated by $\emptyset$.

Log Concept (C^o)		Representation Type (R^t)				
		Isolated (ISO)	Throughput Time (TT)	Case Length (NEPC)[1]	Outcome (OUT)	DFR Performance (DFRP)
1	Variants	$\frac{|\{\sigma \in L_i | var(\sigma)=v_k\}|}{|L_i|}$	$\{Agg(\{tt(\sigma)|\sigma \in L_i \wedge var(\sigma)=v_k\}),$ $Agg(\{tt(\sigma)|\sigma \in L_i \wedge var(\sigma) \neq v_k\})\}$	$\emptyset$	$\{\frac{|\{\sigma \in L_i | var(\sigma)=v_k \wedge \sigma(out)=+\}|}{|\{\sigma \in L_i | var(\sigma)=v_k\}|},$ $\frac{|\{\sigma \in L_i | var(\sigma) \neq v_k \wedge \sigma(out)=+\}|}{|\{\sigma \in L_i | var(\sigma) \neq v_k\}|}\}$	$\emptyset$
2	Directly-Follows Relations (DFRs)	$\{\frac{|\{\sigma \in L_i | dfr_k \in dfr(\sigma)\}|}{|L_i|}$ $Agg(\{\tau_n^{dfr_k}(\sigma)|\sigma \in L_i\})\}$	$\{Agg(\{tt(\sigma)|\sigma \in L_i \wedge dfr_k \in dfr(\sigma)\}),$ $Agg(\{tt(\sigma)|\sigma \in L_i \wedge dfr_k \notin dfr(\sigma)\})\}$	$\{Agg(\{|\sigma||\sigma \in L_i \wedge dfr_k \in dfr(\sigma)\}),$ $Agg(\{|\sigma||\sigma \in L_i \wedge dfr_k \notin dfr(\sigma)\})\}$	$\{\frac{|\{\sigma \in L_i | dfr_k \in dfr(\sigma) \wedge \sigma(out)=+\}|}{|\{\sigma \in L_i | dfr_k \in dfr(\sigma)\}|},$ $\frac{|\{\sigma \in L_i | dfr_k \notin dfr(\sigma) \wedge \sigma(out)=+\}|}{|\{\sigma \in L_i | dfr_k \notin dfr(\sigma)\}|}\}$	$Agg(\{dfr_{perf}(dfr_k, \sigma)|\sigma \in L_i\})$
3	Categorical Case Feature	$\frac{|\{\sigma \in L_i | \sigma(cf)=l_k\}|}{|L_i|}$	$\{Agg(\{tt(\sigma)|\sigma \in L_i \wedge \sigma(cf)=l_k\}),$ $Agg(\{tt(\sigma)|\sigma \in L_i \wedge \sigma(cf) \neq l_k\})\}$	$\{Agg(\{|\sigma||\sigma \in L_i \wedge \sigma(cf)=l_k\}),$ $Agg(\{|\sigma||\sigma \in L_i \wedge \sigma(cf) \neq l_k\})\}$	$\{\frac{|\{\sigma \in L_i | \sigma(cf)=l_k \wedge \sigma(out)=+\}|}{|\{\sigma \in L_i | \sigma(cf)=l_k\}|},$ $\frac{|\{\sigma \in L_i | \sigma(cf) \neq l_k \wedge \sigma(out)=+\}|}{|\{\sigma \in L_i | \sigma(cf) \neq l_k\}|}\}$	$\emptyset$
4	Numerical Case Features	$Agg(\{\sigma(cf)|\sigma \in L_i\})$	$Agg(\{\frac{tt(\sigma)}{\sigma(cf)}|\sigma \in L_i\})$	$Agg(\{\frac{|\sigma|}{\sigma(cf)}|\sigma \in L_i\})$	$\{Agg(\{\sigma(cf)|\sigma \in L_i \wedge \sigma(out)=+\}),$ $Agg(\{\sigma(cf)|\sigma \in L_i \wedge \sigma(out)=-\})\}$	$\emptyset$
5	Categorical Event Feature	$\frac{|\{\sigma \in L_i | l_k \in \sigma(ef)\}|}{|L_i|}$	$\{Agg(\{tt(\sigma)|\sigma \in L_i \wedge l_k \in \sigma(ef)\}),$ $Agg(\{tt(\sigma)|\sigma \in L_i \wedge l_k \notin \sigma(ef)\})\}$	$\{Agg(\{|\sigma||\sigma \in L_i \wedge l_k \in \sigma(ef)\}),$ $Agg(\{|\sigma||\sigma \in L_i \wedge l_k \notin \sigma(ef)\})\}$	$\{\frac{|\{\sigma \in L_i | l_k \in \sigma(ef) \wedge \sigma(out)=+\}|}{|\{\sigma \in L_i | l_k \in \sigma(ef)\}|},$ $\frac{|\{\sigma \in L_i | l_k \notin \sigma(ef) \wedge \sigma(out)=+\}|}{|\{\sigma \in L_i | l_k \notin \sigma(ef)\}|}\}$	$\emptyset$
6	Numerical Event Features	$Agg(\{\overline{\sigma(ef)}|\sigma \in L_i\})$	$Agg(\{\frac{tt(\sigma)}{\overline{\sigma(ef)}}|\sigma \in L_i\})$	$Agg(\{\frac{|\sigma|}{\overline{\sigma(ef)}}|\sigma \in L_i\})$	$\{Agg(\{\overline{\sigma(ef)}|\sigma \in L_i \wedge \sigma(out)=+\}),$ $Agg(\{\overline{\sigma(ef)}|\sigma \in L_i \wedge \sigma(out)=-\})\}$	$\emptyset$

[1] *NEPC* stands for *number of events per case.*

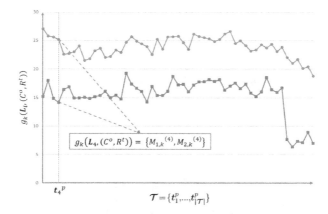

Fig. 1. Visual illustration of the mapping function for one particular dimension k ($\in [1, \ldots, K]$). After having constructed a *log discretization* and having defined the *domain definition*, the associated mapping function is applied to every sublog, thereby yielding one or more time series. In this illustration, $g_k\left(L_i, (C^o, R^t)\right)$ results in $\xi = 2$ measures, and hence two time series are ultimately constructed and visualized.

The dimensions that still need to be selected are the K variants that are to be analyzed, which we denote by $[v_1, \ldots, v_K]$. A good starting point would be to analyze the K most frequently occurring variants in the whole event log L. As shown in Table 1, three different representation types can be used to represent the *Variants* concept.

Directly-Follows Relations. The *Directly-Follows Relations* log concept is concerned with individual DFRs as defined in Definition 6. DFRs, like variants, can be utilized to conceptualize the control-flow perspective. While variants describe complete process executions, DFRs allow for a more granular analysis of the control-flow perspective by focusing on small subsegments comprising of two consecutive activities in process executions. Furthermore, the K dimensions correspond to the K DFRs to be selected by the user, which we denote by $[dfr_1, \ldots, dfr_K]$. A good initial selection strategy could be to select the most important successions of two activities, like for example the K DFRs with the largest number of occurrences in event log L. As shown in Table 1, five different representation types can be used to represent the *Directly-Follows Relations* concept.

Categorical Case Features. Assuming that one or more categorical case features are present in the event log L, and that the practitioner has chosen one particular categorical case feature cf to explore, the *Categorical Case Feature* log concept focuses on the individual categories, also known as levels, of cf. Let us first denote the cardinality of cf as κ. Furthermore, the user might opt for selecting a subset of levels to analyze, or analyzing all levels at once. The K dimensions to be selected correspond to the K ($K \leq \kappa$) levels of cf ultimately chosen by the user, which we define as $[l_1, \ldots, l_K]$. As illustrated in Table 1, the *Categorical Case Feature* log concept can be represented by means of four different representation types.

Numerical Case Features. Assuming the existence of one or more numerical case features in an event log L, the *Numerical Case Features* log concept facilitates the exploration of the dynamics of each individual numerical feature in a concurrent manner. If multiple numerical case features exist, one can choose to explore all of these separate features at once, or only a subset of them. The set of K numerical case features ultimately selected for further analysis is denoted by $[cf_1, \ldots, cf_K]$. Table 1 displays the four different ways in which the *Numerical Case Features* concept can be represented.

Categorical Event Features. Also event features can be categorical. The difference with categorical case features, is that categorical event features can take on different values over the different events belonging to the same trace. For this reason, although highly similar, a separate concept is dedicated to categorical event features. Assuming that one or more categorical event features are present in the event log L, and assuming the user has chosen one particular event feature ef to explore, the *Categorical Event Feature* log concept focuses on the individual levels of ef. Let us first also denote the cardinality of categorical event feature ef as κ. Furthermore, the user might opt for analyzing a subset of levels, or analyzing all levels at once. This corresponds to selecting the K dimensions for the *Categorical Event Feature* concept. The K ($K \leq \kappa$) levels of ef ultimately chosen, is defined as $[l_1, \ldots, l_K]$. As illustrated in Table 1, the *Categorical Event Feature* log concept can be represented by means of four different representation types. Furthermore, since we regard resource information as a categorical event feature, this concept can be utilized to conceptualize both the data and resource perspective. Additionally, also the activity labels can be regarded as a categorical event feature. Therefore, this concept can also be leveraged to examine the dynamics of specific activities over time.

Numerical Event Features. As discussed in Definition 2, the function $\sigma(ef)$ gives us the vector $[efv_1, \ldots, efv_{|\sigma|}]$ which contains all values of numeric event feature ef that occurred in the events of trace σ. The *DyLoPro* framework primarily aims to chart the evolution of trace-level characteristics over time. Consequently, an additional *abstraction method* is needed to project a trace's sequence of numeric event feature values onto a single numeric value. We formally denote this *abstraction operation* as $\overline{\sigma(ef)} : \sigma(ef) \mapsto \mathbb{R}$. Depending on the nature of ef, different abstraction operations can be used, like e.g. taking *the last value*, the *average*, the *minimum*, *maximum* or the *sum* (of all non-null values). After having chosen an abstraction operation, the *Numeric Event Features* concept becomes equivalent with the *Numeric Case Features* concept *(see supra)*.

Now that we have established the concepts, as well as the embodiment of the K dimensions for each of these concepts, we will discuss the five proposed representation types, together with their implementation for the different concepts. Before doing so, it should also be noted that for some of the *log concepts*, each sublog L_i can be further subdivided into two distinct subgroups of cases, for each of the selected dimensions k ($\forall\ k \in [1, \ldots, K]$). In particular, for the *Variants* and *Categorical Case Feature* concepts, a sublog can be further subdivided into cases pertaining to respectively a certain variant v_k or categorical level l_k on the one hand, and all remaining cases on the other hand. Similarly, for the *Directly-Follows Relations* and *Categorical Event Feature* concepts, each sublog can be further subdivided into cases containing at least one occurrence of a certain DFR dfr_k or categorical level l_k, vs. all other cases.

Representation Types

Isolated. The *Isolated* representation type focuses on the evolution of each selected dimension k ($\forall k \in [1, \ldots, K]$). The associated mapping functions, for the *Numerical Case and Event Feature* concepts, map each sublog L_i to the an aggregate measure of that feature, e.g. the mean. For the other four concepts, each sublog is mapped to one[4] measure: the relative fraction of cases corresponding to that dimension, or containing that dimension at least once.

Throughput Time. The *Throughput Time* representation type establishes a link between each concept-specific dimension k ($\forall k \in [1, \ldots, K]$) and the throughput time of cases. The associated mapping functions, for the *Numerical Case and Event Feature* concepts, map each sublog L_i to an aggregated measure of the ratio of units of throughput time needed per unit of that feature. For the other four concepts, each dimension k can be utilized to subdivide the cases of a sublog L_i into two groups. Accordingly, given a sublog, two throughput time measures are computed for each dimension k by computing the aggregate throughput time over the cases for both subgroups. By doing so and by comparing these two throughput time aggregations, possible effects of each dimension on the cases' throughput times can be identified.

Case Length. The *Case Length* representation type establishes a link between each concept-specific dimension k ($\forall k \in [1, \ldots, K]$) and the case length *in number of events per case)*. The associated mapping functions, for the *Numerical Case and Event Feature* concepts, map each sublog L_i to an aggregated measure of the ratio of case length per unit of that feature. The *Case Length* representation type is not applicable to the *Variants* concept, since the case length for a certain variant remains constant. For the other three concepts, we again first subdivide a sublog's cases into two groups for each dimension k as discussed earlier. Consequently, given a sublog, the aggregate case length is computed for both groups of cases. By doing so and by comparing these two case length aggregations, any distinctive relation between a dimension k ($\forall k \in [1, \ldots, K]$) and the number of process steps needed to complete a case can be examined.

Outcome. This paper makes the simplifying assumption of case outcomes being binary. However, the framework can easily be extended to cater for higher dimensional outcomes too. If such an outcome target is present in the event log, then

---

[4] For the *DFRs* concept, two measures are computed. One giving the relative fraction of cases in which dfr_k occurs at least once, and another one giving the aggregated amount of occurrences per case.

possible correlations between a concept-specific dimension k ($\forall k \in [1, \ldots, K]$) and the outcome of a case can be analyzed by means of the *Outcome* representation type. For the *Numerical Case and Event Feature* concepts, this is realized by means of mapping each sublog to the two following measures: the aggregated numerical feature value for cases with a positive outcome, and the aggregated feature value for cases with a negative outcome. For the other four concepts, again after first having subdivided a sublog $\boldsymbol{L_i}$ into two groups of cases for each dimension k *(see supra)*, this is realized by computing the fraction of positives for both of these groups.

DFR Performance. Finally, the *DFR Performance* representation type is only applicable to the *Directly-Follows Relations* concept. Instead of linking the presence or absence of a certain dfr_k ($\forall k \in [1, \ldots, K]$) to trace-level performance characteristics *(such as the throughput time, case length and outcome)*, here, we will focus on a performance measure at the level of individual DFRs themselves, namely the *DFR Performance* measure *(Definition 6)*. As such, for each dfr_k ($k \in [1, \ldots, K]$), a sublog $\boldsymbol{L_i}$ is mapped to an aggregate of DFR Performance for dfr_k, over all its occurrences in $\boldsymbol{L_i}$.

4.3   Stage 3: Time Series Construction and Visualization

After having defined a *log discretization* in the first stage, and subsequently having defined the process domain in the second stage, a *time series construction* procedure is performed. This basically boils down to applying the mapping function resulting from the *domain definition* on each of the chronologically ordered sublogs part of the *log discretization*.

Definition 11 (Time Series Construction). *Given an event log $\boldsymbol{L}$, a corresponding log discretization $D(\boldsymbol{L}) = \{\boldsymbol{L_1}, \ldots, \boldsymbol{L_{|\mathcal{T}|}}\}$, the domain-specific mapping function $g_k(\boldsymbol{L_i}, (C^o, R^t))$, and the chosen dimensionality K, then for each dimension k ($\in [1, \ldots, K]$), a uni- or multivariate time series $\boldsymbol{TS_k}$ can be constructed as follows:*

$$\forall k \in [1, \ldots, K] : \boldsymbol{TS_k} = [g_k(\boldsymbol{L_1}, (C^o, R^t)), \ldots, g_k(\boldsymbol{L_{|\mathcal{T}|}}, (C^o, R^t))]$$

Subsequently visualizing the constructed uni- or multivariate time series $\boldsymbol{TS_k}$ ($\forall k \in [1, \ldots, K]$) allows for easily identifying any interesting patterns, changes or trends over time. This was already graphically illustrated in Fig. 1. By applying the mapping function, which returns two measures $\boldsymbol{M_k} = \{M_{1,k}, M_{2,k}\}$, on each of the consecutive sublogs, a multivariate ($\xi = 2$) time series comprised of two univariate ones, is constructed and visualized. This example could e.g. correspond to the visualization of the *throughput time* dynamics for a certain *categorical case feature* level l_k, and hence uncover the interesting pattern of cases pertaining to l_k continuously having a significantly larger throughput time compared to other cases. In addition, a clear decline in the throughput time of cases pertaining to l_k, and an even sharper decline in the throughput time of all other cases, can be observed towards the end of the event log.

5 Experimental Evaluation

The visualization capabilities offered by the framework are implemented in the associated *DyLoPro* Python library. We evaluated the *DyLoPro* framework and library by means of conducting an extensive analysis on a number of commonly used real-life event logs[5]. In the remainder of this section, we provide and discuss extracts of an extensive analysis of two large real-life event logs, and thereby demonstrate the effectiveness of our framework in identifying interesting patterns prior to the application of other PM techniques.

The BPIC17[6] and BPIC19[7] event logs are two of these large real-life event logs used to evaluate and benchmark a variety of PM techniques in the literature. Thoroughly examining the dynamics in these two event logs prior to applying PM on them, is however hardly done. By applying the *DyLoPro* framework, by means of the identically named Python library, on both event logs, we uncovered a number of interesting patterns. In what follows, we provide and discuss extracts from these analyses.

The *BPIC17* event log records cases of a loan application process of a Dutch financial institute, and consists of 31,413 cases and 1,198,366 events. The log has, inter alia, been extensively utilized to evaluate PPM algorithms for various prediction targets. We shift our focus towards the use of BPIC17 for outcome prediction . More specifically, a loan application can either be accepted, refused or canceled. In the literature, this multi-class classification problem has been broken down into three separate binary classification tasks [9]. For each of these three tasks, the *Outcome* representation type can be leveraged to examine any outcome-related patterns, and the stability of these patterns, over time. Our extensive dynamic profiling of the BPIC17 event log unearthed multiple interesting patterns, but among them, one pattern in particular stands out as especially significant, and is displayed in Fig. 2. There you can see that the numerical event feature *'CreditScore'* has a weekly average of 0 for cases being refused, and a weekly average fluctuating around 500 for cases not being refused by the bank. As this pattern might indicate data leakage, further analysis was conducted. This pointed out that indeed 3,719 out of 3,720 'positively' labeled cases only had a value of 0 for the numeric event feature *'CreditScore'*. This pattern could potentially signify leakage, and therefore warrants further investigation before including it as an input in future research.

The *BPIC19* event log pertains to a purchase handling process of a large Dutch multinational active in the area of coatings and paints. It contains 251,470 cases and 1,587,925 events. Our comprehensive visual analysis has unearthed some remarkable patterns. In particular, the mean weekly throughput time, which fluctuates around 80 days during the initial eight months, suddenly starts decreasing gradually around 09/2018 and converges to zero towards the end of

---

[5] Annotated notebooks with the most interesting visualizations for each event log can be found here: https://github.com/BrechtWts/DyLoPro_CaseStudies.

[6] Data: https://doi.org/10.4121/uuid:5f3067df-f10b-45da-b98b-86ae4c7a310b.

[7] Data: https://doi.org/10.4121/uuid:d06aff4b-79f0-45e6-8ec8-e19730c248f1.

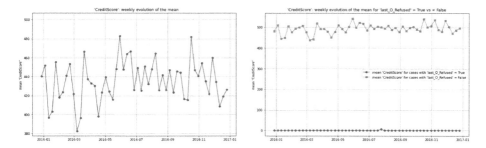

Fig. 2. *DyLoPro* library - BPIC17 event log: Weekly evolution of the mean value for the numeric event feature *'CreditScore' (left)* and its mean value for cases with a positive outcome vs. cases with a negative outcome *(right)*.

the event log. Additionally, we observe a significant and sudden drift in the control-flow perspective around this exact same point in time *(09/2018)*. This is shown in Fig. 3. The left panel contains the weekly amount of cases initialized on its left axis, and the weekly average throughput time on its right axis. The right panel displays the evolution of the weekly relative fraction of cases accounted for by the six most occurring variants. The fraction of cases pertaining to the most frequently occurring variant *(variant 1)* initially remained relatively stable, fluctuating around the range of 25–30% of cases, but suddenly incurred a sudden and steep decline around 09/2018. Similar changes can be observed for the second, fourth and fifth most occurring variants. In contrast, the sixth most frequent variant *(variant 6)* goes from being almost non-existent to being the 'main supplier' of cases around the exact same point in time. The potential inclusion of incomplete cases towards the end of the log would have been a sensible explanation for the observed control-flow and performance drifts. However, the specific sequence of activities in variant 6, i.e. *(Create Purchase Requisition Item, Create Purchase Order Item, Vendor creates invoice, Record Goods Receipt, Record Invoice Receipt, Clear Invoice)*, the fact that activity *'Clear Invoice'* is also the ending activity for four of the five most common variants, the similarity between variants 1 and 6, and the average case length amounting to 6.3 events per case all make this explanation unlikely.

Additionally, Fig. 4 displays multiple time series regarding the two most frequent levels of categorical case feature *'case:Spend area text'*, namely *'Packaging'* and *'Sales'*, and thereby illustrates how *DyLoPro* can also be used to uncover interesting relations between the data perspective and the performance perspective. On average, purchase orders (POs) pertaining to the *'Packaging'* level have a significantly higher throughput time compared to all other cases. For POs pertaining to the *'Sales'* level, the opposite holds. Both patterns initially remain relatively stable over time, but also break down around this exact same point in time, *2018-09*. Similar patterns can be observed for other levels, both for the

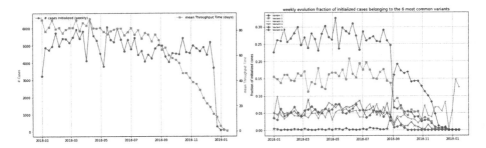

Fig. 3. *DyLoPro* library - BPIC19 event log: Weekly evolution of the number initialized cases and mean throughput time *(left)*, and of the fraction of total cases belonging to the 6 most occurring variants *(right)*.

same categorical case feature, as for others present in BPIC19. These findings illustrate that at least a certain degree of cautiousness is recommended when evaluating and comparing PM techniques on the BPIC19.

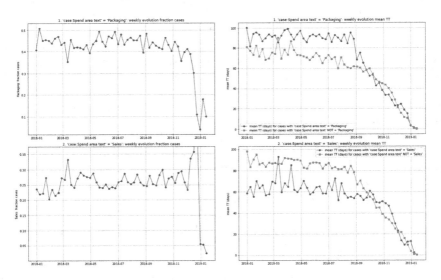

Fig. 4. *DyLoPro* library - BPIC19: Dynamics two most occurring levels of categorical case feature '*case:Spend area text*'. Relative fraction of cases pertaining to each level *(left)* and the mean throughput time of cases belonging to each level, vs. to all other cases *(right)*.

6   Discussion and Future Direction

In this paper, we proposed the *DyLoPro* framework, complemented with the *DyLoPro* library, enabling PM researchers and practitioners to efficiently and comprehensively analyze the dynamics in (real-life) event logs over time. First of all, the framework starts with subdividing the event log into a chronologically ordered set of sublogs. Afterwards, guided by the process perspective of interest, the log domain has to be defined, which in its turn consists of choosing a way to conceptualize the log, and subsequently choosing a way in which this log concept should be represented. The chosen log domain will determine the mapping function used to quantify each of the sublogs. In the third and final stage, time series are constructed and subsequently visualized by applying this mapping function on each of the chronologically ordered sublogs. Furthermore, we briefly demonstrated *DyLoPro* by uncovering two interesting patterns in two often-used public event logs, the BPIC17 and the BPIC19 event logs. We have also applied the *DyLoPro* framework and library on other frequently used real-life event logs, as well as conducted a more elaborate analysis on the BPIC17 and BPIC19 logs. These results are documented in annotated notebooks and can be found here: https://github.com/BrechtWts/DyLoPro_CaseStudies. These first use cases already indicate the potential of this framework and library in terms of enabling PM researchers and practitioners to efficiently and comprehensively explore the dynamics in event logs, prior to applying PM techniques on them. Thereby, biased results because of the stationarity assumption of PM techniques being violated, can be avoided. Additionally, complementing the results of PM with these visuals facilitates interpreting these results.

Future Direction. The framework and proposed mapping functions *(Table 1)* are not meant to be exhaustive, and can be extended in both dimensions. Accordingly, in the near future, we are planning on extending the framework's and corresponding Python Library's capabilities. Moreover, fostering collaboration within the PM community, requests with enhancements to the open source *DyLoPro* Python library will be monitored and accepted if successfully tested.

References

1. van der Aalst, W., et al.: Process mining manifesto. In: Daniel, F., Barkaoui, K., Dustdar, S. (eds.) BPM 2011. LNBIP, vol. 99, pp. 169–194. Springer, Heidelberg (2012). https://doi.org/10.1007/978-3-642-28108-2_19
2. Adams, J.N., van Zelst, S.J., Quack, L., Hausmann, K., van der Aalst, W.M.P., Rose, T.: A framework for explainable concept drift detection in process mining. In: Polyvyanyy, A., Wynn, M.T., Van Looy, A., Reichert, M. (eds.) BPM 2021. LNCS, vol. 12875, pp. 400–416. Springer, Cham (2021). https://doi.org/10.1007/978-3-030-85469-0_25
3. Denisov, V., Fahland, D., van der Aalst, W.M.P.: Unbiased, fine-grained description of processes performance from event data. In: Weske, M., Montali, M., Weber, I., vom Brocke, J. (eds.) BPM 2018. LNCS, vol. 11080, pp. 139–157. Springer, Cham (2018). https://doi.org/10.1007/978-3-319-98648-7_9

4. van Dongen, B.F., de Medeiros, A.K.A., Verbeek, H.M.W., Weijters, A.J.M.M., van der Aalst, W.M.P.: The ProM framework: a new era in process mining tool support. In: Ciardo, G., Darondeau, P. (eds.) ICATPN 2005. LNCS, vol. 3536, pp. 444–454. Springer, Heidelberg (2005). https://doi.org/10.1007/11494744_25
5. Bose, R.P.J.C., van der Aalst, W.M.P., Žliobaitė, I., Pechenizkiy, M.: Handling concept drift in process mining. In: Mouratidis, H., Rolland, C. (eds.) CAiSE 2011. LNCS, vol. 6741, pp. 391–405. Springer, Heidelberg (2011). https://doi.org/10.1007/978-3-642-21640-4_30
6. Jebb, A., Parrigon, S., Woo, S.E.: Exploratory data analysis as a foundation of inductive research. Hum. Resour. Manag. Rev. **27**, 265–276 (2016). https://doi.org/10.1016/j.hrmr.2016.08.003
7. Sato, D.M.V., De Freitas, S.C., Barddal, J.P., Scalabrin, E.E.: A survey on concept drift in process mining. ACM Comput. Surv. **54**(9), 1–38 (2021). https://doi.org/10.1145/3472752
8. Song, M., van der Aalst, W.: Supporting process mining by showing events at a glance. In: Proceedings of 17th Annual Workshop on Information Technologies and Systems (WITS 2007), pp. 139–145 (2007)
9. Teinemaa, I., Dumas, M., La Rosa, M., Maggi, F.M.: Outcome-oriented predictive process monitoring: review and benchmark. ACM Trans. Knowl. Discov. Data **13**(2), 1–57 (2019). https://doi.org/10.1145/3301300
10. Tukey, J.W.: Exploratory Data Analysis. Addison-Wesley, Boston (1977)
11. Yeshchenko, A., Di Ciccio, C., Mendling, J., Polyvyanyy, A.: Comprehensive process drift detection with visual analytics. In: Laender, A.H.F., Pernici, B., Lim, E.-P., de Oliveira, J.P.M. (eds.) ER 2019. LNCS, vol. 11788, pp. 119–135. Springer, Cham (2019). https://doi.org/10.1007/978-3-030-33223-5_11
12. Zerbato, F., Soffer, P., Weber, B.: Initial insights into exploratory process mining practices. In: Polyvyanyy, A., Wynn, M.T., Van Looy, A., Reichert, M. (eds.) BPM 2021. LNBIP, vol. 427, pp. 145–161. Springer, Cham (2021). https://doi.org/10.1007/978-3-030-85440-9_9

Does This Make Sense? Machine Learning-Based Detection of Semantic Anomalies in Business Processes

Julian Caspary, Adrian Rebmann, and Han van der Aa$^{(\boxtimes)}$

University of Mannheim, Mannheim, Germany
julian.yuya.caspary@students.uni-mannheim.de,
{rebmann,han.van.der.aa}@uni-mannheim.de

Abstract. The detection of undesired behavior is a key task in process mining, supported by techniques for conformance checking and anomaly detection. A downside of conformance checking, though, is that it requires a process model as a basis, limiting its applicability, whereas existing anomaly detection techniques look for statistically infrequent behavior, even though infrequency does not necessarily imply undesirability. The recently introduced concept of semantic anomaly detection overcomes these issues by detecting behavior that stands out from a semantic point of view, such as a claim being paid after it has been rejected. In this manner, it detects behavior that is undesirable, while its grounding in natural language analysis allows it to consider behavioral regularities extracted from other processes, alleviating the need to have a process model available. However, the state-of-the-art approach for semantic anomaly detection, a rigid, rule-based approach, is limited in its scope and accuracy. Therefore, we propose a machine learning-based alternative, which uses a classifier trained to recognize whether observed process behavior is normal or anomalous. Our experiments show that this learning-based approach greatly outperforms the state of the art. Users can directly apply our approach to detect semantic anomalies in their own event data by using one of our pre-trained classifiers, even if their data contains so far unseen process behavior.

Keywords: Process mining · Anomaly detection · Natural language processing · Machine learning

1 Introduction

Process mining analyzes data recorded during the execution of business processes in order to gain insights into an organization's operations [2]. A common task in this regard involves the detection of undesired process behavior, since such occurrences can, e.g., reveal compliance issues, operational inefficiencies, or recording errors. Such undesired behavior can be detected using conformance checking techniques [3], though only when a normative process model is available. Alternatively, techniques for anomaly detection [18] can be used to detect behavior that stands out in a statistical sense, i.e., because it is infrequent. However, infrequent behavior does not necessarily mean that it

© The Author(s), under exclusive license to Springer Nature Switzerland AG 2023
C. Di Francescomarino et al. (Eds.): BPM 2023, LNCS 14159, pp. 163–179, 2023.
https://doi.org/10.1007/978-3-031-41620-0_10

is undesired, since it could point to rare, but acceptable situations, whereas, conversely, behavior that is common is not necessarily desirable.

The recently introduced concept of semantic anomaly detection [1] overcomes these limitations by aiming to detect process behavior that stands out from a semantic point of view. It achieves this by considering the natural language labels associated with events, which allows it to recognize behavior that does not make logical sense, such as a claim being *paid* after it has been *rejected* or an order that is *updated* before it is *created*. Given that such semantic issues are often applicable across processes, semantic anomaly detection can exploit behavioral regularities extracted from existing resources, such as process model repositories. In this manner, undesired behavior can be detected, without the need to have a process model for the particular process at hand.

The problem that we address in this paper is that the state-of-the-art approach for semantic anomaly detection [1] is limited in terms of its accuracy and scope. In particular, it can only detect anomalies between pairs of activities that relate to the same business object and has limited generalization capabilities due to its rule-based nature.

Therefore, we propose an alternative approach that uses machine learning (ML) and state-of-the-art natural language processing (NLP) techniques. The core of our approach is formed by a classifier that we trained on data from a large process repository, covering a variety of domains. Our approach uses this classifier to detect *out-of-order* and *exclusion* anomalies, thus being able to recognize when two events should have been performed in a different order or should not have both been executed for the same case. We test classifier architectures that use classical ML techniques with word embeddings and deep learning techniques based on transformers. Our evaluation shows that both architectures greatly outperform the state of the art in terms of precision and recall, allowing our approach to detect a broader range of anomalies in a more accurate manner. Due to its demonstrated generalization capabilities, users can apply our approach directly on their event data, i.e., without requiring any training or labeled examples, even if their data contains process behavior that our classifiers have not seen before.

The remainder of this paper is structured as follows. Section 2 motivates the goal of semantic anomaly detection and highlights the limitations of the state of the art. Section 3 defines essential preliminaries. Section 4 presents our proposed ML-based approach, which we evaluate in Sect. 5. Finally, Sect. 6 discusses related work, whereas Sect. 7 concludes the paper.

2   Motivation

This section illustrates the potential of semantic anomaly detection, before describing the current state-of-the-art approach and its limitations.

Illustration. Consider the following two traces of a claims-handling process:

$t_1 = \langle$*receive claim, accept claim, check claim, pay compensation*$\rangle$
$t_2 = \langle$*receive claim, check claim, reject claim, pay compensation*$\rangle$

Without having any additional information about the process, the activity labels in these traces reveal two clearly undesirable process executions: in trace t_1, a claim was *accepted* before it had been *checked*, rather than the other way around, whereas in trace t_2, compensation was *paid*, even though the claim had been *rejected*.

Such examples demonstrate the potential of semantic anomaly detection based on the natural language of activity labels. This task is particularly easy for humans, who can apply their commonsense and transferable knowledge to the process at hand in order to recognize that both traces show undesirable process behavior. To do this in an automated manner, however, requires an approach to learn such semantic relations between steps in a process, which is notably harder.

State of the Art. The approach by Van der Aa et al. [1] tackles this challenge by establishing a *knowledge base* that captures rules about the semantics of appropriate process executions, extracted from a linguistic resource (*VerbOcean*) and a process model repository. Each rule captures a relation that should not be violated between two actions (i.e., verbs) applied to the same business object in a trace. For example, the knowledge base contains a rule that states that business objects should be *checked* before they can be *accepted*, which can be used to detect the anomaly in trace t_1.

Limitations of the State of the Art. As the first approach of its kind, the approach by Van der Aa et al. [1] demonstrated the potential of semantic anomaly detection, yet also left considerable room for improvement with respect to its scope and accuracy. In particular, the approach proposed in our paper overcomes the following limitations:

Only Intra-object Anomalies. The existing approach can only detect anomalies involving two activities related to the same business object. In this manner, it can detect that *accept claim* should not come before *reject claim* in trace t_1, since both relate to a *claim* object. However, the approach cannot recognize that *pay compensation* should not follow *reject claim*, since these relate to different objects. By contrast, our proposed learning-based approach can detect both intra and inter-object anomalies.

Limited Generalizability. The existing approach is limited in its ability to generalize information on process behavior. Specifically, it can learn individual rules, e.g., that *reject* and *accept* are mutually exclusive actions, and can generalize these to some degree by recognizing synonymous terms, e.g., that a claim can then also not be *rejected* and *approved* (a synonym of accept). However, this is as far as its generalization capabilities go, since the approach does not make connections between the rules it extracts, e.g., in order to recognize that in general *positive outcomes* (accept, approve, confirm, support, etc.) are exclusive to *negative outcomes* (reject, refuse, limit, etc.). By contrast, our proposed learning-based approach incorporates the information from all examples it encounters, to learn such broader behavioral regularities.

No Context-Specific Anomalies. Finally, the existing approach treats all business objects in the same manner. However, many desirable or undesirable behavioral relations are context specific, meaning that they should or should not be allowed for certain business objects. For example, whereas in general it is fine that objects can be *changed* after they have been *created*, e.g., updating a created text document, this does not apply to objects such as a so-called *rush order* in SAP systems, which are special order types that are not allowed to be changed after creation, so that they can be safely processed immediately. Our proposed learning-based approach can make such context-specific distinctions, provided it receives training data that contains examples of when a certain behavioral regularity should and should not hold.

3   Preliminaries

Event Model. Our work takes as input an event log L, which is a collection of traces. A trace $t = \langle e_1, e_2, .., e_n \rangle \in L$ is a sequence of events belonging to the same case of a process. Each event in a trace is associated with a textual label, indicating the activity to which it corresponds. Without loss of generality, we denote traces as sequences of their event labels when convenient, as e.g., shown for traces t_1 and t_2 in Sect. 2.

Process Model. A process model defines desired execution dependencies between the activities of a process. For our purposes, it is sufficient to abstract from specific process modeling languages and focus on the behavior defined by a model. Therefore, we define a model M as a set of activity label sequences that lead the defined process from its start to its final state. We also define $M^F \subseteq M$ as the set of *loop-free* label sequences of M, i.e., the sequences that do not repeat any process fragments.

Eventually-Follows Relation. We use the eventually-follows relation $\prec$ to capture inter-relations between pairs of labels, stemming from recorded traces or allowed process model sequences. Given a trace $t = \langle e_1, e_2, .., e_n \rangle$, we use $e_i \prec_t e_j$ to denote that e_i occurs (directly or indirectly) before e_j in the trace. Similarly, $a_i \prec_M a_j$ holds if a process model M contains an execution sequence in which activity label a_i occurs before a_j, and $a_i \prec_{M^F} a_j$ holds if M contains a loop-free sequence with that relation.

Vector Operations. We use $\mathbf{v} = [v_1\ v_2 \cdots v_n]$ to denote a numerical n-dimensional vector, with $v_i \in \mathbb{R}$ for $1 \leq i \leq n$. The average of two vectors, $\mathbf{v}$ and $\mathbf{w}$, is obtained by dividing the sum of the vectors by two, i.e., $(\mathbf{v} + \mathbf{w})/2$. Finally, we denote the concatenation of two vectors as $[\mathbf{v}\ \mathbf{w}]$.

4   Approach

This section presents our proposed approach for semantic anomaly detection. As shown in Fig. 1, our approach takes as input a trace t and consists of two main components. The first component, the *event-pair extractor*, takes a trace t and extracts a set of eventually-follows pairs of events P_t to be checked for anomalies. Then, the second component, the *anomaly detector*, takes each event pair $e_i \prec e_j \in P_t$ and uses a classifier to determine if events e_i and e_j should be able to follow each other in this particular order, i.e., whether or not this behavior is anomalous. We provide classifiers that we pre-trained on data stemming from a large process model repository, which means that users of our approach do not have to train their own classifier themselves. Based on such a classifier, our approach detects *ordering anomalies*, i.e., cases where e_j should have become before e_i rather than vice versa, e.g., *accept request* followed by *check request*, as well as *exclusion anomalies*, i.e., cases where e_j should not follow e_i because the two are mutually exclusive, e.g., *reject request* followed by *pay compensation*.

4.1   Event-Pair Extractor

Our approach detects anomalies for pairs of events that are in an eventually-follows relation in a trace $t = \langle e_1, e_2, .., e_n \rangle$. We use this abstraction level, instead of a directly-follows relation, because semantic inter-relations between process steps often remain

Fig. 1. Overview of our anomaly detection approach.

applicable even when other process steps occur in between them. For example, the notion that *receive request* should precede *accept request* holds true, irrespective of the occurrence of other steps in between, such as *check request* or *read request documents*.

However, when extracting eventually-follows pairs from a trace, it is important to consider the notion of rework, stemming from loops in a process. This is a crucial factor, because rework can have an important impact on the semantics of a process instance, particularly with respect to which events may or may not follow each other. For example, although *reject request* should normally not be followed by *pay compensation* (cf., trace t_2 in Sect. 2), this does not apply to trace t_3 shown in Fig. 2. There, rework was conducted after initially rejecting the request, which made the subsequent payment of compensation acceptable.[1]

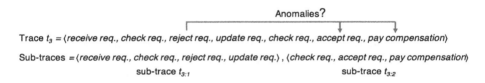

Fig. 2. Illustration of the necessity to identify rework in traces.

Therefore, to prevent the detection of incorrect anomalies, we only check for anomalous behavior within the same cycle of an instance's execution, by avoiding the comparison of behavior occurring in different loops through the process. To do this in the absence of a normative process model (which would render anomaly detection unnecessary), we detect rework at a trace level. Specifically, we split traces into sub-traces, by creating a new sub-trace each time we observe a label that was already present in the current sub-trace. For example, as shown in the lower part of Fig. 2, trace t_3 is split into two sub-traces, where sub-trace $t_{3:2}$ starts when *check request* occurs for the second time. Alternatively, a process model can be discovered for the event log, in order to recognize loops in the process.

Given such sub-traces (a single one for traces without any repetition), we then extract the set of event pairs P_t so that it includes all pairs of events that are in an eventually-follows relation within an identified sub-trace. For sub-trace $t_{3:2}$, this yields *check request* $\prec$ *accept request*, *check request* $\prec$ *pay compensation*, *accept request* $\prec$ *pay compensation*, whereas P_{t_3} comprises $6 + 3 = 9$ event pairs in total. By contrast, without splitting t_3 into its sub-traces, the set would comprise 21 pairs.

---

[1] Note that rework considerations would also apply to directly-follows relations, e.g., in $\langle ..,$ *check, reject, check, accept* $\rangle$ we observe that *check [request]* (directly) follows *reject [request]*, rather than vice versa, yet that this is allowed due to rework being conducted.

4.2 Anomaly Detector

In this component, we apply a classification model to determine for each event pair in the set P_t if it is anomalous or not. We train this classification model on data from a large process model repository, which can then be used when applying our approach on any event log. We test two model architectures for this: An architecture using Support Vector Machines (SVMs), a traditional machine learning technique, in combination with word embeddings, and a transformer-based architecture, a deep learning technique, using a fine-tuned BERT model.

Given that SVMs are simpler and faster to train, whereas transformers often gain better accuracy on complex problems, the comparison of the two architectures allows us to gain insights into the complexity of the problem (i.e., whether or not traditional machine learning suffices), as well as into the benefit of building on a pre-trained language model (i.e., BERT) and a more computationally-intensive method (i.e., deep learning using transformers).

Model Architecture 1: SVM-Based Classification.. Architecture 1 first transforms the natural language labels of an event pair $e_i \prec e_j \in P_t$ into a vector representation by using word embeddings. This vector is then fed into a trained SVM, which will return a classification, i.e., whether or not $e_i \prec e_j$ is an anomaly.

Vector Representation Using GloVe. Given that SVMs (as most machine learning techniques) require a numerical vector as input, we first turn an event pair $e_i \prec e_j$ into a vector representation $\mathbf{v}_{e_i,e_j}$ using GloVe representations [21].

GloVe (short for *Global Vectors*) is a static word representation technique that can be used to create an embedding for a given word, i.e., a vector representation of the word in a high-dimensional space. Such word embeddings are used to capture the meaning of words in a vector, by placing semantically similar words close to each other in the embedding space. Given an event label, e.g., *accept request*, we first use GloVe to establish an embedding of each word, resulting in two 300-dimensional vectors, e.g., $\mathbf{v}_{accept}$ and $\mathbf{v}_{request}$. Then, to obtain vectors of equal length, independent of the number of words in a label, we take the average of all word vectors of the event label, resulting in a single vector representation, e.g., $\mathbf{e_i} = (\mathbf{v}_{accept} + \mathbf{v}_{request})/2$. Finally, to encode an event pair, we concatenate the vectors of the two event labels, i.e., $\mathbf{v}_{e_i,e_j} = [\mathbf{e_i}\ \mathbf{e_j}]$, resulting in a vector of size 600, which accounts for the order in which e_i and e_j were observed (i.e., vectors $\mathbf{v}_{e_i,e_j}$ and $\mathbf{v}_{e_j,e_i}$ are different).

By using embeddings as input for text classification, a classifier can recognize event pairs that have a similar meaning, allowing it to generalize from its training data. For example, if the relation *check application* $\prec$ *approve application* is observed during training, a classifier can recognize that the relation *check request* $\prec$ *accept request* is semantically similar (i.e., has a similar vector representation), and thus recognize that this previously unseen behavioral relation is not an anomaly.

Support Vector Machine. We use the obtained vector representation $\mathbf{v}_{e_i,e_j}$ as input for a two-class SVM, which is a common technique for supervised (traditional) machine learning on textual data [4]. As shown in Fig. 3a, an SVM aims to establish a hyperplane that separates data points belonging to different classes in the feature space, in our case event pairs in the high-dimensional vector space obtained through embedding.

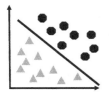 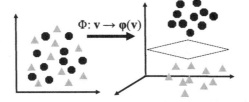

(a) Data separated by a linear hyperplane   (b) Use of a kernel function Φ to separate data   (c) Benefit of parameter C.

Fig. 3. Illustration of SVMs (based on [4]).

When training an SVM (the procedure is described below), we alter two primary settings: the applied kernel function Φ and the choice of the regularization parameter C. A *kernel function* Φ transforms the dimensionality of the data at hand, aiming to represent the data in a higher dimensionality that allows for better separation. Figure 3b shows an example in which data points are not separable in a two-dimensional space, but that can be clearly separated after applying a kernel function that transforms the data into a three-dimensional space. In our experiments, we test linear, polynomial, and Gaussian functions. Furthermore, we alter the *regularization parameter C*, which sets the degree of misclassification allowed when establishing a hyperplane. Particularly, as shown in Fig. 3c, by allowing for a *soft margin* (corresponding to a low C value), some of the training data points may fall outside of the hyperplane, i.e., be misclassified. By allowing for this, the SVM can avoid overfitting to the training data.

Model Architecture 2: BERT-Based Classification. Architecture 2 is a transformer-based architecture that uses a fine-tuned BERT model for anomaly detection.

Transformers and BERT. A *transformer* is a type of neural network architecture that uses self-attention mechanisms to process sequences of data, such as natural language sentences, and learn the relationships between different elements in the sequence [24]. BERT (short for Bidirectional Encoder Representations from Transformers), in turn, is a transformer-based language representation model [6] that has been shown to achieve excellent performance on a broad range of natural language processing tasks.

BERT learns to understand language by processing large amounts of text data (such as the entire English Wikipedia) in an unsupervised manner. This *pre-training* is performed using masked language modeling (Masked LM) and next sentence prediction (NSP). Masked LM trains the model to predict masked tokens based on the context of the surrounding tokens, which allows BERT to learn to represent words in the context of the entire sentence, rather than just based on their local context. In NSP, BERT is trained to predict whether two sentences are consecutive in the input text or not, which helps the model to learn about the relationships between sentences and the broader context of the text. By pre-training on these tasks, BERT learns to represent words and sentences in a way that captures the semantic relationships between them, allowing it to understand natural language text and perform well on a wide range of downstream tasks, such as text classification, question answering, and named entity recognition.

BERT Fine-Tuning. To use BERT for semantic anomaly detection, we fine-tune a pre-trained BERT model on the task at hand. Fine-tuning has the benefit that the classification model can use the language understanding it obtained during pre-training, while requiring much fewer training samples and computation time than would be required when training such a model from scratch.

To perform fine-tuning, we extend BERT's architecture with an additional output layer for two-class classification (whether an input pair is anomalous or not) and then train it in a supervised manner on a collection of positive and negative training samples. Since BERT takes a sequence as input, we provide it event pairs in the following manner: $[[CLS], \ receive, \ request, [SEP], \ check, \ document, \ completeness, [SEP], [PAD], ..., [PAD]]$, where $[CLS]$ is a special token to indicate a classification task, $[SEP]$ is used to indicate the end of an event label, and $[PAD]$ is used to fill the input vector to its maximum length of 128 (since transformers process an entire input sequence of fixed length at once).

Model Training. We train our SVM-based and BERT-based classification models on label pairs extracted from an available process model collection $\mathcal{M}$ (details on the data itself provided in Sect. 5.1). Given a process model $M \in \mathcal{M}$, we extract training samples in the form of allowed and anomalous label pairs based on the model's loop-free eventually-follows relation $\prec_{MF}$. Specifically, we first establish a set of positive label pairs P_M^+, which consists of all pairs of activity labels that can appear in an eventually-follows relation in model M, without any loops in the process. For the example model M_1 in Fig. 4, this yields a set $P_{M_1}^+$ with eight eventually-follows relations, such as *receive request* $\prec^+$ *check request* and *check request* $\prec^+$ *pay compensation*.[2]

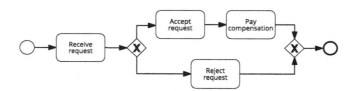

Fig. 4. Exemplary process model used as a basis for training samples.

Then, we establish a set of anomalous samples P_M^- consisting of label pairs not allowed in model M, i.e., that are not included in P_M^+. To provide a balanced training set, we populate P_M^- by randomly selecting pairs that are not in P_M^+, until we have an equal number of positive and negative samples. This would yield a set P_M^- that also consists of eight relations for the example from Fig. 4, e.g., including *reject request* $\prec^-$ *pay compensation* and *accept request* $\prec^-$ *check request*.

Anomaly Detection. Finally, we use a trained classification model to classify each event pair $e_i \prec e_j \in P_t$ as either anomalous or not, resulting in a set of anomalous relations $A_t \subseteq P_t$. Note that, before feeding a label pair into a classifier, we first sanitize the labels using a previously proposed tokenization technique [22], which deals with, e.g., underscores and camel case labels.

---

[2] For clarity, we use $\prec^+$ and $\prec^-$ to denote positive and negative training samples, respectively.

4.3 Approach Output

Our approach yields a set of detected anomalies A_t per trace. When presenting the results to a user, we recognize that a single issue in a process can lead to multiple anomalous label pairs. For example, a trace $t_4 = \langle accept\ claim,\ receive\ claim,\ check\ claim \rangle$ will yield two anomalous relations, i.e., $A_{t_4} = \{accept\ claim \prec receive\ claim,\ accept\ claim \prec check\ claim \}$, which both relate to the premature occurrence of *accept claim*. Furthermore, we verbalize the detected issues using a standard template in order to make them easier to interpret. For the exemplary trace t_4, this then yields the following output:

"Anomaly in t_4: *accept claim* occurred before *receive claim* and *check claim*."

Finally, we aggregate the anomalies detected for all traces in an event log L, resulting in a multi-set of identified issues and their respective frequencies.

5 Experimental Evaluation

This section reports on evaluation experiments conducted to assess the accuracy of our proposed approach, including its two model architectures, and compare it to the state-of-the-art rule-based approach. We describe the data collection Sect. 5.1 and the experimental setup in Sect. 5.2. In Sect. 5.3, we present the evaluation results demonstrating that our ML-based approach accurately detects semantic anomalies and greatly outperforms its rule-based competitor in this regard. Finally, Sect. 5.4 shows an application scenario in which we apply our approach on a real-world event log. The employed implementation, data collection, evaluation pipeline, and raw results are all available in our repository.[3]

5.1 Data Collection

To evaluate our approach, we require a data collection consisting of traces with known anomalies, or, more specifically, event pairs for which a gold standard classification as anomalous or not is available. Since there are no real-world event logs available that include such a gold standard, we instead obtain gold standard data from a large collection of real-world process models from the BPM Academic Initiative (BPMAI) [25].

Specifically, we selected those process models from BPMAI that are in the BPMN notation, have English labels, and that can be turned into a sound workflow net, yielding a total set $\mathcal{M}$ of 2,813 process models. This set comprises process models from a broad range of types and domains, including typical processes related to the handling of orders and requests, as well as more specialized processes, e.g., from software engineering and healthcare domains.

Train-Test Split. To evaluate our approach in an unbiased manner, we established a random split of the process model collection by dividing $\mathcal{M}$ into a training set, $\mathcal{M}_{train}$, comprising 70% (i.e., 1,969) of the models, and a test set, $\mathcal{M}_{test}$, containing the remaining 30% (844). The training set was used for model training, including hyper-parameter

---

[3] https://gitlab.uni-mannheim.de/processanalytics/ml-semantic-anomaly-dection.

optimization (see Sect. 5.2), whereas the test set is exclusively reserved for assessing the performance of our approach. The train-test split is available on our repository.

Characteristics. For each model $M \in \mathcal{M}$, we establish equally-sized sets of (unique) normal and anomalous eventually-follows pairs, i.e., P_M^+ and P_M^-, using the method described in the *Model training* paragraph of Sect. 4.2, this means that 50% of the label pairs in the training and in the test are anomalies, whereas the rest corresponds to regular process behavior.

Table 1 shows the characteristics of the train and test set separately. It shows that nearly half of the label pairs in the test set (10,073) are not included in the training set. With so many unseen label pairs in the test set, a successful anomaly detection approach needs to be able to generalize well from the pairs that it observes during training, making the data collection highly suitable for our purpose.

Table 1 also reports on the number of label pairs that relate to the same business object (BO), such as *create order* $\prec$ *accept order*, since the rule-based state of the art [1] is restricted to such pairs.

Table 1. Characteristics of the data collection. The *Unseen* column refers to labels or label pairs that only occur in the test set, not the training set.

	Training set	Test set	
	Total	Total	Unseen
Process models	1,969	844	–
Unique labels	9,089	4,715	2,684
Total label pairs	53,598	23,770	10,073
Unique label pairs	43,483	21,934	9,906
Total label pairs (same BO)	3,724	1,714	723
Unique label pairs (same BO)	2,711	1,488	694

5.2 Experimental Setup

Implementation and Environment. We implemented our approach in Python (see our repository) and conducted experiments on a machine with 768 GB of RAM, an Intel Xeon 2.6 GHz CPU, and an Nvidia RTX 2080 Ti GPU (used to fine-tune BERT).

Hyper-parameter Optimization. We conducted a hyper-parameter search to identify the most promising configuration for each of our model architectures.

For the SVM-based architecture, we tested different kernel functions, i.e., a linear, a polynomial, and a Gaussian radial basis function (RBF), various values for the degree D of the polynomial kernel function, i.e., $D \in \{2, 4, 6, 8, 10\}$, and different settings for the regularization parameter C, i.e., $C \in \{2^{-5}, 2^{-3}, 2^{-1}, 2, 2^3\}$. Moreover, we tested the effect of reducing the dimensionality of the embedding vectors using Principal Component Analysis (PCA) prior to training the SVM, because this can help reduce the time required for training. The results obtained after using PCA showed that this causes too much information loss, though, and thus leads to considerably worse performance. Therefore, we use the original vector size of 600, obtained by concatenating two 300-dimensional GloVe embeddings, one per label of a pair.

For the BERT-based architecture, we tested two base models for English: *bert-base-cased*, which is pre-trained on text that is case sensitive, and *bert-base-uncased*, which is pre-trained on all lower case text. Furthermore, we tested different learning rates ($5e-5$, $4e-5$, $3e-5$, $2e-5$) and warm-up steps the model performs when fine-tuning (0, 500, and 1000 steps). We use 3 epochs for fine-tuning in order to avoid over-fitting [6].[4]

To select a configuration, we randomly split the models of the training set $\mathcal{M}_{train}$ into a 90% part that is used for the actual training and 10% that are used for validation. We then conducted a train-validation run per configuration, and selected the configuration that yielded the best results:

- *SVM*: an RBF kernel with a C-value of 2, and a vector size of 600.
- *BERT*: *bert-base-uncased* with a learning rate of $5e-5$ and 500 warm-up steps.

We trained an SVM-based and a BERT-based classification model on the entire training set using these optimal configurations, which we use for our experiments on the test set and provide as pre-trained models to users of our approach in our repository.

Baseline. We compare our approach against the rule-based approach by Van der Aa et al. [1], of which the details are described in Sect. 2. It is important to note that this baseline can only detect anomalies for label pairs that share the same business object (as reported in Table 1); the baseline thus automatically classifies pairs with distinct BOs as non-anomalous. We compare our work against the configuration with the best results reported in the original paper, referred to as *SEM4* in their experiments. Most importantly, this configuration uses a semantic similarity threshold to improve the generalizability of the rules learned by the approach.

Measures. We measure the performance of our approach in terms of precision, recall, and F_1-scores, obtained by comparing the predicted classes of label pairs (i.e., anomaly or normal behavior) to the gold standard. Given a class $c \in \{Anomaly, Normal\}$, we denote the number of pairs correctly assigned to c as tp, the number of pairs that are incorrectly assigned to c as fp, and the number of pairs that belong to c in the gold standard, yet, are not assigned to c as fn. Precision (Prec.) is then defined as $tp/(tp+fp)$, recall (Rec.) as $tp/(tp+fn)$, and the F_1-score as the harmonic mean of the two.

5.3 Results

This section presents the results obtained through our evaluation experiments. We first focus on an in-depth analysis of the classification performance of our approach and the baseline, followed by a report on the training and inference times.

Overall Results. Table 2 gives an overview of the main results of our experiments. It shows precision, recall, and F_1-scores per model architecture and the baseline, for different subsets of the event pairs included in the test set.

---

[4] We provide detailed results of the hyper-parameter optimization in our repository.

Overall, the SVM-based model achieves a reasonable F_1-score of 0.69 when considering the entire test set, which shows its general capability to distinguish semantic anomalies from normal behavior. The similar F_1-scores for the individual classes, i.e. 0.68 for the *Anomaly* and 0.70 for *Normal* class, indicate that the model's has learned to recognize anomalous behavior and normal behavior equally well.

Table 2. Results of the evaluation experiments obtained on the test set. Bold numbers indicate the best score for that particular row.

Scope	Class	Support	SVM			BERT			BL [1]		
			Prec.	Rec.	F_1	Prec.	Rec.	F_1	Prec.	Rec.	F_1
All pairs	Anomaly	11,885	0.70	0.66	0.68	**0.76**	**0.74**	**0.75**	0.69	0.01	0.01
	Normal	11,885	0.68	0.72	0.70	**0.75**	0.77	**0.76**	0.51	**0.99**	0.67
	Overall	23,770	0.69	0.69	0.69	**0.76**	**0.76**	**0.76**	0.60	0.50	0.54
All pairs w. same BO	Anomaly	857	0.73	0.71	0.72	**0.81**	**0.76**	**0.78**	0.69	0.07	0.12
	Normal	857	0.72	0.73	0.73	**0.77**	0.82	**0.80**	0.51	**0.97**	0.67
	Overall	1,714	0.72	0.72	0.72	**0.79**	**0.79**	**0.79**	0.60	0.52	0.56
Unseen pairs	Anomaly	5,009	0.64	0.64	**0.64**	0.62	**0.66**	**0.64**	**0.67**	0.01	0.01
	Normal	5,064	**0.64**	0.64	0.64	**0.64**	0.60	0.62	0.50	**0.99**	0.67
	Overall	10,073	**0.64**	**0.64**	**0.64**	0.63	0.63	0.63	0.58	0.50	0.54
Unseen pairs w. same BO	Anomaly	343	0.72	0.73	0.73	**0.79**	**0.76**	**0.77**	0.68	0.07	0.12
	Normal	380	0.75	0.75	0.75	**0.79**	0.82	**0.80**	0.54	**0.97**	0.69
	Overall	723	0.74	0.74	0.74	**0.79**	**0.79**	**0.79**	0.60	0.54	0.57

Our BERT-based model outperforms its SVM-based alternative in all aspects on the entire test set. It achieves a good overall F_1-score of 0.76, which shows that it accurately classifies unseen behavior into semantic anomalies and normal behavior. The better results compared to the SVM-based model suggest that the general language understanding of the transformer in combination with its process-specific fine-tuning improves the performance on our anomaly detection task. At the same time BERT's performance is also stable across classes, achieving comparable scores (0.74–0.77) for all metrics, for both the *Anomaly* and the *Normal* class.

Both our models greatly outperform the baseline (with the exception of recall on the *Normal* class), which achieves an overall F_1-score of 0.54, versus 0.69 and 0.76 of our models. Part of this difference occurs because the baseline, by definition, cannot detect anomalies for event pairs with different business objects, which comprises about 93% of the total pairs. As a result, the baseline assigns the *Normal* class in the vast majority of cases, resulting in a low precision (0.50) but high recall (0.99) for that class, while achieving a recall of only 0.02 for the *Anomaly* class, with a precision of 0.64.

Same BO Pairs. We also computed the results for the subset of label pairs that share the same business object (i.e., intra-object anomaly detection), this, among others, provides a fairer comparison to the baseline. We observe that—in line with expectations—the performance of the baseline improves for this subset, achieving an overall F_1-score of 0.56 compared to 0.54 on the full collection, caused by an increase in recall to 0.07

for the *Anomaly* class (versus 0.01 for the total collection. However, the baseline is still outperformed by both our models, which achieve overall F_1-scores of 0.72 (SVM) and 0.79 (BERT). The performance of the SVM-based model slightly improved from 0.69 overall F_1-score on the entire dataset to 0.72 on the subset of data, whereas the BERT-based model demonstrated larger gains, achieving an overall F_1-score of 0.79 compared to 0.76 for the full collection. We can thus note that even if we only consider data that the baseline is designed to handle, our approach still consistently achieves better results.

Unseen Label Pairs. To be able to assess how well our approach can deal with unseen behavior, we computed the results for the subset of label pairs that only occur in $\mathcal{M}_{test}$ and thus have not been observed by our models during training.

The results obtained for this subset show that both model architectures of our approach can generalize reasonably well to such unseen data, although it is clear that this anomaly detection task is more challenging. We observe that the performance of the BERT-based model drops to an F1-score of 0.63, from 0.76 for the entire test set, whereas the SVM-based model is more stable, achieving an F1-score of 0.64, compared to 0.69 for the entire set.

The main generalization capabilities of our approach become apparent when considering the detection of intra-object anomalies, i.e., by considering unseen label pairs with the same business object. For this subset, both model architectures perform equally well on unseen pairs as on the set including seen pairs, achieving F1-scores of 0.79 (BERT-based) and 0.74 (SVM-based). It should be noted that this subset is relatively small, though, consisting of 723 label pairs.

Overall, these results reveal that anomaly detection can be well-generalized to intra-object relations, e.g., by learning that objects should be *received* before they are *checked*, whereas it is more challenging to learn rules that also apply to unseen activities that relate to different business objects, e.g., just because an *order* must be created before a *delivery*, does not mean that this also applies to two unseen objects.

Post-hoc Analysis. In order to gain deeper insights into the results, we go beyond a quantitative analysis and take a closer look at the correct and incorrect classifications of our approach and the baseline. Specifically, we focus on our BERT-based model, which has demonstrated better performance. We find that our approach is able to correctly identify a wide range of semantically problematic behaviors in the test set. For instance, it finds that *reject credit* should not happen before *assess risk*, that *wait for payment* should only happen after *create invoice*, and that *receive payment* should not be followed by *confirm order*. Note especially that none of these anomalies can be detected by the baseline, since the activities per pair refer to distinct business objects.

Looking at the baseline's results in detail, we observe that, even though it specifically targets the detection of semantic anomalies that involve the same business object, our approach finds additional, relevant intra-object anomalies that the baseline could not detect. For instance, our model correctly detects that *approve application* should not follow *cancel application* and that *evaluate request* should not happen before *prioritize request*, which the baseline does not find. Such cases illustrate the capability of our approach to better consider the meaning of entire activities, not just the actions applied to the same object, as done by the baseline.

However, there is also behavior that our approach fails to classify correctly. For instance, our approach detects *receive payment* followed by *pick shipment* as an anomaly, whereas it is well-imaginable that for some order handling process a shipment is indeed only sent after payment for that shipment was collected. Conversely, our approach did not find that, e.g., *send loan request* followed by *fill out loan request* may be problematic, for instance, if these are executed by the same resource. Our approach currently does not consider such context-dependent anomalies, which would require the incorporation of resource information, a direction for future research.

Computation Time. Table 3 depicts training and inference times of our models and the baseline. The training time refers to the time it takes to train a model using the pairs in $\mathcal{M}_{train}$. The duration refers to the actual training, thus excluding the time it takes to load process models and establish training pairs (which is the same for all approaches).

Table 3. Average training and inference times of our models and the baseline.

	SVM	BERT	BL [1]
Training time	23.4s	1,859.8 s	2.8 s
Inference time per label pair	0.01 s	0.01 s	0.03 s

We find that our SVM-based model requires a training time of 23 s, whereas the BERT-based model requires 1,859 s (~31 min) for fine-tuning in 3 epochs. The knowledge base population of the baseline only takes about 3 s, since it just performs a single pass over the label pairs, storing their counts. As such, there is a trade-off between lower training times and optimized performance. Nonetheless, the performance gain is so strong that we give a clear recommendation for using the BERT-based model. This is especially the case because users do not need to train our approach themselves, but can directly use the fine-tuned model provided in our repository.

With respect to inference time, both our models and the baseline are fast, classifying label pair in less than 0.03 s on average.

5.4 Real-World Application

Finally, we also applied our approach on real-world data: the *permit log* from the BPI 2020 challenge [9], which captures data on work trips conducted by university employees. The process flow concerns the request for and approval of a travel permit, the trip itself, a subsequent travel declaration, as well as associated reimbursements.

Although there is no gold standard available that indicates true anomalies in this process, our approach (using the BERT-based model) detects various interesting situations, as shown in Table 4. The examples correspond to situations in which trips happened before a permit was properly handled or approved (*a1* and *a2*), declarations submitted before a trip rather than after (*a3*), as well as payments that were approved before the respective permit was (*a4*). Still, we also recognize that certain detected anomalies look concerning, but are fine in light of the specifics of the process. This applies to anomaly *a5*, which corresponds to payments occurring before a declaration was actually approved. Although this seems problematic, it is possible in this process for payments related to pre-paid expenses.[5]

Table 4. A selection of anomalies detected for the real-world *permit log*.

ID	Detected anomaly	Frequency
a1	*end trip* occurred before *permit final approved by supervisor*	2,800
a2	*start trip* occurred before *permit submitted by employee*	2,205
a3	*declaration submitted by employee* occurred before *start trip*	4,707
a4	*request for payment approved by administration* occurred before *permit approved by supervisor*	611
a5	*declaration final approved by supervisor* occurred after *request payment* and *payment handled*	4,292

6   Related Work

Various approaches for anomaly detection in process mining have been proposed. Most of these are frequency-based, arguing that uncommon behavior is not of interest or undesirable, as opposed to our semantic approach. Such frequency-based anomaly detection is an inherent part of certain process discovery algorithms [14], which use it to preserve only the most common process behavior. Close to our approach are ML-based techniques, such as works that use autoencoders [13,19], as well as LSTMs (long short-term memory) [12], working in unsupervised or semi-supervised manners. Whereas most approaches only consider control-flow information, others also incorporate additional perspectives, such as BINet [18] for deep learning-based anomaly detection detection, and pattern-based techniques employed in the context of filtering in the process discovery [16] and the repair of event log imperfections [8].

Our work also relates to other NLP applications that focus on distinguishing normal from abnormal relations, primarily in the form of *commonsense reasoning*. Beyond using ML-based techniques, such reasoning can be done based on, e.g., lexical resources, such as WordNet [17] or VerbOcean [5], which capture relations between words, or commonsense knowledge graphs [11,20], which capture common relations between entities. Such reasoning is, for example, employed to improve the quality of actions and state changes extracted from natural language texts [15,23].

7   Conclusion

In this paper, we proposed an ML-based approach for the detection of semantic anomalies in business processes, allowing users to detect undesired behavior without depending on the availability of a normative process model. By building on state-of-the-art NLP techniques to train an anomaly classifier, our approach has learned to distinguish normal from undesired process behavior based on the textual labels associated with

---

[5] Note that such false positives would be avoided when using an object-centric event log, since there would be no relation between the events related to pre-payments and declarations.

recorded events. Our experiments demonstrate that our learning-based approach greatly outperforms an earlier, rule-based approach for semantic anomaly detection in terms of both scope and accuracy.

Still, our work is subject to limitations. Our approach itself is limited by its focus on event pairs. Although this perspective is chosen because it allows us to achieve accurate results and fine-granular anomaly insights (i.e., much more specific than detecting whether or not a trace is anomalous), the event-pair perspective does not allow us to detect missing behavior, e.g., that *check request* was skipped (unlike the baseline approach [1]). Also, the performance of our approach depends on the similarity of behavioral regularities observed during training and those in the event log on which it is applied. Positive points in this regard are that we trained our approach on data from a broad range of domains and that we have demonstrated its capabilities to generalize to unseen data, especially when it comes to intra-object anomalies. Furthermore, in the absence of real-world logs with known anomalies, our experiments are conducted using generated samples. However, these samples are established based on real-world process models, whereas we also show the potential of our work in a real-world application.

In future work, we aim to lift the concept of ML-based semantic anomaly detection to the recent wave of generative large language models, such as ChatGPT and GPT4, once such technology becomes freely accessible, so that experiments can be conducted in a reproducible manner. Here, we would also like to stress that our conceptual idea is independent of a specific language model, so that the same approach can later be updated according to new developments on the NLP side. Furthermore, we also aim to incorporate additional perspectives into the detection of anomalies. Particularly, we aim to encode resource roles and categorical attribute values, allowing our approach to, e.g., consider who performed a certain step and what the outcome of a decision was. Finally, in terms of application scenarios, we plan to integrate our work into analysis pipelines in which semantic correctness plays an important role, such as in the privatization of event data, where the insertion of obvious noise should be avoided [10], and in next activity prediction, where predicted next steps should make semantic sense [7].

Reproducibility: Our employed implementation, data, and obtained results are available through the project repository linked in Sect. 5.

References

1. van der Aa, H., Rebmann, A., Leopold, H.: Natural language-based detection of semantic execution anomalies in event logs. Inf. Syst. **102**, 101824 (2021)
2. van der Aalst, W.M.P.: Process Mining: Data Science in Action, vol. 2. Springer, Cham (2016). https://doi.org/10.1007/978-3-662-49851-4
3. Carmona, J., van Dongen, B., Solti, A., Weidlich, M.: Conformance Checking. Springer, Cham (2018). https://doi.org/10.1007/978-3-319-99414-7
4. Chauhan, V.K., Dahiya, K., Sharma, A.: Problem formulations and solvers in linear SVM: a review. Artif. Intell. Rev. **52**(2), 803–855 (2019)
5. Chklovski, T., Pantel, P.: VerbOcean: mining the web for fine-grained semantic verb relations. In: EMNLP, pp. 33–40 (2004)
6. Devlin, J., Chang, M.W., Lee, K., Toutanova, K.: BERT: pre-training of deep bidirectional transformers for language understanding. In: NAACL, pp. 4171–4186. ACL, Minneapolis, Minnesota (2019)

7. Di Francescomarino, C., Ghidini, C.: Predictive process monitoring. In: W.M.P., Carmona, J. (eds.) Process Mining Handbook. vol. 448. LNBIP, pp. 320–346. Springer, Cham (2022). https://doi.org/10.1007/978-3-031-08848-3_10

8. Dixit, P.M., et al.: Detection and interactive repair of event ordering imperfection in process logs. In: Krogstie, J., Reijers, H.A. (eds.) CAiSE 2018. LNCS, vol. 10816, pp. 274–290. Springer, Cham (2018). https://doi.org/10.1007/978-3-319-91563-0_17

9. van Dongen, B.: BPI challenge 2020 (2020). https://doi.org/10.4121/UUID:52FB97D4-4588-43C9-9D04-3604D4613B51

10. Fahrenkrog-Petersen, S.A., Kabierski, M., van der Aa, H., Weidlich, M.: Semantics-aware mechanisms for control-flow anonymization in process mining. Inf. Syst **114**, 102169 (2023)

11. Havasi, C., Speer, R., Alonso, J.: ConceptNet 3: a flexible, multilingual semantic network for common sense knowledge. In: RANLP. pp. 27–29. John Benjamins Philadelphia, PA (2007)

12. Krajsic, P., Franczyk, B.: Semi-supervised anomaly detection in business process event data using self-attention based classification. Proc. Comput. Sci. **192**, 39–48 (2021)

13. Krajsic, P., Franczyk, B.: Variational autoencoder for anomaly detection in event data in online process mining. In: ICEIS (1), pp. 567–574 (2021)

14. Leemans, S.J.J., Fahland, D., van der Aalst, W.M.P.: Discovering block-structured process models from event logs - a constructive approach. In: Colom, J.-M., Desel, J. (eds.) PETRI NETS 2013. LNCS, vol. 7927, pp. 311–329. Springer, Heidelberg (2013). https://doi.org/10.1007/978-3-642-38697-8_17

15. Losing, V., Fischer, L., Deigmoeller, J.: Extraction of common-sense relations from procedural task instructions using BERT. In: 11th Global Wordnet Conference, pp. 81–90 (2021)

16. Mannhardt, F., de Leoni, M., Reijers, H.A., van der Aalst, W.M.P.: Data-driven process discovery - revealing conditional infrequent behavior from event logs. In: Dubois, E., Pohl, K. (eds.) CAiSE 2017. LNCS, vol. 10253, pp. 545–560. Springer, Cham (2017). https://doi.org/10.1007/978-3-319-59536-8_34

17. Miller, G.A.: WordNet: a lexical database for English. Commun. ACM **38**(11), 39–41 (1995)

18. Nolle, T., Luettgen, S., Seeliger, A., Mühlhäuser, M.: BiNet: multi-perspective business process anomaly classification. Inf. Syst. **103**, 101458 (2019)

19. Nolle, T., Luettgen, S., Seeliger, A., Mühlhäuser, M.: Analyzing business process anomalies using autoencoders. Mach. Learn. **107**(11), 1875–1893 (2018)

20. Omeliyanenko, J., Zehe, A., Hettinger, L., Hotho, A.: LM4KG: improving common sense knowledge graphs with language models. In: Pan, J.Z., et al. (eds.) ISWC 2020. LNCS, vol. 12506, pp. 456–473. Springer, Cham (2020). https://doi.org/10.1007/978-3-030-62419-4_26

21. Pennington, J., Socher, R., Manning, C.: GloVe: Global vectors for word representation. In: EMNLP, pp. 1532–1543. ACL, Doha, Qatar (2014)

22. Rebmann, A., van der Aa, H.: Enabling semantics-aware process mining through the automatic annotation of event logs. Inf. Syst. **110**, 102111 (2022)

23. Tandon, N., Dalvi, B., Grus, J., Yih, W.t., Bosselut, A., Clark, P.: Reasoning about actions and state changes by injecting commonsense knowledge. In: EMNLP, pp. 57–66 (2018)

24. Vaswani, A., et al.: Attention is all you need. In: NeurIPS, vol. 30 (2017)

25. Weske, M., Decker, G., Dumas, M., La Rosa, M., Mendling, J., Reijers, H.A.: Model Collection of the Business Process Management Academic Initiative (2020)

Inferring Missing Entity Identifiers from Context Using Event Knowledge Graphs

Ava Swevels$^{(\boxtimes)}$, Remco Dijkman , and Dirk Fahland

Eindhoven University of Technology, Eindhoven, The Netherlands
{a.j.e.swevels,r.m.dijkman,d.fahland}@tue.nl

Abstract. Complete event data is essential to perform rich analysis. However, real-life systems might fail in recording the (correct) case identifiers the system has operated on, resulting in incomplete event data. We aim to infer missing case identifiers of events by considering the physical constraints of the process which previous work has failed to do. We extended Event Knowledge Graphs (EKGs) with concepts for context and rule-based inference. We use the extended EKGs to model event data in its physical context and define five inference rules to infer identifiers of physical objects in a process. We evaluate the effectiveness of the rules on data from the IC manufacturing industry using conformance checking. Initially, none of the traces were complete. Our method inferred a case identifier for 95% of the events resulting in 88% complete traces.

Keywords: Log repair · Event Knowledge Graph · Modeling · Inference Rule · Contextual Information · Physical Constraints

1 Introduction

Business Process Analytics (BPA) is an area of data analytics that facilitates rich analysis of the way in which business processes work, providing insight into how business processes can be improved. BPA techniques rely on event logs that record the events that happened in the business process, along with (the identifier of) the case to which they belong and the moment in time at which they happened, possibly extended with other information. However, in real-life systems the data in the event logs may be incomplete [20]. Consequently, before event data can be used for BPA, missing data must be added or incomplete events must be removed. If missing data can somehow be *inferred* from the context of the events, that is preferred, because it leads to more usable data to work with. This work focuses on inferring missing case identifiers, also known as the "event-case correlation" problem [5].

Several methods exist to infer missing case identifiers from context. The dominant context information that these methods use for inferring case identifiers is a process model [13,18]. This has some clear drawbacks. Firstly, a process model

© The Author(s), under exclusive license to Springer Nature Switzerland AG 2023
C. Di Francescomarino et al. (Eds.): BPM 2023, LNCS 14159, pp. 180–197, 2023.
https://doi.org/10.1007/978-3-031-41620-0_11

may not exist or may be hard to create, for example in multi-entity processes [11] or in case the data is at a different level of abstraction than the level at which the process is understood. Secondly, inference for cyclic processes either requires further context information, such as constraints on time [4] or data [5], which may not be available either, or it may require time-consuming iterations [6] to do the inference. Thirdly, most existing methods only connect the incomplete information to the context information within the algorithm. Only [6] makes the connection between data and context available for further analysis.

To alleviate these problems, we aim to develop techniques for inferring missing identifiers *without* relying on a process model. This paper focuses on inferring missing case identifiers for *processes with batching* where *context* information is available about the handling of *physical objects*. Since these objects are bound by the laws of physics, information about them can be used in combination with simple physics rules to infer other information. For example, if a physical object is in one place in one event and in another place in another event, it must have been moved in between and a movement event must have involved this object.

We propose to use *knowledge graphs* (KGs) for such inference. A KG combines a graph-based representation of data and its context with an inference mechanism [7]. Specifically, translating an incomplete event log into an (incomplete) *Event Knowledge Graph* (EKG) [11] models which events are known to relate to which entities. The problem now is to infer missing relations between events and entities (instead of a global case) which also allows inference for processes with batching. EKGs [11] currently lack concepts for context and inference.

This paper extends EKGs with context information about the process and a pattern-based inference mechanism over this context to infer missing identifiers (relations between events and entities). We demonstrate how to define context in terms of the physical context of events (locations of activities and their handling of physical objects) and how to define inference rules that encode constraints for objects based on the physical context of events.

We implemented and evaluated the technique on an industrial case study with NXP Semiconductors by showing that it can be used to infer missing entity identifiers. Using our method on event data of 7250 events, where 86% of the events lacked an entity identifier, we could infer an identifier for 95% of the events within 30 s. Subsequent conformance checking against a normative process model validated the correctness of the inferred identifiers.

Against this background, the remainder of the paper is structured as follows. The related work is discussed in Sect. 2. The problem is elaborated on in Sect. 3 along a running example. Section 4 describes a method for inferring identifiers from context using an EKG. Section 5 shows the proposed method by creating inference rules for the running example and industrial use case and validates the inference using conformance checking. The findings are discussed in Sect. 6.

2   Related Work

Data-driven process analysis defines minimal requirements for event logs [14]: each event contains at least an activity, timestamp and a case identifier. However, in practice, collected event data might not meet these requirements.

Table 1. Overview of types of (incomplete) data

Timestamp	Activity	Case Identifier	Context Knowledge used for Inference
-	✓	✓	timing information [12,19]
✓	-	✓	other attributes to activities [2,15,23]
✓	✓	-	acyclic process model [13,18]; process model + add. constraints [4,5]; process model + sim. annealing [6]; surrogate id [17]; activity properties [**this**]

Event data quality issues are assessed in terms of missing or incorrectly recorded attribute values. These typically manifest themselves in systematic "imperfection patterns" [20] due to imperfect data recording mechanisms. Accordingly, specific techniques have been proposed to detect if such data quality 'patterns' exist in an event log [1] and to repair them. Repairing data enables us to apply techniques that require complete data such as rich analysis techniques and traceability.

Assuming correct activity and case attributes have been identified [3,16], we focus on *missing values* for the standard attributes of case/object identifier, activity, and timestamp. Existing literature infers missing values for one of the attributes based on information in the other attributes and *additional context knowledge* as summarized in Table 1.

Missing timestamps can be inferred by using knowledge of duration of process steps, both, for isolated cases [19] and cases processed via shared resources and queues [12]. *Missing activities* can be inferred by knowledge of how events with specific data attributes in a specific behavioral context relate to activities, e.g., by aggregating low-level observations to activities [15,23] or matching text attributes to activity descriptions [2].

Missing case identifiers are typically inferred by leveraging control-flow knowledge. Existing techniques use a given acyclic probabilistic Markov model [13] or first estimate a model of an acyclic process from an incomplete log and then infer case identifiers [18]. Inferring identifiers for cyclic processes not only requires a process model, but also additional timing constraints [4], data constraints [5], or multiple iterations of matching, e.g., through simulated annealing [6] that also generate rules for correlating events to cases based on the activity, event properties, and their immediate context. No process model is required when a surrogate identifier such as the user in a clickstream is present [17]. A common trait of all techniques is that they assume isolated cases (no batching) and the context knowledge is kept separate from the incomplete event data and reconciled within the algorithmic technique itself.

This paper addresses the problem of inferring missing entity identifiers in processes *with batching* but *without* leveraging a process model defining the control-flow or surrogate identifiers. Instead, we focus on processes handling *physical objects* and use knowledge on the activities themselves and the context the activities operate in to infer entity identifiers.

The problem of inferring missing identifiers is a subproblem of *knowledge inference* addressed by *knowledge graphs* (KGs). A KG thereby consist of (i) a semi-structured graph modeling data and/or knowledge, (ii) a set of inference rules over this graph; applying the rules on the initial graph results in (iii) a derived graph describing derived knowledge [7]. "Semi-structured" means a KG can be flexibly extended with new concepts. *Event Knowledge Graphs* [11] allow to model relations between events and (identified) entities of a process, and allow to describe processes with multiple objects and batching. This makes them a suitable model for our problem (inferring missing identifiers in processes with batching, without a given model). As EKGs as defined in [11] lack concepts for describing a process context and an inference mechanism, we will propose these in Sect. 4.

3   Missing Identifiers in Processes with Physical Objects

We illustrate the problem of identifying from incomplete data which physical objects in a process suffered from errors. We show by a simple example that existing inference techniques fail to reliably infer correct entity identifiers for processes with physical objects and batching. We illustrate how we can reliably infer entity identifiers when applying contextual knowledge about activities, physical objects, and how they are constrained. Then, we state the specific problem together with the expected in- and output.

Notation on Event Data. We first recall some notation on event data over multiple identifiers, i.e., object-centric event data, based on [11]. We write Val for the universe of values, including disjoint sets of activity names and timestamps Act, $Time \subseteq Val$. $Time$ is totally ordered by $\leq$.

An *event table with entities* (i.e., object-centric log) $T = (E, Attr, Ent, \#)$ consists of events E, attributes $\{act, time\} \subseteq Attr$, entity type attributes $\emptyset \neq Ent \subseteq Attr$, and partial attribute value function $\# : E \times Attr \nrightarrow Val$ assigning $e \in E$ and $a \in Attr$ value $\#_a(e) = v$ ($\#_a(e) = \perp$ if a is undefined for e) with $\#_{time}(e) \in Time$ and $\#_{act}(e) \in Act$ are defined.

An event e may have a multi-valued attribute $\#_a(e) = \{v_1, \ldots, v_n\}$ (set) or $\#_a(e) = \langle v_1, \ldots, v_n \rangle$ (list) for example for events referring to multiple objects; and we write $v_i \in \#_a(e)$. For uniform notation, we also write $v \in \#_a(e)$ for single-valued $\#_a(e) = v$.

In the generalized setting of object-centric or multi-entity processes [11] a trace is defined in relation to an entity identifier. Let $ent \in Ent$ be an entity type. The entity identifiers of type $ent \in T$ are $ent(T) = \{n \mid n \in \#_{ent}(e), e \in E\} \setminus \{\perp\}$. Let $n \in ent(T)$ be an entity identifier. An *entity trace* of n is a sequence $\pi_n = \langle e_1 \ldots e_k \rangle$ of events $\{e_1 \ldots e_k\} = \{e \in E \mid n \in \#_{ent}(e)\}$ ordered by $\#_{time}(e_i) \leq \#_{time}(e_j)$ for $1 \leq i < j \leq k$. In the following, we specifically discuss (identifiers of) entities that are *physical objects*, e.g., a box.

Running Example. Figure 1 shows a process, modeled as a proclet [12] (bottom), and the *context* it is executed in, i.e., the physical environment (top).

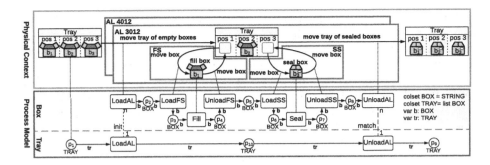

Fig. 1. A process model (proclets [12]) of assembly lines where boxes in a tray are filled, sealed, and labeled (bottom) together with the physical context (top).

There are two *Assembly Lines* at which *Boxes* are filled and sealed. A *Tray* of *Boxes*, i.e. a batch, is loaded into an *Assembly Line* if it is *Empty*, then each *Box* individually passes several stations and finally the *Tray* is unloaded from the *Assembly Line*. The first station is the *Fill Station*, at which a *Box* is loaded, filled and unloaded. The second station is the *Seal Station* at which a *Box* is loaded, sealed and unloaded. *Loading* and *Unloading* of the stations is done by a robot arm that needs to be aware of the *Position* of the *Box* in the *Tray*.

Only when *Sealing*, each *Box* is labeled making it uniquely identifiable. Hence only *Seal* events record an entity identifier for the box. Further, *Load* and *Unload* events register per station the *Position* of a *Box* in the *Tray*. For instance, processing three boxes results in the *incomplete* log shown in Fig. 2 where attribute b records the box identifier and p the position in the tray. No event records, both, b and p. Figure 2(a) visualizes this incomplete log as a *Performance Spectrum* with each color referring to a different box, i.e., blue refers to b_1, orange to b_2 and green to b_3 [9]. Note that no complete traces of boxes can be constructed.

Suppose a *Fill*ing error occurred at e_3. We cannot determine which box was affected by the error on the incomplete log. As a result, an operator has to inspect and possibly even discard the entire *Tray*. To mitigate this problem, we have to obtain complete traces by inferring the likely values for b.

Using control-flow models as context knowledge, e.g., [13,18] allows to infer unknown value $\#_b(e_i)$ from known value $\#_b(e_j)$ when e_j directly precedes or succeeds e_i; e.g., infer $\#_b(e_9) = b_1$ from $\#_b(e_{10}) = b_1$. Existing technique fail to address the physical constraints and batching (boxes in trays) in our example. The method of [18] generates multiple possible solutions of box id values and traces (depending on parameters), two are shown in Fig. 2(b) and (c): besides the analyst having to pick a solution, the method wrongly relates e_1 and e_{20} to only a single box and wrongly claims b_1 went through filling and sealing first, while the log shows that the box at $p = 2$ was filled first and sealed second.

Physical Constraints and Context. In contrast, Fig. 2(d) visualizes the complete traces of the three boxes in line with the process' context and its (physical) constraints. We note three basic (physical) principles that hold for this process:

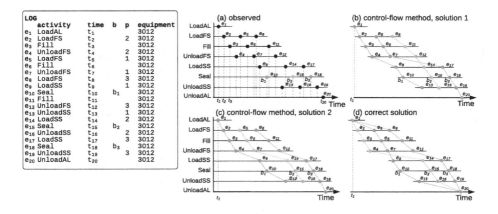

Fig. 2. A small log with missing identifiers: observed events with missing entity identifiers (a), behaviors found with a control-flow method (b, c) and correct behavior (d).

(P1) for a physical object to be present at/removed from a location, the physical load/unload activity of that location must have involved this physical object; (P2) activities are performed at physical locations (stations) and the physical object they operate on must be at that location; (P3) physical objects that are consistently kept at the same (relative) spatial location can consistently be distinguished from each other (e.g. positions in the tray/on a conveyor belt). Applying these principles does *not* require a process model, but context information on activities: (C1) which activities process which kind of physical objects at which locations, (C2) which physical activities move which kinds of physical objects into/out of locations, (C3) relations between locations and (C4) relations between individual physical objects and batches as visualized in Fig. 1.

Applying P1-P3 on the context knowledge shown in Fig. 1 allows us to infer missing identifiers: Activities *Fill* and *Seal* process boxes at distinct locations (the *Fill* and *Seal Station*), while *physical (Un)LoadSS* and *(Un)LoadFS* activities *move* a box into/out of these locations. Physical activities *LoadAL* and *UnloadAL* move a *Tray* of boxes into/out of the *AssemblyLine* containing the locations of the *Seal* and *Fill* station. (1) From $\#_b(e_{10}) = b_1$ we know that b_1 is at the *Seal Station*; by P1, box b_1 must have been moved into/out of the *Seal Station*: $\#_b(e_9) = \#_b(e_{13}) = b_1$. (2) From $\#_b(e_{10}) = b_1, \#_b(e_{15}) = b_2, \#_b(e_{18}) = b_3$ we know that b_1, b_2, b_3 are at the *Seal Station* and thus at the *AssemblyLine*; by P1 they must have been moved into/out of the *AssemblyLine* resulting in multi-valued $\#_b(e_1) = \#_b(e_{20}) = \{b_1, b_2, b_3\}$. (3) Box $\#_b(e_9) = b_1$ is at location $\#_p(e_9) = 1$ in the *Tray*; from $b_1 \in \#_b(e_1) = \#_b(e_{20})$ follows that b_1 at $p = 1$ is at the *AssemblyLine* from time t_1 to t_{20}. By P2, P3 and $\#_p(e_5) = \#_p(e_7) = 1$ we know that e_5 and e_7 must have operated on the box with $p = 1$ at the fill station, resulting in $\#_b(e_7) = b_1$. (4) From $\#_b(e_5) = \#_b(e_7) = b_1$ we know that b_1 is present at the *Fill Station* from t_5 to t_7. From P2 follows that e_6 must operate on a box present at the *Fill Station*, thus $\#_b(e_6) = b_1$. Consistently applying this reasoning infers all identifiers shown in Fig. 2(d).

Problem statement. The problem we address in this paper is how to automate the above inference of missing entity identifiers based on context knowledge of a process with physical objects. We assume as input: (I1) an event table T where each event has a timestamp, activity and the top-level location (e.g. equipment) but may not have an entity identifier, and for each real physical object in the process, there is at least one event $e \in E$ that refers to it, and (I2) context information about the process, e.g., physical constraints C1-C4 above. Given I1 and I2, we want to (O1) infer for each event the (likely) entities involved such that (O2) the resulting entity traces describe a consistent execution of the process matching the (physical) context. The latter can be validated by replaying the log on a process model, though the model itself is not a required input.

4   Inferring Entity Identifiers from Context

We now describe our method for inferring identifiers from context. We first encode the incomplete event data (I1) in an *Event Knowledge Graph* (EKG). Exploiting the flexibility of KGs, we show how to extend the EKG with context information (I2). We then define inference rules over EKGs to infer the missing identifiers (O1). While the concepts for modeling context and inference rules are generic, we demonstrate their application for the concrete problem of inferring identifiers for processes with physical objects stated in Sect. 3.

4.1   Basic Idea

We first illustrate the basic idea on how to define and use context information for inference on a part of the log of Sect. 3 repeated in Fig. 3.

The log has event e_{10} with activity *Seal* for box b_1 in equipment 3012. The physical context described in Fig. 1 shows that *Seal* occurs at the *Seal Station* within equipment *3012*. The log contains events e_9 and e_{13} with activities *LoadSS* and *UnloadSS* which are are not correlated to a box, i.e. $\#_b(e_9) = \#_b(e_{13}) = \bot$. Context (c.f. Figure 1) shows that (1) activities *LoadSS* and *UnloadSS* operate on a box, thus e_9 and e_{13} have incomplete information, and (2) these activities occur at the same location as e_{10} (the *Seal Station* of equipment *3012*). From principle (P1), b_1 must

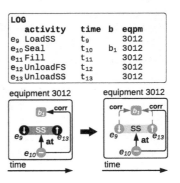

Fig. 3. Example on how to infer missing entity identifiers using context information.

have been loaded into the *Seal Station* before t_{10} and unloaded from *Seal Station* after t_{10}, e_9 and e_{13} are the first preceding load event and succeeding unload event, hence $\#_b(e_9) = \#_b(e_{13}) = b_1$. Even though events e_{11} and e_{12} occur in between e_{10} and e_{13}, they are not considered simply because they occur at a different location.

Figure 3 (bottom) schematically visualizes this reasoning over the context of the events as an inference rule. On the left-hand side, only event e_{10} is correlated to b_1 (blue circle, edge to b_1) while events e_9 and e_{13} are uncorrelated (black circles); e_9 and e_{13} perform activities at the same physical location (bar labeled SS from e_9 to e_{13}) where e_9 loads an object into the location (down arrow) and e_{13} unloads an object (up arrow); e_{10} occurs time-wise between e_9 and e_{13} and performs an activity at location SS (arrow from e_{10} to SS). On the right-hand side, the correlation of e_9 and e_{13} to b_1 is inferred (blue circles, red dashed edges to b_1). In the following subsections, we discuss how to precisely define and apply such inference rules on event data.

4.2 Modeling Context in Property Graphs

The rule and its application we illustrated in Fig. 3 reasoned over events correlated to multiple entities, and their context of activities (with properties) and locations (see C1-C4 in Sect. 3). We now show how to formally model event data with such contextual information using *event knowledge graphs* (EKGs). We first recall the underlying data model of *labeled property graphs* (LPG), the specific model of EKGs, and how the incomplete log of Fig. 3 is modeled as an EKG. Afterwards, we extend the meta-model of EKGs with contextual information and define rules over these extended EKGs.

Existing Models. An LPG G is a directed multi-graph where each node and each edge of G (called relationship) is typed by one or more *labels Lab*. A node n may have multiple labels $\lambda(n) \in 2^{Lab}$; a relationship r always has one label $\lambda(r) \in Lab$. A node or relationship can carry properties, i.e., attribute-value pairs, written $\#_a(n) = v$; see [21, App. A]. We write $n \in \ell$ if $\ell \in \lambda(n)$ and $(n, n') \in \ell$ or $n \ \ell \ n'$ if r is an edge from n to n', $\overrightarrow{r} = (n, n')$, and $\lambda(r) = \ell$.

An Event Knowledge Graph (EKG) is an LPG G with node labels Event, Entity and relationship labels corr ("event correlated to entity"), and df ("event directly followed by event") so that (1) Event and Entity nodes are disjoint, (2) each $e \in$ Event has $\#_{time}(e) \in Time$ and $\#_{act}(e) \in Act$ defined and (3) each df-relationship $r \in$ df, $\overrightarrow{r} = (e, e')$ defines that e' directly follows e from the perspective of entity $r.ent = n \in$ Entity.[1] The meta-model of this basic EKG is shown in Fig. 4 (red-dashed rectangle).

We work with the slightly extended meta-model shown at Fig. 4 (top) proposed in [11] which additionally defines relationships n_1 rel n_2 between $n_1, n_2 \in$ Entity and relationships e observed a describing that $e \in$ Event executed activity $a \in$ Activity.

The standard EKG construction from an event table with entities creates Event, Entity and Activity nodes and corr, df and observed relationships as fol-

---

[1] Formally, let $e, e' \in$ Event be correlated to the same entity $n \in$ Entity, $(e, n), (e', n) \in$ corr: e df e' holds iff $\#_{time}(e) < \#_{time}(e')$ and there is no other event $e'' \in$ Event, $(e'', n) \in$ corr between e and e', i.e. $\#_{time}(e) < \#_{time}(e'') < \#_{time}(e')$.

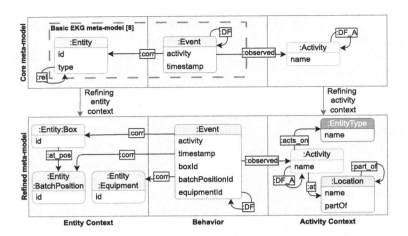

Fig. 4. Schematic meta-model; (a) basic EKG meta-model [10]; (b) extended meta-model; (c) extension by this work; (d) refined meta-model for use case

lows [11]: (1) each event record is translated into an Event node; (2) infer entity node n with $\#_{type}(n) = ET$ if there is an Event node e with $\#_{ET}(e) = n$ and add e corr n (e is correlated to n); (3) infer an n_1 rel n_2 relationship if there is an event node e with e corr n_1 and e corr n_2; (4) infer df-relationships between events correlated to the same entity node n; (5) infer Activity nodes a with $\#_{name}(a) = act\_name$ if there is an Event node e with $\#_{act}(e) = act\_name$ and add e observed a.

In case of incomplete information in the event table, step (2) results in incomplete correlation (corr) relationships and step (3) results in incomplete and false df-relationships.

Adding Context. In every process, each event is related to the activity performed and the entities involved. We therefore consider activity and entities as the natural *context* of an event, while the timestamp is local to the event itself. For a specific process, we can further detail the context of an event based on domain-knowledge of the process. For this, we have to *refine* the Entity and Activity nodes in the meta-model in larger Entity and Activity contexts (see Fig. 4 (bottom)). Next, we discuss this refinement for the processes with physical objects in our problem.

The entity context is refined by introducing a dedicated label for each entity type and for each relationship type in the process. The refined entity context of our example is shown in Fig. 4 (bottom) and defines entity labels for 3 types of entity nodes for Equipment, Box and BatchPosition and an at_pos relation.

The *activity context* of an event is refined by adding *more information* about the activity that is executed, i.e., by adding more related nodes that describe the activity further. Which information is added depends on the process and the required inference. For our use case, we require activity context in the form of how the activities operate on entities and their locations, see C1–C4 in Sect. 3.

Table 2. Records containing contextual data

Location Records		Activity Records			
name	part of	activity	label	entity type	location
AL	-	LoadAL	load	box	AL
FS	AL	UnloadAL	unload	box	AL
SS	AL	LoadFS	load	box	FS
		UnloadFS	unload	box	FS
		Fill	acts_on	box	FS
		..	..	..	..

We introduce two additional node labels EntityType and Location and relationship labels acts_on, at and part_of. Each Activity acts on EntityTypes at a specific Location, e.g., *LoadSS* acts_on a *Box*. Depending on the use case, a more specific label instead of acts_on can be chosen to describe the semantics of the activity, e.g. activity *LoadSS* loads a *Box*. Further, we use part_of relation to express when a location l_1 is physically part of a larger location l_2, e.g., *SS* part_of *AL*.

We adapt the procedure to construct an EKG with a refined entity and activity context as follows. We change to procedure to infer refined Entity nodes (step 2) and relationships (step 3): (2) We infer a node n with labels $\lambda(n) = \{entity, ET\}$ if there is an Event node e with $\#_{ET}(e) = n$; then relationship e corr n is added as usual; (3) We use the label of a refined relationship (instead of generic rel) between two entity nodes whenever one is defined in the schema. Adding the activity context to the EKG requires use-case specific adaptations to the procedure. Assuming context information about activities and locations is specified in tables similar to event tables (see Table 2), we perform the following steps after step 5: (6) For each location record l we treat $\#_{name}(l)$ as a unique identifier and create node $loc \in$ Location and set $\#_{prop}(loc) = \#_{prop}(l)$ for each $prop \in Attr$ in l. (7) Add relationship l part_of l' for $l, l' \in$ Location iff $\#_{name}(l') = \#_{part\_of}(l)$. (8) We create for each unique value $et = \#_{entitytype}(r)$ in Activity records r a new node $et \in$ EntityType. (9) Each Activity Record r provides details for an activity $\#_{activity}(r) = a$ that already exists as a node $a \in$ *Activity* in the EKG. Thus, we can link a to the location node $l = \#_{location}(r)$ created in step (6) by adding relationship a at l. (10) For each activity $a = \#_{activity}(r)$ in Activity Records r, $qualifier = \#_{label}(r)$ defines how a operates on $et = \#_{entitytype}(r)$ and we add a relationship a $qualifier$ et.

Figure 5 shows part of the EKG obtained from the incomplete log (Fig. 2) extended with Activity and Location nodes based on the records and relations in Table 2.

4.3   Inference Rules

We now have an EKG G extended with context information where corr relationships are incomplete due to missing information in the underlying event table.

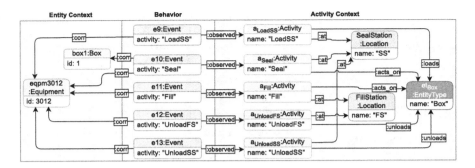

Fig. 5. Partial instance of the EKG

As df-relationships are unreliable due to missing corr relationships, we remove all df-relationships from G. We infer the missing corr relationships based on which we can compute reliable df-relationships. Modeling event context in an EKG enables us to infer missing corr relationships by defining local rules describing contextual patterns in EKGs. We first define the rules in general, then present a first example and then define the semantics of rule application on EKGs.

We define an inference rule as a simple graph-transformation pattern in an EKG G that defines for nodes $n_1, \ldots, n_r$ a "left-hand-side" (LHS) graph pattern specifying the context from which missing relationships can be inferred. The "right-hand side" (RHS) of the rule are then relationships from $n_1, \ldots, n_r$ to other nodes in the LHS; e.g. adding corr relationships between events and entities provides the missing identifiers.

More concretely, an inference rule $IR = (G, R_{inferred}, c, m)$ is an EKG G where we mark one or more relationships $R_{inferred}$ of G as the RHS, i.e., to be inferred, by setting property $\#_{RHS}(r) = True$. All other nodes and relationships of G define the LHS. Further, IR defines an *ordering condition* c and a *minimization condition* m over the properties of the nodes N in G to limit the matches of the LHS; see [21, App. A].

For example Fig. 6 describes an inference rule based on principle (P1) from Sect. 4.1. It infers the unknown entity identifiers of load and unload events f_0 and f_2 from event f_1 occurring in between f_0 and f_2 at the same location.

Figure 6 depicts the LHS of the rule by solid edges. It defines three events ($f_0, f_1, f_2 \in$ Event). The activity operation and location of each event f_i is modeled through the activity f_i observed $a_i, a_i \in$ Activity. Specifically, by the relationship labels r from a_0 and a_2 to $et \in EntityType$ with $\#_{name}(et) = $ 'Box', events f_0 and f_2 observe a physical loading and unloading activity for a box. The events are observed at the

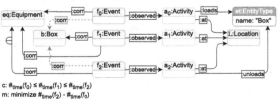

$c: \#_{time}(f_0) \leq \#_{time}(f_1) \leq \#_{time}(f_2)$
$m:$ minimize $\#_{time}(f_2) - \#_{time}(f_0)$

Fig. 6. Inference rule detailing the short-hand notation of Fig. 3.

same location ℓ by the relationships f_i observed a_i at $\ell, \ell \in$ Location, $i = 1, 2, 3$. Furthermore all events are correlated to the same equipment f_i corr $eq \in Equipment, i = 1, 2, 3$, and only f_1 is correlated to box b, i.e., f_1 corr $b \in Box$ while f_0 and f_2 are not correlated to a box, i.e., $N_{incomplete} = \{f_0, f_2\}$.

Ordering condition c: $\#_{time}(f_0) \leq \#_{time}(f_1) \leq \#_{time}(f_2)$ restricts the LHS to events f_0, f_1, f_2 that follow each other in time. Minimization condition m: minimize $\#_{time}(f_2) - \#_{time}(f_0)$ restricts the LHS to only those f_0, f_2 such that no other (un)load events happen in between f_0 and f_2. m also implies that f_0, f_2 are the first preceding load event and first succeeding unload event w.r.t. f_1. Note that the LHS cannot rely on df-relationships to express ordering of the events as df-relationships are incomplete due to incomplete corr relationships.

The RHS of the rule is $R_{inferred} = \{r_0, r_2\}$ with $\vec{r_0} = (f_0, b)$ and $\vec{r_2} = (f_2, b)$ (indicated by red dashed edges in Fig. 6). Subsequently, we use the notation shown in Fig. 3 as short-hand for inference rules, i.e., Fig. 3 denotes the rule of Fig. 6.

An inference rule $IR = (G, R_{inferred}, c, m)$ is *applied* on an (incomplete) EKG G' as follows. Let $LHS(IR)$ be the graph G without $R_{inferred}$. (1) An *instance* of $LHS(IR)$ is a sub-graph G'' of G' that is injectively homomorphic to $LHS(IR)$, so that the ordering condition c holds in G'', i.e., the sub-graph G'' has all nodes, relationships, labels, and properties described in G and c except $R_{inferred}$, and possibly other relationships not described in G. (2) If the minimization condition m is defined, pick only those instances $G''_1, \ldots, G''_k$ of $LHS(IR)$ where m is minimal; otherwise pick all instances. (3) For each picked instance G''_i of $LHS(IR)$ in G', apply IR by extending G''_i by the RHS of IR, i.e., adding the missing relationships $R_{inferred}$; see [21, App. A].

For example, $LHS(IR)$ of Fig. 6 has an instance G'' in the EKG of Fig. 5 as $LHS(IR)$ injectively maps to G'' by $f_0 \mapsto e_9, f_1 \mapsto e_{10}, f_2 \mapsto e_{13}, b \mapsto box1, eq \mapsto eqpm3012, a_0 \mapsto a_{LoadSS}, a_1 \mapsto a_{Seal}, a_2 \mapsto a_{UnloadSS}, \ell \mapsto SealStation$ and $et \mapsto et_{Box}$. Note that G'' has all nodes, labels, properties and relationships specified in $LHS(IR)$. Further, c holds in G'' as $\#_{time}(e_9) \leq \#_{time}(e_{10}) \leq \#_{time}(e_{13})$. G'' also minimizes $m : \#_{time}(e_{13}) - \#_{time}(e_9)$. Applying the RHS adds corr relationships $\vec{r_9} = (e_9, b_1)$ and $\vec{r_{13}} = (e_{13}, b_1)$

The complete inference procedure using a set of rules $\{IR_1, \ldots, IR_k\}$ on an EKG G (with added context) is: (1) remove all df-relationships from G, (2) repeatedly apply each IR_i until no more relationships are added to G, (3) infer the df-relationships (now based on complete corr-relationships).

5   Application and Evaluation

We now use the generic principles for extending EKGs with (use-case specific) context (see Sect. 4.2) and inference rules (see Sect. 4.3) to define concrete inference rules for the problem of inferring identifiers for physical objects defined in Sect. 3. The rules are implemented as queries over the Neo4j graph DB. We report on their use in an industrial use case.

5.1   Inference Rules

Based on the entity and activity context defined in the refined EKG meta-model of Fig. 4, we translated principles P1-P3 of Sect. 3 into five inference rules.

Inference for One Level. Fig. 7 shows two rules to infer missing entity identifiers for an object at a location L using the short-hand notation introduced in Sect. 4.1, see [21, App. C] for EKG notation. Rule A is explained in Sect. 4.1; from an event f_1 with known identifier, we can infer the identifiers for the load and unload events f_0 and f_2 at the same location L. Rule B is simply the reverse of Rule A; based on the load and unload events of a location, we infer the missing identifiers of the events happening at that location by propagating downwards. Note that any rule can correlate any number of entities to an event. For Rule A, this is desired behavior: all physical objects present at location L need to be loaded and unloaded from L. For Rule B, it depends on the context whether this is desired behavior. If f_1 should only be correlated to one of the objects loaded/unloaded during f_0 and f_2, more information is required which will be explained in Rule E.

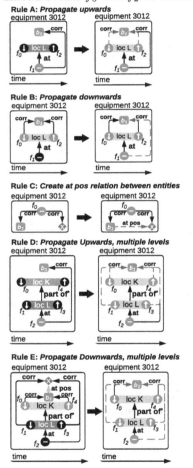

Inference between Entities. Rule C (Fig. 7) is derived from principle (P3); from an event f_0 associated both to a box b_1 and a batch position x, we can infer that box b_1 is at pos x in the tray.

Inference for Multiple Levels. Rules D and E of Fig. 7 use the same principles as Rule A and B respectively for locations containing other locations. As the part of relation is transitive, principles (P1) and (P2) can be also used to infer missing entity identifiers on higher/lower location levels via part of (see Fig. 4) or its transitive closure part of*. Rule D infers missing identifiers through multiple levels by propagating upwards and Rule E downwards through multiple levels.

Rule E also deals with events that should only be correlated to one of the objects loaded/unloaded during f_0 and f_4. As stated before, additional information is required for correct inferences such as the relative spatial position of an object (P3). Hence, events happening at a lower-level location L correlated to position x can be related to entity b_1 at pos x that is loaded into K.

Fig. 7. Rules A, B, C, D and E

These rules allow correlating multiple entities to the same events, which allows to infer, e.g., batching, when physically possible, or reveal ambiguous context information when physically impossible.

5.2 Implementation and Demonstration

We implemented the approach of Sect. 4 and rules A-E of Sect. 5.1 as Cypher queries over the Neo4j graph database; see [21, App. D] and https://github.com/Ava-S/EKG_Inference.

Applying the implementation on an incomplete event table of our running example infers the correlation and traces shown in Fig. 8 as follows: (1) applying Rule D propagates b_1, b_2 and b_3 from *Seal* events upwards to *LoadSS, UnloadSS, LoadAL* and *UnloadAL* events; (2) applying Rule C creates the at_pos relation between Box and BatchPosition nodes using *LoadSS* events which are now correlated to both these entities; (3) applying Rule E propagates entity identifiers from *(Un)LoadAL* events downwards to *LoadFS* and *UnloadFS* events; (4) applying Rule B propagates identifiers from *(Un)LoadFS* events to *Fill* events; see [21, App. B] for the full EKG after inference.

Recall that any rule can correlate any number of entities to an event. Rule D assigns multiple boxes (b_1, b_2 and b_3) to the *LoadAL* event e_1 and *UnloadAL* event e_{20}. Even though Rule D is unaware of batching events, it is still able to infer multiple identifiers to batching events. The resulting traces align with the process in Fig. 1 and the physical constraints.

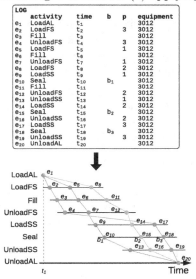

Fig. 8. An incomplete event table of running example.

5.3 Industrial Use Case: NXP's Sawing Process

The proposed method was applied on an industrial use case; the sawing process at NXP Semiconductors (NXP). NXP is a globally operating company that designs, develops, and manufactures Integrated Circuit (IC) chips. We focus on the dicing step in which wafers are sawn into dies (unpackaged chips). The sawing equipment has several sensors installed to record the different events operating on the wafers. The collected data covered a week of manufacturing. Data inspection revealed that some entity identifiers were recorded incorrectly. These identifiers were dropped during data cleaning and needed to be inferred.

Process Description. Multiple wafers are loaded into the sawing equipment together in a rack for wafers. All wafers are then handled in parallel as follows: the

Table 3. Overview of the activity types and whether the wafer identifiers are present.

type	entity	wafer identifier(s)	# activities	# events
act on	wafer	✓	3	≈ 1000
act on	wafer	-	10	≈ 3100
load	(batch of) wafers	-	1	≈ 50
load	wafer	-	5	≈ 1550
unload	wafer	-	5	≈ 1550
Total			24	≈ 7250

wafers are aligned and cut at the cutting station, and then cleaned at the cleaning station. Finally the rack with all cut wafers is unloaded. A rack has multiple slots containing one wafer each; wafers in the same rack are distinguished by their slot position in the rack.

Activities and Events. Table 3 gives an overview of the different activity types in NXP's sawing process, how many activities of this type the process has and their frequency. Only 3/24 activities recorded wafer identifiers, resulting in ≈ 6250/7250 events without wafer identifier.

Inferring Missing Information. We adapted the entity and activity context in the EKG meta-model of Fig. 4 and the inference rules of Sect. 5.1 to match the NXP use case: The Entity:Box and Entity:BatchPosition nodes were replaced respectively with Entity:Wafer and Entity:SlotPosition nodes. We applied the adapted Rules A-E to infer the missing wafer identifiers. Construction of the complete EKG and inference required less than 30 s, resulting in 7225/7250 events correlated to at least one wafer.

Validation. We validated the correctness of the inferred identifiers through alignment-based conformance checking. A proclet reference model of the multi-entity process was created and validated with NXP [22]; the proclet describes for each entity type (e.g. wafer, rack) a life-cycle model as state machine without concurrency. The proclet model was added to the EKG by modeling df_a relationships between Activity nodes (see Fig. 4 and [11]), where a df_a a' describes that activity a can be directly followed by a' for a particular entity, e.g. wafer or a rack. This allowed us to measure whether all df-relationships of a wafer w form a complete trace according to the wafer life-cycle using the technique in [11, Sect 6.4]: we checked whether each df-relationship (e, e') of w has a corresponding df_a-relationship (a, a') for wafers with e observed a and e' observed a'. While for the extracted data, 0 wafers had a complete trace, after inference 88% had complete traces, the remaining 12% of incomplete traces were attributed to non-recorded or double-recorded events.

6 Conclusion

In this paper, we studied the problem of inferring missing entity identifiers in the generalized setting of multi-entity processes (e.g., with batching) that are executed in a physical environment. We showed that the problem can be modeled and solved in Event Knowledge Graphs (EKGs) by extending an EKG over incomplete data with context information about entities and activities, and by defining inference rules over this EKG according to simple principles of physical processes. As a side-effect, the inclusion of context information and inference rules makes EKGs "true" knowledge graphs [7]. Using inference rules over EKGs shows that missing identifiers can be inferred efficiently even when no normative control-flow model is available or applicable (e.g., in case of multi-entity processes or batching). An industrial case study proved that our technique is applicable to industrial processes both in terms of quality of inference and performance. It should be noted that the running example was constructed based on the industry use case.

While the techniques of inference rules over context-extended EKGs are generic, our study only demonstrated context representation and inference rules for multi-entity interactions in the form of batching 1:n synchronization and activities in hierarchical locations. Similar processes with incomplete information exist in healthcare, customer movement, and other production processes [8], which suggest further research on applying the proposed techniques in these domains. For processes with other forms of synchronization or activity context, more research is required to develop a generic, systematic approach for extending EKGs with context information and defining inference rules over incomplete event data. Our approach is also limited in dealing with noise. We assumed each entity identifier to be observed in at least one event and events to not deviate from the intended process. In case all identifiers are missing and we know that a particular activity only happens once for a certain entities, artificial identifiers can be generated. Additionally, in the industrial use case some unload events were not recorded, however the rules were still able to infer the missing identifiers for the preceding load events. Other forms of missing identifiers require further research. In case of deviating events, information about the intended process (i.e., process model) has to be included in the inference. Finally, the proposed inference rule mechanism is not specified to inferring corr edges, suggesting further viable research for inference by integrating event data and process knowledge beyond missing identifiers.

Acknowledgement. The research underlying this paper was partially supported by NXP Semiconductors and by AutoTwin EU GA n. 101092021.

References

1. Andrews, R., Emamjome, F., ter Hofstede, A.H.M., Reijers, H.A.: An expert lens on data quality in process mining. In: ICPM 2020, pp. 49–56 (2020)

2. Baier, T., Di Ciccio, C., Mendling, J., Weske, M.: Matching events and activities by integrating behavioral aspects and label analysis. Softw. Syst. Model. **17**(2), 573–598 (2018)
3. Bala, S., Mendling, J., Schimak, M., Queteschiner, P.: Case and activity identification for mining process models from middleware. In: Buchmann, R.A., Karagiannis, D., Kirikova, M. (eds.) PoEM 2018. LNBIP, vol. 335, pp. 86–102. Springer, Cham (2018). https://doi.org/10.1007/978-3-030-02302-7_6
4. Bayomie, D., Awad, A., Ezat, E.: Correlating unlabeled events from cyclic business processes execution. In: Nurcan, S., Soffer, P., Bajec, M., Eder, J. (eds.) CAiSE 2016. LNCS, vol. 9694, pp. 274–289. Springer, Cham (2016). https://doi.org/10.1007/978-3-319-39696-5_17
5. Bayomie, D., Di Ciccio, C., Mendling, J.: Event-case correlation for process mining using probabilistic optimization. Inf. Syst. **114**, 102167 (2023)
6. Bayomie, D., Revoredo, K., Di Ciccio, C., Mendling, J.: Improving accuracy and explainability in event-case correlation via rule mining. In: ICPM 2022, pp. 24–31. IEEE (2022)
7. Bellomarini, L., Fakhoury, D., Gottlob, G., Sallinger, E.: Knowledge graphs and enterprise AI: The promise of an enabling technology. In: 2019 IEEE 35th International Conference on Data Engineering (ICDE), pp. 26–37 (2019)
8. Blank, P., Maurer, M., Siebenhofer, M., Rogge-Solti, A., Schönig, S.: Location-aware path alignment in process mining. In: EDOC Workshops 2016, pp. 1–8. IEEE (2016)
9. Denisov, V., Fahland, D., van der Aalst, W.M.P.: Unbiased, fine-grained description of processes performance from event data. In: Weske, M., Montali, M., Weber, I., vom Brocke, J. (eds.) BPM 2018. LNCS, vol. 11080, pp. 139–157. Springer, Cham (2018). https://doi.org/10.1007/978-3-319-98648-7_9
10. Esser, S., Fahland, D.: Multi-dimensional event data in graph databases. J. Data Semant. **10**, 109–141 (2021)
11. Fahland, D.: Process mining over multiple behavioral dimensions with event knowledge graphs. In: van der Aalst, W.M.P., Carmona, J. (eds.) Process Mining Handbook. LNBIP, vol. 448, pp. 274–319. Springer, Cham (2022). https://doi.org/10.1007/978-3-031-08848-3_9
12. Fahland, D., Denisov, V., van der Aalst, W.M.P.: Inferring unobserved events in systems with shared resources and queues. Fundam. Inform. **183**(3–4), 203–242 (2021)
13. Ferreira, D.R., Gillblad, D.: Discovering process models from unlabelled event logs. In: Dayal, U., Eder, J., Koehler, J., Reijers, H.A. (eds.) BPM 2009. LNCS, vol. 5701, pp. 143–158. Springer, Heidelberg (2009). https://doi.org/10.1007/978-3-642-03848-8_11
14. Jagadeesh Chandra Bose, R.P., Mans, R.S., van der Aalst, W.M.P.: Wanna improve process mining results?: It's high time we consider data quality issues seriously. In: Proceedings of the IEEE CIDM, pp. 127–134. IEEE (2013)
15. Mannhardt, F., de Leoni, M., Reijers, H.A., van der Aalst, W.M.P., Toussaint, P.J.: From low-level events to activities - a pattern-based approach. In: La Rosa, M., Loos, P., Pastor, O. (eds.) BPM 2016. LNCS, vol. 9850, pp. 125–141. Springer, Cham (2016). https://doi.org/10.1007/978-3-319-45348-4_8
16. de Murillas, E.G.L., Reijers, H.A., van der Aalst, W.M.P.: Case notion discovery and recommendation: automated event log building on databases. Knowl. Inf. Syst. **62**(7), 2539–2575 (2020)

17. Pegoraro, M., Uysal, M.S., Hülsmann, T., van der Aalst, W.M.P.: Uncertain case identifiers in process mining: a user study of the event-case correlation problem on click data. In: Augusto, A., Gill, A., Bork, D., Nurcan, S., Reinhartz-Berger, I., Schmidt, R. (eds.) Enterprise, Business-Process and Information Systems Modeling. BPMDS and EMMSAD 2022. LNBIP, vol. 450, pp. 173–187. Springer, Cham (2022). https://doi.org/10.1007/978-3-031-07475-2_12
18. Pourmirza, S., Dijkman, R.M., Grefen, P.W.: Correlation miner: mining business process models and event correlations without case identifiers. Int. J. Cooper. Inf. Syst. **26**(2), 1742002:1-1742002:32 (2017)
19. Rogge-Solti, A., Mans, R.S., van der Aalst, W.M.P., Weske, M.: Repairing event logs using timed process models. In: Demey, Y.T., Panetto, H. (eds.) OTM 2013. LNCS, vol. 8186, pp. 705–708. Springer, Heidelberg (2013). https://doi.org/10.1007/978-3-642-41033-8_89
20. Suriadi, S., Andrews, R., ter Hofstede, A.H.M., Wynn, M.T.: Event log imperfection patterns for process mining: towards a systematic approach to cleaning event logs. Inf. Syst. **64**, 132–150 (2017)
21. Swevels, A., Dijkman, R.M., Fahland, D.: Inferring missing entity identifiers from context using event knowledge graphs. Technical report, Eindhoven University of Technology (2023). https://doi.org/10.5281/zenodo.7802241
22. Swevels, A.: Creating a digital shadow of a manufacturing process with inferred missing information using an event knowledge graph. Master's thesis, Eindhoven University of Technology (2022)
23. Tax, N., Sidorova, N., Haakma, R., van der Aalst, W.: Mining process model descriptions of daily life through event abstraction. In: Bi, Y., Kapoor, S., Bhatia, R. (eds.) IntelliSys 2016. SCI, vol. 751, pp. 83–104. Springer, Cham (2018). https://doi.org/10.1007/978-3-319-69266-1_5

Process Channels: A New Layer for Process Enactment Based on Blockchain State Channels

Fabian Stiehle[1]([✉]) and Ingo Weber[1,2]

[1] School of CIT, Technical University of Munich, Munich, Germany
{fabian.stiehle,ingo.weber}@tum.de
[2] Fraunhofer Gesellschaft, Munich, Germany

Abstract. For the enactment of inter-organizational processes, blockchain can guarantee the enforcement of process models and the integrity of execution traces. However, existing solutions come with downsides regarding throughput scalability, latency, and suboptimal tradeoffs between confidentiality and transparency. To address these issues, we propose to change the foundation of blockchain-based process enactment: from on-chain smart contracts to *state channels*, an overlay network on top of a blockchain. State channels allow conducting most transactions off-chain while mostly retaining the core security properties offered by blockchain. Our proposal, *process channels*, is a model-driven approach to enacting processes on state channels, with the aim to retain the desired blockchain properties while reducing the on-chain footprint as much as possible. We here focus on the *principled approach of state channels as a platform*, to enable manifold future optimizations in various directions, like latency and confidentiality. We implement our approach prototypical and evaluate it both qualitatively (w.r.t. assumptions and guarantees) and quantitatively (w.r.t. correctness and gas cost). In short, while the initial deployment effort is higher with state channels, it typically pays off after a few process instances—considerably reducing cost. And as long as the new assumptions hold, so do the guarantees.

Keywords: Blockchain · Business Process Enactment · Choreographies · Interorganisational processes · State Channels

1 Introduction

For the enactment of inter-organizational processes, blockchain can guarantee the enforcement of rules and the visibility and integrity of execution traces—without introducing a centralised trusted party. The current state of the art focuses on-chain enactment, where a process model is transformed into a smart contract and executed on the blockchain [1]. However, blockchain execution comes with downsides and suboptimal tradeoffs regarding scalability and confidentiality. [2, Chapter 3]. On-chain process enactment inherits these problems; to address this,

© The Author(s), under exclusive license to Springer Nature Switzerland AG 2023
C. Di Francescomarino et al. (Eds.): BPM 2023, LNCS 14159, pp. 198–215, 2023.
https://doi.org/10.1007/978-3-031-41620-0_12

related work has focused on improving the cost of the on-chain components (e.g., [3–5]). In contrast, we propose a more fundamental change: to move from full on-chain enactment to layer two state channels. Layer-two technologies have emerged as a promising direction to address fundamental challenges of blockchain technology [6]. One of these technologies are state channels. State channels move the bulk of transactions into off-chain channels. In these channels, participants transact directly without the involvement of the blockchain. The blockchain is only used for channel creation and as a dispute resolution and settlement layer. This can greatly reduce the on-chain footprint, enabling new levels of scalability and improved confidentiality, while mostly retaining the core security properties offered by blockchain.

In this paper, we focus on conceptualising the principled approach of enacting processes in state channels. We focus on the fundamental aspects of this new approach: conceiving how to achieve the main functionality and quality attributes, studying where advantages and disadvantages lie, and creating a basis for a new line of research. To this end, we propose *process channels*: a model-driven approach to transform process models into state channel constructions.

To evaluate our approach, we develop a prototype: *Leafhopper*. As we propose a new platform for blockchain-based enactment, we provide a qualitative assessment and investigate the guarantees and assumptions compared to on-chain enactment. We also provide a quantitative evaluation and benchmark our prototype in terms of correctness and gas cost. While gas cost is foremost known as a measure for transaction cost on Ethereum, it is directly linked to the amount, size, computational complexity, and storage requirement of transactions [7]— serving well as a measure to assess the on-chain footprint. We find that, while the initial deployment cost of Leafhopper is higher, it can considerably reduce gas cost. Leafhopper does so without weakening the main security guarantees of the blockchain, given that some additional assumptions hold.

Following open science principles, we make the entire code and data of our prototype and evaluation publicly available – see Footnote 10. Beyond Leafhopper, we also publish *Chorpiler*, the first open-source compiler capable of generating optimised smart contracts from choreographies. In the remainder of the paper, we first discuss the background (Sect. 2) and related work (Sect. 3), before presenting the process channel approach (Sect. 4). Implementation and evaluation are covered in Sect. 5, before Sect. 6 concludes.

2   Blockchain and Layer Two Channels

A blockchain is an append-only store of transactions distributed across a network of nodes [2, Chapter 1]. This data store is called ledger; the ledger and the network of nodes form the defining parts of a particular blockchain system. Participants in the network are identified through their blockchain address, which is derived from a public key. Each transaction is cryptographically signed with the sender's corresponding private key, and is validated according to the blockchain system's protocol. Smart contracts allow to execute user-defined programs on the blockchain. In practice, a blockchain system can provide integrity, immutability, non-repudiation, equal-rights, and full transparency [2, Chapter 1.4]

Layer two channels attempt to scale the underlying blockchain layer by offloading transactions [6]. The blockchain is no longer involved in every minute transaction—these are moved into channels. The idea first emerged in the concept of payment channels (e.g., [8]). Say, you want to pay an online news site $0.10 per article that you read. You create a channel with the news site, where you lock $5.00 as initial funds (or *collateral*). Every time you read an article, you exchange an off-chain transaction with the news site, assigning an additional $0.10 to their account. After 32 articles, you decide to close the channel, with the accrued $3.20 assigned to the news site and the remaining $1.80 refunded to your account. This concept has since been generalised to state channels [9]. Participants wishing to transact first agree on a contract governing the rules of the channel and encode these on the blockchain (e.g., in a smart contract). They then conduct off-chain transactions directly across the channel. For each transaction, they agree on the outcome and cryptographically commit to their agreement. Finally, when they have concluded their contract, they submit the final state to the chain. If at any point a participant (supposedly) violates the rules, e.g., attempts to falsify the outcome of a transaction, or become unavailable, the last unanimously agreed transaction is posted to the blockchain. From this state, participants can then safely resume their interaction on the chain, where the blockchain protocol enforces the honest execution of the contract.

More specifically, a channel constructs off-chain peer-to-peer connections between all channel participants. The off-chain channel replicates an on-chain state machine, e.g., a smart contract, called the *channel contract*. Both can be modelled with state i and state transition function *step*. The transition function takes a set of commands cmd_{i+1} and transitions the state from i to $i+1$. Channels typically transition through the following lifecycle phases [6]:

1. **Establishment**: All participants agree on a channel contract, which encodes the rules of the channel and the initial state. Usually, this phase involves on-chain activity, such as locking collateral or deploying the contract.
2. **Transition(s)**: A participant proposes a new state transition $step_{i+1}$ through a cryptographically signed message. Every other participant verifies that $step_{i+1}$ is in conformance with their local version of the contract. If so, they confirm by signing it and sending their signature to every other participant. Once a participant receives all other signatures on transition $i+1$, they can consider the result of $step_{i+1}$ the new valid state.
3. **Dispute**: If a participant did not receive all signatures on a state transition in time, or they receive conflicting transitions, they must assume that some fault occurred. They must now start a dispute process on the blockchain. To do so, the participant submits the last unanimously signed state transition to the channel contract, which will install the result as new current state. This closes the off-chain channel and starts the dispute window. Should a participant submit a state transition that has already been superseded by a

more recent one, so called *stale state*, another participant must notice and submit the most recent state.

4. **Closure**: After a channel closes and the dispute window elapses, future transitions can be sent directly to the channel contract, which ensures the continued validity of transitions. A channel can also be closed by unanimous vote, installing the last agreed state transition as final state of the contract.

Channels introduce new assumptions that must hold [6]:

1. **Blockchain Reliability**: The security properties of the blockchain hold and valid transactions submitted to the blockchain are eventually accepted.
2. **At Least One Honest Participant** must be present in the channel to contest faulty state transitions.
3. **Always Online**: State channels require participants to remain online during the entire lifecycle of the channel to prevent execution forks [10], in which a malicious actor starts the dispute phase and submits stale state to the blockchain, e.g., $state_{i-1}$. Honest participants must notice such an attempt and submit $state_i$.[1]

Given these assumptions, channels can achieve safety: the integrity of the contract's state is ensured, even when all parties but one are malicious; and liveness: an honest participant can always advance the contract given a valid transition function, even when all other parties try to stall the process [9].

3 Related Work

The current state of the art in blockchain-based business process enactment focuses on-chain enactment [1]. To reduce the on-chain footprint, related work has focused on improving the cost of the on-chain components. García-Bañuelos et al. [3] introduced an optimised generation of smart contracts through petri net reduction. We extend this approach for BPMN choreographies. López-Pintado et al. [4], and Loukil et al. [5] propose an interpreted approach, where a process model is not compiled but executed by an interpreter component on the blockchain. While the deployment becomes more costly, it leads to cost savings over multiple instance runs.

In a recent survey, we pointed to layer two technologies as a promising research direction [1]. To the best of our knowledge, this is the first work to investigate layer two state channels for enacting business processes. The state of the art in state channel[2] constructions focuses on the formalisation and security of protocols for general applications (e.g., [9,12,13]). More in line with our

---

[1] We generally assume a party "looks after themselves", and follows a strategy with the highest payoff.

[2] Hyperledger Fabric uses the terminology of channels for their subnet functionality [11]. The similarity to state channels is weak; like subnets, fabric channels partition the on-chain ledger. State channels construct off-chain channels and use the security guarantees of the on-chain ledger as settlement and dispute resolution layer.

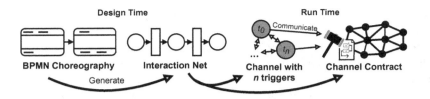

Fig. 1. Overview: From BPMN choreographies to process channels.

work, McCorry et al. [14] present a case study investigating the feasibility and applicability of their state channel construction. They present a template for migrating an existing smart contract to a state channel construction. While they present a state channel architecture for n parties, they chose a two-party game as their case. In two-party games, participants can take turns. However, when more participants are involved, a schedule becomes necessary [13] (see also Sect. 4.2) We present process channels: a model-driven approach, where a process model is used to automatically generate the entire channel setup. The process control-flow naturally enforces a schedule upon the participants. We discuss the particularities of enforcing choreographies in channels and evaluate our approach quantitatively and qualitatively in comparison to a full on-chain baseline. For our evaluation, we chose two commonly used business processes, which allows us to compare our approach to existing related work.

4   Process Channels

In this section, we first give an overview of our approach. We, then, address a particular challenge when designing n-party state channels: the scheduling problem. After, we can describe our model transformation approach, and, lastly, outline the channel's protocol in detail.

4.1   Overview

In Fig. 1, we show an overview of our approach. At its core is the model-driven engineering paradigm. This paradigm has been found to address many challenges of developing blockchain-based applications, specifically, reduce blockchain specific complexity [15]. Channel constructions exacerbate this complexity issue further. More functionality is required, both on and off the chain, thus, also introducing a new trust concern: the security of the channel protocol. To alleviate this issue, we propose to generate channel components from a process model. We base our approach on BPMN choreography diagrams. These fit well the model of state channels, where the focus is on a set of interconnected autonomous participants, initiating messages according to some schedule. From a BPMN choreography model, a compiler component generates an interaction Petri net. This serves as a middle layer presentation, and allows us to apply optimisation techniques before compilation. From the optimised interaction net, we generate the *channel trigger* components. Furthermore, the process channel contract is generated, which is deployed on the blockchain. Each participant deploys a trigger and exposes it to the channel network. Each trigger is interconnected. Additionally, each trigger must be connected to the blockchain, providing access to the channel contract. We assume that each organisation runs a process-aware information system (PAIS) in their (private) organisational network. The PAIS communicates with the trigger to interact with the network. In particular, the responsibilities of each generated component are as follows.

– The **channel contract** handles channel establishment, disputes, and closure, and is deployed on the blockchain. Upon deployment, the blockchain addresses of the participants are bound to their corresponding role. Should a dispute be triggered, the contract validates the submitted state transition by verifying that it has been signed by all participants. The channel also contains process enactment capabilities: it enforces the honest continuation of the contract, should a dispute have occurred.
– The **channel trigger** communicates with the channel network to enact the process model. Each trigger must be configured with: the identify and secret key of its participant, the blockchain addresses of the other participants, the host information of the other triggers, and the address of the channel contract. Once a trigger receives a request from its PAIS, the trigger performs a conformance check to verify the request locally; when successful, it proposes a new state transition to the network. The trigger monitors continuously whether the channel contract has transitioned into the dispute phase. Should a trigger not be able to advance the process, or receive a non-conforming transition request, it starts a dispute phase invoking the channel contract.

4.2   The Scheduling Problem

In an n-party state channel construction, multiple concurrent proposals can deadlock the protocol, where a subset of participants is waiting for the consent to a proposed state transition, while other participants, in turn, are waiting for the consent of a concurrently proposed transition [13]. In 2-party state channels, participants can simply take turns (e.g., in [12]). However, for n-parties, this problem constitutes a leader election problem. A probable solution is the utilisation of a leader election algorithm; however, this would introduce communication overhead and is not done in practice. Instead, this problem is either not addressed in literature, or a specific schedule of turns is enforced over all participants (see [16] for a survey). However, the scheduling of a process is a well understood problem in the world of business process management. A process model inherently contains rules considering the order of events, while a process choreography contains rules regarding the roles and interactions of participants [17, Chapter 6]. We can make direct use of this control-flow information to derive a valid schedule to be enforced within the state channel.

4.3   Optimised Generation

As outlined, channel contract and trigger components require process enactment capabilities. We generate these from a BPMN choreography model. Our approach, hereby, is based on the optimised translation technique presented in García-Bañuelos et al. [3]: a process model is converted into a Petri net, this net is then reduced according to well-established equivalence rules. From the optimised net, code is generated. In the code, the process state is then encoded as efficient bit array. While [3] is based on BPMN process models, we use BPMN choreography models. Thus, our approach is based on interaction Petri nets, which are a special kind of labelled Petri nets. Interaction Petri nets have been proposed as the formal basis for BPMN choreographies [18]. As labels, they store the initiator and respondent information, which are essential for the channel transitions. After conversion, we apply the same reduction rules as in [3]. For this contribution, we limit the scope to choreography tasks, start and end events, and parallel and exclusive gateways. As in [3], this also supports looping behaviour. In contrast to [3], we must restrict enforcement to certain roles: only initiators are allowed to enforce tasks.[3] Here, we can differentiate between *manual* and *autonomous transitions*. Manual transitions correspond to tasks that are initiated by a participant; these must be explicitly executed. Autonomous transitions are the remaining silent transitions. Converting a process model into a Petri net creates silent transitions, and while most of them can be deleted through reduction, some cannot be removed without creating infinite-loops [3]. These transitions must then be performed by the blockchain autonomously, given that the correct conditions are met. Consequently, these transitions are not bound to a role. The differentiation allows a more efficient execution: if the conditions for a manual task are met, it is fired and terminated; further autonomous transitions may be fired, without requiring further manual transitions.

4.4   Channel Protocol

Once the components are generated, they execute the channel protocol. In the following, we outline the protocol of our channel construction, based on the channel lifecycle model introduced in Sect. 2.

Establishment. For the following, we assume all channel triggers are deployed and have established, secure connections. From here, blockchain addresses are exchanged between participants. Any participant can now deploy the channel contract. The deployment initialises the contract, binding all addresses to their role and setting the initial state. The address of the deployed contract is then distributed to all participants; these verify the contract to ensure it was initialised

---

[3] A choreography task can be one-way or two-way: i.e., it optionally includes a response. We assume that a choreography task is one-way; two-way tasks can be regarded as syntactic sugar and adding support for those is no conceptual challenge.

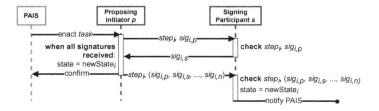

Fig. 2. Sequence diagram for a channel transition (happy path).

with the correct addresses and the correct initial process state.[4] If the contract passes verification, the contract is accepted as channel contract.

Transition. We depict the protocol for a state transition for a sequence flow from state $state_{i-1}$ to $state_i$ in Fig. 2. The PAIS sends an enactment request for a task to its corresponding channel trigger. The task must encode all required information to compute the new process state and can additionally include arbitrary data. The trigger verifies that the enactment of the task is in conformance with its local process state. It then becomes the *proposing initiator p* for this task, and prepares the state transition proposal $step_i$. Each transition proposal is assigned a sequence number i, which is incremented after each successful transition. $step_i$ includes the proposed task and resulting state $newState_i$.[5] The trigger cryptographically signs the proposed transition and sends $step_i$, and its signature $sig_{i,p}$ to all other participants, called the *signing participants*. All signing participants verify that $step_i$ was proposed by the correct initiator by verifying the signature, and that $step_i$ leads to the next conforming state. If all checks pass, each signing participant s signs the new step and sends their signature $sig_{i,s}$ back to the initiator. Once the initiator has collected all signatures $(sig_{i,s}, ..., sig_{i,n})$, it can accept $newState_i$ as new state of the process. It now confirms the transition proposal by sending the set of signatures to all signing participants. These also verify the signatures and, when all checks pass, can also accept $newState_i$. All participants must store the received signatures and corresponding transition proposals, as they are required should a dispute occur.[6]

As we have discussed in Sect. 4.2, a problem of n-party state channels are multiple concurrent state transition proposals. Imposing an order of transitions is, therefore, paramount. Using the control-flow information of the model, it is often trivial to enforce such an order. Consider, a simple sequence flow from

---

[4] This procedure can be made easier by forcing deployment from an agreed upon channel factory contract [2, Chapter 7.4.4].

[5] To prevent the replay of transitions across cases, instances, or blockchains, unique identifiers must also be included, e.g., case ID, instance ID, and chain ID.

[6] To reduce the amount of messages, confirmations can be prepended to a transition proposal. That is, once an initiator has collected all signatures for $step_i.$, it only sends the confirmation to the next initiator. The next initiator prepends the confirmations to the next transition proposal $step_{i+1}$.

task t to task t'. The ordering $t < t'$ naturally follows. This is less so when control-flow branches after gateways. For exclusive gateway branches, it suffices to collapse the possible branches into one during run time. That is, the conditions present on the outgoing sequence flows of the gateway must be based on internal channel state and be part of the transition proposal, so it can be made available to the blockchain in case of a dispute; it can not be based on external state. This requirement can be lifted when all immediately following tasks belong to the same initiator, as it then becomes a (private) choice of a single participant.

Parallel gateways, on the other hand, permit concurrent behaviour and a unique order is not enforceable. Real concurrency is a non-trivial problem in state channel constructions [16]. As each state transition must encode a sequence number and the new global process state, a concurrent execution can deadlock the off-chain protocol. We, thus, require that tasks on parallel branches are serialised at run time into any order chosen by the participants. Should a deadlock still occur, the blockchain contract can always enforce an order during a dispute phase, as the blockchain ledger enforces a unique order of transactions.

Dispute. At any point, the channel contract is in a certain state i. A participant can trigger a dispute by providing all participants' signatures for state transition $step_j$, where $i < j$. Once the contract is in dispute mode, additional state can be submitted until the dispute window elapses. Then, the contract continues on-chain. We describe possible dispute scenarios.

- *Non-Conforming Transition*: Consider a non-conforming transition is proposed in the channel. Assuming there is at least one honest participant, the transition would not acquire unanimous consent. A participant can trigger a dispute submitting the last agreed transition to the blockchain; thus, forcing the continuation of the contract on the blockchain. Faulty participants can, however, also stop to take part in the protocol, which we discuss next.
- *Unavailability:* Consider a transition is proposed to the channel. After a local timeout, the initiator does not receive signatures from all participants. To ensure liveness, a dispute must be triggered to force the transition on-chain. Consider the reverse: after a local timeout, a signing participant does not receive the expected transition proposal or confirmation. The participant can trigger a dispute. Now, the initiator has to perform the transition on-chain or be identified as participant stalling the process, which could be penalised.

Closure. Once participants reach the end event in unanimous consent, they submit the final state to the channel contract. Otherwise, the end event is reached on-chain. In both scenarios, the process terminates and a new case can be instantiated. The contract assigns a new case ID and resets the process state.

5   Implementation and Evaluation

To enable the evaluation of our approach, we developed two prototypes, *Chorpiler* and *Leafhopper*. We perform a quantitative and qualitative evaluation and compare our approach to an on-chain enactment baseline. For the quantitative evaluation, we use process models from well known cases from literature. We verify the correctness of our implementation by replaying process traces and perform benchmarks to assess cost. For the qualitative evaluation, we discuss the required assumptions and provided guarantees of process channel enactment in comparison to full on-chain enactment.

5.1 Implementation and Setup

Chorpiler and Leafhopper. We have developed two tools, Chorpiler and Leafhopper. Chorpiler implements the optimised generation of enactment components, as outlined in Sect. 4. It is capable of generating process channel contracts and on-chain enactment contracts in Solidity, as well as enactment functionality in TypeScript, which is used in the channel triggers. The static trigger component capabilities, e.g., routing, signature verification etc., are implemented in Leafhopper and run on *Node.js*[7]. Leafhopper uses Chorpiler to generate the process channel contract and the enactment capabilities of the channel trigger. For each participant in the choreography, a trigger is deployed. For ease of deployment, each trigger is run in a *Docker* container and the trigger network can be deployed using *Docker Compose*.[8]

Benchmark Setup. We benchmark the supply chain [19] (adapted from [20]) and incident management [21] case, which are well known from related work. To help assess our approach, we compare it to a baseline. The baseline provides the same model support as Leafhopper, but enacts the process completely on-chain, as in related work. For each case, we generate the baseline, channel triggers, and channel contract. The triggers are run in a local network. The smart contracts are deployed to an Ethereum environment; we use the Ethereum simulation tool *Ganache*.[9] Following open science principles, and to enable replicability, we made both our prototypes, evaluation scripts, and data publicly available.[10]

5.2 Quantitative Evaluation

Correctness. We verify that the network only accepts conforming traces and always remains in a stable state, i.e., all triggers report the same state after some finite time. To do so, we follow the methodology outlined in [19]. For each case, we replayed all conforming traces (two for supply chain and four for the incident management case). After, we generated 2000 non-conforming traces; to do so, a conforming trace was randomly manipulated by one of the following

---

[7] See *Node.js*, https://nodejs.org/en, accessed 2023-03-17.
[8] See *Docker Compose*, https://docs.docker.com/compose, accessed 2023-03-17.
[9] See *Ganache*, https://trufflesuite.com/ganache, accessed 2023-03-17.
[10] Leafhopper is available at https://github.com/fstiehle/leafhopper. The repository includes instructions and scripts to automate the replication of our evaluation. Chorpiler is available at https://github.com/fstiehle/chorpiler.

Table 1. Gas cost of Leafhopper in relation to the baseline.

Case	Baseline			Leafhopper				
	Deployment	Avg. Exec.		Deployment	Exec. Best Case	Avg. Exec.		
		Case	Task			Bad Case	Worst Case	On-Chain Task
Supply Chain	396.732	347.076	34.708	772.282	88.319	310.186	495.207	38.691
Incident Mgmt.	408.954	190.509	31.752	784.823	88.319	199.545	328.889	34.774

operations: add an event, remove an event, and swap the position of two events.[11] We replayed these traces from a local script which, for each event, connects to the corresponding initiator to propose the task.[12] All events were classified correctly w.r.t. conformance, and after each trace replay the channel was in a stable state.

Cost Analysis. We compare the cost of our baseline to Leafhopper. For the baseline, we replayed, for each case, all conforming process variants (two for supply chain and four for the incident management case) and recorded the gas cost of all interactions with the blockchain. As gas costs are deterministic, multiple runs of the same variant are not required.

Cost in Leafhopper is more difficult to assess and is driven by the cost for the channel establishment (deployment of the contract) and successful closure (unanimous submission of the final state) or dispute. To study the cost behaviour, we performed, analogous to our baseline, for each conforming process variant the following benchmarks:

1. A **best case** run with no disputes, where the channel is unanimously closed.
2. A **bad case** run, where a dispute is triggered after half of the process. As a result, the other half must be completed on the blockchain.
3. A **worst case** run, where a dispute with stale state is made immediately after the start event. An honest participant then submits the last agreed-upon state, and the entire remaining process must be completed on-chain.

Table 1 shows the recorded gas costs for our benchmark experiments.[13] Compared to our baseline, Leafhopper incurs around twice the cost for *deployment* due to the implemented channel capabilities. However, the *best case* execution considerably improves upon the avg. execution cost of the baseline. Furthermore, we can see that the best case execution cost is fixed. It requires one round of signature verification. The cost, hence, does not depend on the complexity of

---

[11] We removed any coincidentally created conforming traces. In total we replayed 1812 non-conforming traces to the incident mgmt. and 1933 to the supply chain case.

[12] Normally, the local trigger would also verify the request and only forward valid requests. We disabled this functionality to allow us to simulate a faulty component.

[13] While our baseline is based on [3], it incurs increased gas cost (compare with Table 2), as it additionally implements role enforcement (c.f. Sect. 4.3).

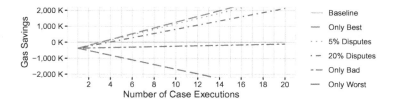

Fig. 3. Gas savings compared to baseline, when the initial deployment is re-used.

Table 2. Gas cost of Leafhopper in relation to other approaches.

Case	Approach	Deployment	Avg. Case Execution	Leafhopper Overhead			
				Deployment	Best Case	Bad Case	Worst Case
Supply Chain	García-Bañuelos et al. [3]	298.564	272,186	1.6	−0.7	0.1	0.8
	Caterpillar [22]	1,100,590	566,861	−0.3	−0.8	−0.5	−0.1
	ChorChain [23]	2,802,543	1,156,734	−0.7	−0.9	−0.7	−0.6
	CoBuP [5]	4,832,706	254,661	−0.8	−0.6	0.2	0.9
Incident Mgmt.	García-Bañuelos et al. [3]	345.743	166,345	1.3	−0.5	0.2	1.0
	Caterpillar [4]	1,119,803	324,420	−0.3	−0.7	−0.4	0
	ChorChain [23]	3,278,656	1,028,505	−0.8	−0.9	−0.8	−0.7
	CoBuP [5]	4,639,652	249,378	−0.8	−0.6	−0.2	0.3

the process or its tasks, only on the number of participants—five participants for both cases.

The *bad case* execution cost is still lower for the supply chain case and slightly higher for the incident management case compared to the avg. execution cost of the baseline. The incident management case is on average shorter (six versus ten tasks of the supply chain case) but has the same number of participants. Thus, the fixed state verification cost has a higher impact on the total cost of the shorter process. The *worst case* cost is considerably higher than the baseline's average. This is expected, as it constitutes two state submissions and the enactment of the entire remaining process on the chain. Should a dispute occur, Leafhopper exhibits slightly (around 10%) higher cost for enacting a singular task on the blockchain. This is the result of having to determine whether a dispute is currently active.

Cost Under Different Dispute Scenarios. To put the cost of Leafhopper into context, we assess the relative overhead of Leafhopper compared to the baseline when considering different dispute rates. That is, the number of expected disputes over multiple case runs. We illustrate this in Fig. 3, where we show how the average cost savings for a case execution on Leafhopper develops for different dispute scenarios, when the initial deployment is re-used. We depict five scenarios, only best case runs, 5% disputes, 20% disputes, only bad, and only worst case runs. For the 5% and 20% runs, we assume an equal share of bad and worst cases. Both percentages are taken from an industry survey, where 5%

constitutes the best and 20% the worst measured average contract dispute rate by industry sector.[14]

We can see that, when considering only best cases, the deployment costs of Leafhopper is amortised after three case runs. The 5% dispute rate follows closely, while the 20% run requires 4 cases to break even. Furthermore, we can see that less than one best case run can more than amortise one worst case run.

Cost Compared to Related Work. In Table 2, we compare the cost of Leafhopper to selected approaches from literature by reporting its relative overhead.[15] We chose García-Bañuelos et al. [3] for its efficient implementation technique; Caterpillar [22] for providing the most complete feature set; and the choreography-based approaches ChorChain [23] (compiled) and CoBuP [5] (interpreted).

The results of this comparison are in line with our above analysis. Leafhopper's best case provides vast cost improvements (between 1/10 to 1/2 of the cost). The medium case overhead ranges from big improvement (1/10) to slightly worse (20% more expensive) depending on the efficiency and feature support of the approach. The worst case cost ranges from double the cost to still considerably (3/10) cheaper. Notably, Caterpillar and ChorChain exhibit the highest cost. However, both provide more features. ChorChain additionally implements answer and response patterns and does not implement net reduction and encodes the process state as simple array type, leading to increased cost. This shows the potential of Leafhopper to improve the cost of more complex implementations with its fixed state verification cost.

5.3   Qualitative Evaluation

To gain a more holistic understanding of the proposed approach, we now move to a higher level of abstraction and compare the process channel approach to our baseline on-chain approach, which enacts the entire process on-chain, on the basis of relevant quality attributes. As such, we assess which guarantees the approaches provide, and which assumptions must hold. Xu et al. [2, Chapter 1.4] identify the main non-functional properties that blockchain provides: immutability, non-repudiation, integrity, transparency, and equal rights. We summarise our assessment in Table 3.

Assumptions. Both approaches assume the reliability of the blockchain. In addition, process channels require at least one honest participant in a channel; otherwise, colluding participants can install arbitrary state. In contrast, an on-chain approach can still rely on other validators in the blockchain network to verify

---

[14] IACCM: *Are you in an adversarial industry? Insights for contract negotiators and managers.* 2014. https://wp.me/pa5oX-RH, accessed 2023-03-28.

[15] Due to the different feature sets being supported, these approaches incur different gas costs; cost should not be understood as the only yardstick to compare approaches by. Since our approach in this paper is quite different from full on-chain approaches, we find this comparison worthwhile reporting.

Table 3. Assumptions and guarantees of on-chain and process channel enactment.

Assumptions	On-Chain	Process Channel
Blockchain Reliability	✓	✓
At Least One Honest Participant	✕	✓
Explicit Role-Binding	✕	✓
Participants are Always Available	✕	✓
Security of Off-Chain Protocol	✕	✓
Guarantees		
Integrity	✓	✓ [§]
Immutability	✓	✓ [†]
Non-Repudiation	✓	✓ [†*]
Transparency	✓	✓ [§]
Equal Rights	✓	✓ [§]

[*]Stalling is a non-attributable fault
[†]Requires storage of proof
[§]Requires access to channel

a transaction. Process channels also require explicit role-binding: roles must be bound to (trigger) hostnames and blockchain accounts. This information must be propagated through the channel. Additionally, participants joining the network must deploy trigger components. This inhibits process flexibility. Also, participants in the channel must be always available, they cannot go offline in-between tasks. They must sign transitions and watch for disputes. Some usage scenarios, e.g., energy-constrained wireless devices, may be ruled out by this requirement. Finally, channels introduce additional components and, thus, additional attack surface.

Guarantees. As only channel participants see and verify transitions, integrity and transparency can only be demonstrated to participants with access to the channel. Also, equal rights can only hold within the channel. Immutability and non-repudiation require that transition proposals are stored durably by a participant. Otherwise, a faulty participant can submit stale state. Additionally, under certain circumstances a participant can stall the process without being identified as doing such. It is undecidable whether an initiator is stalling the process by not sending the next transition proposal or whether a signing participant has refused to sign it [14]. While the process will be forced to continue on-chain, the channel contract cannot attribute who was at fault, limiting the use of penalties.

5.4   Discussion

In Sect. 5.3, we analysed the assumptions and guarantees of off-chain enactment. When these assumptions are met, the channel can offer comparable guarantees to an on-chain approach. However, additional complexity is introduced by requiring dedicated channel components, which require a certain degree of redundancy, as they must stay online. While choreographies already require participants to handle messages reliably, process channels increase this reliability requirement. All participants must stay online to advance the choreography and prevent execution forks. Additionally, process data is no longer durably stored on the blockchain; where such durability is required, it needs to be ensured by the participants off-chain. Finally, the closed nature of the channel hinders flexibility. Third parties can only verify the honest execution of the process if given access to all messages in the channel.

In Sect. 5.2, we reported on the gas cost of our prototype Leafhopper. While gas costs are an especially limiting factor when enacting processes on public blockchains, they can also limit scalability of private and permissioned deployments, as gas costs are directly linked to the number, computational complexity, and storage requirement of transactions. We found that, for processes which are repeated more than 3–4 times without disputes, Leafhopper can significantly reduce the gas cost of on-chain enactment. In Leafhopper, costs are highly dependent on how often a dispute occurs. If we assume the off-chain protocol will resolve trivial faults, such as temporary connectivity issues, the blockchain will be mainly involved in resolving permanent faults, such as malicious acts or long-term crashes, which constitute a contract breach. For our benchmarked cases, a worst case run was amortised after only one dispute-free run. In the case of interorganisational processes, organisations generally form a collaboration to work towards a common business goal; and under typical industry dispute rates, Leafhopper was able to significantly reduce cost. Nonetheless, Leafhopper's cost is case specific and not as predictable as on-chain enactment. This introduces additional uncertainties. We have provided an analysis to help gauge the cost of Leafhopper. Still, future studies are required to address these uncertainties and aspects not covered in our current evaluation, like latency.

Beyond cost, process channels have the potential to improve further dimensions such as latency and confidentiality. On-chain approaches are highly dependent on the underlying latency of the blockchain. Process channels can reduce this reliance. Blockchain latency differs greatly between different blockchain platforms and the required trust assumptions of the application; related work reports latency ranging from seconds (e.g., [19,24]) to minutes and more (e.g., [25]). Furthermore, blockchain consensus can be highly probabilistic, resulting in high latency outliers (see e.g., [24,25]). Similarly, channels reduce the exposure of data. While confidentiality is breached during a dispute phase, there is potential to design the dispute phase in a confidentiality-preserving manner; for example, by utilising zero-knowledge proofs. While these are usually costly operations, they would only be required in the case of a dispute, making their use viable [26]. There are further topics not addressed and left to future work.

- *Length of Dispute Window*: The choice of the dispute window is an important factor. It must be chosen so that an honest participant has time to react to stale state. Thus, it must be multiples of the underlying blockchain latency.

There is currently no consensus in literature on how to determine this. In the BPM context, the particular business case may also influence this choice.

– *Dispute Phase Design*: In a choreography, a participant that stops to collaborate will stall the process indefinitely. A more advanced dispute design could penalise faulty participants and replace them. There is potential to design dispute processes, specific to business cases, to incentivise honest participation. However, such a design is limited by non-attributable faults.

– *Channel Networks:* In channel networks, multiple channels are supported by one *root contract*. In our current design, the channel smart contract is application specific. Exploring a design where a contract can support multiple processes could pave the way toward a network of cost efficient, blockchain-based choreographies.

6   Conclusion

In this paper, we propose to address challenges in inter-organizational process enactment by moving to a layer two approach: blockchain-based state channels. With this approach, we aim to reduce the on-chain footprint. The quantitative evaluation shows a significant reduction in gas cost for common settings. The qualitative evaluation shows that the blockchain properties largely remain intact when moving to our channel approach—as long as the assumptions are met, such as having at least one honest participant per channel.

Moving communication and state into channels may, in the future, prove useful to achieve lower latency and improved confidentiality—but those aspects were out of scope for this paper, where we focused on the principled approach and extensive evaluation. Future work will, thus, address latency and confidentiality. Beyond that, we outlined a multitude of other research opportunities, such as the design of process channel networks, where multiple channels are supported by a singular on-chain contract.

References

1. Stiehle, F., Weber, I.: Blockchain for business process enactment: a taxonomy and systematic literature review. In: Marrella, A., et al. (eds.) BPM 2022, vol. 459, pp. 5–20. Springer, Cham (2022). https://doi.org/10.1007/978-3-031-16168-1_1
2. Xu, X., Weber, I., Staples, M.: Architecture for Blockchain Applications. Springer, Cham (2019). https://doi.org/10.1007/978-3-030-03035-3
3. García-Bañuelos, L., Ponomarev, A., Dumas, M., Weber, I.: Optimized execution of business processes on blockchain. In: Carmona, J., Engels, G., Kumar, A. (eds.) BPM 2017. LNCS, vol. 10445, pp. 130–146. Springer, Cham (2017). https://doi.org/10.1007/978-3-319-65000-5_8
4. López-Pintado, O., Dumas, M., García-Bañuelos, L., Weber, I.: Interpreted execution of business process models on blockchain. In: EDOC, pp. 206–215. IEEE (2019)
5. Loukil, F., Boukadi, K., Abed, M., Ghedira-Guegan, C.: Decentralized collaborative business process execution using blockchain. WWW **24**(5), 1645–1663 (2021). https://doi.org/10.1007/s11280-021-00901-7

6. Gudgeon, L., Moreno-Sanchez, P., Roos, S., McCorry, P., Gervais, A.: SoK: layer-two blockchain protocols. In: Bonneau, J., Heninger, N. (eds.) FC 2020. LNCS, vol. 12059, pp. 201–226. Springer, Cham (2020). https://doi.org/10.1007/978-3-030-51280-4_12

7. Buterin, V.: A Next-Generation Smart Contract and Decentralized Application Platform (2014). https://ethereum.org/en/whitepaper. Accessed 29 Mar 2023

8. Poon, J., Dryja, T.: The bitcoin lightning network: scalable off-chain instant payments (2016). https://lightning.network/lightning-network-paper.pdf. Accessed 29 Mar 2023

9. Miller, A., Bentov, I., Bakshi, S., Kumaresan, R., McCorry, P.: Sprites and state channels: payment networks that go faster than lightning. In: Goldberg, I., Moore, T. (eds.) FC 2019. LNCS, vol. 11598, pp. 508–526. Springer, Cham (2019). https://doi.org/10.1007/978-3-030-32101-7_30

10. McCorry, P., Bakshi, S., Bentov, I., Meiklejohn, S., Miller, A.: Pisa: arbitration outsourcing for state channels. In: AFT 2019, pp. 16–30. ACM (2019)

11. Androulaki, E., Barger, A., Bortnikov, et al.: Hyperledger fabric: a distributed operating system for permissioned blockchains. In: EuroSys, pp. 1–15 (2018)

12. Dziembowski, S., Faust, S., Hostáková, K.: General state channel networks. In: ACM SIGSAC CCS, pp. 949–966 (2018)

13. Dziembowski, S., Eckey, L., Faust, S., Hesse, J., Hostáková, K.: Multi-party virtual state channels. In: Ishai, Y., Rijmen, V. (eds.) EUROCRYPT 2019. LNCS, vol. 11476, pp. 625–656. Springer, Cham (2019). https://doi.org/10.1007/978-3-030-17653-2_21

14. McCorry, P., Buckland, C., Bakshi, S., Wüst, K., Miller, A.: You sank my battleship! a case study to evaluate state channels as a scaling solution for cryptocurrencies. In: Bracciali, A., Clark, J., Pintore, F., Rønne, P.B., Sala, M. (eds.) FC 2019. LNCS, vol. 11599, pp. 35–49. Springer, Cham (2020). https://doi.org/10.1007/978-3-030-43725-1_4

15. Di Ciccio, C., et al.: Blockchain support for collaborative business processes. Informatik Spektrum **42**(3), 182–190 (2019)

16. Negka, L.D., Spathoulas, G.P.: Blockchain state channels: a state of the art. IEEE Access **9**, 160277–160298 (2021)

17. Weske, M.: Business Process Management: Concepts, Languages, Architectures, 3rd edn. Springer, Heidelberg (2019). https://doi.org/10.1007/978-3-642-28616-2

18. Decker, G., Weske, M.: Local enforceability in interaction petri nets. In: Alonso, G., Dadam, P., Rosemann, M. (eds.) BPM 2007. LNCS, vol. 4714, pp. 305–319. Springer, Heidelberg (2007). https://doi.org/10.1007/978-3-540-75183-0_22

19. Weber, I., Xu, X., Riveret, R., Governatori, G., Ponomarev, A., Mendling, J.: Untrusted business process monitoring and execution using blockchain. In: La Rosa, M., Loos, P., Pastor, O. (eds.) BPM 2016. LNCS, vol. 9850, pp. 329–347. Springer, Cham (2016). https://doi.org/10.1007/978-3-319-45348-4_19

20. Fdhila, W., Rinderle-Ma, S., Knuplesch, D., Reichert, M.: Change and compliance in collaborative processes. In: IEEE SCC, pp. 162–169 (2015)

21. OMG: BPMN 2.0 by Example, Version 1.0 (2010). https://www.omg.org/cgi-bin/doc?dtc/10-06-02. Accessed 29 Mar 2023

22. López-Pintado, O., García-Bañuelos, L., Dumas, M., Weber, I., Ponomarev, A.: Caterpillar: a business process execution engine on the Ethereum blockchain. Softw. Pract. Exp. **49**(7), 1162–1193 (2019)

23. Corradini, F., Marcelletti, A., Morichetta, A., Polini, A., Re, B., Tiezzi, F.: Engineering trustable and auditable choreography-based systems using blockchain. ACM TMIS **13**(3), 1–53 (2022)

24. Corradini, F., et al.: Model-driven engineering for multi-party business processes on multiple blockchains. Blockchain Res. Appl. **2**(3), 100018 (2021)
25. Prybila, C., Schulte, S., Hochreiner, C., Weber, I.: Runtime verification for business processes utilizing the Bitcoin blockchain. FGCS **107**, 816–831 (2020)
26. Zhang, Y., Long, Y., Liu, Z., Liu, Z., Gu, D.: Z-channel: scalable and efficient scheme in zerocash. Comput. Secur. **86**, 112–131 (2019)

Action-Evolution Petri Nets: A Framework for Modeling and Solving Dynamic Task Assignment Problems

Riccardo Lo Bianco[1]([✉]), Remco Dijkman[1], Wim Nuijten[1,2], and Willem van Jaarsveld[1]

[1] Eindhoven University of Technology, Eindhoven, The Netherlands
{r.lo.bianco,r.m.dijkman,w.p.m.nuijten,w.l.v.jaarsveld}@tue.nl
[2] Eindhoven Artificial Intelligence Systems Institute, Eindhoven, The Netherlands

Abstract. Dynamic task assignment involves assigning arriving tasks to a limited number of resources in order to minimize the overall cost of the assignments. To achieve optimal task assignment, it is necessary to model the assignment problem first. While there exist separate formalisms, specifically Markov Decision Processes and (Colored) Petri Nets, to model, execute, and solve different aspects of the problem, there is no integrated modeling technique. To address this gap, this paper proposes Action-Evolution Petri Nets (A-E PN) as a framework for modeling and solving dynamic task assignment problems. A-E PN provides a unified modeling technique that can represent all elements of dynamic task assignment problems. Moreover, A-E PN models are executable, which means they can be used to learn close-to-optimal assignment policies through Reinforcement Learning (RL) without additional modeling effort. To evaluate the framework, we define a taxonomy of archetypical assignment problems. We show for three cases that A-E PN can be used to learn close-to-optimal assignment policies. Our results suggest that A-E PN can be used to model and solve a broad range of dynamic task assignment problems.

Keywords: Petri Nets · Dynamic Assignment Problem · Business Process Optimization · Markov Decision Processes · Reinforcement Learning

1 Introduction

During the execution of a business process, tasks become executable and resources become available to execute these tasks. As resources are assigned to tasks, they become unavailable to execute other tasks. Consequently, continuously assigning the right task to the right resource is essential to run a process efficiently. This problem is known as dynamic task assignment. The dynamic task assignment problem can be seen as a particular case of the *dynamic assignment problem*, which, according to [1], is the problem of assigning a fixed number of

© The Author(s), under exclusive license to Springer Nature Switzerland AG 2023
C. Di Francescomarino et al. (Eds.): BPM 2023, LNCS 14159, pp. 216–231, 2023.
https://doi.org/10.1007/978-3-031-41620-0_13

individuals to a sequence of tasks, such as to minimize the total cost of the allocations, which may include setup costs, travel costs, or other time-varying costs. This problem has been extensively studied in business process optimization [2] as well as related areas, such as manufacturing [3]. For the sake of brevity, we will employ the term "assignment problem" to indicate the general dynamic (task) assignment problem.

To solve an assignment problem, it must first be modeled mathematically. Markov Decision Processes (MDPs) are a common technique for modeling assignment problems [4], and they are the standard interface for Reinforcement Learning (RL) algorithms [5]. The basic definition of MDP involves a single agent interacting with an environment to maximize a cumulative reward, which is a global signal of the goodness of the actions chosen by the agent during a (possibly infinite) sequence of system states. In the context of business process optimization, the environment is the business process that must be executed, and the agent decides which task to assign to which resource. The reward is calculated based on what we want to optimize in the process, such as the total time resources spend working, the total cost of employing the resources, or the time customers spend waiting. While MDPs provide a good formalism for modeling the agent's behavior, they consider the environment, in our case the business process, as a black box that provides rewards for the decisions taken by the agent without exposing its internal behavior. Moreover, they do not have an agreed-upon syntax and lack any type of graphical representation. On the other hand, (Colored) Petri Nets [6] are a well-known formalism for modeling a business process but have no inherent mechanisms for modeling and calculating the best decision in a given situation. Also, frameworks exist for many mathematical optimization techniques, such as linear programming and constraint programming, where problems can be modeled and solved without additional effort. However, no such framework exists for dynamic task assignment problems.

To fill this gap, this paper presents a unified and executable framework for modeling assignment problems. We use the term "unified" to refer to the capability of expressing both the agent and the environment of the assignment problem in a single standardized notation, thus simplifying the modeling of new problems. We use the term "executable" to refer to the possibility of using the models to train and test decision-making algorithms (specifically RL algorithms) without additional effort. To this end, we propose a new artifact in the form of a modeling language with a solid mathematical foundation, namely A-E Petri Net (A-E PN), which draws from the well-known Petri Net (PN) formalism to model assignment problems in a readable and executable manner. This paper pays particular attention to embedding the A-E PN formalism in the RL cycle, such that RL algorithms can be trained and used to solve assignment problems without additional effort.

The proposed artifact is evaluated by modeling and solving a set of archetypical assignment problems. A taxonomy of assignment problem variants is proposed, and an example for each of the three main variants is modeled through A-E PN. An RL algorithm is trained on each instance, achieving close-to-optimal results. Apart from modeling each assignment problem as an A-E PN, no additional effort is required to achieve these results, empirically demonstrating that A-E PN constitutes a unified and executable framework for modeling and solving assignment problems.

Against this background, the remainder of this paper is structured as follows. Section 2 is dedicated to a review of relevant literature. Section 3 introduces Timed-Arc Colored Petri Nets (T-A CPN). Section 4 is devoted to the formal definition of Action-Evolution Petri Net and the description of the integration of A-E PN in the classic RL loop. In Sect. 5, an essential taxonomy of assignment problem variants is presented. A problem instance for each variant is modeled through A-E PN, and a RL algorithm is trained on each instance, obtaining close-to-optimal results. Section 6 discusses the proposed method's benefits and limitations and delineates the next research steps.

2   Related Work

To the best of our knowledge, this paper presents the first attempt at defining a unified and executable framework for assignment problems. In contrast, the relation between (generalized stochastic) Petri Nets and Markov Chains is well studied [7], but Markov Chains cannot be used to model and optimize (task assignment) decisions. Since Markov Decision Processes can be seen as an extension to Markov Chains, the idea of extending Petri Nets to model Markov Decision Processes follows naturally. Several attempts at this exist in the literature, but none focus on the assignment problem. An overview of existing frameworks for modeling and solving dynamic optimization problems is presented in Table 1, listing, for each framework, the Petri Net variant employed, the scope of applicability, and whether the framework is unified and executable. The current work is presented in the last line.

Table 1. Comparison of existing frameworks for dynamic optimization.

Reference	PN	Scope	Unified	Executable
[8]	FPN	Problems expressible as finite MDPs	Yes	Yes*
[9]	DPN	Problems expressible as finite MDPs	No	Yes
[10]	GSPN	A single power management problem	Yes	No
[11]	TCPN	A single manufacturing scheduling problem	Yes	No
[12]	TCPN	Manufacturing scheduling problems	Yes	No
This paper	A-E PN	Assignment problems	Yes	Yes

*No executable example is provided.

In [8], the authors define a CPN variant: Factored Petri Net (FPN). In FPNs, the transition probabilities are defined explicitly, and a reward is attached to each network state. A limitation of [8] is that actions must be input marks from a single source transition (a transition without input arcs), while our framework allows actions to be defined anywhere in the Petri net, thus allowing for more modeling flexibility.

In [9], the authors propose the Decision Petri Net (DPN) formalism. In DPN, the network is partitioned into a probabilistic network, in which transition probabilities are determined on arcs, and a non-deterministic network, corresponding to the actions that can be taken at a given moment by the decision maker. In our framework, we remove the need for two separate subnets and model the agents as tokens in the network, obtaining a unified representation. Both [8], and [9] require the number of states in the system to be finite, whereas our approach does not rely on states enumeration.

In [10], the authors propose a model for a power-managed distributed computing system that is based on the Generalized Stochastic Petri Net (GSPN) formalism and provide a translation to the equivalent continuous-time MDP. The work demonstrates the expressive power of PN variants, but the resulting model is not executable. Also, the paper presents a single case study, while our approach is demonstrated to be generally applicable to modeling and solving problems with different characteristics.

In [11], a manufacturing scheduling problem is modeled using Timed Colored Petri Nets (TCPN). The search for an optimal policy is implemented using Q-learning, where each action corresponds to a complete schedule, which is a path from the initial marking to a final marking of the TCPN representing the system, whereas in our case, an action corresponds to a single assignment, which allows for more flexible modeling of decisions. Moreover, [11] only covers a single case study, relying heavily on problem-specific heuristics.

In [12], the authors provide an example usage of TCPN in the context of manufacturing systems, focusing on reinforcement learning as solving approach. While [12] highlights the relationship between TCPN and RL, TCPNs are used only to describe the environment and not to train or test solving algorithms. In contrast, our work provides a unified and executable framework.

3   Preliminaries

This section provides the formal definition of Colored Petri Net (CPN) and Timed-Arc Colored Petri Net (T-A CPN), which will be used to define the new formalism.

Colored Petri Net (CPN) [6] is an extension of Petri Nets (PN) in which tokens have different characteristics called colors. In the remainder of this section, we rely on the CPN definition provided in [13].

Definition 1 (Colored Petri Net). *A CPN is defined as a tuple $CPN = (\mathcal{E}, P, T, F, C, G, E, I)$, such that:*

- *$\mathcal{E}$ is a finite set of types called color sets. Each color set must be finite and non-empty.*
- *P is a finite set of places.*

- T *is a finite set of transitions, such that* $P \cap T = \emptyset$
- $F \subseteq P \times T \cup T \times P$ *is a finite set of arcs.*
- $C : P \to \mathcal{E}$ *is a color function that maps each place p into a set of possible token colors. Each token on p must have a color that belongs to the type* $C(p)$, *which is called the place's color set.*
- G *is a guard function. It is defined from* T *into expressions such that for each* $t \in T$, $G(t)$ *is a Boolean expression and* $Type(Var(G(t))) \subseteq \mathcal{E}$, *where* $Type(x)$ *denotes the type of x and* $Var(f)$ *denotes the set of free variables in the function f.*
- E *is an arc expression function. It is defined from* F *into expressions such that for each* $f \in F$, $Type(E(f)) = C(P(f))_{MS}$ *and* $Type(Var(E(f))) \subseteq \mathcal{E}$ *where* $P(f)$ *is the place of f. This means that each evaluation of the arc expression must yield a multi-set (indicated by the MS subscript) over the color set attached to the corresponding place.*
- I *is an initialization function. It is defined from* P *into expressions such that* $\forall p \in P : Type(I(p)) = C(p)_{MS}$. *The initialization function determines the network's initial marking.*

Definition 2 (Marking). *A marking of a CPN is a function M, such that for each place* $p \in P$, *it defines a multi-set of colors* $C(p) \to \mathbb{N}$, *which maps each possible color of the place to the number of times it occurs.*

For a place p with colors $C(p) = \{c_1, c_2\}$, we also write $M(p) = c_1^n c_2^m$ to denote that p has n) token with color c_1 and m tokens with color c_2. Since a marking is a multi-set, multi-set operations, such as $\geq$, $+$, and $-$, are available on markings.

Definition 3 (Binding). *For a transition t, the variables* $Var(t) = Var(G(t)) \cup \{Var(E(f)) | f \in F, T(f) = t\}$ *represent the set of variables from the guard function and the expressions on its arcs, where* $T(f)$ *is the transition of arc f.*

A binding of a transition $t \in T$ *is a function Y that maps each* $v \in Var(t)$ *to a color, such that* $\forall v \in Var(v) : Y(v) \in Type(v)$ *and* $G(t)\langle Y \rangle$ *evaluates to true, where* $f\langle Y \rangle$ *denotes the evaluation of a function f with its free variables bound as Y.*

For a transition t with variables $Var(t) = \{v_1, v_2\}$, we also write $Y(t) = \langle v_1 = c_1, v_2 = c_2 \rangle$ to denote that the binding Y assigns color c_1 to variable v_1 and color c_2 to variable v_2.

We now define the behavior of a CPN through its firing rules.

Definition 4 (CPN Firing Rules).

1. *A transition t is enabled in marking M for binding Y if and only if* $\forall (p, t) \in F : M(p) \geq E((p,t))\langle Y \rangle$.
2. *An enabled transition can fire, changing the Marking M into a marking* M', *such that* $\forall p \in P : M'(p) = M(p) - E((p,t))\langle Y \rangle + E((t,p))\langle Y \rangle$.

The standard CPN definition assumes that the effect of a firing is always instantaneous. To account for time, we will refer to a modified version of the Timed-Arc Petri Net (T-A PN) formulation [14]. Our version defines a global clock, updated according to a next-event time progression. This is also the time management paradigm implemented in CPN Tools [15], a widely adopted software for Petri Nets modeling.

Definition 5 (Timed-Arc Colored Petri Net). *A T-A CPN is defined by a tuple $TACPN = (\mathcal{E}, P, T, F, C, G, E, I)$, where P, T, F, C, G, I are as in Definition 1, and $\mathcal{E}$ and E are adapted as follows:*

- *$\mathcal{E}$ is a finite set of timed types called timed color sets. A color of a timed color set has both a value v and a time τ, we also denote this as $v@\tau$.*
- *E is an arc expression function. It is defined from F into tuples of two elements. For a given $f \in F$, $E(f)_0$ is defined the same as E in Definition 1 and $E(f)_1$ is a scalar increment, thus $\forall f \in F : Type(E(f)_1) = \mathbb{N}$, that indicates the generated tokens' time with reference to the global clock. The second tuple element is ignored for arcs outgoing from places and incoming to transitions since the scalar increment is only used when producing new tokens.*

Note that each color now has a time and consequently, each color in a marking and in a binding has time. For example, we can refer to the marking of a place p with $M(p) = c_1@2^1 c_1@3^5$ as the marking that has one token with color c_1 at time 2 and five tokens with color c_1 at time 3. With some abuse of notation, we will allow arc expression functions $E(f)_0$, to ignore the time element of colors and leave it unaffected, and we will denote with $c@e$ that an expression e only changes the time element of a timed color.

We also extend the concept of marking to account for the presence of a global clock, which we need further on in the paper to define the transition rules for A-E PN.

Definition 6 (Timed Marking). *A timed marking is defined as the tuple $TM = (M, \tau)$, where M is a marking and τ is the current value of the global clock.*

The T-A CPN firing rule can then be expressed as follows:

Definition 7 (T-A CPN Firing Rules).

1. *Let t be a transition that is enabled in marking M for binding $Y = \langle v1 = c_1@\tau_1, v2 = c_2@\tau_2, \ldots, v_n = c_n@\tau_n \rangle$ as in Definition 4 (using only E_0 for E). The enabling time of the transition, denoted τ_E, is $max(\tau_1, \tau_2, \ldots, \tau_n)$.*
2. *An enabled transition t is time-enabled in timed marking (M, τ), if its enabling time τ_E is less than or equal to τ, and there exists no transition t' that is enabled in marking M for some binding Y' with enabling time $\tau'_E \leq \tau_E$.*
3. *A transition t that is time-enabled in timed marking (M, τ) for binding Y with enabling time τ_E can fire, changing the timed marking to (M', τ_E), where M' is constructed, such that $\forall p \in P : M'(p) = M(p) - E((p, t))_0 \langle Y \rangle + E((t, p))_0 \langle Y \rangle @\tau_E + E((t, p))_1$.*

4. When there exists no t in timed marking (M, τ), for which there is a binding Y, such that t is time-enabled, the global clock τ is increased until there is.

In practice, point 4 can be performed by evaluating bindings that are enabling but not time-enabling. The binding that leads to the lowest enabling time reveals the minimal increase of the global clock, making it possible to update the global clock using a next-event time progression.

4   Action-Evolution Petri Nets

This section extends the definition of T-A CPN to provide a model that can automatically learn close-to-optimal task assignment policies. This extension is called Action-Evolution Petri Nets (A-E PN). The new elements are first described informally, then a formal definition is provided. Finally, the definition is incorporated into the RL cycle, allowing for automated learning of close-to-optimal task assignment policies.

4.1   Tags and Rewards

The overall objective of A-E PN is to mimic the behavior of an agent that observes changes in the environment and acts upon those changes when possible. We will thus extend the CPN definition provided in the background section to distinguish two separate types of transitions:

- **Actions:** transitions that represent actions taken by the agent. In the context of assignment problems, the firing of an action transition represents a single assignment.
- **Evolutions:** transitions that represent events happening in the system independently of the actions taken by the agent. The firing of an evolution transition represents a single event in the environment, for example, the arrival of a new order.

This distinction is expressed by associating every transition with a *transition tag*, that can be either A (action) or E (evolution), through a *transition tag function* L. We also extend the concept of marking to embed a *network tag* l, which can assume a single value in $\{A, E\}$: only transitions associated with a tag of the same type as the one in the network tag are allowed to fire. The network tag l must be updated every time no transitions with the same tag are available for firing. The *tag update function* S performs the update by changing the network's tag from A to E or vice versa: $S(l) = A$, if $l = E$; $S(l) = E$, if $l = A$. We use the term *tag time frame* to refer to the period between changes in the network tag.

The objective of the RL cycle is the maximization of a cumulative reward over a (possibly infinite) horizon. To track rewards in A-E PN, we introduce a *transitions reward function* $\mathcal{R}$ that associates a reward to the firing of any transition, and we embed the total reward accumulated by firing transitions, which we call *network reward* ρ, in the network's marking. In general, a reward can be produced by any change in the environment, regardless of whether an action or an evolution produced such change. For this reason, a reward is produced due to the firing of any transition, regardless if the transition is tagged as an action or an evolution. To comply with the classic RL cycle, rewards associated with evolutions are accumulated and awarded to the last action taken, eventually after a normalization operation (see Subsect. 4.3).

To further clarify the basic mechanisms of A-E PN, the example in Fig. 1 provides an overview of a sequence of firings.

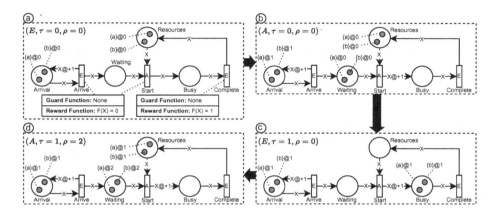

Fig. 1. A sequence of firings in a simple task assignment problem.

The network shows the evolution of a system with two types of tasks, a and b, and two employees, one that can undertake only task a and one that can undertake only task b. A task of each type arrives at every clock tick, and an employee is assigned to a task of the same type. Assignments take one clock tick to complete, and a reward of 1 is produced every time an assignment is completed. The parentheses on the top right corner contain the components of the tagged marking that are not directly represented as network elements. Guard functions and reward functions are associated with single transitions. Timed tokens and arcs follow the notation introduced in Definition 5. The initial marking is presented in the dotted square a, in which only E transitions are enabled. After two firings of transition *Arrive*, consuming both tokens in the *Arrival* place (in any order), no evolution transitions are available, so the tag is updated, and the system transitions to state b. Notice that the transition from e to a does not produce a clock update, since actions are available to be taken at time 0. In b, transition *Start* is enabled. In this case, the RL agent would have two available actions: pairing task a with resource a, or pairing task b with resource b. In this case, both actions will be taken sequentially, in any order, leading to tagged marking c, while in the general case, choices would have to be made by a decision algorithm on which assignments to make. In c, the network tag is again E, and two transitions are associated with time-enabled steps: *Arrive* and *Complete*. The firing of *Arrive* produces two new tokens at time 1 in the *Waiting* place, while the firing of *Complete* places two tokens back in the *Resources* place at time 1 and generates a network reward increment of 2 units in state d.

4.2   Formal Definition of Action-Evolution Petri Net

To provide a formal definition of A-E PN, we must adapt three definitions from T-A CPN: the net itself, the marking, and the firing rules.

Definition 8 (Action-Evolution Petri Net). *Let $\mathcal{T} = \{A, E\}$ be a finite set of tags representing actions and evolutions, and $S : \mathcal{T} \to \mathcal{T}$ a network tag update function. An Action-Evolution Petri Net (A-E PN) is as a tuple $AEPN = (\mathcal{E}, P, T, F, C, G, E, I, L, l_o, \mathcal{R}, \rho_0)$, where $\mathcal{E}, P, T, F, C, G, E, I$ follow Definition 5, and:*

- *$L : T \to \mathcal{T}$ is a transition tag function that maps each transition t to a single tag. Only transitions associated with the same tag as the network can fire.*
- *$l_0 \in \mathcal{T}$ is a singleton containing the network's initial tag, usually equal to E.*
- *$\mathcal{R} : T \to (f : \mathbb{R})$ associates every transition with a reward function. The function can take timing properties or numbers of tokens (representing completed cases) as parameters, thus allowing for flexibility in modeling reward.*
- *$\rho_0 \in \mathbb{R}$ is the initial network reward, usually equal to 0.*

Definition 9 (Tagged Marking). *A tagged marking is a tuple $TM = (M, l, \tau, \rho)$, where the tuple (M, τ) is a timed marking, as in Definition 6, $l \in \mathcal{T}$ is the network tag at the current time τ, and $\rho \in \mathbb{R}$ is the total reward accumulated until the current time τ.*

Definition 10 (A-E PN Firing Rule).

1. *A transition t is tag-enabled in a tagged marking (M, l, τ, ρ) for binding Y if and only if t is enabled in M according to Definition 1, and $L(t) = l$.*
2. *Let t be a transition that is tag-enabled in tagged marking (M, l, τ, ρ) for binding $Y = \langle v1 = c_1 @ \tau_1, v2 = c_2 @ \tau_2, \ldots, v_n = c_n @ \tau_n \rangle$. The enabling time of the transition, denoted τ_E, is $max(\tau_1, \tau_2, \ldots, \tau_n)$.*
3. *An enabled transition t is tag-time-enabled in tagged marking $TTM = (M, l, \tau, \rho)$, if its enabling time τ_E is less than or equal to τ, and there exists no transition t' that is enabled in tagged marking TTM for some binding Y' with enabling time $\tau'_E \leq \tau_E$.*
4. *A transition t that is tag-time-enabled in tagged marking (M, l, τ, ρ) for binding Y with enabling time τ_E can fire, changing the tagged marking to (M', l, τ_E, ρ'), where M' is constructed, such that $\forall p \in P : M'(p) = M(p) - E((p,t))_0 \langle Y \rangle + E((t,p))_0 \langle Y \rangle @ \tau_E + E((t,p))_1$ and $\rho' = \rho + \mathcal{R}(t)$.*
5. *When there exists no t in tagged marking $TTM = (M, l, \tau, \rho)$, for which there is a binding Y, such that t is time-enabled, the set of all transitions is partitioned in two disjoint sets: $T_{current} = \{t \in T | L(t) = l\}$ and $T_{next} =$*

$\{t \in T | L(t) \neq l\}$. Let $\tau_{current}$ be the minimum value for which a transition in $T_{current}$ is time-enabled (according to Definition 7), and let τ_{next} be the minimum value for which a transition in T_{next} is time-enabled. Note that $\tau_{current}$ and τ_{next} can be undefined.

- If $\tau_{current}$ is defined, and $\tau_{current} \leq \tau_{next}$ or τ_{next} is undefined, only the global clock is updated, leading to a new tagged marking $TTM' = (M, l, \tau_{current}, \rho)$.
- If τ_{next} is defined, and $\tau_{current} > \tau_{next}$ or $\tau_{current}$ is undefined, both the global clock and the network tag are updated, leading to a new tagged marking $TTM' = (M, S(l), \tau_{next}, \rho)$.

4.3   Extending the Reinforcement Learning Loop

Having completely defined the characteristics of the A-P PN formalism, we can clarify how it can be used to learn optimal task assignment policies (i.e. mapping from observations to assignments) by applying it in a Reinforcement Learning (RL) cycle. Figure 2 shows the RL cycle. In every step in the cycle, the agent receives an observation (a representation of the environment's state), then it produces a single action that it considers the best action for this observation. The action leads to a change in the environment's state. The environment is responsible for providing a reward for the chosen action along with a new observation. Then the cycle repeats, and a new decision step takes place. The MDP formulation is the standard framework for training an agent to take actions that lead to the highest cumulative reward.

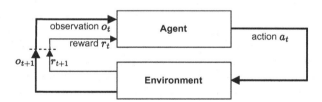

Fig. 2. A common representation of the RL training cycle [5].

In recent years, the embedding of neural networks in RL algorithms gave birth to the field of Deep Reinforcement Learning (DRL), achieving breakthroughs in settings such as playing board games [16] and robotic manipulation [17], as well as successful applications in domains like industrial process control [18], and healthcare [19]. With the proliferation of robust DRL algorithms, the main hurdle in modeling new problems is the definition of the environment, which is usually represented as a black box, as in Fig. 2, thus leaving the implementation of the system's dynamics entirely to the modeler. The lack of a standardized interface makes the creation of new environments time-consuming and dependent on the modeler's coding skills. Moreover, even introducing small changes potentially requires substantial effort once the environment has been modeled. These observations motivate the effort to provide a unified and executable framework. In Fig. 3, the classic RL cycle is extended to account for the presence of A-E PN. The main element is the A-E PN, which acts as a simulator for the whole process.

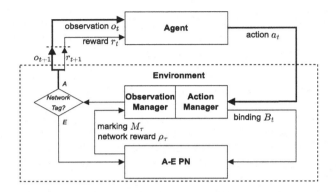

Fig. 3. The reinforcement learning cycle with A-E PN

The A-E PN communicates with the agent through two sub-components: *observation manager* and *action manager*. The observation manager is invoked every time the tagged marking changes, regardless if due to a firing or not. The new reward is stored, and the network tag is evaluated: if the tag is E, no action is required, and the control is given back to the A-E PN, which can fire a new E transition. If the tag is A, the accumulated rewards are added up, and the result is divided by $1 + (\tau_{t+1} - \tau_t)$. The resulting value is returned to the agent as r_{t+1}. The reward value takes into account the possible misalignment between clock ticks (τ) and RL steps (t), given by the fact that multiple actions can happen at the same τ. The observation manager also returns to the agent the new observation o_{t+1}. For the set of experiments presented in the next section, the observation is built as a vector containing, for each place, the number of tokens of each color in the place's color set. The action manager is invoked every time the agent chooses an action a_t, which it transforms into the corresponding binding B_t (associated with an action transition) to be fired.

5   Evaluation

This section aims to show that A-E PN constitutes a unified and executable framework for expressing dynamic task assignment problems with different characteristics: in fact, all the examples were modeled using a single notation (except for color-specific functions on arcs, guards, and rewards) and a RL algorithm was trained on each problem, without any additional development effort.

We provide a (non-exhaustive) taxonomy of assignment problem variants based on [20]. We distinguish three archetypes of assignment problems.

- **Assignment Problem with Compatibilities:** resources are assigned to tasks according to a measure of compatibility. Two problem subclasses can be formulated:
  - **Assignment Problem with Hard Compatibilities:** resources can only be assigned to tasks if they are compatible. The dynamic task assignment problem in Subsect. 5.1 falls into this subclass.
  - **Assignment Problem with Soft Compatibilities:** resources can always be assigned to tasks, but different assignments result in different system behaviors. An example of such a problem is if multiple resources can perform a task, but some will be faster at it than others.
- **Assignment Problem with Multiple Assignments:** the same resource can be assigned to multiple tasks, or the same task can be assigned to multiple resources. Two problem subclasses can be formulated:
  - **Assignment Problem with Resource Capacity:** resources have a maximum capacity of tasks that they can undertake before being considered full. In the simple case each resource can only be busy with a single task at a time. The dynamic bin packing problem in Subsect. 5.2 provides a more elaborate example.
  - **Assignment Problem with Task Capacity:** tasks have a minimum capacity of resources to be assigned to them before processing. In the simple case each tasks needs exactly one resource.
- **Assignment Problem with Dynamic Resources' Behavior:** resources have dynamic behavior. Two problem subclasses can be formulated:
  - **Assignment Problem with Action-Dependent Dynamic Resources' Behavior:** resources change their attribute values as the consequence of taking actions. The dynamic order-picking problem in Subsect. 5.3 falls into this category.
  - **Assignment Problem with Action-Independent Dynamic Resources' Behavior:** resources change their attribute values as the consequence of evolutions in the environment. For example, resources may take breaks or go on holidays.

In the following sections, one example is detailed for each archetype. An example for each subclass is implemented in the provided Python package.

5.1 Dynamic Task Assignment Problem with Hard Compatibilities

Let us consider a system that solves a task assignment problem, similar to the one presented in Fig. 1. At every clock tick, two tasks arrive: one has type $r1$ and the other $r2$. Two resources are available for the assignment: one can only undertake tasks of type $r1$, while the other can undertake tasks of type $r1$ or $r2$. Once a task is assigned to a resource, completion always takes one clock tick, after which the resource becomes available for a new assignment. A resource cannot work on multiple tasks at the same time. A network reward of 1 is returned every time a task is assigned to a resource and every time an assignment completes, leading to a theoretical maximum reward of 200 over 100 clock ticks. The problem can be fully expressed in terms of A-E PN, as reported in Fig. 4.

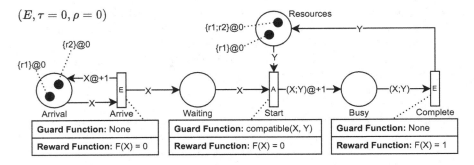

Fig. 4. A-E PN initial marking for the dynamic task assignment problem

5.2 Dynamic Bin Packing Problem

In this scenario, we model a dynamic version of the bin packing problem where items (the problem tasks, characterized by their *weight*) arrive sequentially and they must be allocated to two bins (the problem resources, characterized by the total weight of objects in the bin *curr* and the bin's total capacity *tot*) that are emptied at every clock tick (except for the first, which is used to generate the objects to be put in the bins). The fullness of the bins before being emptied gives the measure of goodness of the object's allocation, quantified as the weight of objects in the bin divided by the total bin capacity. This problem showcases how tokens' colors can be used to model non-trivial reward functions. In the example reported, three objects arrive in the system at every clock tick, one of weight 1 and two of weight 2. Two initially empty bins are available, one with capacity 2 and one with capacity 3. The optimal allocation would give a reward of 2, leading to a theoretical maximum reward of 200 over a 100 clock ticks horizon. The A-E CPN formalization of the problem is reported in Fig. 5.

5.3 Dynamic Order-Picking Problem

In this section, we present an example of action-dependent resource behavior (i.e. the agent taking decisions on the actions that it performs). The example is a simple order-picking problem in which a single agent (the resource) moves on a squared grid of size 2, trying to pick orders (the tasks). The agent's and the orders' colors are characterized by two parameters representing the coordinates on the grid (infinite capacity is assumed). The agent starts in position (0,0) and can move left, right, up, or down, but not over a diagonal. If an order is in the same position as the agent, the latter can use an action to pick the order. A

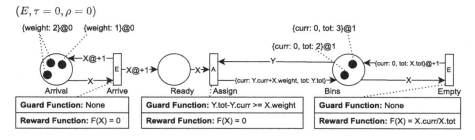

Fig. 5. A-E PN initial marking for the dynamic bin packing problem

single order arrives at every clock tick, always in position $1,1$, and the order stays on the grid for exactly one clock tick, according to a time-to-live (TTL) parameter. The agent's objective is to pick as many orders as possible, so it gets a reward of 1 every time an order is picked, leading to a theoretical maximum reward of 98 over a 100 clock ticks horizon (at least two orders will be lost due to the agent moving to position $(1,1)$. The problem is formulated in terms of A-E PN in Fig. 6.

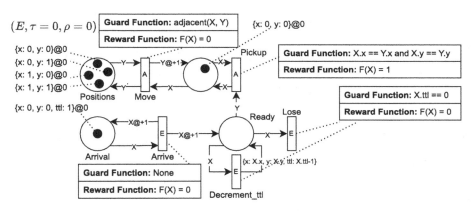

Fig. 6. A-E PN initial marking for the dynamic order-picking problem

5.4   Experimental Results

All experiments were implemented in a proof-of-concept package[1], relying on the Python programming language and the widely adopted RL library Gymnasium [21]. Proximal Policy Optimization (PPO) [22] with masking was used as the training algorithm. Specifically, the PPO implementation of the *Stable Baselines* package [23] is used. Note, however, that the mapping from each A-E PN to PPO was automated and requires no further effort from the modeler. The PPO algorithm was trained on each example for (10^6 steps with 100 clock

---

[1] The code is publicly available in https://github.com/bpogroup/aepn-project.

ticks per episode, completed in less than 2300 seconds on a mid-range laptop, without GPUs), always using the default hyperparameters. The experimental results were computed on (network) rewards obtained by the trained agent and following a random policy over 1000 trajectories, each of duration 100 clock ticks. In Table 2, the average and standard deviations of rewards obtained by the trained PPO are compared to those of a random policy on each of the three presented problem instances, with reference to the maximum attainable reward. In all cases, PPO shows to be able to learn a close-to-optimal assignment policy.

Table 2. The results for the three presented problem instances.

Instance	Random	PPO	Optimal
Task Assignment	186.894 ± 2.084	199.852 ± 0.398	200
Bin Packing	186.746 ± 1.941	199.963 ± 0.186	200
Order Picking	6.046 ± 2.585	96.776 ± 2.019	98

6   Conclusions and Future Work

This paper presented a framework for modeling and solving dynamic task assignment problems. To this end, it introduced a new variant of Petri Nets, namely Action-Evolution Petri Nets (A-E PN), to provide a mathematically sound modeling tool. This formalism was integrated with the Reinforcement Learning (RL) cycle and consequently with existing algorithms that can solve RL problems. To evaluate the general applicability of the framework for modeling and solving task assignment problems, a taxonomy of archetypical problems was introduced, and working examples were provided. A DRL algorithm was trained on each implementation, obtaining close-to-optimal policies for each example. This result shows the suitability of A-E PN as a unified and executable framework for modeling and solving assignment problems.

While the applicability of the framework was shown, its possibilities and limitations are yet to be fully explored. This will be done in future research by expanding the provided taxonomy of assignment problems and considering different problem classes.

Acknowledgement. The research that led to this publication was partly funded by the European Supply Chain Forum (ESCF) and the Eindhoven Artificial Intelligence Systems Institute (EAISI) under the AI Planners of the Future program.

References

1. Kuhn, H.W.: The Hungarian method for the assignment problem. Nav. Res. Logist. Q. **2**(1–2), 83–97 (1955)
2. Gülpınar, N., Çanakoğlu, E., Branke, J.: Heuristics for the stochastic dynamic task-resource allocation problem with retry opportunities. Eur. J. Oper. Res. **266**(1), 291–303 (2018)

3. Hu, L., Liu, Z., Hu, W., Wang, Y., Tan, J.: Petri-net-based dynamic scheduling of flexible manufacturing system via deep reinforcement learning with graph convolutional network. J. Manuf. Syst. **55**, 1–14 (2020)
4. Spivey, M.Z., Powell, W.B.: The dynamic assignment problem. Transp. Sci. **38**(4), 399–419 (2004)
5. Sutton, R., Barto, A.: Reinforcement Learning: An Introduction, p. 352. The MIT Press, Cambridge (2014)
6. Jensen, K.: A brief introduction to coloured Petri Nets. In: Brinksma, E. (ed.) TACAS 1997. LNCS, vol. 1217, pp. 203–208. Springer, Heidelberg (1997). https://doi.org/10.1007/BFb0035389
7. Bause, F., Kritzinger, P.: Stochastic Petri nets: an introduction to the theory. ACM SIGMETRICS Perform. Eval. Rev. **26**(2), 2–3 (1998)
8. Eboli, M.G., Cozman, F.G.: Markov decision processes from colored Petri nets. In: da Rocha Costa, A.C., Vicari, R.M., Tonidandel, F. (eds.) SBIA 2010. LNCS (LNAI), vol. 6404, pp. 72–81. Springer, Heidelberg (2010). https://doi.org/10.1007/978-3-642-16138-4_8
9. Beccuti, M., Franceschinis, G., Haddad, S.: Markov decision Petri net and Markov decision well-formed net formalisms. In: Kleijn, J., Yakovlev, A. (eds.) ICATPN 2007. LNCS, vol. 4546, pp. 43–62. Springer, Heidelberg (2007). https://doi.org/10.1007/978-3-540-73094-1_6
10. Qiu, Q., Wu, Q., Pedram, M.: Dynamic power management of complex systems using generalized stochastic Petri nets. In: Proceedings of the 37th Conference on Design Automation - DAC 2000, pp. 352–356. ACM Press (2000)
11. Drakaki, M., Tzionas, P.: Manufacturing scheduling using colored Petri nets and reinforcement learning. Appl. Sci. **7**(2), 136 (2017)
12. Riedmann, S., Harb, J., Hoher, S.: Timed coloured Petri net simulation model for reinforcement learning in the context of production systems. In: Behrens, B.-A., Brosius, A., Drossel, W.-G., Hintze, W., Ihlenfeldt, S., Nyhuis, P. (eds.) WGP 2021. LNPE, pp. 457–465. Springer, Cham (2022). https://doi.org/10.1007/978-3-030-78424-9_51
13. Jensen, K., Rozenberg, G.: High-Level Petri Nets: Theory and Application. Springer, Heidelberg (1991). https://doi.org/10.1007/978-3-642-84524-6
14. Jacobsen, L., Jacobsen, M., Møller, M.H., Srba, J.: Verification of timed-arc Petri nets. In: Černá, I., et al. (eds.) SOFSEM 2011. LNCS, vol. 6543, pp. 46–72. Springer, Heidelberg (2011). https://doi.org/10.1007/978-3-642-18381-2_4
15. CPN Tools. https://cpntools.org/
16. Silver, D., et al.: Mastering the game of Go without human knowledge. Nature **550**(7676), 354–359 (2017)
17. Kalashnikov, D., et al.: QT-Opt: scalable deep reinforcement learning for vision-based robotic manipulation, November 2018
18. Nian, R., Liu, J., Huang, B.: A review on reinforcement learning: introduction and applications in industrial process control. Comput. Chem. Eng. **139**, 106886 (2020)
19. Yu, C., Liu, J., Nemati, S.: Reinforcement learning in healthcare: a survey, April 2020. arXiv:1908.08796 [cs]
20. Pentico, D.W.: Assignment problems: a golden anniversary survey. Eur. J. Oper. Res. **176**(2), 774–793 (2007)
21. Gymnasium. https://gymnasium.farama.org/
22. Schulman, J., Wolski, F., Dhariwal, P., Radford, A., Klimov, O.: Proximal Policy Optimization Algorithms, August 2017. arXiv:1707.06347 [cs]
23. SB3-contr. https://github.com/Stable-Baselines-Team/stable-baselines3-contrib

Context-Based Activity Label-Splitting

Sebastiaan J. van Zelst[1,2]($\boxtimes$) (ID), Jonas Tai[2], Moritz Langenberg[2],
and Xixi Lu[3] (ID)

[1] Fraunhofer Institute for Applied Information Technology,
Sankt Augustin, Germany
`sebastiaan.van.zelst@fit.fraunhofer.de`
[2] RWTH Aachen University, Aachen, Germany
`{jonas.tai,moritz.langenberg}@rwth-aachen.de`
[3] Utrecht University, Utrecht, The Netherlands
`x.lu@uu.nl`

Abstract. The information systems used in companies store event data describing the historical execution of the processes they support. Process mining covers the automated analysis of such data, generating insights that may ultimately lead to process improvement. A core branch of process mining is process discovery, dealing with event-data-based automated discovery of process models. In practice, the same activity may often be executed in a significantly different context, e.g., in a vaccination program, multiple vaccine doses are typically provided at different points in time. Process discovery algorithms assume that all executions of the same activity are to be mapped onto the same modeling element. Consequently, the presence of repeated activity executions under different contexts typically leads to underfitting discovered process models. To this end, activity label-splitting algorithms have been proposed to relabel the recordings of the same activity occurring in significantly different execution contexts. Yet, the state-of-the-art label-splitting algorithm adopts a trace-level-mapping strategy, yielding inferior results in the presence of loop constructs and infeasible computation time. Therefore, this paper proposes a novel label-splitting preprocessing technique that overcomes these issues. Our experiments confirm that our newly proposed label-splitting algorithm outperforms the state-of-the-art.

Keywords: Process mining · Process discovery · Label-splitting

1 Introduction

Most business processes executed in companies in various domains are supported by multiple, often interconnected, information systems. Among storing documents and artifacts, many such information systems, e.g., Enterprise Resource Planning (ERP) systems and Manufacturing Execution Systems (MES), store a digital representation of the historical execution of the processes they support.

© The Author(s), under exclusive license to Springer Nature Switzerland AG 2023
C. Di Francescomarino et al. (Eds.): BPM 2023, LNCS 14159, pp. 232–248, 2023.
https://doi.org/10.1007/978-3-031-41620-0_14

Such data is referred to as *event data*. The intrinsic value of event data is confirmed by the successful application of event-data-driven analysis techniques in various domains, i.e., referred to as *process mining* [1].

In process mining, *process discovery*, which aims at the *automated discovery* of business process models describing the process as recorded, is one of the most prominent tasks [1]. Many process discovery algorithms have been proposed and studied in the literature [5]. While the discovery algorithms have made considerable progress and shown their values in real life, many other challenges remain largely unsolved. One of these challenges is the accurate handling *duplicated tasks* [1,5,19].

A *duplicated task* is manifested as a process task executed at different stages of the process, representing different activities. For example, the patient consultations at the beginning and end of a treatment trajectory are both called *consultations* but refer to different activities. When modeling such a process, a process analyst typically uses two task nodes to represent such duplicated tasks. However, when such tasks are executed, the corresponding events are recorded with the same label, i.e., *consultation*. Most existing discovery algorithms then consider these events to belong to the same activity and discover an overgeneralized loop to capture the behavior in the event log. To tackle this challenge, *activity label-splitting* techniques have been proposed [19,23]. Label-splitting techniques aim to detect the groups of events that refer to the same activity but are executed in a different context and, therefore, should be treated by the process discovery algorithm as conceptually different activities.

It is shown that label-splitting algorithms can significantly improve the quality of subsequently discovered process models [19], yet a relatively limited amount of work has been done in the area. The label-splitting technique proposed in [19] has shown that splitting the labels can lead to discovering more precise process models in some cases. However, the proposed approach uses a brute-force algorithm to find an optimal mapping of the events with the same labels between different traces, also called *trace mapping*, to detect candidate events for label-splitting. When two events have the same label in every trace in the log, this approach has to search 2^N possible mappings to find an optimal solution. As a result, the technique has a high time complexity (thus a poor running time) and has difficulties handling processes with loops. Other techniques use additional contextual information (such as the timestamps of the events) [23]. As a result, these approaches cannot handle an arbitrary log.

Therefore, we propose a novel label-splitting framework that is robust to looping behavior executed in the process and can handle real-life logs. The key artifact of our proposal is an *event graph* connecting all events that describe the same label and are candidates for label-splitting. Several techniques can be applied to detect clusters of equally labeled events with similar contexts, e.g., community detection [13]. The proposed event graph ignores the process instances in which the events occur and focuses on the execution context of the events (e.g., preceding and succeeding activities); the events of the same loop occur in a similar context and, thus, will be clustered automatically. As a result, the approach can handle event logs from the processes with loops. We conducted an extensive range of quantitative experiments to assess our proposed framework. The results of our experiments show that our proposed approach consistently outperforms the state-of-the-art label-splitting preprocessing method.

The remainder of this paper is structured as follows. In Sect. 2, we discuss related work. Section 3 presents background concepts and the notation used in this paper. We present our main contribution in Sect. 4. In Sect. 5, we present the evaluation of our approach. Finally, Sect. 6 concludes this work.

2   Related Work

In this section, we discuss related work. We primarily focus on label-splitting. For a general overview of process mining, we refer to [1]. For an overview of existing process discovery algorithms, we refer to [5]. In terms of event data preprocessing techniques, next to label-splitting, we primarily identify two significant fields of study, i.e., *outlier and noise detection* [14], and *event abstraction* [27].

Various label-splitting methods exist that refine imprecise labels as a *preprocessing step*. Lu et al. [18,19] propose a label-splitting algorithm that refines event labels based on their context similarity by creating a mapping between process traces. The goal is to maximize the pairs of mapped events with similar contexts. For this, they use a cost function based on various aspects like neighbors of the events and location of the events in the trace is used. However, mapping complete traces cannot express the relationship between events within the same trace and leads to various issues in practice, most significantly for traces with loops. Our approach proposes to use an event graph that connects all events of the same label. This allows our approach to cluster the events of a loop and, thus, tackle this limitation.

Tax et al. [23] propose using the timestamps of events to perform label-splitting. The assumption is that when the events occurred at different times of the day, this may suggest the events carry a different meaning (e.g., eating during the morning versus eating during the evening). This leads to good results on the event logs (such as smart devices) that satisfy this assumption, yet, it does not apply to arbitrary process event logs, as this method requires a correlation between the time and execution of events in different contexts. Our approach does not have such assumptions and is generally applicable to any event log. In addition, our approach can be extended to also take additional context (such as timestamps) into account as features, which are used as input for clustering or community detection algorithms.

Another type of approach to label-splitting is the extension of existing process discovery algorithms. For example, Fodina [7] is an extension of the heuristic miner algorithm [25] that introduces a simple label-splitting based on the local context of events. There has also been an effort to extend the α-algorithm [4] to enable it to deal with duplicate tasks [17]. Another class of process discovery algorithms that can apply label-splitting are genetic process discovery algorithms [2,9]. This class of algorithms uses an *evolutionary computational paradigm* to gradually learn process models, naturally supporting adding several modeling elements with the same label. Some *region-based process mining algorithms* also support label-splitting [3,10]. However, the emphasis of label-splitting in these algorithms is to enable the discovered model to describe all observed behavior in the input data. Consequently, these approaches need to take more contexts into account, which can lead to excessive label refinement. Another technique suggested by Vazquez-Barreiros et al. [24] discovers duplicate tasks in an already mined heuristic net or causal net. Finally, Yang et al. [26] use hidden Markov models to discover workflow models and split states during discovery. A general downside of all methods that integrate label-splitting within process discovery is that these methods are tied to their respective discovery algorithms and can not be used as a general preprocessing method. In contrast, our approach is independent of any discovery algorithm and can be seen as a preprocessing step of the event log.

3   Background

This section presents the background concepts used in this paper. After briefly presenting notational conventions, we introduce the notion of *event data*.

A sequence σ of length n over a set X is a function $\sigma \colon \{1, \ldots n\} \rightarrow X$. We write $\sigma = \langle x_1, x_2, \ldots x_n \rangle$, where $x_i = \sigma(i)$ for $1 \leq i \leq n$. The length of a sequence σ is denoted as $|\sigma|$. Given $1 \leq i < j \leq |\sigma|$, we let $\mathsf{sub}(\sigma, i, j) = \langle \sigma(i), \ldots, \sigma(j) \rangle$, i.e., the strict sub-sequence of σ ranging form index i to j. We let X^* denote the set of all possible sequences over X. Given a sequence $\sigma \in X^*$, we let $\mathsf{elem}(\sigma) = \{x | 1 \leq i \leq |\sigma| \wedge (\sigma(i) = x)\}$ to return all elements in σ.

We define an *event log* as follows. Consider Table 1, presenting a simplified example of an event log. Each row refers to an *event*, recording the execution of an activity, e.g., the first row represents a recording of the "Open Expense Report" activity. Each event has a *unique event identifier*. Similarly, each event has a unique *case identifier*, representing the process instance for which the activity was executed. Finally, a *timestamp* is recorded, recording the activity execution time. We formally the notion of event logs as follows.

Definition 1 (Event, Case, Event Log). *Let $\mathcal{C}$ denote the universe of cases, let $\mathcal{E}$ denote the universe of events, and let Σ denote the universe of activity labels. An event $e \in \mathcal{E}$ is a data tuple, recording the historical execution of an activity. We assume that at minimum, an event describes:*

- *An activity attribute, accessed by $\pi_{act}(e) \in \Sigma$,*
- *A timestamp attribute, accessed by $\pi_{time}(e) \in \mathbb{R}^+$.[1]*

A case $c \in \mathcal{C}$ records an instance of the process and describes a collection of events, i.e., $\pi_{events}(c) \subseteq \mathcal{E}$. An event log L is a collection of cases, i.e., $L \subseteq \mathcal{C}$.

---

[1] We assume that, for $t_0, \Delta \in \mathbb{R}^+$, every timestamp t can be represented as $t = t_0 + \Delta$.

Table 1. Simple event log describing recorded process behavior. The event log captures at what point in time an activity was executed for a specific case.

Event ID	Case ID	Activity	Timestamp
1	1	Open Expense Report	26-10-2022 9:40 AM
2	1	Attach Receipts	26-10-2022 9:42 AM
3	1	Send Report	26-10-2022 9:43 AM
4	2	Open Expense Report	26-10-2022 10:21 AM
5	2	Attach Receipts	26-10-2022 10:27 AM
6	2	Write Supporting Motivation	26-10-2022 10:35 AM
7	2	Send Report	26-10-2022 10:42 AM
8	2	Receive Revision Request	26-10-2022 5:25 PM
9	2	Write Supporting Motivation	27-10-2022 9:45 AM
10	2	Send Report	27-10-2022 9:53 AM
11	1	Receive Confirmation	27-10-2022 11:13 AM
12	1	Close Report	27-10-2022 11:14 AM
13	2	Receive Confirmation	28-10-2022 11:18 AM
14	3	Open Expense Report	29-10-2022 11:22 AM
15	3	Attach Receipts	29-10-2022 11:28 AM
16	3	Write Supporting Motivation	29-10-2022 11:36 AM
17	3	Send Report	29-10-2022 11:43 AM
18	3	Receive Revision Request	29-10-2022 3:20 PM
19	3	Write Supporting Motivation	31-10-2022 3:55 PM
20	3	Send Report	31-10-2022 4:27 PM
21	3	Receive Confirmation	31-10-2022 5:16 PM
⋮	⋮	⋮	⋮

We write $\hat{c}$ as a shorthand notation for $\pi_{\texttt{events}}(c) \subseteq \mathcal{E}$. In the context of this paper, we assume that a total order is deterministically available for a case, i.e., $\texttt{seq}(c) \in \hat{c}^*$, s.t., $\texttt{elem}(\texttt{seq}(c)) = \hat{c}$, $|\texttt{seq}(c)| = |\hat{c}|$ and:
$\forall 1 \leq i < j \leq |\texttt{seq}(c)| \; (\pi_{\texttt{time}}(\texttt{seq}(c)(i)) \leq \pi_{\texttt{time}}(\texttt{seq}(c)(j)))$. We also assume that events occur uniquely in one case in an event log, i.e., $\forall c, c' \in L (\hat{c} \cap \hat{c}' \neq \emptyset \implies c = c')$.

4   Event-Graph-Based Label-Splitting

In this section, we present our novel proposed framework for activity label-splitting. In Sect. 4.1, we present a motivating example, which we use as a running example in the remainder of the paper. In Sect. 4.2, we present an overview of our proposed framework. Section 4.3 presents the construction of the event graph, i.e., the foundational artifact of our approach. Section 4.4 briefly discusses graph clustering. We present the relabeling mechanism in Sect. 4.5.

(a) Ground truth process model (in BPMN modeling formalism) describing the normative behavior of the running example process. Our proposed approach successfully leads to *rediscovery* of the ground truth model.

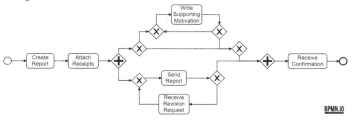

(b) Process model automatically discovered on the running example event data, using the Inductive Miner algorithm [15].

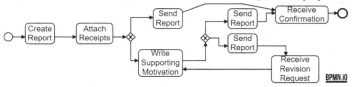

(c) Process model automatically discovered on the running example event data, using the Inductive Miner algorithm [15] in combination with the state-of-the-art label-splitting algorithm [19].

Fig. 1. Example used in this paper, describing a simple compensation request.

4.1 Motivating Example

To motivate our proposed approach and to ease the readability of this paper, we explain the steps of our framework using a motivating running example. We consider a (simplified) reimbursement process. Consider Fig. 1a, in which we depict a process model (using the BPMN modeling formalism) describing the reimbursement process. Firstly, a report is created. Then, receipts are attached to the report. If the total sum of the reimbursement claim is below $500, the report is directly submitted. Subsequently, the report is automatically accepted, and a confirmation is sent out to the applicant. If the sum of the claim is above $500, the applicant writes a supportive motivation, after which the report is submitted. The applicant either receives a confirmation or a revision request.

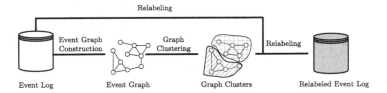

Fig. 2. Schematic overview of the proposed approach. The event log is converted to an event graph in which equally-labeled similar events are connected. Community detection is applied to detect events that depict similar behavioral contexts.

The applicant must revise and resubmit the supporting motivation if a revision request is received. Table 1 captures three executions of the process described.

State-of-the-art approaches for process discovery inaccurately handle event data describing the behavior of Fig. 1a. When applying the Inductive Miner algorithm [15] on a noise-free event log based on the model in Fig. 1a, we obtain the process model depicted in Fig. 1b. Since the discovery algorithm maps all occurrences of the **send report** activity on the same model element, the model is severely underfitting, i.e., it describes many more execution sequences than the process's reference model. When applying the state-of-the-art label-splitting algorithm [19], we obtain the process model depicted in Fig. 1c.[2] Whereas the model discovered by applying the label-splitting algorithm of [19] is *language equivalent* to the ground truth model (cf. Fig. 1a), it does have conceptual quality issues. The model falsely suggests that two distinct "Send Report" activities are possible after writing the supporting motivation, yielding a different outcome. As modeled in the ground truth model, the decision point of confirmation or requiring another revision is made after receiving the report. In contrast, our newly proposed algorithm can, when splitting the "Send Report" activity, rediscover *rediscover* the ground truth model (cf. Fig. 1a).

4.2 Overview

This section presents an overview of our proposed framework for activity label-splitting. Consider Fig. 2, in which we schematically present the basic steps of our framework. The input of our framework is an *event log*. We assume that some activity label $a \in \Sigma$, i.e., we aim to split label a, has been determined in advance by a domain expert. As a first step, the event log is converted into an *event graph* where all events e in the log that describe activity a ($\pi_{\mathrm{act}}(e) = a$) form the vertices of the graph. If they are significantly similar, given some arbitrary context, two vertices are connected. Generally, the context and similarity function are parameters of the approach. Examples include, among others, the resource executing the event, the activities preceding and succeeding the event,

---

[2] The default implementation of the algorithms falsely splits the event data on "Write Supporting Motivation". The model in Fig. 1c is closest to the ground truth model (Fig. 1a) and is obtained by using a custom parameterization of the algorithm.

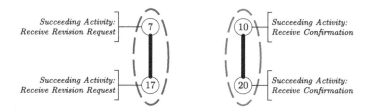

Fig. 3. Example event graph, based on the running example event log (cf. Table 1). The vertices contain the event ids, and the similarity context used is the succeeding activity within the case that the event belongs to. Vertices 7 and 17 have a similar context. Similarly, vertices 10 and 20 have a similar context.

etc. We apply graph clustering on the graph to detect groups of equally labeled events that occur in a different context. All events in the same cluster obtain a "fresh" activity label.

4.3 Event Graph Construction

In this section, we describe the first step of our approach, i.e., *event graph construction*. Events that have the same activity label, i.e., $e, e' \in \mathcal{E}$ s.t. $\pi_{\texttt{act}}(e) = \pi_{\texttt{act}}(e')$, form the vertices of the graph. Two events vertices are connected if, for some *context-based symmetrical similarity function*, their corresponding events are significantly similar. Generally, such a similarity function can be any contextual data feature recorded for the events, e.g., the two activities may be executed by the same resource, the two activities may require the same input document, etc. From a formal perspective, we require the similarity metric to be symmetric.

As a simple example, reconsider Table 1. Observe that the activities "Write Supporting Motivation" and "Send Report" are executed twice for both case 2 and case 3. We decide to apply label-splitting on the "Send Report" activity, hence, the events describing said activity form the vertices in the event graph (events 7, 10, 17, and 20). For simplicity, assume that we use each event's direct succeeding activity as a context (within the same case). For events 7 and 17, followed by events 8 and 18, respectively, the succeeding activity is "Receive Revision Request". Similarly, events 10 and 20 are followed by events 13 and 21, respectively, which both describe the "Receive Confirmation" activity. If we only connect those events in the event graph with the exact same context, i.e., succeeding activity, we obtain the graph depicted in Fig. 3. We define the notion of an event graph as follows.

Definition 2 (Event Graph). *Let $V \subseteq \mathcal{E}$ be a collection of events. Let $\varphi \colon \mathcal{E} \times \mathcal{E} \to [0,1]$ be a context-based symmetric similarity function on the universe of events and let $t_s \in [0,1]$. We let $G_{(\varphi, t_s)} = (V, E)$ be an undirected graph, referred to as the (φ, t_s)-driven event graph of L, where $\{v, v'\} \in E$ iff $\varphi(v, v') \geq t_s$.*

The exact characterization of the similarity function φ is a parameter of our approach, i.e., it depends on the attributes available in the event data as well as the nature of the process and its corresponding logging. Hence, we refrain from providing formal definitions of instantiations of φ, yet, since we assume that at least an activity and timestamp attribute are available, we provide an example instantiation based on control flow. Additionally, note that, in certain scenarios, the φ-value can be used as a weight function on the edges of the event graph.

Let $\sigma = \mathsf{seq}(c)$ for some $c \in \mathcal{C}$ denote a sequence of events (recall $\mathsf{seq}(c) \in \mathcal{E}^*$), let $1 \leq i \leq |\sigma|$ and let $k \in \mathbb{N}$. The sequence $\mathsf{sub}(\sigma, \max(1, i-k), \max(1, i-1))$ describes the preceding k events of the ith event in σ. Similarly, $\mathsf{sub}(\sigma, \min(i+1, |\sigma|), \min(i+k, |\sigma|))$ describes the succeeding k events of the ith event in σ. Clearly, given some $c' \in \mathcal{C}$ with $\sigma' = \mathsf{seq}(c')$ and $1 \leq j \leq |\sigma'|$, we compare the k preceding events of event $\sigma(i)$ in σ with the k preceding events of event $\sigma'(j)$ in σ', e.g., by computing the *edit distance* between $\mathsf{sub}(\sigma, \max(1, i-k), \max(1, i-1))$ and $\mathsf{sub}(\sigma', \max(1, j-k), \max(1, j-1))$. The same can be applied for the k succeeding events. Both distances can subsequently normalized and a weighted average can be computed. Several variations of the above scheme are possible. For example, the event graph in Fig. 3 uses $k = 1$ and ignores the preceding activities.

4.4   Graph Clustering

The second step of our approach entails *global graph clustering* [21]. Any algorithm that computes a *partitioning* of the vertices of an undirected graph based on the graph's topological structure is applicable. For example, *connected components* is used by the approach of [19]. However, we found that *community detection* leads to better results which is why we decided to focus on this clustering method.

Community Detection; Community detection algorithms [13] detect clusters in which the intra-connectivity of vertices of a cluster is high, and the inter-connectivity to vertices in a different cluster is low. Applying community detection allows for detecting clusters, even if the graph is connected. For example, consider the schematic example graph in Fig. 2. The graph itself is connected, yet, two separate communities are identifiable.

Observe that, in the context of our running example, most clustering techniques find the two vertex clusters visualized in Fig. 3, i.e., events 7 and 17, and, events 10 and 20 are grouped together.

4.5   Relabeling

In the final step of our framework, we *relabel* the events that form the clusters in the event graph. In particular, all events belonging to the same cluster obtain the same activity label. Let $G_{(\varphi,k)} = (V, E)$ be a (φ, k)-driven event graph of some event log L, similarity function φ, and threshold k. Further, assume that a clustering algorithm of choice resulted in a partitioning $\mathcal{V} = \{V_1, V_2, \ldots, V_n\}$. To relabel the events in the event log, we return a relabeling function $\lambda\colon V\to\Sigma$, s.t., $\lambda(e) \neq {}'\pi_{\mathtt{act}}(e)$, $\forall 1 \leq i \leq n\, (\forall e, e' \in V_i\, (\lambda(e) = \lambda(e')))$, and $\forall 1 \leq i < j \leq n (\forall e \in V_i e' \in V_j(\lambda(e) \neq \lambda(e')))$. The process discovery algorithm that is used on the event log L uses $\lambda(e)$ as a replacement for $\pi_{\mathtt{act}}(e)$. As an example instantiation of λ, assume the activity label that we aim to split is label $a \in \Sigma$. Given the partitioning $\mathcal{V} = \{V_1, V_2, \ldots, V_n\}$, for $e \in V_1$, we let $\lambda(e) = a_1$, for $e' \in V_2$, we let $\lambda(e') = a_2$, etc.

In practice, the label-splitting algorithm typically outputs two artifacts, i.e., an event log L' and the label function λ. In event log L', the $\pi_{\mathtt{act}}(e)$ values (e.g., `Activity` column in Table 1) are simply overwritten by the λ function. After process model discovery is applied on L', the $\lambda(e)$ values occurring in the discovered model are replaced by their original $\pi_{\mathtt{act}}$ value. The framework can be iteratively applied, i.e., if the new activity labels used for the identified event clusters are unique.

5   Evaluation

In this section, we present the evaluation of our approach. In Sect. 5.1, we discuss the implementation of our approach, followed by the experimental setup in Sect. 5.2. The results are presented in Sect. 5.3.

5.1   Implementation

A public implementation of our framework is available[3]. The implementation supports three types of contexts based on the k preceding/succeeding activities (cf. Sect. 4.3). Based on the length-k preceding and succeeding activity sequences, we support the computation of *normalized edit distance*. Additionally, the sequences can be further abstracted using either the set abstraction (`elem`-function defined in Sect. 3) or the Parikh vector representation [20] (multiset representation counting the occurrences of each activity in the activity sequences). The implementation uses *the Louvain community detection algorithm* [6] for community detection.

The implementation supports *variant compression*. In the compression, all events occurring in cases that describe the same sequence of activities, e.g., cases 2 and 3 in Table 1, are represented by a single unique vertex (observe that all these events have the same control-flow context). The arc weights between the vertices are equal to the sum of the arc weights in the uncompressed event graph. Additionally, self-loops are added to the vertex to represent the similarity of the equal events. Consider Fig. 4, showing a visual example of the application of variant compression. The compression equals the initial data structure used in the Louvain community detection algorithm and is, as such, particularly useful in combination with said community detection algorithm.

---

[3] https://github.com/jonas-tai/python-label-refinement.

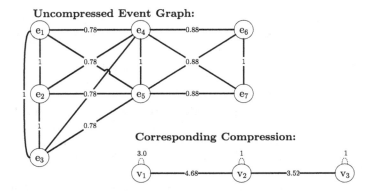

Fig. 4. Example of variant compression. Events $\{e_1, e_2, e_3\}$, $\{e_4, e_5\}$, and $\{e_6, e_7\}$ are part of the same *case variant*, respectively. Vertex v_1 represents $\{e_1, e_2, e_3\}$, vertex v_2 represents $\{e_4, e_5\}$, and vertex v_3 represents $\{e_6, e_7\}$.

5.2   Experimental Setup

In this section, we describe the experimental setup of our experiments. We conduct two sets of experiments, i.e., an experiment using several *synthetic event logs* (part of [19]) with a known ground-truth and an experiment using several *real event logs*, i.e., without any known ground-truth.

We are primarily interested in the *precision* of the discovered process model after applying label-splitting. The precision of the discovered models describes the additional amount of behavior described by the model compared to an event log. Typically, models discovered on the raw data are underfitting and have low precision. As such, we investigate the increased precision of the discovered process models due to label-splitting.

For the synthetic data, we know precisely which events belong to the same "activity cluster". Therefore, we can use *Adjusted Rand Index* (ARI) of the discovered event clustering and the ground truth clustering to measure the general quality of the detection mechanism (i.e., the ARI measures the similarity of two clusterings).

Experiments with Synthetic Data. We compare our approach with and without variant compression to the state-of-the-art approach presented in [19] (we use the same set of data as presented in [19]). In our experiments, we found that the use of *community detection* instead of *connected components* leads to better results. To show that our method outperforms the approach of [19] independent of the clustering method, we substitute the use of *connected components* by *community detection* in their implementation. In our approach, we use 11 different similarity thresholds, varying from 0.0 to 1.0, for including edges in the constructed graph, five context sizes, and three metrics to measure the similarity of events. We evaluate the approach of [19] with similar configurations of their unfolding threshold t_u, used to determine if two events with the same

Table 2. Parameter space for experiments on the synthetic event logs.

Algorithm	Parameters
Context-based with Variant Compression	Similarity threshold $t_s = \{0, 0.1, \ldots, 0.9, 1\}$
	Context size $k = \{1, 2, 3, 4, 5\}$
	Distance metric $dm = \{$edit distance,
	set distance,
	multi-set distance$\}$
Case-Mapping-based with Community Detection	Unfolding threshold $t_u = \{0, 0.1, \ldots, 0.9, 1\}$
	Variant threshold $t_v = \{0, 0.1, \ldots, 0.9, 1\}$

Table 3. Parameter space used for the experiments on the real life event logs.

Algorithm	Parameters
Context-based with Variant Compression	Similarity threshold $t_s = \{0, 0.25, 0.5, 0.75, 1\}$
	Context size $k = \{1, 3, 5\}$
	Distance metric $dm = \{$edit distance,
	set distance,
	multi-set distance$\}$
Case-Mapping-based with Community Detection	Unfolding threshold $t_u = \{0, 0.25, 0.5, 0.75, 1\}$
	Variant threshold $t_v = \{0, 0.25, 0.5, 0.75, 1\}$

label in one case get different labels, i.e., if they are part of a loop or not, and variant threshold t_v, used to prune edges on the graph structure created by the algorithm to compare case mappings, parameters. A detailed list of the used parameter space for each of the algorithms is depicted in Table 2. We use the *Inductive Miner* [15] algorithm for process discovery without embedded noise filtering. As such, the algorithm guarantees that the process model describes all event data in the input (referred to as *perfect fitness*). Since a ground truth is available, we know what labels are candidates to be used for the label-splitting approach.

Experiments with Real-Life Event Data. We use publicly available event logs in combination with our proposed algorithm in experiments with real-life event data. We use 5 different event logs, i.e., the *BPI Challenge logs* from 2012 [11], 2013 (*Closed Problems log*) [22] and *2017* [12] (referred to as BPIC12, BPIC13 and BPIC17, respectively). Additionally, we use the *road fines* event log [16] and the environmental permits event log [8]. Due to the size of the event graphs, we primarily focus on the results of the variant compression, i.e., as presented in Sect. 5.1. The values for the parameters are listed in Table 3. We again use the *Inductive Miner* [15] algorithm for process discovery, in this case, with embedded noise filtering. We investigated several activity labels as

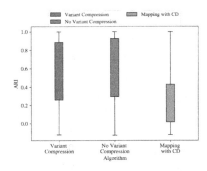
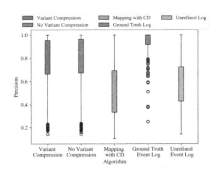

(a) The ARI scores obtained. Our approach (with and without compression) and the existing technique (Mapping with CD) [19]. The effect of compression is negligible, our approach achieves a significantly higher average ARI score.

(b) The precision scores obtained. Our approach tends to outperform the existing technique (Mapping with CD) [19] as well as applying process discovery without label-splitting.

Fig. 5. Results obtained for the experiments with synthetic data.

a candidate for splitting. We tested our approach on the three most frequently occurring activities. We selected the best-performing candidate, as we did not have a domain expert to pick the best candidate for every event log manually. For the road fines log, we selected the *Payment* activity for splitting due to insides from a manually created model [16]. This model shows that the *Payment* activity can be executed in different contexts, making it a prime candidate for label-splitting.

5.3 Results

Here, we present the results of the experiments. We further divide this section based on the results on *synthetic event data* and *real event data*, respectively.

Results on Synthetic Event Data. As indicated, in total, we used 270 different event logs from [18].

Consider Fig. 5, in which we present the corresponding results. In terms of ARI score (Fig. 5a), our approach outperforms the existing state-of-the-art label-splitting algorithm. Secondly, applying variant-based compression has a negligible effect on the overall ARI score. In Fig. 5b, the precision scores of the discovered process models is presented. The results are in line with the results obtained for the ARI score, i.e., in general, our technique outperforms the approach presented in [19]. Clearly, using the ground-truth event log generally leads to models of near-perfect precision. Yet, the median result of applying process discovery with our proposed label-splitting as a preprocessing step is well over 0.8. On average, the approach presented in [19] seems to have little effect compared to applying discovery on the unrefined event data.

Table 4. An overview of the results of the experiments with real event logs.

Event Log	IM Noise Threshold	Min Cases per Variant	Unrefined Log		Variant Compression				
			Precision	F1-Score	Max Precision	Average Precision	Max F1-Score	Average F1-Score	Average Runtime
BPI12	0.1	3	0.31	0.48	0.58	0.33	0.73	0.48	36 s
BPI13	–	1	0.8	0.89	0.98	0.87	0.99	0.93	49 s
BPI17	0.1	3	0.37	0.53	0.54	0.4	0.68	0.56	492 s
Road Fines	0.1	1	0.56	0.72	0.83	0.68	0.89	0.79	2 s
Permit	0.1	1	0.19	0.31	0.36	0.25	0.52	0.4	0.2 s

Results on Real Event Data. Here, we present the results of applying our proposed label-splitting algorithm on real event data. The variant-compression-based version of our framework is the only algorithm that finished within a reasonable time for all event logs. The algorithm presented in [19] only finished within the time-out set for the *permits event log* (13 s vs. 0.2 s of our approach, lower precision results: 0.15 average precision vs. 0.25 of our approach), *road traffic fines event log* (2823 s seconds vs. 2 s of our approach, lower precision results: 0.6 average precision vs. 0.68 of our approach) and *BPI challenge 2013 event log* (575 s vs. 49 s of our approach, equal precision results: 0.87 average precision vs. 0.87 of our approach).

In Table 4, we present an overview of the results of our algorithm (using different parameter configurations) and compare them with the results applied to the raw event data. To reduce the number of events in the event data, for BPI12 and BPI13, we enforce a minimum of 3 cases describing the same variant to be included in the event log. We use the Inductive Mining algorithm as a discovery algorithm with a noise threshold of 0.1 (except for BPI13, where no threshold is used). The noise threshold allows for ignoring small portions of noise, i.e., generally a large share of the behavior in the event log. We observe that our algorithm increases the precision scores and, similarly, the $F1$-score. The increase in precision is most significant in the Road Fines event log, where, on average, an increase of 0.12 is measured. The compression yields a relatively small graph for some logs, and the algorithm terminates within a few seconds. Clearly, we observe higher runtime values for larger graphs (e.g., BPI17).

Finally, consider Fig. 6, in which we present box plots of the precision obtained for the different instantiations of our algorithm on the real event logs. We also present the result of [19] for the *permits event log* ("Case Mapping" in the figure). We observe that the median value of our results outperforms the results of the process discovery algorithm on the unrefined event log for all logs except for BPI12, where it is slightly lower. For the road fines log, all obtained precision values exceed the result on the raw event log. Notably, the approach presented in [19] does not always improve the quality of discovered process models, compared to using the raw event data, inline with the conclusion in [19].

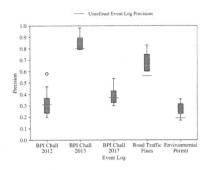
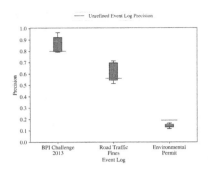

(a) The precision scores obtained with our approach with variant compression.

(b) The precision scores obtained with the state-of-the-art technique from [19].

Fig. 6. Results obtained for the experiments with real life event data. The red line indicates the precision on the unrefined event log. (Color figure online)

6 Conclusion and Discussion

Label-splitting, i.e., an established preprocessing technique in process mining, generally allows for better results from process discovery algorithms. However, the state-of-the-art label-splitting algorithm performs poorly on real event data and often has infeasible runtime. Therefore, this paper presented a novel label-splitting framework based on *event similarity graphs*. Our experiments show that, compared to the state-of-the-art label-splitting algorithm, our approach yields process models with higher precision and has better runtime performance.

One interesting finding of our evaluation is that the best parameter configuration for our algorithms highly depends on the input event log. One solution is to run our algorithm over a range of parameter configurations. By using various optimizations in the implementation, e.g., only recalculating the graph edges after the first iteration, we found this to be a feasible solution in our evaluation of real event data.

Future Work. Using variant-compression allows us to obtain a feasible runtime for the algorithm. We plan to perform experiments with different community detection algorithms and investigate whether the compression can be adopted. Secondly, we aim to investigate *detection mechanisms* for label-splitting, i.e., indicating what activity labels may be good candidates for splitting.

References

1. van der Aalst, W.M.P.: Process Mining - Data Science in Action, 2nd edn. Springer, Heidelberg (2016). https://doi.org/10.1007/978-3-662-49851-4
2. van der Aalst, W.M.P., de Medeiros, A.K.A., Weijters, A.J.M.M.: Genetic process mining. In: Ciardo, G., Darondeau, P. (eds.) ICATPN 2005. LNCS, vol. 3536, pp. 48–69. Springer, Heidelberg (2005). https://doi.org/10.1007/11494744_5

3. van der Aalst, W.M.P., Rubin, V.A., Verbeek, H.M.W., van Dongen, B.F., Kindler, E., Günther, C.W.: Process mining: a two-step approach to balance between under-fitting and overfitting. Softw. Syst. Model. **9**(1), 87–111 (2010)

4. van der Aalst, W.M.P., Weijters, T., Maruster, L.: Workflow mining: discovering process models from event logs. IEEE Trans. Knowl. Data Eng. **16**(9), 1128–1142 (2004)

5. Augusto, A., et al.: Automated discovery of process models from event logs: review and benchmark. IEEE Trans. Knowl. Data Eng. **31**(4), 686–705 (2019)

6. Blondel, V.D., Guillaume, J.L., Lambiotte, R., Lefebvre, E.: Fast unfolding of communities in large networks. J. Stat. Mech. Theory Exp. **2008**(10), P10008 (2008)

7. vanden Broucke, S.K.L.M., Weerdt, J.D.: Fodina: a robust and flexible heuristic process discovery technique. Decis. Support Syst. **100**, 109–118 (2017)

8. Buijs, J.: Receipt phase of an environmental permit application process (WABO), CoSeLoG project, March 2022. https://doi.org/10.4121/12709127.v2

9. Buijs, J.C.A.M., van Dongen, B.F., van der Aalst, W.M.P.: Quality dimensions in process discovery: the importance of fitness, precision, generalization and simplicity. Int. J. Cooperative Inf. Syst. **23**(1), 1440001 (2014)

10. Carmona, J., Cortadella, J., Kishinevsky, M.: A region-based algorithm for discovering Petri nets from event logs. In: Dumas, M., Reichert, M., Shan, M.-C. (eds.) BPM 2008. LNCS, vol. 5240, pp. 358–373. Springer, Heidelberg (2008). https://doi.org/10.1007/978-3-540-85758-7_26

11. van Dongen, B.: BPI Challenge 2012, April 2012. https://doi.org/10.4121/uuid:3926db30-f712-4394-aebc-75976070e91f

12. van Dongen, B.: BPI Challenge 2017, February 2017. https://doi.org/10.4121/uuid:5f3067df-f10b-45da-b98b-86ae4c7a310b

13. Fortunato, S.: Community detection in graphs. Phys. Rep. **486**(3), 75–174 (2010)

14. Koschmider, A., Kaczmarek, K., Krause, M., van Zelst, S.J.: Demystifying noise and outliers in event logs: review and future directions. In: Marrella, A., Weber, B. (eds.) BPM 2021. LNBIP, vol. 436, pp. 123–135. Springer, Cham (2022). https://doi.org/10.1007/978-3-030-94343-1_10

15. Leemans, S.J.J., Fahland, D., van der Aalst, W.M.P.: Discovering block-structured process models from event logs - a constructive approach. In: Colom, J.-M., Desel, J. (eds.) PETRI NETS 2013. LNCS, vol. 7927, pp. 311–329. Springer, Heidelberg (2013). https://doi.org/10.1007/978-3-642-38697-8_17

16. de Leoni, M.M., Mannhardt, F.: Road Traffic Fine Management Process, February 2015. https://doi.org/10.4121/uuid:270fd440-1057-4fb9-89a9-b699b47990f5

17. Li, J., Liu, D., Yang, B.: Process mining: extending α-algorithm to mine duplicate tasks in process logs. In: Chang, K.C.-C., et al. (eds.) APWeb/WAIM -2007. LNCS, vol. 4537, pp. 396–407. Springer, Heidelberg (2007). https://doi.org/10.1007/978-3-540-72909-9_43

18. Lu, X., Fahland, D., van den Biggelaar, F.J.H.M., van der Aalst, W.M.P.: Detecting deviating behaviors without models. In: Reichert, M., Reijers, H.A. (eds.) BPM 2015. LNBIP, vol. 256, pp. 126–139. Springer, Cham (2016). https://doi.org/10.1007/978-3-319-42887-1_11

19. Lu, X., Fahland, D., van den Biggelaar, F.J.H.M., van der Aalst, W.M.P.: Handling duplicated tasks in process discovery by refining event labels. In: La Rosa, M., Loos, P., Pastor, O. (eds.) BPM 2016. LNCS, vol. 9850, pp. 90–107. Springer, Cham (2016). https://doi.org/10.1007/978-3-319-45348-4_6

20. Parikh, R.: On context-free languages. J. ACM **13**(4), 570–581 (1966)

21. Schaeffer, S.E.: Graph clustering. Comput. Sci. Rev. **1**(1), 27–64 (2007)
22. Steeman, W.: BPI Challenge 2013, closed problems, April 2013. https://doi.org/10.4121/uuid:c2c3b154-ab26-4b31-a0e8-8f2350ddac11
23. Tax, N., Alasgarov, E., Sidorova, N., Haakma, R., van der Aalst, W.M.P.: Generating time-based label refinements to discover more precise process models. J. Ambient Intell. Smart Environ. **11**(2), 165–182 (2019)
24. Vázquez-Barreiros, B., Mucientes, M., Lama, M.: Enhancing discovered processes with duplicate tasks. Inf. Sci. **373**, 369–387 (2016)
25. Weijters, A., van der Aalst, W.M.P., De Medeiros, A.: Process mining with the heuristicsminer algorithm. BETA publicatie: working papers, Technische Universiteit Eindhoven (2006)
26. Yang, S., et al.: Medical workflow modeling using alignment-guided state-splitting HMM. In: 2017 IEEE International Conference on Healthcare Informatics, ICHI 2017, Park City, UT, USA, 23–26 August 2017, pp. 144–153. IEEE Computer Society (2017)
27. van Zelst, S.J., Mannhardt, F., de Leoni, M., Koschmider, A.: Event abstraction in process mining: literature review and taxonomy. Granul. Comput. **6**(3), 719–736 (2021)

Verifying Resource Compliance Requirements from Natural Language Text over Event Logs

Henryk Mustroph[1]([✉]), Marisol Barrientos[1], Karolin Winter[2],
and Stefanie Rinderle-Ma[1]

[1] TUM School of Computation, Information and Technology,
Technical University of Munich, Garching, Germany
{henryk.mustroph,marisol.barrientos,stefanie.rinderle-ma}@tum.de
[2] Department of Industrial Engineering and Innovation Sciences, Eindhoven
University of Technology, Eindhoven, The Netherlands
k.m.winter@tue.nl

Abstract. Process compliance aims to ensure that processes adhere to requirements imposed by natural language texts such as regulatory documents. Existing approaches assume that requirements are available in a formalized manner using, e.g., linear temporal logic, leaving the question open of how to automatically extract and formalize them for verification. Especially with the constantly growing amount of regulatory documents and their frequent updates, it can be preferable to provide an approach that enables the verification of processes with requirements in natural language text instead of formalized requirements. To this end, this paper presents an approach that copes with the verification of resource compliance requirements, e.g., which resource shall perform which activity, in natural language over event logs. The approach relies on a comprehensive literature analysis to identify resource compliance patterns. It then contrasts these patterns with resource patterns reflecting the process perspective, while considering the natural language perspective. We combine the state-of-the-art GPT-4 technology for pre-processing the natural language text with a customized compliance verification component to identify and verify resource compliance requirements. Thereby, the approach distinguishes different resource patterns including multiple organizational perspectives. The approach is evaluated based on a set of well-established process descriptions and synthesized event logs generated by a process execution engine as well as the BPIC 2020 dataset.

Keywords: Compliance Requirements Verification · Resource Mining · Natural Language Text · Process Descriptions · Event Logs

1 Introduction

Business process compliance is the task of ensuring that processes obey the rules, guidelines, and constraints imposed on them. Those compliance requirements

ⓒ The Author(s), under exclusive license to Springer Nature Switzerland AG 2023
C. Di Francescomarino et al. (Eds.): BPM 2023, LNCS 14159, pp. 249–265, 2023.
https://doi.org/10.1007/978-3-031-41620-0_15

are typically outlined in extensive regulatory documents such as legislative texts or ISO norms [12]. Compliance requirements need to be verified, i.e., they are checked against the actual execution of a process captured by an event log. In order to enable verification, the compliance requirements are typically formalized using, e.g., linear temporal logic [2,15,26]. However, as regulatory documents change frequently, the compliance requirements have to be re-formalized equally frequently [27]. Therefore, it can be desirable to enable direct compliance verification between requirements provided in natural language and event logs.

Business process compliance refers to multiple perspectives beyond control flow, i.e., time, resources, and data [15]. In recent work, we focus on verifying quantitative temporal compliance requirements over event logs [4]. In this paper, we consider the resource perspective which captures legally relevant concepts such as the segregation and binding of duties [25], reframed as *Responsibility deviation Requirements* in [22]. In general, process activities are carried out by resources, which can be classified into categories human and non-human [21]. Human resources refer to individuals involved in the process, while non-human resources are typically machines, robots, or computer-based systems. The presented approach can extract both, compliance requirements referring to human and non-human resources, but the main focus will be on human resources. In the case of human resources, in this paper four different types of resources are distinguished following [21]: *organizations*, such as company X, represent a larger grouping of resources. *Organizational units* such as departments or teams are subgroups within an organization and are responsible for specific activities. *Roles* are used to define the specific responsibilities and tasks assigned to different resources within an organization. The role of a software developer, for example, contains responsibilities such as developing and maintaining software applications. Finally, *users* are specific individuals. In the case of human resources, users can be identified by their name or an ID, while non-human resources may be identified by their unique identifiers or serial numbers.

The identification of resources paired with activities from natural language text has been addressed by, e.g., [5,11,20]. Yet, a) the organizational structure reflected by the resources, b) the different compliance requirement patterns associated with each resource, and c) the compliance verification of natural language texts over event logs have not explicitly been considered. Moreover, we see the extraction of resource activity pairs as a pre-processing step, and rely for this on the model GPT-4 [19]. Therefore, the main contribution is not how to identify and extract those pairs from natural language text, but how to use the output w.r.t. compliance verification over event logs. In order to address this question, we analyze in Sect. 2 i) the process perspective through workflow resource patterns [21], mining organizational structures from event logs [10] and the eXtensible event stream (XES) standard [1], and ii) the compliance perspective by identifying resource compliance requirements [3,15,24,25]. We establish a mapping between both perspectives and extract a summary of resource compliance patterns that are feasible to detect from a natural language processing perspective. Moreover, we identify which assumptions an event log needs to fulfill to enable compliance verification. Based on those findings, in Sect. 3 we design an approach consisting of five steps divided into a pre-processing component and the actual compliance verification component. The approach is evaluated in Sect. 4 followed by a discussion of evaluation results and limitations in Sect. 5. Section 6 presents related work and Sect. 7 concludes the paper with a summary and outlook on future work.

2 Resource Compliance Requirements Pattern Elicitation

In the following, we provide an analysis of which resource compliance requirements patterns exist and how they are reflected by the process and event log perspective. This analysis constitutes the fundamentals for the resource compliance requirements verification approach, presented in Sect. 3. First, Sect. 2.1 summarizes how resources are represented from the process and event log perspective by reviewing the literature on workflow resource patterns [21], mining organizational structures from event logs [10] and the XES standard [1]. Second, in Sect. 2.2, we analyze literature on resource compliance requirements [3, 15, 23–25], and related work. This body of literature was selected based on a literature search conducted on DBLP[1] using keywords "resource" or "organizational" or "role" and "mining", "organization" or "resource" and "compliance" or "requirements", and "resource-aware process verification". Table 1 presents a mapping between the resource patterns considered in [21] and those identified papers. A detailed analysis is provided in a spreadsheet, which can be accessed at https://www.cs.cit.tum.de/bpm/data/.

Table 1. Mapping of Resource Patterns; ✓ = mentioned; ✗ = not mentioned

Pattern	Description	Process		Compliance					
		[21]	[10]	[24]	[15]	[25]	[3]	[23]	[28]
1. Performed by	A_1 must be performed by resource Re	✓	✓	✓	✓	✓	✓	✓	✓
2. Not Performed by	A_1 can not be performed by resource Re	✗	✓	✗	✗	✗	✓	✗	✗
3. Dynamic SoD (4-Eyes-Principle)	A_1 and A_2 must be performed by different resources, independently of roles.	✓	✓	✓	✓	✓	✓	✓	✓
4. Static SoD	A_1 and A_2 must be performed by different resources with different roles	✓	✓	✓	✓	✓	✓	✓	✓
5. Multi-segregated	A set of activities $(A_1, A_2, A_3, ..., A_n)$ must be performed by (m) different resources	✗	✗	✗	✗	✓	✗	✗	✓
6. Dynamic Bonded	A_1 and A_2 must be performed by different resources with the same role	✓	✓	✓	✓	✓	✗	✗	✗
7. Static Bonded	A_1 and A_2 must be performed by the same resource with the same role	✓	✓	✓	✓	✓	✗	✓	✗
8. Multi-bonded	A set of activities $(A_1, A_2, A_3, ..., A_n)$ must be performed by the same resource with the same role	✓	✗	✓	✓	✓	✗	✓	✗
9. Automatic	A_1 is automatically executed	✓	✗	✗	✗	✗	✗	✗	✗

2.1 Process Perspective

In [21], 43 patterns describing the distribution and execution of activities in workflow systems are proposed. These patterns are classified into five categories:

---

[1] https://dblp.org/, last access: 2023-06-19.

Creation patterns limit how activities are executed, determine which resources can perform an activity, and match activities with capable resources. *Push* and *Pull* are distribution patterns where activities are allocated by a central authority or chosen by resources respectively. *Detour* and *Visibility* allow resources to modify ongoing activities or view available ones, while *Multi-Resource* sets limits on how many activities a resource can perform at the same time. These patterns ensure efficient and effective execution while meeting business requirements. Our approach takes into account the *Creation* patterns, assuming that a resource will be able to perform the same activities throughout the entire process. In all the *Creation* patterns, the requirements only involve the relationship between a resource and the activities' performance. To represent these requirements, we introduce the concept of Resource-Activity Requirement (R-AR) which capture the patterns presented in Table 1. Algorithm 1 determines how R-ARs are built.

The sub-patterns related to *Allocation-based Creation* are omitted from Table 1. These sub-patterns, encompassing diverse allocation strategies like assigning tasks based on hierarchy or deferring assignments to future activities, are viewed as ancillary data rather than essential for forming resource-activity pairs. For example, we focus on whether *Resource R* executed *Activity A*, rather than whether *Resource R* was the original assignee. This approach aids to avoid unnecessary complexity and potential biases, thereby centering our research on fundamental resource-activity pairings. Of all the literature analyzed, [10] adheres most closely to the resource patterns presented in [21]. Nonetheless, we suggest further research to extract control flow aspects from process descriptions, which could extend our approach and incorporate cross-perspective patterns.

As we aim at compliance verification over event logs we consider the XES standard [1], which contains an organizational extension describing human actors. Therein, three elements are distinguished, first resource which contains the "name, or identifier, of the resource having triggered the event." [1], a role which reflects the "role of the resource having triggered the event, within the organizational structure." [1] and a group that represents the "group within the organizational structure, of which the resource having triggered the event is a member." [1] Considering the terminology we chose for this paper, a resource refers to a rather generic term compared to the term resource as used within the XES standard. A user corresponds to resource, group corresponds to organizational unit and we additionally add organization in order to include the perspective that multiple organizations and interactions between them could be described in one natural language text. Overall, the event log must contain basic elements as specified by the XES standard, including trace names, event labels, event IDs and fields denoting the resource type as well as a field detailing the organization. The event log may not contain errors like duplicate events.

2.2   Compliance Requirements Perspective

The first aspect that can be observed in Table 1, is that the papers that focus on studying resource compliance verification do not consider the *Creation automatic activity execution* sub-pattern, even though it involves the system as a resource.

We decided to maintain the automatic execution sub-pattern [21] because there may be cases where the system is the only one permitted to execute an activity, and no one else can. In contrast, [3,10] introduced the *Not Performed by* pattern, which is not covered in [21]. Similarly, the *Multi-segregated* pattern is only discussed in [25,28]. As there is little consensus among the papers regarding the definition of resource (e.g. using other terms like *agent*, or considering only *users* and *roles*), we decided to define the resource concept and its types in Sect. 1. In [24], the authors distinguish between resources and agents and consider both resource-aware and data-aware compliance requirements. This paper presents examples of cross-perspective patterns, such as *If attribute X has value v then resource Re must execute the activity* pattern. From a resource perspective, both [15,24] cover the same patterns. However, [24] discusses authorization, whereas [15] addresses the same issues but refers to *users* and *roles*. In [25], the same resource patterns as in [15,24] are addressed, with the addition of the *Multi-segregated* pattern. Both [3,23] focus on less than 10 specific resource examples, which overlap with each other and do not introduce any new patterns. Additionally, [28] focuses on the *Segregation of Duties* (SoD) when studying resource patterns, as their primary objective is task-based authorization constraints. In this paper, they include aspects that previous papers do not handle but we are, such as *The same activity can only be performed twice by the same resource,* and *A₁ and A₂ must be performed by different resources.*

3   R-AR Verification Approach

The resource compliance requirements verification approach is presented in Fig. 1, and illustrated based on a running example depicted in Fig. 2, and Fig. 3. The approach consists of a pre-processing component (cf. Sect. 3.1) and a compliance verification component (cf. Sect. 3.2). It takes as input a natural language text, e.g., a process description, and an event log, which must fulfill the assumptions as detailed in Sect. 2.1. Consistency in naming resources and activities within the process description and within the log is assumed, but synonyms of both, resources and activities, across the description and log are handled by the approach. Note that the textual document is always considered as the ground truth, i.e., compliance of a given event log is verified against the textual document.

The approach consists of five steps and allows for user intervention for compensating errors after steps 1, 2, and 3. These step's results are saved for possible necessity, as the approach's outcome hinges on their quality. In general, the approach is independent of the domain but with all intermediate results available, domain knowledge can be easily integrated, e.g., in order to tailor the GPT-4 prompt towards a specific dataset. Moreover, several parameters must be set by the user such as thresholds for similarity mappings. Further details are provided in the explanations of the single steps. The output of the compliance verification component consists of two files. The first one, generated after step 3, details on activity matching results for transparency allowing for tracing whether errors

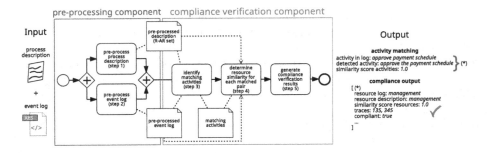

Fig. 1. Overview of Resource Compliance Requirements Verification Approach

occurring later on were caused by the previous activity matching. The second file contains the compliance verification results. Both files are ordered by unique event log pairs, i.e., resource-activity pairs, for easier comprehension.

3.1 Pre-processing Component

The pre-processing component handles a process description in natural language text and in parallel the event log data. The intermediate pre-processed outputs serve as input for the second component, the compliance verification component.

Step 1 – Pre-process Process Description. Consider the running example, Fig. 2, depicting a process description along with examples of resource compliance patterns. The intended outcome of the running example's pre-processing is a collection of Resource-Activity Requirements (R-ARs). Each R-AR includes a resource (marked in bold), and activity (underlined), along with the necessary metadata (cf., output of Algorithm 1) used to identify the appropriate resource pattern for each R-AR.

1. Whenever **Elite Holdings** receives a customer request, it demands a solvency check from **Miracle Credit**.
2. At **Miracle Credit** exactly **two clerks** perform a solvency check.
3. **Miracle Credit** hands back the results of the solvency check to **Elite Holdings**.
4. If the solvency check results is negative, a **clerk from the customer advisory** informs the customer and deletes the customer's request.
5. If the solvency check result is positive, **Anna** or **Hans,** bot not both, develop a payment schedule.
6. Afterward, the schedule is sent to the **manager**.
7. Both **he** and **another clerk from the management** must approve the payment schedule.
8. Approve payment schedule may never be executed by **Anna** or **Hans**
9. If the payment schedule has been approved, an email is sent to the customer **automatically**, otherwise, the **customer advisory** calls the customer to suggest an alternative.
10. In both cases, the request must be closed.

Fig. 2. Running Example – Natural Language Text

Examples of these patterns include *Performed by*, as in ④ *a clerk from customer advisory informs the customer*, and *Not Performed by*, which identifies resources that are not allowed to execute an activity, such as ⑧ *approve payment schedule may never be executed by Anna or Hans.* Additionally, more complex requirements involving multiple resources may be present, such as a *Multi-segregated*, where ⑦ *he and another clerk from the management must approve the schedule.* In this case while pre-processing the description we need to consider anaphora resolution in order to know which resource is meant by *he.* Note that resources are not limited to humans; automated tasks can also occur during process execution, such as ⑨ *sending an email to the customer (Automatic Pattern).* Moreover, the running example shows the different granularity levels of a resource: organization (*Elite Holdings, Miracle Credit*), organizational units (*customer advisory, management*), roles (*clerk, manager*), and particular users (*Anna, Hans, system*).

The extraction of R-ARs from process descriptions is carried out using the GPT-4 model from OpenAI [19]. The prompt schema is showcased in Algorithm 1. To ensure reproducibility, the original prompt is also provided as input for the evaluation of our implementation. The prompt provides fundamental information such as the meaning of a resource and an organization, along with examples for each. These examples are not taken from our dataset. Details about the required format, necessary parameters, and the type of each parameter are also included in the prompt. The delivery of this information is thoroughly explained within the prompt. For example, if the activity concerns two resources, the aim is to treat each as a separate resource instead of a collective group. In cases of lengthy process descriptions, even if there is not a specific question about the control flow, the model takes the initiative to propose a control flow concept, which could be exploited if the control flow is also analyzed in future work.

Algorithm 1: Process Description Pre-processing Prompt Schema

```
Input: Process Description
Output: Set of R-ARs with fields: "role", "user", "org_unit", "organization", "activity", "inclusion", "exclusion",
        "min", "max", "equals", "anaphora", "is_performed"
Function SetFields():
    Initialize: "role", "user", "org_unit", "organization", "activity"
    Set "inclusion" and "exclusion"
        // capture relations between R-ARs (e.g., conflicts between R-ARs)

    Set "min", "max", "equals"  //limit the number of resources performing an activity

    Set "anaphora"  //indicate the original resource which references the pronoun

    if activity is not performed then
    |    Set "is_performed" = false
    else
    |    Set "is_performed" = true
    end
```

Step 2 – Pre-process Event Log. Figure 3 illustrates an excerpt of an event log reflecting the process description in Fig. 2. For brevity, the trace ID in Fig. 3 is omitted and it is assumed that all depicted events belong to the same trace and that the trace is complete. The events that comply with the process description are E1, E2 and E7 (Fig. 2 ①, ② and ⑨). Events E3, E4, E5 and E6 (Fig. 2 ③, ⑤, ⑦) do not comply. In event E3, the organization should be referred to as *Miracle Credit* not *Elite Holdings*. Events E4 and E5 must not occur simultaneously in one trace. Additionally, the activity in event E6 should be carried out by employees holding the role of manager, as well as by a specific employee in the clerk role, rather than being exclusively assigned to one unique employee with the clerk role.

Event ID	concept:name	org:resource	org:role	org:unit	organization
① E1	demand solvency check	unknown	unknown	unknown	Elite Holdings
② E2	perform solvency check	Peter	clerk	unknown	Miracle Credit
③ E3	hand back results	unknown	unknown	unknown	Elite Holdings
⑤ E4	develop payment schedule	Anna	unknown	unknown	Elite Holdings
E5	develop payment schedule	Hans	unknown	unknown	Elite Holdings
⑦ E6	approve payment schedule	unknown	clerk	management	Elite Holdings
⑨ E7	send email to customer	system	unknown	unknown	Elite Holdings
...					

Fig. 3. Running Example – Event Log Trace containing Violations

The event log, i.e., also this trace, serves as input for the event log pre-processing step. This step constructs an object-oriented representation of the event log's structure, which includes an attribute, event, trace, and event log class. Each class contains methods to extract the most crucial information from the event log. To generate a pre-processed event log output for persistent storage, an event log is parsed into a data frame using PM4Py [7]. The data frame is then converted into an event log object from the custom-created class. The event log class features a method generating an output file containing every distinct event in the log. Thereby, a distinct event is defined as a unique resource-activity pair, where a resource is a combination of a user, role, organizational unit, and organization. An example of a pre-processed event is *Event = {"activity": "perform solvency check", "user": "Peter", "role": "clerk", "org_unit": "unknown", "organization": "Miracle Credit"}*

3.2 Compliance Verification Component

After pre-processing the process description and the event log, the actual compliance verification can be carried out. This requires a mapping between the R-ARs from process descriptions and event logs. The component unfolds in the following three steps.

Step 3 – Identify Matching Activities. In this step of the overall approach, activities in the event log output file are compared for similarity with those in the description output file. To make the approach more resilient in real-world settings, we accept variations in how activities are phrased between the event log and the process description, as we only expect terminology consistency within each of these individually. All the activities in the pre-processed log are compared with all activities in the description output, identifying matches based on high similarity scores. Each match's score is then evaluated against a predefined threshold to decide whether to accept it. If the score falls short of this threshold, we denote the log activity as unmatched. This safeguards against finding matched resource-activity pairs in subsequent steps. The threshold and similarity score for matching pairs range from zero to one. If activities are few and similar, scores are near one; diverse activities yield scores near zero. An expert, familiar with the data, should set a threshold to eliminate misclassified matches. This threshold is adjustable to accommodate various activities and the selected similarity measure, impacting synonym levels in pre-processed files. Three similarity measures - TF-IDF[2], BERT[3], and spaCy[4] evaluate activity likenesses in the log and description. The user initially selects one, alongside the threshold for valid activity matches. The output of this step contains a measurement-type section containing the compliance verification input information, e.g., the set threshold for activity similarity matching and the activity matching result information. *Activity Matching Output: Measure Types: {"similarity measure": "TF-IDF", "threshold": 0.65, ...}, "activity_matching_output": [{ "Activity Log": "accept order", "Detected Activity": "accept the order", "Similarity Score": 1.0}...]*

Step 4 – Determine Resource Similarity. In the previous step, each activity from the pre-processed event log was matched with an activity from the pre-processed process description. In this step, we evaluate whether the resource performing each activity for a specific event in the event log is similar to its counterpart in the process description. For that, the same similarity measures are used as within step 3. This semantic similarity needs to be executed since resources can be subject to different naming conventions, and wording. The threshold used for comparing resources can differ from the one used for activities, reflecting the importance of knowledge of the mentioned resource types in the text and event log. Furthermore different types of resources to be checked for similarity to ensure event log compliance can be selected. These resources can be organizational units or users with specific roles, depending on what is stored in the log and the information granularity provided in the text document. If the chosen resource structure type is not specified for a particular event in the log, a mechanism is in place to automatically build the resource to be checked using the available resource values in the log, as long as they are defined. For instance, if the initial choice was to check the organizational unit as the resource structure but the value is undefined for a specific event in the log, the mechanism will instead construct the resource based on the user, role and organization information.

Step 5 – Generate Compliance Verification Results. The last step creates, based on the outcome of the resource checks, the compliance output file. Hence, the implemented approach verifies the minimum criteria necessary to classify a resource-activity pair as compliant or non-compliant. Two main distinctions are made for these checks. When a resource, whether human or non-human, is expected to perform an activity (patterns 1 and 9), the resulting similarity

---

[2] https://scikit-learn.org/stable/install.html, last access: 2023-06-19.

[3] https://www.sbert.net, last access: 2023-06-19.

[4] https://spacy.io/usage/linguistic-features, last access: 2023-06-19.

score from step 4 must exceed a predefined threshold. On the other hand, if a specific resource should not perform an activity (pattern 2), the similarity score should be lower than the threshold. The resulting output file maintains the same granularity, naming convention, and structure for both approaches. The order of events in the output file corresponds to the pre-processed event log file, as the compliance of the process is verified for each distinct event by comparing it to the description. Similar to activity matching, the output includes a section for measurement types and appears as follows: *Compliance Matching Output: Measure Types: ..., default compliance check output: ["Matched Activity": "accept the order", "Resource Log": "sales department", "Resource Description": "member of the sales department", "Similarity Score of Matched Resources": 0.65, "Corresponding Traces": ..., "Compliant": true, "Non-Compliant Reason": " ...].*

Furthermore, based on the information given in all resulting files from the overall process, users can manually verify if patterns incorporating multiple resource-activity pairs at once (patterns 3–8) are also compliant, if necessary.

4   Evaluation

The approach has been implemented as a prototype in Python 3 and can be accessed publicly at the following location: https://www.cs.cit.tum.de/bpm/software/. All input, intermediate files, like JSON files from GPT-4 prompt execution, and pre-processed event logs, and output files are available, as well, via the above link. In Sect. 4.1, details on the datasets used in the evaluation are provided while the evaluation results for the pre-processing, as well as compliance verification component, are described separately in Sect. 4.2. All the synthesized and modified event logs are available at https://www.cs.cit.tum.de/bpm/data/.

4.1   Dataset Preparation

The evaluation features synthetic as well as real-world datasets. In the following, we describe how the synthetic datasets were generated and how the real-world dataset was prepared.

Synthetic Datasets. First of all, we take into account the process description of the *Running Example* (RE), which was initially introduced in Sect. 3. This example was meticulously designed to allow the evaluation of an extensive set of patterns, which are comprehensively detailed in Table 1. Additionally, the PET dataset, a well-known collection of 45 process descriptions [6], with a total number of resource activity pairs of 449, was included in the analysis. On average each process description of the PET dataset contains 10 resource activity pairs. From this extensive dataset, two specific descriptions were singled out for consideration: *Bicycle Manufacturing* (BM) and *Schedule Meetings* (SM). On average, each of these two descriptions contain 10.50 resource activity pairs. For each of these three process descriptions a model and event logs were generated using the Cloud Process Execution Engine[5] (CPEE) [16]. These event logs did initially not

---

[5] https://cpee.org, last access: 2023-06-19.

contain any resource violations. To test the compliance verification component of our approach, resource compliance violations were introduced into the logs. This was done by randomly selecting resource activity pairs and modifying the resource executing the activity. This process of modeling, log generation, and log alteration was overseen by one of our authors who was not involved in the technical implementation process.

Real-World Dataset. In addition to the synthetic datasets, we had a look at real-world event logs used within the Business Process Intelligence Challenges (BPIC). Among those, we identified the BPIC 2020 to be suitable since it contains resources not only in the form of IDs but verbal (e.g., *budget owner*) and comes at the same time with a detailed textual process description[6]. This dataset, collected between 2017 and 2019, comprises a total of 270,216 events recorded across five logs. In BPIC, resources are classified as *STAFF MEMBER* or *SYSTEM*. The first type of resource can have a role while the system not. The column values for event and case *ID* were adjusted to match the terminology used in the implementation for pre-processing the event log. The names of the roles (e.g., *DIRECTOR*) were removed from the original label of the activity (e.g. *PERMIT REJECTED by DIRECTOR*). Offered as supplementary material, the script specifically designed for this task could be employed as a blueprint, enabling the adaptation of an event log from any domain, with a different data structure, to suit our approach.

4.2 Results

Owing to the significant impact of the pre-processing quality on the compliance verification results, we first provide a succinct overview of the results for the pre-processing component before diving deeper into the evaluation of the compliance verification component.

Pre-processing Component. To assess the performance of the pre-processing of process descriptions, a gold standard file for each process description is generated containing the desired set of R-AR results. We utilize this gold standard to compare the performance of GPT-3.5 and GPT-4 in retrieving the set of R-ARs. To accomplish this, the GPT models receive as input each process description together with the prompt, which was built following the steps outlined in Algorithm 1. All intermediate files generated during this evaluation process are saved and stored in our repository for future reference and analysis. Table 2 displays the precision and recall values for the results obtained from both the GPT-3.5 and GPT-4 model.

A R-AR is deemed successfully detected if it contains both the activity and resource, representing a pair in the process description, and matches any of the patterns outlined in Table 1. By looking at Table 2, on average, the precision scores improved by 40% when using GPT-4 compared to its predecessor. Additionally, the recall demonstrated a substantial increase of 135%. Most GPT-3.5

---

[6] https://doi.org/10.4121/uuid:52fb97d4-4588-43c9-9d04-3604d4613b51 last access: 2023-06-19.

errors stem from anaphoras, scattered information, passive voice, and activity boundary detection issues.

In [5], a GPT-3 model was utilized. However, this model had limitations in identifying the resource responsible for an activity if the word *perform* was not explicitly mentioned in the sentence. In contrast, our pre-processing approach was designed to handle more complex cases and successfully identi-

Dataset	Precision		Recall	
	GPT-3.5	GPT-4	GPT-3.5	GPT-4
RE	0.92	1.00	0.92	1.00
BM	0.62	0.90	1.00	1.00
SM	0.50	0.89	0.17	1.00
BPIC	0.76	1.00	1.00	1.00

Table 2. Evaluation Results Pre-processing Component

fied resources in all instances, even those involving anaphora, embedded conditions, and other related factors such as excluding and including activities. Misclassifications in the results produced by GPT-4 often stem from the model trying to assume too much information, which can result in false positives when the model tries to infer activities that were not clearly specified. As outlined in [18], they suggest improving the quality of the PET dataset by employing data augmentation methods. Another issue with models like GPT-4, such as generating labels with new words or synonyms, can be mitigated by narrowing the prompt, as indicated in [13]. This approach not only helps avoid these issues but also leads to more faithful and reasonable texts, reducing the hallucination, i.e., AI creating information without input basis, that occurs in natural language generation.

Compliance Verification Component. The evaluation of the compliance verification component is depicted in Table 3. It depicts the intermediate results for the matched activities (step 3), the extraction of resource similarity (step 4), and the final compliance verification outcomes (step 5). For the synthetic datasets, RE, BM, and SM, which had less contextual detail, TF-IDF was applied, and activity (step 3) and resource-activity (step 4) thresholds were between 0.60 and 0.80. For the compliance verification of BPIC, we used the BERT model to account for varying naming conventions and synonyms. To manage BERT's tendency to assign high similarity scores to dissimilar pairs, we set activity and resource-activity thresholds at 0.8 and 0.9. This approach maintained accuracy by ensuring context-specific matches, despite the presence of synonyms.

Upon examining the output of step 1 of BPIC, 17 unique resource activity pairs were identified. However, step 2 revealed a larger number, presenting 55 unique resource activity pairs. Once we processed the results of the compliance verification component, we deduced that only 7 pairs from the original event log were accurately represented in the process description. These pairs included (*declaration final approved, director*), (*declaration approved, administration*), (*declaration rejected, employee*), (*declaration submitted, employee*), (*request payment, automatic*), (*permit rejected, employee*), and (*request for payment rejected, employee*). The BPIC process description lacks explicit information about the remaining 48 pairs, preventing further interpretations without making extensive assumptions. This challenge that the BPIC dataset presents, is partially solved by the use of thresholds, and it is not present in the synthetic dataset because the total number of unique pairs in the log is very close to the total number of resource-activity pairs identify in the process description.

Considering the results for step 3 in Table 3, it is notable to see that the approach was able to match in all the cases the activities presented in the process description. The small proportion of mismatches (c.f., precision of BM or SM) from the activities presented in the event log was due to the

		RE	BM	SM	BPIC
Step 3	Precision	0.94	0.87	0.81	1.0
	Recall	1.0	1.0	1.0	1.0
Step 4 & 5	Precision	0.64	1.0	1.0	0.71
	Recall	1.0	0.92	1.0	1.0

Table 3. Evaluation Results of Steps 3–5

presence of events not described in the process description. The results of steps 4 and 5 are evidence of the impact that has the good handling of the granularity of a resource. The RE contained a wider variability of granularity which ended up being more challenging and had a direct impact on the extraction of non-compliant traces.

In RE, BM, and SM, the majority of non-compliant traces were due to a conflict with another resource with different granularity, as indicated in the process description, performing an activity (e.g., *Elite Holdings* vs *Hans from Elite Holdings*). While in the BPIC dataset, non-compliant traces were primarily characterized by three factors. Firstly, traces that contained *missing* as a resource keyword were flagged. Secondly, traces that involved staff members who, by hierarchical implication, were permitted to perform the activities of their subordinates were noted. Lastly, instances, where system automation carried out activities intended for human staff members, were also marked as non-compliant. Future enhancements should thus address these distinct issues accordingly. The ensuing section will discuss the limitations of the current study and potential avenues for further research.

5   Discussion and Limitations

In order for the presented approach to achieve optimal results, it is essential that the pre-processing of the natural language text delivers all necessary information, such as activities, the different levels of granularity of a resource, and other fields, as shown in Algorithm 1. We utilize state-of-the-art GPT-4 technology to accomplish this as it is undoubtedly powerful and provides the means to boost the performance of our approach. However, there are certain drawbacks to consider which also led to the decision to develop a customized compliance verification component without GPT-4.

Reproducibility. GPT-4 is a black box model, which makes ensuring reproducibility challenging due to its dependence on finely-tuned prompts. To address this issue, we provide both the prompt we used and the output from GPT-4, which can then be incorporated into the compliance verification component. If we had used GPT-4 also for compliance verification we see a further challenge in describing the pattern-based check accurately in a prompt for GPT-4. Crafting a prompt that consistently produces reliable results for compliance verification might require a significant amount of effort and expertise.

Reliability, Explainability, and Transparency. Furthermore, GPT-4 is a third-party service and may not always be accessible when needed (e.g., having a cap of 25 messages every 3 h). During the evaluation, we also recognized a certain instability in the results it produced, e.g., slower service, loss of history. GPT-4, being a large transformer model, is considered a black box meaning its results might be difficult to understand. In particular, the incorrect classification of events into compliant or non-compliant in the compliance verification would provide a lack of transparency without being able to explain the process. Our aim is to provide a transparent step-by-step resolution of compliance verification results, which is not easily achievable with GPT-4.

Technical Feasibility and Costs. As users must pay for each executed prompt, it might become expensive and also impractical to process a large event log containing thousands of events which is another reason we want to keep the usage of GPT-4 to a minimum. The final goal is to develop a fully automated compliance verification approach, however, this is difficult to offer when relying on a commercial product.

Suggestions for Improvement. One option for a customized pre-processing component could be built based on existing work, e.g., [5,11,20]. The task of identifying resources could be, e.g., follow a rule-based approach incorporating a custom-trained named entity recognition (NER) model. A second option is to explore OpenAssistant [14], a lightweight open-source project to collaborate on large language models. Further improvements include the implementation of a hierarchy compliance resolution verification system that evaluates compliance for role and department hierarchies. For instance, this would enable a manager to carry out tasks usually done by a junior developer. The method will also evaluate varying organizational structures like user-role and role-department, merging the best combinations for each scenario. Additionally, it could investigate handling cases with multiple process logs but only one process description, looking into how these logs can be combined and compliance ensured.

6   Related Work

Related work on how resources are handled from a process, compliance and event log perspective has been discussed in Sect. 2. In order to provide a holistic view of the topic, we outline additional related work in the following.

For the pre-processing component, literature on the identification and extraction of resource information from natural language text constitutes a further line of related work. This task has been addressed by approaches that aim at process model generation from natural language text, like, e.g., [11]. Other approaches employ semantic role labelling in order to extract resources [20] or pre-trained language models and in-context learning in order to extract business process entities and their relations from natural language texts [5]. The latter makes use of GPT-3 [8]. However, when using GPT-3.5 in the pre-processing component, we could not come up with satisfying results. Only the latest released GPT-4 model [19] was capable of delivering the necessary quality for the pre-processing component. Another line of research is focusing on extracting access control policies from natural language text, e.g., [17]. They also make use of semantic role labelling like [20] but as demonstrated in Sect. 2, resource compliance patterns are more diverse. Moreover, none of the mentioned approaches has envisioned compliance verification over event logs as it is the aim of this paper.

In order to enable compliance verification, by now, compliance requirements need to be formalized as, e.g., LTL formulas manually. Recent efforts have focused on automatically extracting LTL formulas from natural language texts, cf., e.g., [9] for a comprehensive state-of-the-art analysis. However, according to this survey "a general enough solution, that is capable of translating free, natural English texts into unbounded, general LTL formulas is still missing." [9].

7   Conclusion and Future Work

In this paper, an approach for resource compliance requirements verification over event logs has been presented. Compared to existing work, resource compliance requirements do not need to be formalized in, e.g., LTL formulas, but can be kept as natural language text. The approach consists of a pre-processing component that makes, i.a., use of recent advances in deep learning, in particular GPT-4. The compliance verification component constitutes the main contribution of this paper and encounters several steps to achieve resource compliance verification. Each step of the approach was evaluated quantitatively using precision and recall on multiple synthetic as well as a real-world dataset, the BPIC 2020 dataset. The evaluation results are promising and provide clear pointers for future work. In particular, we plan to implement a customized pre-processing component for the requirements extraction from natural language text, e.g., using OpenAssistant [14], and compare it to the current solution which uses GPT-4. This allows us to cope with limitations arising due to possible downtimes of GPT-4, the necessity to have access to GPT-4, and investing time to fine-tune the employed prompt. Moreover, we plan to incorporate further perspectives like control flow, data, and time to come up with a holistic compliance verification approach.

Acknowledgements. This work has been partly funded by SAP SE in the context of the research project "Building Semantic Models for the Process Mining Pipeline" and by the Deutsche Forschungsgemeinschaft (DFG, German Research Foundation) – project number 277991500.

References

1. IEEE standard for eXtensible event stream (XES) for achieving interoperability in event logs and event streams. IEEE Std 1849–2016, pp. 1–50 (2016). https://doi.org/10.1109/IEEESTD.2016.7740858

2. van der Aalst, W.M.P., de Beer, H.T., van Dongen, B.F.: Process mining and verification of properties: an approach based on temporal logic. In: On the Move to Meaningful Internet Systems, pp. 130–147 (2005). https://doi.org/10.1007/11575771_11

3. van der Aalst, W.M.P., van Hee, K.M., van der Werf, J.M.E.M., Kumar, A., Verdonk, M.: Conceptual model for online auditing. Decis. Support Syst. **50**(3), 636–647 (2011). https://doi.org/10.1016/j.dss.2010.08.014

4. Barrientos, M., Winter, K., Mangler, J., Rinderle-Ma, S.: Verification of quantitative temporal compliance requirements in process descriptions over event logs. In: Indulska, M., Reinhartz-Berger, I., Cetina, C., Pastor, O. (eds.) CAiSE 2023. LNCS, vol. 13901, pp. 417–433. Springer, Cham (2023). https://doi.org/10.1007/978-3-031-34560-9_25

5. Bellan, P., Dragoni, M., Ghidini, C.: Extracting business process entities and relations from text using pre-trained language models and in-context learning. In: Enterprise Design, Operations, and Computing, pp. 182–199 (2022). https://doi.org/10.1007/978-3-031-17604-3_11

6. Bellan, P., Ghidini, C., Dragoni, M., Ponzetto, S.P., van der Aa, H.: Process extraction from natural language text: the PET dataset and annotation guidelines. In: Proceedings of the Sixth Workshop on Natural Language for Artificial Intelligence (NL4AI 2022), vol. 3287, pp. 177–191. CEUR-WS.org (2022). http://ceur-ws.org/Vol-3287/paper18.pdf

7. Berti, A., van Zelst, S.J., van der Aalst, W.M.P.: Process mining for python (PM4Py): bridging the gap between process- and data science. CoRR abs/1905.06169 (2019). http://arxiv.org/abs/1905.06169

8. Brown, T.B., et al.: Language models are few-shot learners. In: Annual Conference on Neural Information Processing Systems (2020). https://proceedings.neurips.cc/paper/2020/hash/1457c0d6bfcb4967418bfb8ac142f64a-Abstract.html

9. Brunello, A., Montanari, A., Reynolds, M.: Synthesis of LTL formulas from natural language texts: state of the art and research directions. In: 26th International Symposium on Temporal Representation and Reasoning, TIME, LIPIcs, vol. 147, pp. 17:1–17:19 (2019). https://doi.org/10.4230/LIPIcs.TIME.2019.17

10. Cabanillas, C., Ackermann, L., Schönig, S., Sturm, C., Mendling, J.: The RALph miner for automated discovery and verification of resource-aware process models. Softw. Syst. Model. **19**(6), 1415–1441 (2020). https://doi.org/10.1007/s10270-020-00820-7

11. Friedrich, F., Mendling, J., Puhlmann, F.: Process model generation from natural language text. In: Mouratidis, H., Rolland, C. (eds.) CAiSE 2011. LNCS, vol. 6741, pp. 482–496. Springer, Heidelberg (2011). https://doi.org/10.1007/978-3-642-21640-4_36

12. Hashmi, M., Governatori, G., Lam, H.-P., Wynn, M.T.: Are we done with business process compliance: state of the art and challenges ahead. Knowl. Inf. Syst. **57**(1), 79–133 (2018). https://doi.org/10.1007/s10115-017-1142-1

13. Ji, Z., et al.: Survey of hallucination in natural language generation. ACM Comput. Surv. **55**(12), 248:1–248:38 (2023). https://doi.org/10.1145/3571730

14. Köpf, A., et al.: Openassistant conversations - democratizing large language model alignment. CoRR abs/2304.07327 (2023). https://doi.org/10.48550/arXiv.2304.07327

15. Ly, L.T., Maggi, F.M., Montali, M., Rinderle-Ma, S., van der Aalst, W.M.P.: Compliance monitoring in business processes: functionalities, application, and tool-support. Inf. Syst. **54**, 209–234 (2015). https://doi.org/10.1016/j.is.2015.02.007

16. Mangler, J., Rinderle-Ma, S.: Cloud process execution engine: architecture and interfaces (2022). https://doi.org/10.48550/ARXIV.2208.12214
17. Narouei, M., Takabi, H., Nielsen, R.: Automatic extraction of access control policies from natural language documents. IEEE Trans. Dependable Secur. Comput. **17**(3), 506–517 (2020). https://doi.org/10.1109/TDSC.2018.2818708
18. Neuberger, J., Ackermann, L., Jablonski, S.: Beyond rule-based named entity recognition and relation extraction for process model generation from natural language text. CoRR abs/2305.03960 (2023). https://doi.org/10.48550/arXiv.2305.03960
19. OpenAI: GPT-4 technical report (2023)
20. Quishpi, L., Carmona, J., Padró, L.: Extracting decision models from textual descriptions of processes. In: Business Process Management, pp. 85–102 (2021). https://doi.org/10.1007/978-3-030-85469-0_8
21. Russell, N., van der Aalst, W.M.P., ter Hofstede, A.H.M., Edmond, D.: Workflow resource patterns: identification, representation and tool support. In: Pastor, O., Falcão e Cunha, J. (eds.) CAiSE 2005. LNCS, vol. 3520, pp. 216–232. Springer, Heidelberg (2005). https://doi.org/10.1007/11431855_16
22. Sai, C., Winter, K., Fernanda, E., Rinderle-Ma, S.: Detecting deviations between external and internal regulatory requirements for improved process compliance assessment. In: Indulska, M., Reinhartz-Berger, I., Cetina, C., Pastor, O. (eds.) CAiSE 2023. LNCS, vol. 13901, pp. 401–416. Springer, Cham (2023). https://doi.org/10.1007/978-3-031-34560-9_24
23. Semmelrodt, F., Knuplesch, D., Reichert, M.: Modeling the resource perspective of business process compliance rules with the extended compliance rule graph. In: Bider, I., et al. (eds.) BPMDS/EMMSAD -2014. LNBIP, vol. 175, pp. 48–63. Springer, Heidelberg (2014). https://doi.org/10.1007/978-3-662-43745-2_4
24. Taghiabadi, E.R., Gromov, V., Fahland, D., van der Aalst, W.M.P.: Compliance checking of data-aware and resource-aware compliance requirements. In: Meersman, R., et al. (eds.) OTM 2014. LNCS, vol. 8841, pp. 237–257. Springer, Heidelberg (2014). https://doi.org/10.1007/978-3-662-45563-0_14
25. Türetken, O., Elgammal, A., van den Heuvel, W., Papazoglou, M.P.: Capturing compliance requirements: a pattern-based approach. IEEE Softw. **29**(3), 28–36 (2012). https://doi.org/10.1109/MS.2012.45
26. Voglhofer, T., Rinderle-Ma, S.: Collection and elicitation of business process compliance patterns with focus on data aspects. Bus. Inf. Syst. Eng. **62**(4), 361–377 (2019). https://doi.org/10.1007/s12599-019-00594-3
27. Winter, K., van der Aa, H., Rinderle-Ma, S., Weidlich, M.: Assessing the compliance of business process models with regulatory documents. In: Dobbie, G., Frank, U., Kappel, G., Liddle, S.W., Mayr, H.C. (eds.) ER 2020. LNCS, vol. 12400, pp. 189–203. Springer, Cham (2020). https://doi.org/10.1007/978-3-030-62522-1_14
28. Wolter, C., Schaad, A.: Modeling of task-based authorization constraints in BPMN. In: Alonso, G., Dadam, P., Rosemann, M. (eds.) BPM 2007. LNCS, vol. 4714, pp. 64–79. Springer, Heidelberg (2007). https://doi.org/10.1007/978-3-540-75183-0_5

From Text to Performance Measurement: Automatically Computing Process Performance Using Textual Descriptions and Event Logs

Manuel Resinas[1][(✉)] [iD], Adela del-Río-Ortega[1] [iD], and Han van der Aa[2] [iD]

[1] SCORE Lab, I3US, Universidad de Sevilla, Seville, Spain
{resinas,adeladelrio}@us.es
[2] University of Mannheim, Mannheim, Germany
han.van.der.aa@uni-mannheim.de

Abstract. Process performance measurement assesses how well a process is running, covering various dimensions such as time, cost, and quality. This task involves the definition of measurable Process Performance Indicators (PPIs), which in many cases are calculated based on data recorded in an event log. An inhibitor of effective performance analysis is that establishing PPI definitions measurable from event logs is highly complex, because it requires process analytical expertise, as well as in-depth knowledge about the structure and contents of the available event data. Given that managers typically do not have such knowledge, this means that those stakeholders that are generally most interested in measuring process performance cannot do so in a convenient manner. Recognizing this, we bridge this gap by proposing an approach for the measurement of process performance based on textual descriptions and event logs, which combines state-of-the-art natural language processing techniques with matching strategies that are tailored to the task at hand. Evaluation experiments using textual descriptions provided by both industry and academic users demonstrate the accuracy of our approach.

Keywords: Process performance measurement · process mining · natural language processing · matching

1 Introduction

Process Performance Measurement is the practice of evaluating various dimensions of business processes, such as time, cost, and quality, to determine if business processes are achieving strategic and operational goals, and to assist in their

This work has been partially supported by projects PID2021-126227NB-C21/ AEI/10.13039/501100011033/ FEDER, UE; TED2021-131023B-C22/ AEI/10.13039/501100011033/ Unión Europea NextGenerationEU/PRTR, and US-1381595 (Junta de Andalucía/FEDER, UE).

© The Author(s), under exclusive license to Springer Nature Switzerland AG 2023
C. Di Francescomarino et al. (Eds.): BPM 2023, LNCS 14159, pp. 266–283, 2023.
https://doi.org/10.1007/978-3-031-41620-0_16

optimization. It includes the definition, collection, visualization and analysis of Process Performance Indicators (PPIs), which are quantifiable metrics used to evaluate the efficiency and effectiveness of one or more business processes [4]. PPIs can be calculated from different sources, with recorded execution data, stored event logs, being among the main ones [15] and the focus of this paper.

The definition of PPIs calculated from event logs consists of two primary parts: (1) establishing a formal definition of the manner in which process performance should be measured, according to a certain metamodel for PPIs [13,16,21], and (2) linking that definition to the data structure and contents of a specific event log. For example, to measure the "*average time until reimbursement*" in a travel reimbursement process, part (1) involves the recognition that this corresponds to a PPI that should compute the average (an aggregate measure) time (a measure type) between *receiving a request* (start moment) and the *reimbursement being paid* (end moment). Part (2), in turn, requires one to recognize that this measure should be linked to the time between "*receive request*" and "*payment handled*" activities in a particular event log.

The problem here is that both parts of this task involve expertise from users regarding process performance measurement and process mining (how to properly define PPIs, how event logs work, etc.), as well as in-depth knowledge about the data in an event log (which activity or attribute value corresponds to an occurrence of interest, e.g., that the moment of reimbursement, non-straightforwardly, corresponds to a "*payment handled*" activity). Given that such expertise and domain knowledge are rarely (both) held by managers, process performance currently cannot be conveniently measured by those stakeholders most interested in it. Instead, managers need to involve process analysts and domain experts to obtain the insights they desire, which can lead to considerable hindrance in terms of additional effort and delays, as well as possibly incorrect measurements caused by miscommunication [1].

In this work, we overcome this barrier by proposing an approach for measuring process performance based on event logs and textual descriptions. In this manner, our work allows managers to conveniently and quickly obtain useful insights by describing desired performance measures in a textual manner, such as "*the fraction of requests that are rejected*" or "*the maximum time between receiving and delivering orders that were approved*". With this goal, our work complements recent work on using natural language querying in process mining [2,11], with an approach tailored to the task of process performance measurement.

Our approach builds on a language model fine-tuned to the task of extracting entities from PPI descriptions. To align these extracted entities to the contents of an event log, we propose various matching functions, as well as heuristics to infer missing information. Evaluation experiments using real-world event logs and PPI descriptions collected from industry and academic users highlight the potential of our approach, yet also reveal the challenging nature of the task.

In the remainder, Sect. 2 describes the challenges of automated transformation, whereas Sect. 3 provides essential definitions. Section 4 describes our automated approach to compute PPIs described in natural language. Section 5 presents a quantitative evaluation of our approach. Streams of related work are described in Sect. 6 and we conclude our paper in Sect. 7.

2 Problem Illustration

This section illustrates the main challenges associated with the transformation of textual PPI descriptions into measurable definitions. As a basis for this, we use the well-known *domestic declarations* event log from the 2020 BPI Challenge [5] and the PPI descriptions in Table 1. This process involves the submission, approval or rejection, and payment of travel declarations by employees.

Table 1. Exemplary PPI descriptions

ID	Description
ppi1	The average duration between submission and payment of a declaration
ppi2	The average time it takes for a declaration to be paid after its submission
ppi3	The amount of time until reimbursement
ppi4	The percentage of rejected requests
ppi5	The number of denied declarations as a fraction of the total submitted ones
ppi6	The number of declarations above 100 euros
ppi7	The total amount paid per year

C1: High Flexibility of Natural Language. Textual PPI descriptions can describe the same measure in many different ways. For example, *ppi1* to *ppi3* all describe the time between *submission* and *payment* of declarations, using different structures (e.g., starting point first or last) and terminology (e.g., *duration* versus *time*). Similarly, whereas *ppi4* immediately indicates that this is a fractional measure, this information comes later in *ppi5*, which starts in the exact same manner (*"The number of [..]"*) as is common for count measures, such as *ppi6*. These examples are only the tip of the iceberg, though, which means that a transformation approach must be able to deal with highly variable input.

C2: Differences Between Description and Data. When describing a measure of interest, users do not necessarily account for the way that a process is recorded, which can result in large differences between the contents of a textual description and an event log. This often results in the use of synonyms (*denying* versus *rejecting*), though differences may also be process specific. For example, *ppi3* refers to the time until *reimbursement*, yet there is no activity in the event log that contains this term. Rather, the moment of *reimbursement* corresponds to the *Payment handled* activity. Therefore, when matching information extracted from a PPI description to the contents of an event log, an approach must be able to find challenging correspondences.

C3: Missing Information. Finally, PPI descriptions may leave certain information implicit that is required to define a measure. Common examples are: missing aggregation functions (does *ppi3* refer to the average, total, or individual time until payment?), missing starting points of time measures (what is the starting point of *ppi3*?), and missing denominators of fractions (e.g., in *ppi4*). A transformation approach needs to be able to make the right choices in such situations, in order to still be able to compute a value for the desired measure.

3   Measurable PPI Definitions

This section presents the formal PPI definitions that our approach uses. These definitions are inspired by the PPINOT metamodel [4], which we specifically adapted to the way in which PPIs are commonly described in natural language, allowing for easier and more accurate transformation from textual description to a formal PPI. The values for PPIs defined in this manner can be automatically measured using a PPI computation tool (cf., [15]).

Scope. Our work covers a broad range of PPI definitions, allowing users to combine the following aspects. Each PPI definition should correspond to one of three types of base measures: *count*, *time*, and *data*. These can be complemented with additional operators such as *aggregation functions* (e.g., *min.*, *max.*, *average*), *group-by conditions* (e.g., *request per year or department*), *negation* (e.g., *requests not accepted*), and *filters* (such as *above 100 euros* or in the form of a fraction, such as the *fraction of rejected declarations*).

Definitions. We formalize measurable PPIs through the following definitions.

Definition 1 (Universes). *We define the following universes:*

- $\mathcal{U}_{att}$ *and* $\mathcal{U}_{val}$ *are the universes of* attribute names *and* values *in an event log, including an* activity *attribute to refer to activities and their names. For each* $a \in \mathcal{U}_{att}$, *we use* $dom(a) \subseteq \mathcal{U}_{val}$ *to refer to values that* a *can take.*
- $\mathcal{U}_{agg}$ *is the universe of* aggregation functions. *In this paper we consider* $\mathcal{U}_{agg} = \{avg, max, min, sum, perc\}$.
- $\mathcal{U}_{op}$ *is the universe of* operations. *In this paper* $\mathcal{U}_{op} = \{==, \neq, >, <, \geq, \leq\}$.
- $\mathcal{U}_{case}$ *is the universe of* possible conditions *that refer to a case. In this paper,* $\mathcal{U}_{case} = \{begin, end\}$, *which refer to the beginning and end of a case.*
- $\mathcal{U}_{mval}$ *is the universe of possible values of base measures computed on an event log. This includes integers (for count measures), time intervals like* 7 days *for time measures, and* $\mathcal{U}_{val}$ *for data measures.*

We next define the different kinds of conditions used to specify measures:

Definition 2 (Conditions). *We define two sets of conditions:*

- Instant conditions $\mathcal{C}_I = \mathcal{C}_E \cup \mathcal{C}_C$ *comprise event and case conditions. Event conditions* $\mathcal{C}_E = \mathcal{U}_{att} \times \mathcal{U}_{op} \times \mathcal{U}_{val}$ *are tuples that relate an attribute name to a value using a comparison operator. Case conditions* $\mathcal{C}_C = \mathcal{U}_{case}$ *refer to either the beginning or end of a case.*

- Measure conditions $\mathcal{C}_M = \mathcal{U}_{op} \times \mathcal{U}_{mval}$ *are tuples that define a boolean expression based on a comparison against a possible measure value* $(\mathcal{U}_{mval})$.

An example of an event condition is $(activity, ==, submit\ declaration)$, which occurs when a *submit declaration* activity is completed, while to select cases of *declarations above 100 euros*, we specify $(>,€100)$ as a measure condition over a measure value that refers to a case attribute. Using instant and measure conditions, we can define the base measures supported by our approach.

Definition 3 (Base measures). Base measures $\mathcal{M}_B = \mathcal{M}_C \cup \mathcal{M}_T \cup \mathcal{M}_D$, *comprise three types:*

- Count measures $\mathcal{M}_C = \mathcal{C}_I \times (\mathcal{C}_M \cup \{\bot\})$ *are tuples that include an instant condition that specifies when to count, and an optional measure condition that is applied to the result of the count ($\bot$ as the absence of a condition).*
- Time measures $\mathcal{M}_T = \mathcal{C}_I \times \mathcal{C}_I \times (\mathcal{C}_M \cup \{\bot\})$ *are tuples that include two instant conditions specifying when the time measure starts and stops, respectively, and an optional measure condition ($\bot$ as the absence of a condition).*
- Data measures $\mathcal{M}_D = \mathcal{U}_{att} \times (\mathcal{C}_M \cup \{\bot\})$ *are tuples that include the attribute whose value we want to obtain, and an optional measure condition.*

For example, the count measure for *ppi4* is $((activity, ==, reject), (>, 0))$,[1] the time measure for *ppi3* is $(begin, (activity, ==, payment\ handled), \bot)$, and the data measure for *ppi7* is $(amount, \bot)$.

Finally, we define *aggregated measures*, which expand the expressiveness of base measures with aggregation, group-by, and filtering options:

Definition 4 (Aggregated measures). *The set of* aggregated measures $\mathcal{M}_A = \mathcal{M}_B \times \mathcal{U}_{agg} \times (\mathcal{U}_{att} \cup \{None\}) \times (\mathcal{C}_E \cup \bot)$ *is the set of tuples such that* $a = (b, agg, att, c)$ *means that the aggregation function agg is applied over the base measure b, grouping by attribute att, and filtering the cases that meet condition c. If att = None, this means that no grouping is applied; if c =$\bot$, this means that no condition is applied.*

For example, the full measure for *ppi1* is $(((activity, ==, submit), (activity, ==, payment\ handled), \bot), average, None, \bot)$, i.e., the average time between the two activities, and for *ppi7* we get $((amount, \bot), sum, 'year', \bot)$, to group the data measure (sum of the *amount* attribute) per year.

4   Approach

Figure 1 depicts an overview of our approach. The input is a textual description of a PPI and the output is the result of evaluating this PPI against a given event log. The approach consists of four main steps. Step 1 focuses on the extraction of relevant entities from the textual PPI description (tackling challenge C1). Step

---

[1] $(>, 0)$ is used to count the cases for which this activity happens at least once.

2 matches the extracted entities against the contents of the event log in order to start establishing a measurable PPI definition (challenge C2). Then, for cases in which a user left out certain required information (challenge C3), Step 3 uses various heuristics to fill in the gaps and thereby complete the PPI definition. Finally, Step 4 uses the established definition in order to compute the desired PPI, thus directly measuring process performance for the event log.

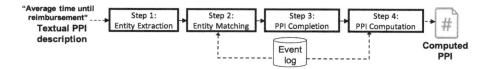

Fig. 1. Overview of our approach

4.1 Step 1: Entity Extraction

In this first step, our approach takes a textual PPI description P as input and extracts the entities necessary to establish a PPI definition. We first discuss the kinds of entities we extract, before describing the extraction technique itself, and the data augmentation strategy we used to compensate for the small size of the available training data.

Entity Types. Our approach aims to extract a number of entity classes, which we each denote with a specific tag. The tags for base measures are:

- Count measures: We use a count entity CE to capture what should be counted, e.g., *requests for reimbursement* or *accepted orders*.
- Time measures: We use start and endpoints (TSE and TEE) when present in descriptions, e.g., *"The amount of time until reimbursement"* contains an end point (*"reimbursement"*). However, descriptions of time measures may also use a single entity to refer to the range from start to end, for which we use a TBE tag, such as for *"The time to approve declarations"*.
- Data measures: We use the DMA tag to refer to the description of the attribute to be measured, such as *the total amount* in *ppi7*.

On top of these classes for the measure types, we extract aggregation functions (AGR), such as *average* or *maximum*, group-by clauses (GBC), such as *per year* in *ppi7*, measure conditions, which are composed of an operator (CCI) and a measure value (MEV), such as *above 100 euros* in *ppi6*, and filters (FDE), such as *pre-approved* in *the total amount of requests that are pre-approved.*

Extraction Technique. We apply a two-stage approach for entity extraction, using an annotated training dataset of PPI descriptions $\mathcal{D}_T$ (see below) and a pre-trained language model [20] as a basis for both stages. In this paper, we use DistilBERT [18] as the pre-trained language model, but the proposal is independent of the language model used.

In the first stage, we use *text classification* to categorize a PPI description P according to its base type (count, time, or data). To this end, we fine-tune the pre-trained language model, with a linear layer on top of its pooled output, using the gold-standard measure types of the training collection $\mathcal{D}_T$. We use the resulting fine-tuned model to infer the measure type of unseen PPI descriptions.

In the second stage, we use *token classification* to identify the entities of interest in a description P. As a basis for this, we use the gold-standard tags of the descriptions in $\mathcal{D}_T$. Since entities can span multiple words, tags are assigned per *chunk*, e.g.: `The\O,average\AGR, time between\O submission\TSE and\O payment of a declaration\ TEE`.[2] Based on these gold-standard tags, we fine-tune the pre-trained language model for token classification, using a linear layer on top of its hidden-states output. We do this separately for count, time, and data measures, so that we obtain language models that are specifically fine-tuned to extract information from a given measure type, as identified by the aforementioned text classifier.

Given a textual PPI description P, we represent the output after token classification as a sequence $\Phi_P \backslash T_P = \langle \phi_1 \backslash t_1, \ldots, \phi_m \backslash t_m \rangle$. Each $\phi_i \backslash t_i$ represents a chunk of text (ϕ_i) and its assigned tag (t_i). Each chunk, ϕ_i consists of one or more consecutive words from P and each word is assigned to exactly one chunk.

Data Augmentation. We only had a collection of 165 PPI descriptions available, including 129 from prior research [1]. However, the fine-tuning of language models requires a considerable amount of training data, especially when dealing with such diverse kinds of input and entities as in our work (challenge C1).

We address this challenge through data augmentation. Specifically, we greatly extend the initial set of PPI descriptions with automatically generated ones. To this end, we handcrafted textual patterns based on the descriptions in the initial set, making sure that a wide variety of different patterns is included. For instance, a typical text pattern for time measures is `[Agg] time from [cond1] to [cond2]`. Then, we used *Chatito*[3], a text generation tool, to generate distinct training phrases by combining all alternatives provided for each pattern. In this manner, we ended up with a total of 12,036 annotated descriptions in $\mathcal{D}_T$.

Using this augmented set to train the aforementioned text and token classification techniques, our entity extraction step can deal with highly flexible input.

4.2 Step 2: Entity Matching

We next set out to establish an actual measure M_P, according to the concepts defined in Sect. 3. To illustrate this step, we use *ppi1* as an example, which Step 1 identifies as a *time measure* with the tagged sequence $\Phi_P \backslash T_P$: $\langle average \backslash \texttt{AGR}, submission \backslash \texttt{TSE}, payment\ of\ a\ declaration \backslash \texttt{TEE} \rangle$.

To establish M_P, we start with the structure of an aggregated measure (b, agg, att, c), as defined in Definition 4, and expand its base measure b with the structure of the identified measure type. Since *ppi1* is a time measure, b's structure is $(s, e, (op, v))$ (cf. Definition 3), yielding $M_P = ((s, e, (op, v)), agg, att, c)$.

---

[2] Tag O indicates that a chunk does not belong to any entity from the tag set.
[3] https://rodrigopivi.github.io/Chatito/.

Overall Procedure. Given a tagged sequence $\Phi_P \backslash T_P$, each tagged chunk $\phi \backslash t$ will be used to find a value for an element of M_P. The correspondence between a chunk $\phi \backslash t$ and an element ϵ follows from the tag t. For instance, $average \backslash \texttt{AGR}$ corresponds to agg, whereas $submission \backslash \texttt{TSE}$ and $payment\ of\ a\ declaration \backslash \texttt{TEE}$, respectively, correspond to the time measure's start (s) and end (e).

For a chunk $\phi \backslash t$, we use a *matching function* $\texttt{match}(\phi \backslash t)$ to identify the right value for its corresponding element ϵ in M_P, from a target domain D. For instance, $\texttt{match}(\phi \backslash \texttt{AGR})$ identifies the value for its corresponding element agg from its domain, which is $\mathcal{U}_{agg}$ (cf., Definition 4). As detailed next, we propose six instantiations of $\texttt{match}(\phi \backslash t)$, for different tags t. The general procedure is the same, though: $\texttt{match}$ evaluates the similarity between the chunk ϕ and elements of the target domain D using a similarity measure, returning the element with the highest score.

Matching AGR Tags. Chunks with $\texttt{AGR}$ tags correspond to aggregation functions, which means that the matching function's target domain is $\mathcal{U}_{agg} = \{avg, max, min, sum, perc\}$. Therefore, to match $\phi \backslash \texttt{AGR}$, we consider the similarity between ϕ and each of the possible values of $\mathcal{U}_{agg}$. Additionally, we also consider synonyms for each of them (e.g., *average, mean, total average*).

This is formalized as $\texttt{match}(\phi \backslash \texttt{AGR}) = argmax_{d \in \mathcal{U}_{agg}} \texttt{sim}_{comb}(\phi, d)$, where $\texttt{sim}_{comb}$ combines both the syntactic and semantic similarity of text chunk ϕ to the respective domain values d of the tag and its synonyms as follows:

$$\texttt{sim}_{comb}(\phi, d) = w_1 \texttt{semSim}(\phi, d) + w_2 \texttt{synSim}(\phi, d)$$

The semantic similarity ($\texttt{semSim}(\phi, d)$) of ϕ with a domain value d is the average of the semantic similarity of d and its synonyms. This semantic similarity is the cosine distance between vectors obtained using an algorithm like word2vec[4]. The syntactic similarity ($\texttt{synSim}(\phi, d)$) is determined using the mean of the well-known Damerau-Levenshtein $dl(\phi, d)$ and the Jaro-Winkler $j(\phi, d)$ distances, where we consider the synonym of ϕ with the highest score.

Matching CCI Tags. This tag corresponds to an operator (e.g., *above* or *greater than*). Therefore, its target domain is $\mathcal{U}_{op}$. The approach followed is exactly the same as for aggregation functions: $\texttt{match}(\phi \backslash \texttt{CCI}) = argmax_{d \in \mathcal{U}_{op}} \texttt{sim}_{comb}(\phi, d)$.

Matching DMA and GBC Tags. These tags corresponds to the attribute name used in data measures (e.g., *amount* in "*the total amount paid per year*"), and *group by* clauses (e.g., "*orders per customer type*"), respectively. For these tags, the function $\texttt{match}$ again uses $\texttt{sim}_{comb}$, this time to compare ϕ to L's attribute names in $\mathcal{U}_{att}$, which is their target domain. However, for the $\texttt{GBC}$ tag, we restrict the target domain to a subset of $\mathcal{U}_{att}$, so that it only includes attributes with a relatively low number of distinct values (we use 15 as a guideline). The rationale is that we expect users to be interested in groupings with a limited number of categories (e.g., per *customer_type*), as opposed to grouping by attributes that have unique values per case (e.g., per *order_ID*).

---

[4] We use the $\texttt{en\_core\_web\_lg}$ model provided by spacy (https://spacy.io/).

Matching CE and TBE Tags. The target domain of these tags is the set of instant conditions $\mathcal{C}_I$ of the event log L, which are composed of three parts: attribute, operator, and value. Therefore, given a chunk $\phi_i \backslash$TBE like *reimbursement* in *ppi3*, its matched value is (*activity*, ==, *payment handled*). We perform this matching at once, for just a single chunk ϕ, since users generally do not specify conditions in an attribute-operator-value manner, but rather use a shorthand. For instance, "*reimbursement*" does not explicitly state that this condition refers to an *equals to* operator and that the relevant attribute is an event's *activity*.

A challenge is that the target domain is huge (spanning numerous combinations of attributes and their values), so we apply three actions to reduce it. First, we only consider categorical attributes whose number of categories is lower than a threshold (we use 100). Second, we only consider the attributes whose value changes across the events of at least one case. The reason is that, if the attribute does not change, it is not useful to use it as a condition in count or time metrics. Finally, we restrict the operator of instant conditions to *equals to*, whereas we support *not equals to* in the PPI completion step (Sect. 4.3). This restriction of the operator does not apply to the operator identified by the CCI tag.

Let $\mathcal{S} \subseteq \mathcal{C}_I$ be the subset of conditions after applying these three actions. Then, to match a text chunk ϕ to an instance condition, we define $\text{match}(\phi \backslash t) = argmax_{d \in \mathcal{S}} \text{conSim}(\phi, d)$, where $t \in \{\text{CE}, \text{TBE}\}$. Here we quantify the condition similarity conSim between ϕ and a condition $d = (a, ==, v)$ as follows:

$$\text{conSim}(\phi, d) = (1 - w_{att})\text{valSim}(\phi, a, v) + w_{att} \sum_{v_i \in dom(a)} \frac{\text{valSim}(\phi, a, v_i)}{|dom(a)|}$$

The first part of the equation computes the value similarity (valSim) between the text chunk ϕ and the attribute-value pair of the condition. The second part of the equation considers the average similarity between the chunk and all values of the attribute domain ($dom(a)$). This allows us to prioritize those attributes whose domain is closer to ϕ. w_{att} represents the weight given to the latter. $\text{valSim}(\phi, a, v)$ itself is computed as follows:

$$\text{valSim}(\phi, a, v) = (1 - w_c)\text{indSim}(\phi, v) + w_c\text{indSim}(\phi, a + v)$$

Here, the first part of the equation computes the individual similarity (indSim) between the text chunk and the value of the attribute v, whereas the second part of the equation computes the same similarity but considering both the name of the attribute and its value ("$< a > < v >$"). This is done to account for cases where the text chunk either omits or includes the name of the attribute.

Finally, $\text{indSim}(\phi, v)$ can be computed using any established similarity measures. In this paper, we combine four similarity measures with a weighted sum: sim_{comb} as defined above, sim_{is}, which uses a standard measure for bag-of-words-based similarity we applied in a previous work [1]. sim_{emb}, which uses the cosine distance between the sentence embeddings of the chunks [14]. sim_{bert}, which uses pre-trained natural language inference models as zero shot text classifiers [22].

Matching TSE and TEE Tags. These tags refer to the conditions that determine the beginning (TSE) and end (TEE) of a time measure such as "*the duration* *between submission and payment of a declaration.*" When both of these tags are extracted from P, we can extend the approach used for CE and TBE with the following two heuristics to improve its performance. First, the *from* and *to* conditions are likely to be related to the same attribute (e.g., if the first condition refers to an activity, the second condition tends to refer to an activity as well). Second, the *from* condition should commonly occur before the *to* condition in the case. Therefore, given a text chunk ϕ, we compute the similarity for each possible pair of conditions $d_i = (a_i, ==, v_i)$ and $d_j = (a_j, ==, v_j)$ as follows:

$$\texttt{pairSim}(\phi, d_i, d_j) = (1 - w_{mh})\frac{\texttt{conSim}(\phi, d_i) + \texttt{conSim}(\phi, d_j)}{2}$$
$$+ w_{mh}\frac{\texttt{same}(a_i, a_j) + \texttt{cr}(d_i, d_j)}{2} \tag{1}$$

where $\texttt{same}(a_i, a_j)$ has a value of 1 if $a_i = a_j$ and 0 otherwise, and $\texttt{cr}(d_i, d_j)$ is the ratio of cases where the *from* condition occurs before the *to* condition from all cases where both conditions occur. To penalize only those conditions where the condition ratio is significantly low, we apply a logistic function normalized between 0.5 and 1 as follows: $2\left(\frac{1}{1 + e^{-k \times cr}} - 0.5\right)$, where k is a parameter that determines the steepness of the curve. We use $k = 10$ in our implementation.

Matching MEV Tag. This tag refers to a measure value used in a measure condition, such as "*100 euros*" in *ppi6*. Its target domain is the set of possible values of base measures computed on an event log L ($\mathcal{U}_{mval}$). The matching is performed using the context provided by the measure type identified in Step 1. For count measures, it just involves parsing an integer. For time measures, we use a syntactic parsing of time deltas. For data measures, it depends on the domain of the attribute used in it. For instance, in *The number of declarations with amount above 100 euros*, we would match *100* with the domain of *amount*, which is an integer. If the domain is categorical, we use $\texttt{sim}_{comb}$.

4.3 Step 3: PPI Completion

After the previous step, a PPI definition M_P has been built according to the information that is explicitly provided in the textual description P. However, as described in challenge C3, some details may have been left implicit, which means that certain mandatory slots in M_P may still be empty. In this step, we fill these remaining slots based on several heuristics, which reflect common-sense interpretations of the missing pieces of information.

Missing Time Points. Descriptions such as "*The amount of time until reimbursement*" only describe a single point in time (the end, here), even though a time measure requires both a start and an end. Therefore, we complete time measures with an unspecified start point by setting it to the earliest timestamp of a case, and use the last timestamp for missing endpoints.

Default Conditions. Users can provide descriptions of count measures like "*the percentage of rejected requests.*" Such percentage aggregations require measure conditions in order to be properly defined. Therefore, we add a measure condition > 0 to the result of the count, in our example: $((activity, ==, declaration\ rejected\ by\ employee), (>, 0))$, to capture that this activity should occur at least once for a case to be considered in the aggregation's numerator.

Default Aggregations. Users may provide descriptions for time measures such as "*The amount of time until reimbursement*" (*ppi3*). Although these technically do not indicate that this is an aggregate measure, we recognize that a user is likely not interested in the time of individual instances, which is why we automatically set the aggregation function to *average* here. Similarly, for a count measure such as "*the number of declarations*", we employ a *sum* aggregation by default.

Applying Negation. Instant conditions can include negations, e.g., *the number of requests that are not paid*, which can be recognized using available *dependency parsers*.[5] In these cases, we must ensure that the corresponding negated measure is properly defined. For time measures, this is straightforward, since we can just change the operator of the identified instant condition from *equals to* to *not equals to*. However, if we negate the condition of a count measure, such as $(activity, \neq, payment\ handled)$, we would get a situation in which all non-payment activities are counted. Therefore, we instead insert a measure condition, i.e., $(=, 0)$, which ensures that all cases for which the *payment handled* activity did not occur are counted.

4.4 Step 4: PPI Computation

Step 3 yields a complete PPI definition. To actually compute a value for it, this definition can be translated into the input for a PPI computation tool, like ppinot4py [15] or the Celonis Process Query Language (PQL) [19]. In this paper, we use ppinot4py because there is a direct correspondence between the elements of the PPI definition and the PPINOT model so no transformation is needed.

5 Evaluation

To test our approach, we conducted an evaluation in which we compare PPI definitions obtained by our approach to a manually created gold standard.[6]

5.1 Evaluation Data

We collected two data sets of textual PPI descriptions for real-world event logs, whose main characteristics are summarized in Table 2. To allow for a high external validity of our evaluation, the data was obtained from different sources. The first one was gathered during different BPM courses with undergraduate and

---

[5] We again use the Spacy library for this.

[6] More information, our prototype, and links to the materials can be found at https://github.com/isa-group/ppinat.

master students and includes 52 PPI descriptions related to the event log of a public traffic fine management process (TF) [12]. The second one was collected using an online questionnaire, through which industry and academic users from different countries provided 53 PPI definitions, all related to the domestic declarations process (DD) [5]. Participation was voluntary and anonymous.

Table 2. Information about the datasets in the test collection

Dataset	Participants	PPIs	Reason to exclude			PPIs tested		
			(1)	(2)	(3)	Time	Count	Data
Traffic fines (TF)	18	52	7	7	18	12	8	0
Declarations (DD)	14	53	12	9	2	11	17	2

(1) Information not in the log. (2) Not supported. (3) Ambiguous information.

In both datasets, a pre-processing step was needed and, as shown in Table 2, some PPI textual descriptions from the original data collection had to be excluded. These PPI descriptions could not be manually transformed into structured definitions to be computed against the information available in the corresponding event logs. Although most PPIs are time-related, the diversity of the PPI descriptions provided by the participants is noteworthy.

5.2 Experimental Setup

Implementation. To conduct the evaluation, we implemented the presented approach in Python, available in our repository.

Hyperparameter Search. The entity-matching step uses various weights to operationalize the matchers. To find the appropriate settings, we perform an exhaustive grid search considering the following values: $w_{\texttt{sim}_x} \in \{0, 0.25, 0.5, 0.75, 1\}$, $w_{att} \in \{0, 0.1, 0.2\}$, $w_c \in \{0, 0.5, 1\}$, and $w_{mh} \in \{0, 0.25, 0.5\}$, where w_{att}, w_c, w_{mh} are the weights defined in Sect. 4.2, and $w_{\texttt{sim}_x}$ captures the weights of the similarity measures used in $\texttt{indSim}$, testing a total of 945 combinations.

In this manner, we selected a configuration with the following weights: $w_{\texttt{sim}_{is}} = 0.25, w_{\texttt{sim}_{emb}} = 0.5, w_{\texttt{sim}_{bert}} = 0.25, w_{\texttt{sim}_{comb}} = 0, w_c = 0.5, w_{att} = 0.2$ and $w_{mh} = 0.25$. We report on the impact of these parameters below in the discussion of the results of Step 2.

Evaluation Measures. To assess the quality of our approach we use the well-known *precision* and *recall* measures to compare generated PPI definitions to a manually created *gold standard*. The gold standard, available in the repository, was created by the three authors, who independently established measurable definitions for the gathered PPI descriptions based on their understanding of the respective event logs. The few differences were then resolved through a joint discussion. Here, *precision* reflects the fraction of slots that our approach filled correctly according to the gold standard, whereas *recall* represents the fraction of slots filled in the gold standard that were also correctly filled by our approach.

5.3   Results

In this section we report on the overall results obtained using our approach, followed by an assessment of its individual steps and configurations.

Overall Results. Table 3 summarizes the results obtained in our evaluation. The *Approach* column reports on the results obtained by applying our full approach on the data, which shows that it obtains a precision and recall of above 0.70 for both datasets.

Table 3. Evaluation results obtained for the two datasets and various configurations

Dataset:	Domestic declarations (DD)						Traffic fines (TF)					
Config.:		*Approach*		*Perfect*		*No comp.*		*Approach*		*Perfect*		*No comp.*
Slot type	n	prec.	rec.	prec.	rec.	prec.	rec.	n	prec.	rec.	prec.	rec.	prec.	rec.
Aggreg	29	0.93	0.93	1	1	0.88	0.76	18	0.89	0.89	0.95	0.95	0.75	0.33
Cond	12	0.8	1	0.92	1	1	0.08	3	0.75	1	1	1	0	0
From	11	0.72	0.72	0.82	0.82	0.6	0.27	12	0.58	0.58	0.75	0.75	0.33	0.08
To	11	0.45	0.45	0.54	0.54	0.6	0.27	12	0.67	0.67	0.83	0.83	0.67	0.17
When	16	0.44	0.44	0.47	0.47	0.44	0.44	6	0.83	0.83	0.87	0.87	0.83	0.83
Total	82	0.70	0.72	0.78	0.80	0.67	0.44	52	0.72	0.75	0.86	0.86	0.64	0.27

As expected, the best results are obtained for the matching of *aggregation* slots, which have a small, fixed target domain. *Condition* slots also get a high precision and recall in both datasets, but in this case the majority of true positives comes from the PPI completion of Step 3 (see below). Regarding *from, to,* and *when* slots, their precision and recall changes significantly from one dataset to the other, which suggests that they are heavily domain-dependent. Note that we omitted slot types (group-by, data, and filters) with $n \leq 2$.

Step 1: Impact of Entity-Extraction Quality. We evaluate if and how any mistakes made by the parser used in Step 1 affect the overall result of our approach. To do this, we also computed results obtained when using perfectly extracted entities (from the gold standard) as input for Step 2. As shown through the *Perfect* column in Table 3, we then obtain a precision of 0.78 and recall of 0.80 for DD, and precision and recall of 0.86 for the TF dataset, showing that our matching strategies are accurate. Compared to the results obtained with our full approach, we observe differences between 0.08 in DD and 0.14 in TF. This improvement is especially apparent for *from* and *to* slots, where more precise extraction leads to clear improvements for entity matching.

Step 2: Matching Configurations. Next, we assess the impact of the various parameter settings and heuristics used in Step 2 to match extracted entities to the elements of an event log. The results obtained during hyperparameter search (available in the repository) show that the configuration of the matchers considerably affects the overall result quality. Out of the 945 matcher configurations, the best configured matchers outperform the worst matchers by 0.11 for DD and 0.14 for TF in terms of precision and recall. To further examine the effects of the weights, we compared the values of the top 25% and the worst 25% configurations. The best configurations typically combine at least two similarity metrics (out of the four sim_x options) for matching extracted entities to instant conditions for the *from*, *to*, and *when* slots. Moreover, virtually all configurations in the top 25% use a similarity metric based on language models (sim_{bart} or sim_{emb}), frequently combining both.

With respect to the matching heuristics, we find that top-25% configurations typically consider the order in which entities matched to *from* and *to* slots appear in traces (by using $w_{mh} > 0$), whereas the worst configurations tend to omit this consideration (using $w_{mh} = 0$). Furthermore, when matching *from*, *to*, and *when* slots, the majority of the top-5% configurations consider attribute names (on top of attribute values), by setting $w_c = 0.5$. By contrast, there is no clear difference between configurations that assign positive weights to w_{att}, which also lets a matcher consider the entire domain of an attribute. Nevertheless, the positive impact of w_{mh} and w_c thus highlight the importance of considering the contents of an event log, in terms of event order and attribute names, during matching.

Step 3: Impact of PPI Completion. We assess the relevance of the *PPI completion* step by comparing the results of our full approach to those obtained when omitting this step (*No comp.* in Table 3). The results reveal a significant drop in recall, i.e., of 0.28 in DD and 0.49 in TF. This shows that the proposed PPI completion heuristics help to identify many missing default values for *aggregation*, *condition*, *from*, and *to* slots. Precision also decreases (e.g., from 0.72 to 0.64 for TF), because without Step 3, negations are not properly interpreted.

Runtime Efficiency. We tested our approach on an Intel i9 PC with 64 GB of RAM, a 2TB SSD hard drive, and a consumer GPU GeForce RTX 3080 Ti. The average execution time for steps 1 to 3 for each PPI defined for DD and TF is 0.66 and 0.91 seconds, respectively. The initialization time, which is executed just once for each log and involves loading the log and computing the embeddings of the attribute names and values, is 10 seconds for DD and 42 seconds for TF.

5.4   Discussion

A post-hoc analysis of the results obtained reveals that the approach faces several challenges related to the entity-extraction and entity-matching steps.

Entity-Extraction Challenges. The usage of a state-of-the-art token classification technique allows our approach to deal with highly flexible input (challenge C1), and performs well with previously unseen terms. However, it is occasionally infeasible to distinguish different entity types based on just a description. For instance, in the TF dataset, the description *"Percentage of fines with an increment"* corresponds to a percentage aggregation over a count entity CE (counting the cases with an *"Add Penalty"* activity). By contrast, the description *"Percentage of fines appealed with vehicle class A"*, involves a *filter entity* FDE on top of the count entity CE, even though the two descriptions are structurally identical. Here, one might envision a post-processing step that evaluates all different alternatives and picks the one that fits best with the event log or even fine-tuning the token classification for the specific domain at hand.

Matching Challenges. The entity-matching step also faces the problem of lack of domain knowledge to resolve its task. For instance, in the DD dataset there are several activities with the text "approval" in them (e.g., *Declaration approved by Administration* and *Declaration final approved by supervisor*). If the PPI description does not provide indications about what kind of approval it refers to, e.g., *"Percentage of declarations are approved"*, it is near impossible for the entity matching to tell to which approval activity it should be matched.

A related problem occurs when confounding words appear in the PPI description and the entity-matching step gives them more relevance than the actual key words of the description. For instance, the DD dataset has two activities called *Request payment* and *Approve request*. When trying to match a PPI description like *"Average time to approve the request for reimbursement"*, the matcher has to decide which part of the text is more relevant: *"approve the request"*, which matches best with *Approved request* or the confounding *"request for reimbursement"*, which matches best with *Request payment*. This problem may be addressed by recognizing when a PPI description refers to the overall goal of a process (e.g., *request for reimbursement*) and using that information to reduce the relevance of these confounding elements for matching.

6   Related Work

Process Performance Measurement. Most process mining tools support the computation of some types of PPIs. In most cases, however, they just support a predefined set of metrics, mainly related to time. There are some exceptions to this, e.g. the *PQL* language to define customized PPIs in *Celonis* [19]. The main drawback in this case is that the computation results are not designed to be used outside the tool platform and integrated with other tools or workflows. Recently, *ppinot4py* was presented as a library that can be used to compute a wide variety of custom PPIs [15]. Yet, its main limitation is that users need to know low-level details of the log involved as well as technical aspects of the definition of PPIs. There is another thread of works that proposes user-friendly approaches to define custom PPIs, in the form of graphical representations [4,9] or templates [17]. A caveat is that the PPIs from these approaches cannot be directly computed over an event log.

NLP Interfaces for Data Base Querying. Highly related to our approach is the existing work on defining queries on tables or databases using natural language. There are two main threads in this regard. The first thread uses natural language text and a database schema as input to generate an SQL query that can be directly computed on the database. An example of this is the work presented in [8], but many more can be found in a recent survey [10]. The second thread does not generate an intermediate query model, but uses deep learning to learn the appropriate output, given a natural language query and a database table. Some examples are [6,7]. Although these works address a similar problem to the one presented in this paper, the nature of processes and the event logs that collect their information differs significantly from that of databases, which prevents us from using the same approaches off-the-shelf.

NLP Interfaces for Process Mining. Finally, there is developing interest in using NLP to facilitate process mining tasks. For instance, Kobeissi et al. [11] present an intent-based natural language interface to allow users to perform queries on an event log. However, their work focuses on queries about event related data, with performance queries being out of scope. In addition, [11] needs to be adapted for each event log, unlike our proposal, which is log-independent and does not require human intervention to hand-craft the training set for each new event log. Barbieri et al. [2] present an architecture to support a conversational, process mining oriented interface to existing process mining tools, but only focused on questions over process execution data. This preliminary work is extended in [3]. It introduces a taxonomy for natural language questions for process mining and also provides support to queries over process behavior and process mining analyses. However, their implementation and evaluation is performed with general questions applicable to any event log, leaving those associated with selected, domain-specific event logs for future work.

7 Conclusion

In this paper, we presented a first approach to calculate process performance indicators against a given event log based on textual descriptions. Our work builds on a fine-tuned language model to extract relevant entities, tailored techniques to match these entities to the contents of an event log, as well as completion heuristics to deal with incomplete descriptions provided by users. The evaluation performed yielded promising results, although there are also clear open challenges. It is worth noting that the heuristics used for completing PPIs have been useful in improving the performance of our approach. In addition, the publicly available dataset we established for this work is valuable in itself, as it allows other researchers and us to continue advancing in this direction.

In future work, we aim to improve both the scope and accuracy of our work. For this, we naturally aim to use the exponentially increasing potential of large language models (LLMs) such as GPT-4. A first direction, aiming for accuracy improvements, is to incorporate LLMs directly into our proposed approach by using their functionality to replace parts of our work that currently rely on other NLP technologies, such as the entity-extraction step and the computation of semantic similarity scores. Next to that, we aim to use the conversational capabilities of LLMs to facilitate interaction between a user and an approach such as ours, in order to guide users through the step-by-step definition of PPIs using textual input. In such an interactive setting, the agent can ask clarifying questions where appropriate and, furthermore, allow users to iteratively build up highly expressive performance measures, thus improving both the accuracy and scope of our work.

Acknowledgments. We thank Maria Isabel Ramos and Javier Vilariño for their support in the implementation.

References

1. van der Aa, H., Leopold, H., del Río-Ortega, A., Resinas, M., Reijers, H.A.: Transforming unstructured natural language descriptions into measurable process performance indicators using hidden markov models. Inf. Syst. **71**, 27–39 (2017)
2. Barbieri, L., Madeira, E.R.M., Stroeh, K., van der Aalst, W.M.P.: Towards a natural language conversational interface for process mining. In: Munoz-Gama, J., Lu, X. (eds.) ICPM 2021. LNBIP, vol. 433, pp. 268–280. Springer, Cham (2022). https://doi.org/10.1007/978-3-030-98581-3_20
3. Barbieri, L., Madeira, E.R.M., Stroeh, K., van der Aalst, W.M.: A natural language querying interface for process mining. J. Intell. Inf. Syst. (2022). https://doi.org/10.1007/s10844-022-00759-9
4. del-Río-Ortega, A., Resinas, M., Durán, A., et al.: Visual PPINOT: a graphical notation for process performance indicators. Bus. Inf. Syst. Eng. **61**(2), 137–161 (2019)
5. van Dongen, B.: BPI challenge 2020 domestic declarations. https://doi.org/10.4121/uuid:52fb97d4-4588-43c9-9d04-3604d4613b51
6. Eisenschlos, J.M., Gor, M., Müller, T., Cohen, W.W.: MATE: multi-view attention for table transformer efficiency (2021)
7. Herzig, J., Nowak, P.K., Müller, T., Piccinno, F., Eisenschlos, J.: TaPas: weakly supervised table parsing via pre-training. In: ACL, pp. 4320–4333 (2020)
8. Hui, B., et al.: Improving text-to-SQL with schema dependency learning. arXiv preprint arXiv:2103.04399 (2021)
9. Janiesch, C., Matzner, M.: BAMN: a modeling method for business activity monitoring systems. J. Decis. Syst. **28**(3), 185–223 (2019)
10. Katsogiannis-Meimarakis, G., Koutrika, G.: A survey on deep learning approaches for text-to-SQL. VLDB J. **32**, 905–936 (2023)
11. Kobeissi, M., Assy, N., Gaaloul, W., Defude, B., Haidar, B.: An intent-based natural language interface for querying process execution data. In: ICPM, pp. 152–159. IEEE (2021)
12. de Leoni, M., Mannhardt, F.: Road traffic fine management process. https://doi.org/10.4121/uuid:270fd440-1057-4fb9-89a9-b699b47990f5 (2015)
13. Popova, V., Sharpanskykh, A.: Modeling organizational performance indicators. Inf. Syst. **35**(4), 505–527 (2010)
14. Reimers, N., Gurevych, I.: Sentence-BERT: sentence embeddings using Siamese BERT-networks. In: EMNLP. ACL, November 2019
15. Resinas, M., del Río-Ortega, A., Ruiz-Cortés, A.: PPINOT computer and ppinot4py: Two libraries to compute process performance indicators. In: ICPM (Demo track) (2021)

16. del Río-Ortega, A., Resinas, M., Cabanillas, C., Ruiz-Cortés, A.: On the definition and design-time analysis of process performance indicators. Inf. Syst. **38**(4), 470–490 (2013)
17. del Río-Ortega, A., Resinas, M., et al.: Using templates and linguistic patterns to define process performance indicators. Ent. Inf. Sys. **10**(2), 159–192 (2016)
18. Sanh, V., Debut, L., Chaumond, J., Wolf, T.: DistilBERT, a distilled version of BERT: smaller, faster, cheaper and lighter. arXiv preprint arXiv:1910.01108 (2019)
19. Vogelgesang, T., Ambrosy, J., Becher, D., Seilbeck, R., Geyer-Klingeberg, J., Klenk, M.: Celonis PQl: a query language for process mining. In: Polyvyanyy, A. (ed.) Process Querying Methods, pp. 377–408. Springer, Cham (2022). https://doi.org/10.1007/978-3-030-92875-9_13
20. Wang, H., Li, J., Wu, H., Hovy, E., Sun, Y.: Pre-trained language models and their applications. Engineering (2022)
21. Wetzstein, B., Ma, Z., Leymann, F.: Towards measuring key performance indicators of semantic business processes. In: Abramowicz, W., Fensel, D. (eds.) BIS 2008. LNBIP, vol. 7, pp. 227–238. Springer, Heidelberg (2008). https://doi.org/10.1007/978-3-540-79396-0_20
22. Yin, W., Hay, J., Roth, D.: Benchmarking zero-shot text classification: datasets, evaluation and entailment approach. In: EMNLP, pp. 3914–3923 (2019)

Agent Miner: An Algorithm for Discovering Agent Systems from Event Data

Andrei Tour[1]([✉])[iD], Artem Polyvyanyy[1][iD], Anna Kalenkova[2][iD],
and Arik Senderovich[3][iD]

[1] The University of Melbourne, Parkville, VIC, Australia
`atour@student.unimelb.edu.au`, `artem.polyvyanyy@unimelb.edu.au`
[2] The University of Adelaide, Adelaide, SA, Australia
`anna.kalenkova@adelaide.edu.au`
[3] York University, Toronto, ON, Canada
`sariks@yorku.ca`

Abstract. Process discovery studies ways to use event data generated by business processes and recorded by IT systems to construct models that describe the processes. Existing discovery algorithms are predominantly concerned with constructing process models that represent the control flow of the processes. Agent system mining argues that business processes often emerge from interactions of autonomous agents and uses event data to construct models of the agents and their interactions. This paper presents and evaluates Agent Miner, an algorithm for discovering models of agents and their interactions from event data composing the system that has executed the processes which generated the input data. The conducted evaluation using our open-source implementation of Agent Miner and publicly available industrial datasets confirms that our algorithm can provide insights into the process participants and their interaction patterns and often discovers models that describe the business processes more faithfully than process models discovered using conventional process discovery algorithms.

1 Introduction

Process discovery is a subarea of process mining that studies ways to construct models that faithfully describe processes of a system based on event data the system has generated [26]. Constructed models assist analysts in understanding the system and, consequently, deciding how to improve it. The state-of-the-art process discovery algorithms build models that describe the control flow of the processes. This focus on control flow has at least two limitations. Firstly, the resulting models are not well suited for analyzing the behavior of individual process participants and their interactions, as activities and interactions performed by a specific actor are often scattered across a discovered control flow model. Secondly, control flow models discovered from large data arrays are often too

© The Author(s), under exclusive license to Springer Nature Switzerland AG 2023
C. Di Francescomarino et al. (Eds.): BPM 2023, LNCS 14159, pp. 284–302, 2023.
https://doi.org/10.1007/978-3-031-41620-0_17

complex, the phenomenon known as *spaghetti models* [26]. Models of interacting agents, or *agent systems*, where agents are the process participants, address the former limitation by their very definition and, importantly, do *not* necessarily grow in complexity with the growth of the amount of data they represent [29].

Agent system mining is a type of process mining that studies ways to derive and use knowledge about systems composed of interacting agents based on the events these agents generate [24]. This paper presents and evaluates an agent system discovery algorithm. Concretely, this paper makes these contributions:

- It presents Agent Miner, a divide-and-conquer algorithm for discovering models of agents and their interactions from event data. The algorithm "divides" the input collection of events into several special subsets and "conquers" these subsets using conventional discovery algorithms, like Inductive Miner [11] and Split Miner [2], to construct an agent system that has generated the data.
- It presents the results of an evaluation of Agent Miner based on our open-source implementation of the algorithm that shows that Agent Miner can discover agent and interaction models from publicly available industrial event data that i) provide an additional perspective on the system that has generated the data and ii) often describe the processes more faithfully than the corresponding process models constructed using conventional control flow discovery algorithms; thus, addressing the two stated limitations.
- It demonstrates the value of agent system mining and invites the community to more intensive explorations of the agent-based paradigm in process mining.

The remainder of this paper proceeds as follows. The next section presents a motivating example. Then, Sect. 3 gives basic notions required for understanding the subsequent discussions. Section 4 presents our new discovery algorithm, while Sect. 5 discusses the results of its evaluation. Finally, Sect. 6 surveys related work before Sect. 7 gives concluding remarks on this work.

2   Motivating Example

As a motivating example, we propose a hypothetical health surveillance process sketched in Fig. 1. The process starts with a check (see the *check* activity in the

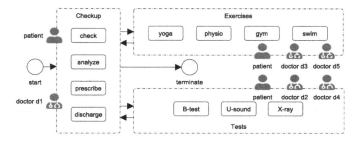

Fig. 1. A schematic visualization of the health surveillance process.

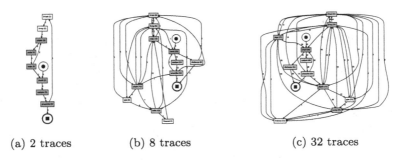

(a) 2 traces   (b) 8 traces   (c) 32 traces

Fig. 2. DFGs that describe the health surveillance process (readability not intended).

figure) of the patient by doctor *d1*. If the analysis (*analyze*) of the results of the check indicates a risk of developing ill-health, the doctor prescribes (*prescribe*) medical tests and preventive therapy. The possible tests are blood test (*B-test*), ultrasound (*U-sound*), and *X-ray*. The therapy includes *yoga*, *physio*, *gym*, and swimming (*swim*). The tests and the therapy are performed independently; any subset of tests and exercises can be prescribed. Doctors *d2* and *d4* perform the tests, while the therapy is conducted by doctors *d3* and *d5*. Once the tests and the therapy are completed, doctor *d1* rechecks (*check*) the patient. If the new check shows good results, the patient is discharged (*discharge*), and the process terminates; otherwise, further tests and therapy are prescribed. The information system that supports the process recorded a *log* of events that stem from managing 1 024 patients. Each recorded event has four attributes that specify the *timestamp* of the event occurrence, *activity* that triggered the event, the patient *case* the event relates to, and the doctor, or *agent*, that performed the activity.

Figure 2 shows three directly-follows graphs (DFGs) [26] that describe control flow dependencies between the activities of the health surveillance process discovered from 2, 8, and 32 traces from the log, where a *trace* is a sequence of all events with the same case attribute ordered by their timestamps. The complexity of the DFGs, defined as the number of nodes and arcs, grows as they represent more data, where the model in Fig. 2c is an example spaghetti model.

Figure 3a and Fig. 3b show the interaction net and one agent net that describes one of the three *agent types* discovered by Agent Miner from the events of the 32 traces used to construct the DFG in Fig. 2c captured as Petri nets [19]. Besides providing an alternative, modular perspective on the process, that is, an explicit representation of process participants and their interactions, these models, similar to the DFGs in Fig. 2, describe control flow dependencies between the activities. These dependencies can be captured explicitly in an integrated Petri net, called the Multi-Agent System (MAS) net, obtained by refining labeled transitions of the interaction net with the corresponding agent nets. Interestingly, this MAS net is smaller and represents the traces encoded in the event data more faithfully than the Petri net constructed from the same 32 traces using Inductive Miner (IM) [11]—a conventional process discovery algorithm. The MAS net has 155 nodes and arcs, while the IM net shown in Fig. 4 has 175 nodes and arcs.

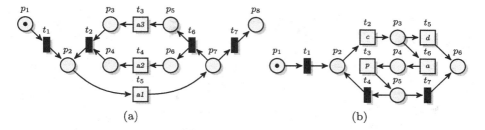

Fig. 3. Petri nets that describe: (a) interactions of the three agent types $a1$ (doctor $d1$), $a2$ ($d2$ and $d4$), and $a3$ ($d3$ and $d5$) and (b) agent type $a1$ from the health surveillance process with transition labels check (c), analyze (a), prescribe (p), and discharge (d).

The precision of the MAS net and the IM net is 0.37 and 0.16, respectively. To support model comparison, both algorithms were configured to discover perfectly fitting models (recall of 1.0). A model has good precision if it does not replay traces not recorded in the log and a good recall if it replays many traces from the log, where the values closer to 1.0 indicate better models. Precision and recall were measured for the 32 traces used to construct the models using the entropy-based approach [16,17], the only existing measures that guarantee that better discovered models result in better measurements [23].

A MAS net often does not require an increase in size to represent more data well due to an ability of an agent system to simulate the non-decreasing complexity of the behavior of a system [24,29]. For instance, the same MAS net we discovered from the 32 traces has precision of 0.43 and recall of 1.0 if measured for all 1 024 traces. This example confirms that Agent Miner can construct modular, fitting, and precise models of process participants and their interactions. Finally, while the interaction and agent nets enable dedicated analyses of the respective artifacts, like identification of repetitive work handovers and verification of safeness and liveness of individual agent nets, the information about agents is scattered in the IM net, see the *U-sound* activity by agent $a2$ amidst activities *swim* and *yoga* by agent $a3$ in Fig. 4, which hinders the analysis of individual agents and their interactions. The input log, models, and measurements discussed in this section are publicly available [25].

3 Preliminaries

This section introduces Petri nets (Sect. 3.1) and event logs (Sect. 3.2).

3.1 Petri Nets and Workflow Nets

Petri nets formalism suits well for describing models of distributed systems [19].

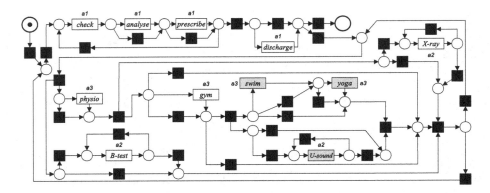

Fig. 4. The IM net discovered from 32 traces of the example health surveillance process.

Definition 3.1 (Petri nets).

A *(labeled) Petri net*, or a *net*, N is a quintuple $(P, T, F, \Lambda, \lambda)$, where P is a finite set of *places*, T is a finite set of *transitions*, $F \subseteq (P \times T) \cup (T \times P)$ is the *flow relation*, Λ is a set of *labels*, such that $\tau \in \Lambda$ is the *silent label* and sets P, T, and Λ are pairwise disjoint, and $\lambda : T \to \Lambda$ is the *labeling function*.   ⌟

If $\lambda(t) = \tau, t \in T$, t is *silent*; otherwise t is *observable*. Observable transitions represent activities from the problem domain, and silent transitions encode internal actions of the system. A *marking* M of a net encodes its state and is a multiset over its places. Figure 3a shows a Petri net $N = (P, T, F, \Lambda, \lambda)$, with eight places ($P = \{p_1, \ldots, p_8\}$) and seven transitions ($T = \{t_1, \ldots, t_7\}$). In the graphical notation, places are drawn as circles, while transitions as squares or rectangles. Transitions t_1, t_2, t_6, and t_7 are silent, shown as black rectangles. The labeling function assigns labels *a1*, *a2*, and *a3* to transitions t_5, t_4, and t_3, respectively. The flow relation is shown as directed arcs. A marking is denoted as an arrangement of black dots, called *tokens*, inside of the corresponding places. Marking M shown in Fig. 3a is the multiset $[p_1]$; see one black dot in place p_1.

Let $n \in P \cup T$ be a place or transition, then by $\bullet n = \{x \in (P \cup T) \mid (x, n) \in F\}$ we denote its *preset* and by $n\bullet = \{x \in (P \cup T) \mid (n, x) \in F\}$ we denote its *postset*. Let $N = (P, T, F, \Lambda, \lambda)$ be a net. A transition $t \in T$ is *enabled* in a marking M of N, denoted by $(N, M) [t\rangle$, if every *input place* of t contains at least one token, i.e., $\forall p \in \bullet t : M(p) > 0$; by $M(p)$ we denote the multiplicity of p in M. An enabled transition $t \in T$ can occur. An *occurrence* of t leads to a fresh marking $M' = (M \setminus \bullet t) \uplus t\bullet$ of N, denoted by $(N, M) [t\rangle (N, M')$.

Workflow nets are special nets used for modeling workflows [26]. A *workflow net* is a Petri net $(P, T, F, \Lambda, \lambda)$ with a dedicated *initial place* $i \in P$, $\bullet i = \emptyset$, a dedicated *final place* $f \in P$, $f\bullet = \emptyset$, and every place and transition on the vertex sequence of some directed walk from i to f in graph $(P \cup T, F)$. Marking $[i]$ is the initial marking of a workflow net. A workflow net is *safe* if every marking reachable from $[i]$ via a sequence of transition occurrences is a set. It is *sound* if every transition of the net can occur in some sequence of transition occurrences that starts in $[i]$, every sequence of transition occurrences that starts in $[i]$ can be extended to "put" a token in the final place, and once there is a token in the final place no other places hold tokens [26]. Figure 3a and Fig. 3b show safe and sound workflow nets with final markings $[p_8]$ and $[p_6]$, respectively.

3.2 Events, Event Logs, and Traces

An *event log*, or *log*, represents real-world processes recorded by an information system. In process mining, events are often organized into time-ordered sequences, called *traces*. As explained in Sect. 2, in our work, each event has at least four attributes: *timestamp*, *activity*, *case*, and *agent* (either instance or type). Let $\mathcal{A}$ be the universe of attribute names, with $\{timestamp, activity, case, agent\} \subseteq \mathcal{A}$. Let $\mathcal{V}$ be the universe of attribute values. Then, an *event* is an attribute function $e : \mathcal{A} \to \mathcal{V}$ that maps attribute names to attribute values. By $\mathcal{E}$ we denote the universe of events. An *event selection* $S \subseteq \mathcal{E}$ is a finite set of events. Without loss of generality, we assume that events have unique timestamps. Table 1 defines event selection $S =$

Table 1. An event selection.

event	timestamp	case	activity	agent
e^a	30 Mar 16:34	case1	check	d1/a1
e^b	30 Mar 16:35	case1	analyze	d1/a1
e^c	31 Mar 16:35	case2	check	d1/a1
e^d	01 Apr 08:22	case2	analyze	d1/a1
e^e	01 Apr 16:05	case1	prescribe	d1/a1
e^f	03 Apr 11:55	case1	B-test	d4/a2
e^g	03 Apr 16:59	case1	X-ray	d4/a2
e^h	06 Apr 10:02	case1	physio	d3/a3
e^i	06 Apr 11:01	case1	swim	d3/a3
e^j	07 Apr 15:55	case1	yoga	d3/a3
e^k	07 Apr 11:11	case1	physio	d3/a3
e^l	10 Apr 13:13	case1	swim	d3/a3
e^m	10 Apr 15:05	case1	yoga	d3/a3
e^n	11 Apr 09:12	case1	physio	d3/a3
e^o	11 Apr 10:05	case1	swim	d3/a3
e^p	13 Apr 11:03	case1	yoga	d3/a3
e^q	13 Apr 14:57	case1	check	d1/a1
e^r	16 Apr 12:11	case1	analyze	d1/a1
e^s	17 Apr 10:03	case1	prescribe	d1/a1
e^t	17 Apr 16:36	case2	prescribe	d1/a1

$\{e^a, \ldots, e^t\}$. Each row of the table describes one event. For example, e^t $= \{(timestamp, 17 \text{ Apr } 16:36), (activity, prescribe), (case, case2), (agent, a1)\}$ is the event from the last row of Table 1; the agent attribute specifies agent type.

Let X be a finite non-empty set. A *partition* of X is a set Π of disjoint subsets of X such that the union of the subsets equals X; the subsets are *parts* of Π. Partition Π can be defined by an equivalence relation $\sim \subseteq X \times X$ such that if two events $e_i, e_j \in X$ are equivalent under $\sim$, that is, it holds that $(e_i, e_j) \in \sim$, then e_i and e_j belong to the same part of Π. We use notation $\Pi = X/\sim$ to denote that partition Π is defined by the equivalence relation $\sim$ over X. Against this background, we define a trace of an event selection as follows.

Definition 3.2 (Traces induced by partitions).
The *trace* σ of event selection S induced by part π of partition $\Pi = S/\sim$ is the ordered by timestamps sequence of all and only events in π. ⌟

We refer to Π and $\sim$ as the *trace partition* and the *trace relation* that induce σ. The *trace set* of S induced by $\Pi = S/\sim$ is the set of traces Σ that for each part π of Π contains the trace of S induced by π, and contains no other traces.

Definition 3.3 (Event logs).
An *event log*, or *log*, of event selection S induced by partition $\Pi = S/\sim$ is a triple (S, Σ, ν), where $\Sigma \subset S^*$ is the trace set induced by Π, and $\nu : S \to \mathcal{V}$ is a naming function that assigns names to events in S. ⌟

Multiple event logs induced by different partitions and naming functions can be defined. For example, in classical process discovery, traces are induced by case attributes. Let $\mathcal{C}$ be a universe of cases. The *case trace set* of event selection S is the trace set of S defined by relation $\sim^{\mathbf{c}} = \left\{(e_i, e_j) \in S \times S \mid e_i(case) = e_j(case)\right\}$, denoted by $\Sigma^{\mathbf{c}}$; we refer to traces in $\Sigma^{\mathbf{c}}$ as *case traces*. For instance, the case trace set of the event selection in Table 1 consists of two case traces, namely $\langle e^a, e^b, e^e, e^f, e^g, e^h, e^i, e^j, e^k, e^l, e^m, e^n, e^o, e^p, e^q, e^r, e^s \rangle$ and $\langle e^c, e^d, e^t \rangle$.

The naming function of an event log specifies the names of the events used by the discovery algorithms to identify activity names in the constructed process models. In process discovery, events are often identified by their activity attributes, that is, $\nu = \{(e, e(activity)) \mid e \in S\}$. In general, other naming functions can be used. We will use this feature in the subsequent sections.

4   Agent Miner

In this section, we introduce the core notions required to define the Agent Miner algorithm (Sect. 4.1) and present the algorithm (Sect. 4.2).

4.1   Agent Logs and Nets

A trace of an *agent trace set* is a sequence of events that refer to the same case, are performed by the same agent (identified by the agent attribute), and are not interrupted by an event from the same case performed by a different agent.

Definition 4.1 (Agent trace sets).
The *agent trace set* of event selection S is the trace set of S induced by the partition of S defined by relation $\sim = \left\{(e_i, e_j) \in S \times S \mid (e_i(agent) = e_j(agent)) \wedge (e_i(case) = e_j(case)) \wedge (\neg \exists e_k \in S : (((e_i(timestamp) < e_k(timestamp) < e_j(timestamp)) \vee (e_j(timestamp) < e_k(timestamp) < e_i(timestamp))) \wedge (e_i(agent) \neq e_k(agent) \wedge e_i(case) = e_k(case))))\right\}$. ⌟

The set $\{\langle e^a, e^b, e^e \rangle, \langle e^c, e^d, e^t \rangle, \langle e^f, e^g \rangle, \langle e^h, e^i, e^j, e^k, e^l, e^m, e^n, e^o, e^p \rangle, \langle e^q, e^r, e^s \rangle\}$ is the agent trace set of the event selection in Table 1. For instance, in trace $\langle e^a, e^b, e^e \rangle$, all events are from *case1*, performed by agent *a1*, and, though interrupted by events e^c and e^d, the latter events are from a different case. Note that, by definition, relation $\sim$ from Definition 4.1 is an equivalence relation.

Next, we define several useful logs. Traces of an interaction log are composed of events that allow identifying all handovers of work between agents.

Definition 4.2 (Interaction logs).
The *interaction log* of event selection S is the triple $(\bar{S}, \Sigma, \nu)$, where $\bar{S}$ is the *agent event selection* composed of the first events of all the traces in the agent trace set Δ of S, that is, $\bar{S} = \{\delta(1) \mid \delta \in \Delta\}$, Σ is the case trace set of $\bar{S}$, and $\nu = \{(e, e(agent)) \mid e \in \bar{S}\}$ names each event by the corresponding agent. ⌟

The interaction log of the event selection from Table 1 is, therefore, the event log $(\bar{S}, \Sigma, \nu)$, where $\bar{S} = \{e^a, e^c, e^f, e^h, e^q\}$, $\Sigma = \{\langle e^a, e^f, e^h, e^q \rangle, \langle e^c \rangle\}$, and $\nu = \{(e^a, a1), (e^c, a1), (e^f, a2), (e^h, a3), (e^q, a1)\}$. For instance, trace $\langle e^a, e^f,$ $e^h, e^q \rangle$ in Σ suggests that agent *a1* starts *case1*; note that $e^a(agent) = a1$ and $e^a(case) = case1$. Then, agent *a2* takes over the work on *case1*. Agent *a2* then passes the work on *case1* to agent *a3*, who later hands work back to agent *a1*.

Traces in the agent trace set done by the same agent compose its agent log.

Definition 4.3 (Agent logs).
The *agent log* of agent a and event selection S is the triple (S^a, Σ, ν), where $S^a = \{e \in S \mid e(agent) = a\}$, Σ is the set of traces in the agent trace set Δ of S performed by a, that is, $\Sigma = \{\delta \in \Delta \mid \forall i \in [1 .. |\delta|] : \delta(i)(agent) = a\}$, and $\nu = \{(e, (a, e(activity))) \mid e \in S^a\}$ is the naming function that names an event by the pair comprising its agent and activity attributes. ⌟

The agent log of agent *a1* and the event selection from Table 1 is defined by the triple (S^{a1}, Σ, ν), where $S^{a1} = \{e^a, e^b, e^c, e^d, e^e, e^q, e^r, e^s, e^t\}$, $\Sigma = \{\langle e^a, e^b, e^e \rangle, \langle e^c, e^d, e^t \rangle, \langle e^q, e^r, e^s \rangle\}$, and ν maps events in S^{a1} to their names and contains, for instance, it holds that $\{(e^a, (a1, check)), (e^d, (a1, analyze))\} \subset \nu$.

Next, we discuss several classes of workflow nets used by Agent Miner. An interaction net describes the structure of interactions between agents.

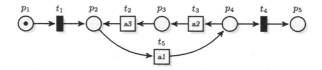

Fig. 5. An i-net.

Definition 4.4 (Interaction nets).
An *interaction net*, or an *i-net*, of event selection S is a workflow net $(P, T, F, \Lambda, \lambda)$, where $\Lambda = \{a \in \mathcal{V} \mid \exists e \in S : e(agent) = a\} \cup \{\tau\}$. ⌟

Fig. 3a and Fig. 5 show two i-nets of the event selection from Table 1. They describe alternative ways the agents could have interacted to generate the event data. For example, the i-net in Fig. 5 suggests that agent *a1* starts the interaction, and then any number of sequences of interactions of *a1* with agent *a2*, then of *a2* with agent *a3*, and finally of *a3* again with agent *a1* can occur.

A MAS net describes how agents perform activities in a collaborative setting.

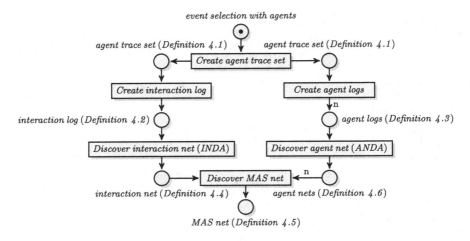

Fig. 6. The Agent Miner algorithm.

Definition 4.5 (Multi-Agent System nets).
A *Multi-Agent System (MAS) net* of event selection S is a workflow net $(P, T, F, \Lambda, \lambda)$, with $\Lambda = \{(a, b) \in \mathcal{V} \times \mathcal{V} \mid \exists e \in S : e(agent) = a \wedge e(activity) = b\} \cup \{\tau\}$. ⌟

A MAS net of events of a single agent is an agent net of this agent.

Definition 4.6 (Agent nets).
A MAS net of event selection S such that the agent attribute of every event in S is equal to a, that is, $\forall e \in S : e(agent) = a$, is an *agent net* of a. ⌟

4.2 Algorithm

Figure 6 defines the Agent Miner algorithm as a workflow net. It is parameterized by two conventional control flow discovery algorithms: an Agent Net Discovery Algorithm (ANDA) and an Interaction Net Discovery Algorithm (INDA). Agent Miner takes an event selection as input and produces an interaction net, agent nets, and a MAS net. Similar to standard process discovery algorithms, the latter explains how the discovered interaction and agent nets represent the traces identified by the case attribute. The algorithm has six steps detailed below.

Create Agent Trace Set. The agent trace set of the input event selection is created by associating each event with an agent trace, see Definition 4.1.

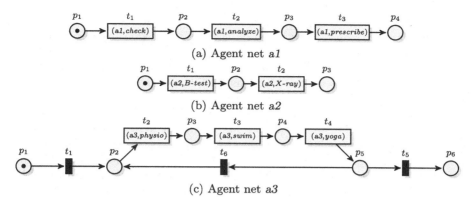

(a) Agent net *a1*

(b) Agent net *a2*

(c) Agent net *a3*

Fig. 7. Three agent nets.

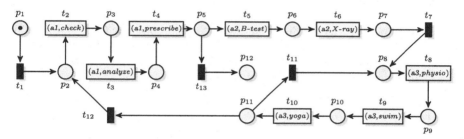

Fig. 8. A MAS net.

Create Interaction Log. The interaction log is created by selecting the first events of all traces in the agent trace set and constructing the case trace set of this event selection, see Definition 4.2.

Discover Interaction Net. In this step, INDA is used to discover an i-net, see Definition 4.4, from the interaction log. For example, the i-net in Fig. 5 was discovered from the interaction log of the event selection in Table 1.

Create Agent Logs. Multiple (n) agent logs are created, one for each agent encountered in the input event selection, by extracting the corresponding agent traces from the agent trace set, see Definition 4.3.

Discover Agent Net. In this step, ANDA is used to discover n agent nets (one net is discovered from each agent log), see Definition 4.6. Figure 7 shows three agent nets discovered from the agent logs of the event selection in Table 1.

Discover MAS Net. Finally, a MAS net, see Definition 4.5, is constructed by "embedding" the agent nets into the i-net and applying the Fusion of Series Places reduction [14] to the refined i-net. Figure 8 shows the embedding of the agent nets in Fig. 7 into the i-net in Fig. 5. The embedding is performed by refining each observable transition in the i-net with the corresponding agent net. The resulting MAS net describes how agents (doctors) interact to support the execution of cases (treatment of patients) in the health surveillance process introduced in Sect. 2 discovered based on the event selection from Table 1.

Thus, Agent Miner is a divide-and-conquer algorithm that "divides" the input event selection into interaction and agent logs and "conquers" these logs using ANDA and INDA. In this work, as ANDA, we used DFG translation to Petri nets (DFG-PN) [12], while as INDA, we used Inductive Miner (IM) [11] and Split Miner (SM) [2] and removed iterations of observable transitions in the obtained i-nets. This approach follows the agent system paradigm, where agents are sequential machines, and parallelism emerges through collaborations of agents. If the algorithms guarantee that the constructed models are safe and sound workflow nets, which holds for DFG-PN and IM, every constructed MAS net is guaranteed to be a safe and sound workflow net (cf. Theorem 2 in [18]).

5   Evaluation

The Agent Miner algorithm (Fig. 6) and its evaluation pipeline (Fig. 9), including the code, the discovered models, and detailed results, are publicly available [25]. This section presents and discusses the design (Sect. 5.1) and results (Sect. 5.2) of an evaluation of Agent Miner using real-world datasets.

5.1   Design

Figure 9 presents our five-step evaluation pipeline. We executed this pipeline for each real-world dataset used in this evaluation. The steps are explained below.

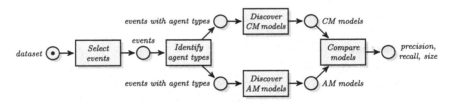

Fig. 9. Evaluation pipeline.

Select Events. In this step, we select events from the dataset to ensure events have required attributes and remove infrequent case traces. Specifically, every event must have a *case*, *timestamp*, *activity*, and *agent* attribute. Moreover, each selected event must be part of a frequent case variant (a trace within *vff*% of the most common case traces, where *vff* is the *variant frequency filter* parameter).

Identify Agent Types. For the event selection, we identify agent instances using agent attribute values of events and group them into agent types by clustering them based on distances between agent instance DFGs, where an agent instance DFG is constructed from the corresponding agent log (Definition 4.3).

For a pair of agents (a_1, a_2), we calculate the distance between the DFGs of a_1 and a_2 as:

$$1 - \max(|DF_{a_1} \cap DF_{a_2}|/|DF_{a_1}|, |DF_{a_1} \cap DF_{a_2}|/|DF_{a_2}|),$$

where DF_{a_1} and DF_{a_2} are the directly-follows relations (sets of edges) in the corresponding DFGs. Thus, the more one DFG subsumes the directly-follows relations of the other DFG, the more likely the corresponding agent instances belong to the same agent type. Once the agent types are identified, the agent attribute values of all the events in the input event selection are updated from instances to types. For example, this step leads to the identification of three agent types $a1$ (doctor $d1$), $a2$ (doctors $d2$ and $d4$), and $a3$ (doctors $d3$ and $d5$) in the motivating example from Sect. 2.

Discover CM Models. We refer to a discovery algorithm based on the conventional process mining paradigm as a Conventional Miner (CM) and a model discovered by a CM algorithm as a CM model. We use two state-of-the-art CM algorithms, Inductive Miner IMf variant (IM) [11] and Split Miner (SM) [2], to construct CM models from the case traces induced by the event selection, that is, traces identified by the case attributes, see Definition 3.3. Given an event selection S, to discover CM models, we use two naming functions: i) activity-only labeling (AOL) defined by $\nu_{aol} = \{(e, e(activity)) \mid e \in S\}$ and ii) agent and activity labeling (AAL) defined by $\nu_{aal} = \{(e, (e(agent), e(activity))) \mid e \in S\}$. Thus, for each event dataset, we construct four CM models. For example, the model in Fig. 4 is the IM model constructed using the ν_{aol} naming function. Hence, agent attributes of events were not used to construct the model, and then observable transitions were annotated with agent info. For each CM algorithm and naming function, we construct ten CM models, one for each configuration of the *threshold* parameter of the algorithm, *noise threshold* for IM and *frequency threshold* for SM (in the range from 0.0 to 0.9 in 0.1 increments). This approach ensures that the discovered CM models show a range of quality characteristics.

Discover AM Models. For the event selection over agent types, we use the Agent Miner (AM) algorithm (refer to Sect. 4) to discover agent nets, the model of agent interactions, and the MAS net that defines the semantics of the resulting agent systems. DFG translation to Petri nets (DFG-PN) [12] algorithm is used as an ANDA. Inductive Miner (IM) [11] and Split Miner (SM) [2] are used as INDAs. For each configuration of ANDA and INDA, we run AM ten times with ten different parameter pairs (ff_i, th_i), with $i \in [1..10]$, $ff_i = i \times 0.1$, and $th_i = 1 - i \times 0.1$. Parameter ff_i is the *activity frequency filter* parameter used by the ANDA. Parameter th_i is the *threshold* parameter used by the INDA, *noise threshold* for IM and *frequency threshold* for SM. The lower the value of i, the more filtered the event selection is and the less the discovered models are fit to the input data.

Compare Models. To compare the discovered CM and AM models, we calculate their *recall* and *precision* with respect to the case trace set of the original event selection, and *size* (as the number of nodes and arcs). To compute precision and recall, we use the entropy-based measures [16,17]. These measures fulfill all the desired properties for these classes of measures [23]. For example, the entropy-based precision measure guarantees that a model that describes fewer traces not in the event log has a better precision score. For CM models constructed using AOL and AAL naming functions of events, we measure their precision and recall with respect to the event logs that identify events using the corresponding naming functions. Note that labels of observable transitions of AM models are composed of agent-activity pairs, see, for example, the MAS net in Fig. 8. Hence, to compare MAS nets to CM models constructed using the AOL naming function, we also measure precision and recall of MAS nets after rewriting their transition labels to only mention activity names.

Table 2. Size, recall, and precision of IM nets and MAS nets discovered by Agent Miner (Inductive Miner as INDA) that rely on activity-only labeling of observable transitions.

BPIC dataset	Variant frequency filter (vf%)	Inductive Miner						Agent Miner (MAS nets)					
		lowest size			greatest precision			lowest size			greatest precision		
		size	recall	prec.	size	recall	prec.	size	recall	prec.	size	recall	xprec.
2011	10%	592	0.72	**0.41**	592	0.72	**0.41**	**466**	**0.92**	**0.35**	**466**	**0.92**	0.35
2012	80%	420	**1.00**	0.06	454	**0.88**	0.32	**333**	0.80	**0.18**	**333**	0.80	0.18
2013	80%	**54**	**0.70**	0.53	105	**0.78**	0.62	69	0.62	0.64	69	0.62	**0.64**
2014	10%	**229**	**0.85**	0.28	335	0.55	**0.45**	231	0.78	**0.36**	**231**	**0.78**	0.36
2015	80%	199	**0.95**	**0.35**	199	**0.95**	**0.35**	122	**0.95**	**0.34**	122	**0.95**	0.34
2017	80%	**216**	**0.85**	0.22	241	0.91	0.26	340	0.76	0.15	597	**0.94**	0.15
2018	10%	**350**	**0.88**	0.13	350	**0.88**	0.13	535	0.85	0.09	535	0.86	0.09
2019	10%	270	0.61	0.24	375	0.54	**0.46**	**191**	**0.65**	**0.34**	**191**	**0.65**	0.34
2020	80%	204	0.72	0.18	220	0.69	**0.25**	**200**	**0.75**	**0.19**	**200**	**0.75**	0.19

5.2 Datasets and Results

To evaluate Agent Miner, we used publicly available real-world Business Process Intelligence Challenge (BPIC) datasets and assumed they stem from agent systems. These datasets are widely used to evaluate conventional process discovery algorithms. We assumed that the resource attributes of events specify agent instances that triggered them and selected all BPIC datasets that specify events with resource attributes, leading to nine selected datasets.[1]

Initially, we used the *vff%* parameter of 80% and completed the evaluation pipeline (cf. Figure 9) for datasets BPIC 2012, 2013, 2015 (Municipality 1), 2017,

---

[1] BPIC 2011 (https://doi.org/10.4121/uuid:d9769f3d-0ab0-4fb8-803b-0d1120ffcf54),
BPIC 2012 (https://doi.org/10.4121/uuid:3926db30-f712-4394-aebc-75976070e91f),
BPIC 2013 (https://doi.org/10.4121/uuid:a7ce5c55-03a7-4583-b855-98b86e1a2b07),
BPIC 2014 (https://doi.org/10.4121/uuid:3cfa2260-f5c5-44be-afe1-b70d35288d6d),
BPIC 2015 (https://doi.org/10.4121/uuid:31a308ef-c844-48da-948c-305d167a0ec1),
BPIC 2017 (https://doi.org/10.4121/uuid:5f3067df-f10b-45da-b98b-86ae4c7a310b),
BPIC 2018 (https://doi.org/10.4121/uuid:3301445f-95e8-4ff0-98a4-901f1f204972),
BPIC 2019 (https://doi.org/10.4121/uuid:d06aff4b-79f0-45e6-8ec8-e19730c248f1), and
BPIC 2020 (https://doi.org/10.4121/uuid:52fb97d4-4588-43c9-9d04-3604d4613b51).

and 2020 (Travel Permit Data). Due to performance reasons, to process the other datasets, we lowered the $vff\%$ parameter to 10%. Table 2 summarizes the quality measurements for workflow nets discovered using Inductive Miner and MAS nets constructed by Agent Miner, using activity-only labeling (AOL) of transitions in the discovered nets. For each dataset, the table lists size, recall, and precision values for the models that scored the lowest size and the greatest precision.

These results confirm that the quality of the MAS nets discovered by Agent Miner is comparable to the quality of CM models. This observation is remarkable for at least two reasons. First, as stated above, we had no background knowledge of whether the datasets stem from agent-driven business processes. Still, for most datasets, we discovered interesting (in terms of size, recall, and precision) MAS nets. Second, in addition to high-quality MAS nets, Agent Miner constructs agent nets and i-nets that can be used as a starting point for analysis and improvement of ways the process participants work individually and together.

To support the above conclusions, Fig. 10 shows three types of trade-offs as two-dimensional Pareto fronts for the nets discovered from the BPIC 2015 dataset. Each point in the plots denotes two quality measurements for one net generated by the evaluation pipeline. The Pareto fronts indicate the nets with better quality measurements. The MAS nets discovered by Agent Miner complement workflow nets discovered by Inductive Miner to result in more saturated Pareto fronts, that is, fronts with more models of interesting qualities. For the BPIC15 dataset, MAS nets 1 to 8 discovered by Agent Miner belong to the recall/precision Pareto front and demonstrate combinations of these quality measurements better than most CM models discovered by Inductive Miner. MAS net 8 belongs to the size/precision Pareto front and is better in terms of size/precision than most of the CM models. MAS nets 1 to 5, 7, and 8 demonstrate a better combination of size/recall measurements than almost all CM models. Overall, the Pareto fronts contain more AM models than CM models, except in the size/precision case, when the front is represented by one AM model and one CM model. Similar to the BPIC 2015 results, the Pareto fronts for the BPIC 2013 dataset, presented in Fig. 11, include points for the MAS nets. The Pareto front plots for all the datasets are included in the evaluation results [25].

Agent Miner uses an additional event attribute that specifies the *agent* that triggered the event. Hence, it is reasonable to expect it to construct comparable or better models than conventional algorithms, which do not require this attribute. To obtain generalizable conclusions, for all the datasets, we analyzed Pareto fronts and performed paired samples t-tests to establish whether CM and AM models are of the same qualities, for CM models discovered using both activity-only labeling (AOL) and agent and activity labeling (AAL). When measuring precision and recall of AOL and AAL models, both AM and CM models, events in the logs were identified correspondingly. The null hypotheses used in the t-tests are such that means of quality measurements for AM and CM models are equal. Table 3 summarizes comparisons of Pareto fronts for IM nets and AM nets (IM as INDA), while Table 4 shows the results of the tests, where results are for size (s), recall (r) and precision (p) measurements. In Table 3, AM, CM,

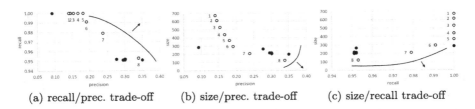

(a) recall/prec. trade-off   (b) size/prec. trade-off   (c) size/recall trade-off

Fig. 10. BPIC 2015 Pareto fronts: IM–AOL nets ("•") and AM–AOL MAS nets ("○").

and AM&CM entries stand for situations when the Pareto front is composed of AM only, CM only, or AM and CM models, respectively. In Table 4, for the significance level of 0.05, AM, CM, and AM&CM entries stand for situations when AM models are significantly better, CM models are significantly better, and the null hypothesis was not rejected, respectively. We accept the AM and AM&CM entries as situations when Agent Miner provides additional value to CM models and highlight them in bold (the majority of the entries).

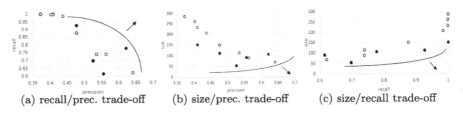

(a) recall/prec. trade-off   (b) size/prec. trade-off   (c) size/recall trade-off

Fig. 11. BPIC 2013 Pareto fronts: IM–AOL nets ("•") and AM–AOL MAS nets ("○").

Table 3. Comparison of Pareto fronts.

BPIC data	AM vs CM fronts (AOL)			AM vs CM fronts (ALL)		
	r/p	s/p	s/r	r/p	s/p	s/r
2011	AM	AM&CM	AM	AM&CM	CM	AM&CM
2012	AM&CM	AM&CM	AM&CM	AM&CM	AM&CM	AM
2013	AM&CM	AM&CM	AM&CM	AM&CM	AM&CM	AM&CM
2014	AM&CM	AM&CM	CM	AM&CM	AM&CM	AM&CM
2015	AM&CM	AM&CM	AM&CM	AM&CM	CM	CM
2017	AM&CM	CM	CM	AM&CM	AM&CM	CM
2018	CM	CM	CM	CM	CM	CM
2019	AM&CM	AM&CM	AM&CM	CM	CM	AM&CM
2020	AM&CM	AM&CM	AM&CM	AM&CM	AM&CM	AM&CM

Table 4. Results of t-tests.

BPIC data	AM vs CM tests (AOL)			AM vs CM tests (ALL)		
	s	r	p	s	r	p
2011	CM	CM	CM	CM	AM	AM&CM
2012	CM	CM	AM&CM	AM&CM	AM	AM&CM
2013	CM	AM&CM	AM&CM	AM&CM	AM	AM&CM
2014	CM	CM	AM&CM	CM	AM	AM&CM
2015	CM	AM&CM	AM&CM	CM	AM	AM&CM
2017	CM	AM&CM	AM&CM	AM&CM	AM&CM	AM&CM
2018	CM	CM	AM&CM	AM&CM	AM&CM	AM&CM
2019	CM	CM	AM	CM	AM	AM&CM
2020	CM	CM	AM&CM	CM	AM	CM

The results confirm that Agent Miner discovers interesting models to complement models constructed by conventional discovery algorithms that also rely on event attributes that specify agents that triggered the events and describe this information in the constructed models. The results for Split Miner, CM models, and AM models (Split Miner as INDA), demonstrate similar conclusions as for IM models presented above.

6 Related Work

This section reviews business process modeling with agents, process discovery addressing agents, and agent system modeling in a broad context.

Agent Modeling in Business Process Management. The relationship between business process management and agent-based modeling was first explored in the '90s by Jennings et al. [7]. In an extension of their work, dubbed ADEPT, the authors propose to model a business process as a negotiation system between agents, similar to our interaction nets [8]. However, their model of negotiating agents is conceptual in its nature, while our model is formal and executable. Several authors advocate for an agent-based perspective on business process modeling [6,22]. In these works, the process is seen as such composed of interacting agents [7]. However, they do not provide automated discovery of models from data.

Process Discovery. Traditional process discovery techniques assume a case perspective when learning models from data while often ignoring additional perspectives such as resources. Several process discovery techniques extended traditional methods beyond the case perspective. Rozinat et al. [20] propose a multi-perspective approach for mining simulation models from event data that includes the resource perspective. The resulting models are executable and can be used for performance analysis of the underlying system. However, resources are considered static entities and not active agents. Van der Aalst et al. [1] address the modeling of behavior and availability of resources. Yet, resources play a secondary role, with cases being the dominant perspective that defines business processes. Klijn et al. [9] develop a technique for querying event logs to uncover interactions between process entities. Yet, they do not provide a formal model that can be evaluated for its correctness or goodness-of-fit. Discovering functional architecture models (FAM) that comprise interacting modules that internally perform various activities was proposed in [28]. Unlike the modules in FAMs, our agents are dynamic, decentralized, and may interact not only with other agents, but also with the environment. Fettke and Reisig presented an approach to system mining called Heraklit [5]. Heraklit proceeds by constructing distributed runs of participating agents and then combining them into the overall system, whereas Agent Miner constructs agent nets and an additional net that explicitly describes agents' interactions.

Tour et al. [24] have shown that by shifting process mining paradigm from case-based to agent-based, one can discover less complex models of business processes. Within this agent paradigm, Nesterov et al. [15] have recently proposed a process discovery solution. Their algorithm constructs sound generalized workflow nets that capture agents' behavior. In contrast to our approach, Nesterov et al. assume to know the interaction patterns between agents and aim to implement this given interaction pattern, while we discover the interactions from the data. Moreover, their algorithm can only discover pre-defined workflow patterns, while Agent Miner is generic and can model any pattern its ANDA component can discover. Furthermore, we evaluated Agent Miner over real-life logs, while Nesterov et al. tested their approach over synthetically generated data.

Our approach can also be viewed as a log-decomposition process discovery approach [27]. Such techniques propose to localize the event log to discover different models tailored to the data in these local logs. In these techniques, the log decomposition is usually driven by the case attribute, whereas we propose an agent-driven log decomposition.

Multi-agent System Modeling. Multi-agent systems (MASs) have been studied extensively in the past; cf. [4] for a recent overview. Here, we focus on approaches that model multi-agent systems using Petri nets, our formalism of choice. A seminal paper by Moldt et al. [13] proposed to use colored Petri nets (CPNs) to model agent systems. Their model captures three crucial components in a MAS: communication, independence (between agents), and intelligence. However, given the richness of CPNs, formal results on correctness of constructed models were not obtain. Celaya et al. [3] model a multi-agent system using elementary Petri nets to capture interactions between agents and use Petri net analysis to ensure the models are deadlock-free. In our work, we discover MAS models from event data rather than relying on human expertise.

7 Conclusion

This paper presents and evaluates Agent Miner, a divide-and-conquer algorithm for discovering models of agents and their interactions from event data. The algorithm "divides" the input data into special parts and then "conquers" the parts using conventional process discovery algorithms. The constructed agent and interaction models provide a new, modular perspective on the data, suitable for analyzing process participants and their interactions. These artifacts can be integrated into a model that describes process control flow. Such integrated models are often smaller and represent the event data more faithfully than corresponding process models constructed using conventional discovery algorithms. The configuration of Agent Miner used in the evaluation reported in this paper ensures that the obtained integrated models are safe and sound.

Agent Miner has several limitations representing areas of interest for future work. First, the interaction logs are constructed by taking the first event in each agent trace of each agent. This approach ignores the information on the duration of agent activities and interactions. For instance, information on the durations of activities between interactions can be obtained by considering the first and the last event from agent traces between those interactions. Consequently, one can apply lifecycle-aware process discovery [10] or queue mining techniques [21] to infer agent interactions. Next, Agent Miner associates each event with only one agent. One can relax this limitation and study the effects of multiple agents sharing the same event. Finally, the evaluation approach used in this article is limited to the traditional model quality measures used in process mining. The use of new agent-specific quality measures for discovered models may highlight additional benefits of Agent Miner and other agent system mining algorithms in the context of agent-based business process management.

Acknowledgments. Andrei Tour was supported via an "Australian Government Research Training Program Scholarship." Artem Polyvyanyy was in part supported by the Australian Research Council project DP220101516.

References

1. van der Aalst, W.M.P., Nakatumba, J., Rozinat, A., Russell, N.: Business process simulation. In: Brocke, J.V., Rosemann, M. (eds.) Handbook on Business Process Management 1. International Handbooks on Information Systems, pp 313–338. Springer, Berlin (2010). https://doi.org/10.1007/978-3-642-00416-2_15
2. Augusto, A., Conforti, R., Dumas, M., La Rosa, M., Polyvyanyy, A.: Split miner: automated discovery of accurate and simple business process models from event logs. Knowl. Inf. Syst. **59**(2), 251–284 (2019)
3. Celaya, J.R., Desrochers, A.A., Graves, R.J.: Modeling and analysis of multi-agent systems using Petri nets. In: 2007 IEEE International Conference on Systems, Man and Cybernetics, pp. 1439–1444 (2007)
4. Dorri, A., Kanhere, S., Jurdak, R.: Multi-agent systems: a survey. IEEE Access **6**, 28573–28593 (2018)
5. Fettke, P., Reisig, W.: Systems mining with HERAKLIT: the next step. In: Di Ciccio, C., Dijkman, R., del Río Ortega, A., Rinderle-Ma, S. (eds.) Business Process Management Forum. BPM 2022. LNBIP, vol. 458, Springer, Cham, pp. 89–104 (2022). https://doi.org/10.1007/978-3-031-16171-1_6
6. Halaška, M., Šperka, R.: Is there a need for agent-based modelling and simulation in business process management? Organizacija **51**(4), 255–2569 (2018)
7. Jennings, N.R., Faratin, P., Johnson, M.J., Norman, T.J., O'Brien, P., Wiegand, M.E.: Agent-based business process management. Int. J. Coop. Inf. Syst. **5**(2&3), 105–130 (1996)
8. Jennings, N.R., Norman, T.J., Faratin, P.: ADEPT: an agent-based approach to business process management. ACM SIGMOD Record **27**(4), 32–39 (1998)
9. Klijn, E.L., Mannhardt, F., Fahland, D.: Aggregating event knowledge graphs for task analysis. In: Montali, M., Senderovich, A., Weidlich, M. (eds.) Process Mining Workshops. ICPM 2022. LNBIP, vol. 468, pp. 493–505. Springer, Cham (2023).https://doi.org/10.1007/978-3-031-27815-0_36
10. Leemans, S.J.J., Fahland, D., van der Aalst, W.M.P.: Using life cycle information in process discovery. In: Reichert, M., Reijers, H.A. (eds.) BPM 2015. LNBIP, vol. 256, pp. 204–217. Springer, Cham (2016). https://doi.org/10.1007/978-3-319-42887-1_17
11. Leemans, S.J.J., Fahland, D., van der Aalst, W.M.P.: Discovering block-structured process models from event logs containing infrequent behaviour. In: Lohmann, N., Song, M., Wohed, P. (eds.) BPM 2013. LNBIP, vol. 171, pp. 66–78. Springer, Cham (2014). https://doi.org/10.1007/978-3-319-06257-0_6
12. Leemans, S.J.J., Poppe, E., Wynn, M.T.: Directly follows-based process mining: exploration & a case study. In: ICPM, pp. 25–32. IEEE (2019)
13. Moldt, D., Wienberg, F.: Multi-agent-systems based on coloured Petri nets. In: Petri Nets. LNCS, vol. 1248, pp. 82–101. Springer (1997)

14. Murata, T.: Petri nets: properties, analysis and applications. Proc. IEEE **77**(4), 541–580 (1989)
15. Nesterov, R., Bernardinello, L., Lomazova, I., Pomello, L.: Discovering architecture-aware and sound process models of multi-agent systems: a compositional approach. Softw. Syst. Model. **22**(1), 351–375 (2022)
16. Polyvyanyy, A., et al.: Entropia: a family of entropy-based conformance checking measures for process mining. In: ICPM Tools. CEUR Workshop Proceedings, vol. 2703, pp. 39–42. CEUR-WS.org (2020)
17. Polyvyanyy, A., Solti, A., Weidlich, M., Di Ciccio, C., Mendling, J.: Monotone precision and recall measures for comparing executions and specifications of dynamic systems. ACM Trans. Softw. Eng. Methodol. **29**(3), 17:1–17:41 (2020)
18. Polyvyanyy, A., Weidlich, M., Weske, M.: Connectivity of workflow nets: the foundations of stepwise verification. Acta Informatica **48**(4), 213–242 (2011)
19. Reisig, W.: Understanding Petri Nets: Modeling Techniques, Analysis Methods. Springer, Case Studies (2013)
20. Rozinat, A., Mans, R.S., Song, M., van der Aalst, W.M.P.: Discovering simulation models. Inf. Syst. **34**(3), 305–327 (2009)
21. Senderovich, A.: Queue mining. In: Sakr, S., Zomaya, A.Y. (eds.) Encyclopedia of Big Data Technologies, pp 1351–1358. Springer, Cham (2019). https://doi.org/10.1007/978-3-319-77525-8_101
22. Sulis, E., Di Leva, A.: An agent-based model of a business process: the use case of a hospital emergency department. In: Teniente, E., Weidlich, M. (eds.) BPM 2017. LNBIP, vol. 308, pp. 124–132. Springer, Cham (2018). https://doi.org/10.1007/978-3-319-74030-0_8
23. Syring, A.F., Tax, N., van der Aalst, W.M.P.: Evaluating conformance measures in process mining using conformance propositions. Trans. Petri Nets Other Models Concurrency **14**, 192–221 (2019)
24. Tour, A., Polyvyanyy, A., Kalenkova, A.: Agent system mining: vision, benefits, and challenges. IEEE Access **9**, 99480–99494 (2021)
25. Tour, A., Polyvyanyy, A., Kalenkova, A., Senderovich, A.: Agent miner: implementation and evaluation results. Technical Report, Melbourne University (2023). https://doi.org/10.26188/21127273
26. van der Aalst, W.M.P.: Process mining – Data Science in Action. Springer (2016)
27. van der Aalst, W.M.P., Kalenkova, A., Rubin, V., Verbeek, E.: Process discovery using localized events. In: Devillers, R., Valmari, A. (eds.) PETRI NETS 2015. LNCS, vol. 9115, pp. 287–308. Springer, Cham (2015). https://doi.org/10.1007/978-3-319-19488-2_15
28. van der Werf, J.M.E., Kaats, E.: Discovery of functional architectures from event logs. In: PNSE@Petri Nets, pp. 227–243 (2015)
29. Wolfram, S.: Computation theory of cellular automata. Commun. Math. Phys. **96**, 15–57 (1984)

Interactive Multi-interest Process Pattern Discovery

Mozhgan Vazifehdoostirani[1(✉)], Laura Genga[1], Xixi Lu[2], Rob Verhoeven[3,4,5],
Hanneke van Laarhoven[4,5], and Remco Dijkman[1]

[1] Eindhoven University of Technology, Eindhoven, The Netherlands
`m.vazifehdoostirani@tue.nl`
[2] Utrecht University, Utrecht, The Netherlands
[3] Netherlands Comprehensive Cancer Organisation (IKNL),
Utrecht, The Netherlands
[4] Amsterdam UMC location University of Amsterdam, Amsterdam, The Netherlands
[5] Cancer Center Amsterdam, Cancer Treatment and Quality of Life,
Amsterdam, The Netherlands

Abstract. Process pattern discovery methods (PPDMs) aim at identifying patterns of interest to users. Existing PPDMs typically are unsupervised and focus on a single dimension of interest, such as discovering frequent patterns. We present an interactive multi-interest-driven framework for process pattern discovery aimed at identifying patterns that are optimal according to a multi-dimensional analysis goal. The proposed approach is iterative and interactive, thus taking experts' knowledge into account during the discovery process. The paper focuses on a concrete analysis goal, i.e., deriving process patterns that affect the process outcome. We evaluate the approach on real-world event logs in both interactive and fully automated settings. The approach extracted meaningful patterns validated by expert knowledge in the interactive setting. Patterns extracted in the automated settings consistently led to prediction performance comparable to or better than patterns derived considering single-interest dimensions without requiring user-defined thresholds.

Keywords: Process Pattern Discovery · Multi-interest Pattern Detection · Process Mining · Outcome-Oriented Process Patterns

1 Introduction

Process pattern discovery methods (PPDMs) aim to discover process patterns that are *of interest* for the human analyst, where a process pattern corresponds to a set of process activities (possibly annotated with additional data) with their ordering relations. The interest of a pattern is usually computed according to one or more functions. Previous studies highlighted how these techniques often uncovered interesting behaviors that would otherwise remain hidden in start-to-end process models [23].

© The Author(s), under exclusive license to Springer Nature Switzerland AG 2023
C. Di Francescomarino et al. (Eds.): BPM 2023, LNCS 14159, pp. 303–319, 2023.
https://doi.org/10.1007/978-3-031-41620-0_18

Several approaches have been proposed to discover process patterns from a given event log [7,13,23] and employed them in several applications, for instance, event abstraction [20], or trace classification [26]. However, most of these approaches focus on a single interest dimension. In particular, they usually aim to detect *frequent* patterns, which often leads to the generation of a multitude of non-interesting patterns and possibly the missing of interesting but infrequent ones [22]. As pointed out by recent studies in the pattern mining field [12], the concept of *interest* of a pattern is often linked to multiple dimensions, some of which may be in conflict with each other. These considerations also hold in the process domain since processes emerge from the interplay of multiple factors, highlighting the need for multi-dimensional thinking in process analysis [11]. Few PPDMs introduced a broader notion of interest, either by allowing the user to define cut-off thresholds for several metrics [23], which are then aggregated to rank the obtained set of patterns, or by directly using a composite metric during the pattern generation phase [9,22]. However, these solutions offer limited support in dealing with a multi-dimensional notion of pattern interest. Defining appropriate cut-off thresholds for different and conflicting metrics is a non-trivial decision that strongly impacts the obtained results. Furthermore, aggregating multiple dimensions in a single one leads to a single ranked collection of patterns which depends on the aggregation setting and hides the interplay of the different dimensions. To deal with this complexity, the detection of process patterns should be expressed as a *multi-objective* problem.

Beside the *multi-objective* challenge, most PPDMs are unsupervised and suffer from pattern explosion in real-life event logs. Previous studies showed that leveraging expert domain knowledge can avoid or mitigate the pattern explosion issue [2,18]. A semi-supervised PPDM [18] was proposed for users to manually select and extend patterns. However, the approach still relies exclusively on frequency-based metrics. Also, the burden of the selection and extension of discovered patterns is left to the user as a manual task without much guidance.

In this work, we introduce the IMPresseD framework (**I**nteractive **M**ulti-interest **Process** **P**attern **D**iscovery) for process pattern discovery. IMPresseD is designed to derive interesting and easily interpretable patterns for the end users by combining different strategies. First, the framework allows users to define different *interest functions* to measure the interest of patterns, supporting customizable multi-dimensional analysis goals. In this way, the user has more control over the measures of relevance that they use, which is expected to lead to patterns that are indeed considered meaningful by end users. Multi-optimization strategies are used to allow the user to go over far fewer patterns than the ones obtained by threshold-dependent strategies to identify the relevant ones. The framework supports an in-depth analysis of the pattern characteristics, which also considers the characteristics of the process executions in which the pattern occurs. Finally, the approach is iterative and interactive. At each step, the user is presented with the process patterns that are best according to the user-defined interest functions, and they can select the ones to expand further.

To showcase the framework's usefulness, we also discuss how to use it with a concrete analysis goal, i.e., *deriving process patterns affecting the process outcome*. This is inherently a complex problem for which different aspects need to be considered. Furthermore, to the best of our knowledge, most outcome-oriented pattern detection approaches do not support a multi-dimensional analysis. Given this concrete analysis goal, we carried out a two-fold evaluation to validate our approach. First, we use a real-world case study in healthcare to show the capability of the proposed framework in supporting domain experts in extracting meaningful patterns in an interactive setting. Then, we evaluate our approach in a quantitative experiment to assess the predictive power of the automatically discovered patterns. We compare the results of our approach with the ones obtained by using a single metric and using the entire pattern set without filtering. The obtained results show that the discovered set of patterns consistently ranked within the top positions, while patterns mined by adopting single metrics led to a more unstable performance. Furthermore, the proposed framework returned a set of patterns significantly smaller than the entire pattern set while preserving a comparable predictive power.

Summing up, the paper contributes to the literature by introducing:

- a multi-interest and interactive process pattern discovery framework;
- tailored interest functions for discovering process patterns affecting the outcome of the process.

The remainder of this paper is organized as follows. Section 2 reviews the relevant related work. Section 3 provides basic concepts used throughout the paper. Section 4 introduces the proposed framework, together with a concrete instantiation of the interest function to support outcome-oriented pattern discovery. Section 5 presents and discusses the evaluation. Finally, Sect. 6 draws the conclusion and delineates some ideas for future studies.

2   Related Work

Most previous PPDMs take an event log as input and generate patterns based on user-defined thresholds on a set of predefined measures of interest. These approaches vary depending on the type of patterns they aim to extract. Early work focused on discovering sequences of event traces, such as identifying sequences that fit predefined templates [6] or using a sequence pattern mining algorithm [14]. More recent research has focused on patterns representing more complex control-flow relationships, for instance, episodes representing eventually-flow relations [16], or graphs representing both sequential and concurrent behaviors [9,15]. Patterns that represent a more comprehensive set of control-flow relationships, including sequences, concurrency, and choice, are considered in the approach proposed by Tax et al. [23]. This approach has been extended to allow the extraction of patterns based on a more general set of utility functions [22]. Taking into account the context in which patterns are observed,

Acheli et al. extended previous work to discover contextual behavioral patterns, allowing for insights into the aspects that influence a process conduction [1].

Although these unsupervised PPDMs can uncover interesting patterns, they offer little or no support for multi-dimensional analysis goals involving possibly conflicting dimensions. Furthermore, they do not incorporate user knowledge, which often results in the return of uninteresting patterns. A possible mitigation strategy to this problem consists in keeping "humans in the loop", as observed by previous authors. For instance, Benevento et al. showed potential improvements in the quality and clarity of the process models by employing interactive process discovery in modeling healthcare processes compared to traditional automated discovery techniques [3,4]. Within the PPDMs domain, a semi-supervised approach is proposed for discovering process patterns which involves the user in the pattern extraction process [18]. However, this approach only exploits frequency-based interest functions based on user-defined thresholds.

3   Preliminaries

In this section, we recall the basic concepts needed to introduce our framework.

Definition 1 (Event). *Let $\mathcal{AC}$ be the universe of activities, $\mathcal{C}$ be the universe of case identifiers, $\mathcal{T}$ be the time domain, and $\mathcal{D}_1, \mathcal{D}_2, ..., \mathcal{D}_m$ be the sets of additional attributes with $i \in [1, m]$, $m \in \mathbb{Z}$. An event is a tuple of $e = (a, c, t, d_1, \ldots, d_m)$, where $a \in \mathcal{AC}$, $c \in \mathcal{C}$, $t \in \mathcal{T}$ and $d_i \in \mathcal{D}_i$.*

Definition 2 (Trace, event log). *A trace $\sigma = \langle e_1, \cdots, e_n \rangle$ is a finite non-empty sequence of events $e_1, \cdots, e_n$ in which their timestamp does not decrease. Let $\mathcal{S}$ denote the universe of all possible traces, an event log can be defined as $L = \{\sigma_1, \sigma_2, \cdots, \sigma_n\} \subseteq \mathcal{S}$ which is a set of traces.*

We use E_σ for the set of events in trace σ. We define $\pi_{act}(e)$, $\pi_{time}(e)$, $\pi_{case}(e)$, and $\pi_{d_i}(e)$ to return the activity, timestamp, case identifier and the attribute d_i associated with e, respectively.

A well-known issue of log traces is that they flatten the real ordering relations among process events, hiding possible concurrency [17]. Since we intend to discover patterns representing both sequential and concurrent relations, we convert log traces in so-called *partially ordered traces*. It is possible to derive partially ordered traces from fully ordered traces by using a conversion oracle function obtained from expert knowledge or data analysis [10,19].

Definition 3 (Partially ordered trace). *Given a conversion oracle function φ and a log trace σ, a partially ordered trace $\varphi(\sigma) = (E_\sigma, \prec_\sigma)$ is a Directed Acyclic Graph (DAG), where E_σ and $\prec_\sigma \in E_\sigma \times E_\sigma$ corresponds to the set of nodes and edges, respectively. We define matrix $A_{\varphi(\sigma)}$ as an upper triangular adjacency matrix that specifies directed edges from e to e', with $e, e' \in E_\sigma$. Also, $R_{\varphi(\sigma)}$ is the reachability matrix derived from $A_{\varphi(\sigma)}$ to represent all possible paths from e to e' of length l such that $2 \leq l \leq |\sigma| - 1$. For each pair of events $e, e' \in E_\sigma$, such that $e \neq e'$, we define the following ordering relations:*

- *if $R_{\varphi(\sigma)}(e,e') \neq 0$, e' eventually follows e',*
- *if $A_{\varphi(\sigma)}(e,e') \neq 0$, e' directly follows e,*
- *if $R_{\varphi(\sigma)}(e,e') = 0$ and $R_{\varphi(\sigma)}(e',e) = 0$, then e is concurrent with e'.*

Definition 4 (Process pattern). *A process pattern $P = (N, \mapsto, \alpha, \beta)$ is a DAG, where:*

- *N is a set of nodes,*
- *$\mapsto$ is set of edges over N*
- *α is a function that assigns a label $\alpha(n)$ to any node $n \in N$,*
- *β is a foundational pattern for P, which means pattern P is extended from pattern β.*

Also, we denote that if $| N | = 1$, then β is NULL, i.e., a single node is considered a pattern without any foundational pattern.

Examples of process patterns can be found in Fig. 2. $P_\theta^1, ..., P_\theta^5$ all share the same foundational pattern θ, represented by the single node b. In turn, pattern ω is the foundational pattern for P_ω^1 and P_ω^2 in the middle column. Given a process pattern, an *instance* of the pattern is an occurrence of the pattern in a log trace.

Definition 5 (Pattern instances set). *Let $P = (N, \mapsto, \alpha, \beta)$ be a pattern, $\varphi(\sigma) = (E_\sigma, \prec_\sigma)$ a partially ordered trace, $A_\mapsto$ be an upper triangular adjacency matrix over N, and $R_\leadsto$ be the reachability matrix of size $| N | - 1$ derived from $A_\mapsto$. Given a subset $E' \subseteq E_\sigma$ of nodes in $\varphi(\sigma)$, such that there is a bijective function $I : E' \to N$, then we define the pattern instances of P in $\varphi(\sigma)$ as $PI(P, \varphi(\sigma)) = \{E' \mid \forall e, e' \in E', A_{\varphi(\sigma)}(e,e') = A_\mapsto(I(e), I(e')) \wedge R_{\varphi(\sigma)}(e,e') = R_\leadsto(I(e), I(e')) \wedge \pi_{act}(e) = \alpha(I(e))\}$. The pattern instances set of pattern P over event log L is defined as $PIS(P, L, \varphi) = \bigcup_{\sigma \in L} PI(P, \varphi(\sigma))$.*

4 IMPresseD Framework

Given an event log, the objective of the IMPresseD framework (Fig. 1) is to discover the set of process patterns that are best according to multiple interest functions defined by the user. The framework includes the following steps.

Step 1 Converting all traces in the event log into partially ordered traces using a conversion oracle derived from expert knowledge or data analysis.
Step 2 Defining the interest functions which fit the users' notion of pattern interestingness based on their analysis goal. Analytical dashboards to visualize the discovered patterns and the computed interest functions are also defined.
Step 3 Extracting patterns of length-1, i.e., individual activities.
Step 4 Measuring the interestingness of each discovered pattern through the set of interest functions defined at *Step 2*.
Step 5 Returning the set of patterns that are the best according to the interest functions (i.e., non-dominated patterns in the Pareto front).

Step 6 If the user is satisfied with the current set of patterns or there is no extension possible, the procedure ends. Otherwise, the user selects pattern(s)
to extend (i.e., the foundational patterns), and the procedure goes to
Step 7.

Step 7 Building all extensions of the foundational patterns and going to *Step 4.*

In the remainder of this section, we delve into the pattern *selection* (Steps 4
and 5) and *extension* (Step 7). Finally, we show an instantiation of the interest
functions (Step 2) using an analysis goal for the discovery of process patterns
affecting the process outcome.

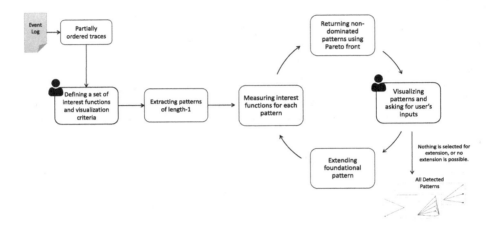

Fig. 1. Overview of the IMPresseD framework

4.1   Pattern Selection

Let $\mathcal{P}_i = \{P^1, P^2, ..., P^k\}$ be the set of all patterns discovered in i^{th} iteration of
the method and let $\mathcal{I} = \{\mathcal{I}_1, \mathcal{I}_2, ..., \mathcal{I}_m\}$ be the set of interest functions, where
$\forall \mathcal{I}_k \in \mathcal{I}, \mathcal{I}_k : \mathcal{P}_i \to \mathbb{R}$. The pattern selection module aims to return the set
of patterns $(P^* \subseteq \mathcal{P}_i)$ that optimize the pre-defined interest functions. This
corresponds to solving a *multi-objective optimization problem (MOP)*.

Several approaches have been proposed in the literature to solve a MOP.
Note, however, that a feasible solution optimizing all objective functions simultaneously usually does not exist. Therefore, the goal is to find the so-called
Pareto Front, which involves a set of patterns that are not dominated by any
other pattern in terms of the multiple interest functions. Informally, solutions on

the Pareto front are such that no objective can be improved without worsening at least one of the other objectives. In this paper, we use the algorithm proposed by [5] to filter out dominated patterns. For any pair of patterns P^l, $P^j \in \mathcal{P}_i$, we say that P^l dominates P^j if and only if: a) $\forall \mathcal{I}_k \in \mathcal{I}, \mathcal{I}_k(P^l)$ is no worse than $\mathcal{I}_k(P^j)$; b) $\exists \mathcal{I}_k \in \mathcal{I}, \mathcal{I}_k(P^l)$ is strictly better than $\mathcal{I}_k(P^j)$.

4.2   Pattern Extension

Informally, extending a pattern P means generating a new pattern P' by adding new nodes and edges to P according to a set of *extension rules* applied on partially ordered traces involving at least one instance of the pattern. Formally, let $\varphi(\sigma) = (E_\sigma, \prec_\sigma)$ be a partially ordered trace, and $P = (N, \mapsto, \alpha, \beta)$ be a pattern, in a way that $\mid PI(P, \varphi(\sigma)) \mid > 0$ and $E' \in PI(P, \varphi(\sigma))$. An extension operator is a function Ext_f that takes as input pattern P and an instance pattern E' and returns a new pattern P' according to the extension rule f. Specifically, $Ext_f(P, E') = (N \cup V_f, \mapsto \cup \mapsto_f, \alpha \cup \alpha', P)$, where V_f is the set of nodes in $\varphi(\sigma)$ satisfying the ordering relation expressed by the extension rule f, $\mapsto_f$ is the set of edges linking the nodes of V_f with the nodes in E', and α' is the labelling function for nodes in V_f. In this paper, $f \in \{\mapsto, \mapsto', \mid\mid, \rightsquigarrow, \rightsquigarrow', dc\}$, which represents respectively: (1) *direct following*, (2) *direct preceding*, (3)*concurrent*, 4)*eventually following*, (5)*eventually preceding*, (6) *direct context* relations.

Given $A_{\varphi(\sigma)}$ as adjacency matrix and $R_{\varphi(\sigma)}$ as reachability matrix of size $\mid \sigma \mid -1$ over E_σ, we define $V_{\mapsto} = \{e \in E_\sigma \mid \forall n \in E', e \notin E', A_{\varphi(\sigma)}(n, e) = 1\}$. In a similar way, $V_{\rightsquigarrow} = \{e \in E_\sigma \mid \forall n \in E', e \notin E', A_{\varphi(\sigma)}(n, e) = 0, R_{\varphi(\sigma)}(n, e) > 0\}$ and $V_{\mid\mid} = \{e \in E_\sigma \mid \forall n \in E', e \notin E', A_{\varphi(\sigma)}(n, e) = 0, R_{\varphi(\sigma)}(n, e) = 0, R_{\varphi(\sigma)}(e, n) = 0\}$. Note, $V_{\mapsto'}$ and $V_{\rightsquigarrow'}$ can be derived by changing the order of e and n in the definition of $V_{\mapsto}$ and $V_{\rightsquigarrow}$, respectively. Finally, dc is defined as $Ext_{dc}(P, E') = Ext_{\mapsto}(P, E') \cup Ext_{\mapsto'}(P, E') \cup Ext_{\mid\mid}(P, E')$.

Figure 2 illustrates some examples of pattern extensions. The black dotted boxes in each column of the figure highlight the instance found in the partially ordered trace of a pattern P we want to extend. For instance, single node b is an instance of pattern θ, and its corresponding extensions are patterns $P_\theta^1, .., P_\theta^5$. , where the number in the red dotted box reflects the ordering of the rule set. For instance, the first (second) rule represents the *directly following (preceding)* relation, which results in patterns $P_\theta^1(P_\theta^2)$; the third rule involves nodes *concurrent* to b, i.e., in this case, only node c^1.; and so on. Users can select all or a subset of rules f to explore all possibilities for the extension of a selected foundational pattern in each iteration of IMPresseD. Therefore, the set of all extended patterns from the foundational pattern P using a subset of rules called F' is defined as $\mathcal{P}_P = \bigcup_{\sigma \in L} \bigcup_{E' \in PI(P, \varphi(\sigma))} \bigcup_{f \in F'} Ext_f(P, E')$.

---

[1] Please note that we added the black dots only for the sake of clarity in the visualization of a concurrent pattern and they do not belong to the extended pattern.

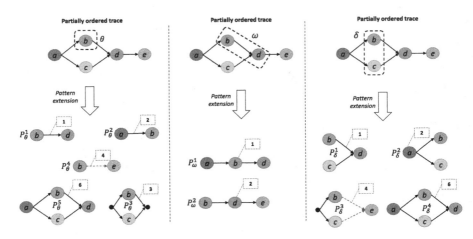

Fig. 2. Pattern extension procedure example

4.3   Interest Functions for Outcome-Oriented Pattern Detection

To show a concrete example of the use of the IMPresseD framework, this section outlines tailored interest functions for the outcome-oriented pattern discovery goal. We designed these functions by analyzing related literature and through discussions with domain experts.

While previous studies in outcome-oriented pattern discovery have focused on identifying patterns that are highly correlated with the outcome [21], we argue that correlation should not be the only dimension of interest. Our discussions with healthcare experts revealed that ignoring the frequency measure may lead to identifying too rare patterns that are often less interesting. In addition, frequent patterns that are not highly correlated may still be worth exploring. For example, a particular treatment "A" may be highly frequent but not highly correlated. However, when studying different extensions of "A", some interesting correlated patterns may emerge. Hence, in our analysis, we define frequency-based interest besides correlation-based interest. Moreover, it is well-known that potential confounding variables may play an important role in determining the outcome of a treatment process [25]. For example, let treatment pattern P_1 be detected as a pattern that negatively affects the treatment outcome. We may find that P_1 is only delivered to elderly patients. This questions the reliability of the relation between P_1 and the treatment outcome since the patients' age may actually be the real factor leading to worse treatment results. To mitigate the effect of confounding variables, we consider the distance between cases with or without a specific pattern as the third interest dimension.

Following these observations, we established three dimensions of interest to support outcome-oriented pattern discovery with their corresponding interest functions.

Frequency Interest. evaluates the frequency of occurrence of a pattern in the event log. In this study, we define *frequency interest function* as the percentage of cases that have at least one pattern instance P as:

$$CC(P, L, \varphi) = \frac{|\{\sigma \in L \mid PI(P, \varphi(\sigma)) > 0\}|}{|L|}$$

Outcome Interest. measures the effect of each pattern on the process outcome. For continuous outcome values, we use a correlation-based function. For the categorical outcomes, we use an information-gain-based function.

Let Φ be a set of values representing possible outcomes. The outcome of a process is defined as a function $f : \mathcal{S} \to \Phi$, that maps the set of all possible input traces to the set of all possible outcome values. Then we define $\mathcal{OV} = (f(\sigma))_{\sigma \in L}$ as the outcome vector for event log L. Let $PC\text{-}freq(P, \varphi(\sigma)) = |PI(P, \varphi(\sigma))|$ be the frequency of pattern P in trace $\varphi(\sigma)$, we define $\mathcal{FV} = (PC\text{-}freq(P, \varphi(\sigma)))_{\sigma \in L}$ as the frequency vector of pattern P for event log L. Then, the *outcome interest function* is defined as $OI(P, L, \varphi) = \rho(\mathcal{OV}, \mathcal{FV})$, where for **continues outcome** ρ is the Spearman correlation coefficient, while for **categorical outcome** ρ is the information gain.

Case Distance Interest. is designed to mitigate the impact of confounding variables. Here, we consider initial case attributes as potential confounding variables. Let $\mathcal{AT}$ be a set of user-defined case attributes, $AT_{\sigma_i} = (\pi_{d_j}(e_1))_{d_j \in \mathcal{AT}}$ is a vector of initial case attributes corresponding to trace σ_i. Let $C_P = \{\sigma \in L \mid | PI(P, \varphi(\sigma)) | > 0\}$ be the set of cases including an instance of the pattern P and $C_{\bar{P}} = \bigcup_{\sigma \in L} \{\pi_{case}(\sigma)\} - C_P$ be the set of cases without P. Then we define the *Case distance function* as $CD(P, L, \mathcal{AT}) = \sum_{\sigma_i \in C_P} \sum_{\sigma_j \in C_{\bar{P}}} \frac{1}{|L|} dist(AT_{\sigma_i}, AT_{\sigma_j})$. Let $dist_{Euc}$ be the *Euclidean* distance for numerical features, and $dist_{Jac}$ be the *Jaccard* distance for m categorical feature, and F_{normal} be a normalization function, then $dist = \frac{F_{normal}(dist_{Euc}) + dist_{Jac}}{m+1}$ as defined in [8].

Ideally, there must be $CD(P, L, \mathcal{AT}) = 0$ to ensure that the pattern P is not influenced by any confounding variable. However, in real-life scenarios, some differences between case attributes are inevitable. To assist users in analyzing which case attributes might have an effect on the outcome, we present a dashboard that visualizes the differences in selected case attributes. This enables the user to pinpoint specific case attributes that may be important for pattern P or explore the reasons behind each process behavior if it is related to the case dimension. An example of this dashboard is presented in Fig. 4.

5   Implementation and Evaluation

This section aims to demonstrate the usefulness of the IMPresseD framework for a concrete analysis goal defined by expert users (i.e., detecting process patterns affecting the process outcome) through two forms of evaluation. We have implemented an open-source tool in Python for outcome-oriented pattern discovery goals, which is publicly available through GitHub[2].

The first evaluation (user-based evaluation) aims to show the usefulness of the proposed framework in supporting the user in dealing with pattern discovery in an interactive and multi-interest setting. In the second evaluation (quantitative evaluation), we performed a comparative analysis using different sets of patterns in a fully automated setting to evaluate their predictive capabilities.

5.1   User Based Evaluation

Evaluation Setup. The goal of this evaluation is to determine whether our framework is able to discover patterns confirming expert knowledge of the treatment process. To this end, we asked two expert users from the medical domain to use the IMPresseD tool on historical data to discover treatment process patterns affecting patients' survival time. We then asked the users to validate the discovered patterns using their own medical knowledge.

As interest functions, we maximize $CC(P, L, \varphi)$ and $OI(P, L, \varphi)$ based on the Spearman correlation, and minimize $CD(P, L, \mathcal{AT})$. Regarding the visualization dashboard, we opted for *distribution plot* for the numerical features (e.g., age, albumin level, etc.) and *pie chart* for the categorical features (e.g., gender, morphology, etc.) based on expert suggestion. We also visualized the Kaplan-Meier curve, as it is a very common graphical representation of the survival probability for a group of patients based on their observed survival times. The Log-rank test is also included to check the significance of the difference in survival time between cases with or without a particular pattern.

Dataset. We used an event log provided by the Netherlands Cancer Registry (NCR) regarding the treatment process for patients with metastatic stomach or esophageal cancer. These patients can usually not be cured and receive palliative care to increase the quality of the remaining lifetime and possibly extend it. Therefore, the outcome of the treatment process is the patient survival time.

We did some data preprocessing according to the domain experts. Specially, we removed cases where there were logging errors (e.g., patients for which the survival time was not known), as well as exceptional cases or outliers, like patients who received one or multiple treatment(s) abroad. Similarly, patients with too deteriorated health are removed from the dataset, as they are not fit enough to receive any treatment. At the end of preprocessing, the event logs consisted of 957 cases, 32 distinct treatment codes, and 368 process variants. We also used domain knowledge as a conversion oracle for transforming each trace into a partially ordered trace. In particular, two groups of treatments are considered to be parallel: 1) systematic treatments starting within three days from each other, and 2) all treatments which start and end on the same day.

---

[2] https://github.com/MozhganVD/InteractivePatternDetection.

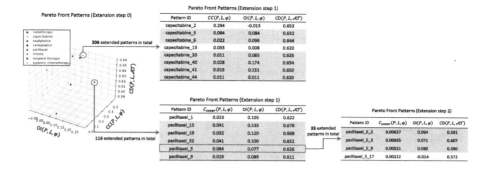

Fig. 3. Non-dominated pattern in three iterations of discovery algorithm

Results. In the first iteration of the algorithm (extension step 0), we obtained 8 non-dominated treatments as shown in Fig. 3 (3D graph left side). The framework allows users to assess every single non-dominated treatment in a three-dimensional view.

Users can select each of the recommended single treatments by the Pareto front as a foundational pattern and apply the extension functions discussed in Sect. 4.2. The experts selected *capecitabine* and *paclitaxel* as interesting patterns to extend. In the second iteration (extension step 1), we identified 18 patterns out of the 206 extended patterns from *capecitabine* and 14 patterns out of the 116 extended patterns from *paclitaxel* in the Pareto front, indicating that the use of defined interest functions and Pareto front enables users to concentrate on a maximum of 10% of the total discovered patterns in this step. The expert decided to filter out patterns with a minimum frequency of 10 patients, thus focusing on the 8 and 6 most frequent patterns from *capecitabine* and *paclitaxel* within the Pareto front, reported in Fig. 3. The users decided to stop after one extension step for *capecitabine*, while a second extension step was carried out for *paclitaxel*. For each pattern, values of each interest function are reported. Furthermore, we also generate a dashboard showing its control-flow structure and corresponding case data. The main goal of the dashboard is to allow users to compare different case attributes corresponding to the cases with and without patterns. These dashboards enable users to investigate the reasons behind each process behavior. An example is shown in Fig. 4. This pattern depicts a treatment pathway that commences with *oxaliplatin* and *capecitabine*. After some time, *oxaliplatin* is stopped, and *capecitabine* is continued. The significant difference between the survival curves of patients with and without this pattern suggests the efficacy of these treatment combinations. Cases with and without patterns are quite similar according to the selected attributes, though the dashboard shows that this pattern was never prescribed to patients with a tumor morphology labeled as "other" (which was in line with experts' expectations).

Patterns colored green in tables inside the Fig. 3 are the patterns marked as interesting by expert users (i.e., patterns validated by medical knowledge). The users considered two patterns from extending *capecitabine* not interesting

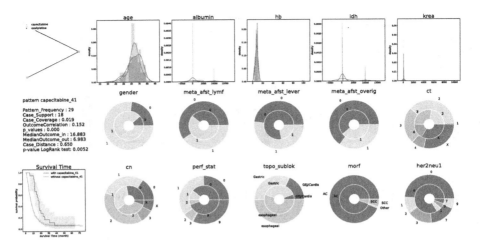

Fig. 4. An example of dashboard visualization for a pattern extended from *capecitabine*. Note: the inner ring of pie charts and red color in distribution plots correspond to the cases with the shown pattern in the dashboard.

because of a too low correlation with the outcome, leading to very similar survival time for patients with and without the pattern (MedianOutcome_in/out in the dashboard), which does not allow them to say anything about the relationship with the outcome. As regards patterns extended from *paclitaxel*, in the first step only one pattern was marked as not interesting. The reason is that the user expected an additional treatment which, however, was not possible to detect in combination with the discovered patterns. Further investigations are needed to determine why the occurrence of this particular treatment in the dataset does not fit with experts' expectations. Note that for the second extension, with foundational pattern *paclitaxel_5*, the users were especially interested in extensions involving radiotherapy. The last extended pattern in the extension step 2 did not involve radiotherapy and was hence marked as not interesting. Overall, the detected patterns confirmed the effectiveness of the previously known combination of treatments, providing valuable evidence-based insight. Only a few patterns were marked as not interesting. Both users found the visualization dashboard very helpful in understanding the detected patterns and in uncovering potential relations with the case attributes. We would like to point out that without using the Pareto front, users would have to either try different thresholds or explore all the extended patterns manually.

5.2   Quantitative Evaluation

Evaluation Setup. The goal of the quantitative evaluation consists in assessing the predictive capabilities of patterns detected employing multi-interest functions compared to patterns detected utilizing a single dimension or without any filtering. If the multi-interest functions obtain a predictive performance in line

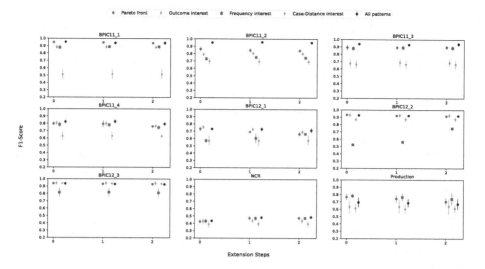

Fig. 5. Quantitative evaluation results

with the other strategies, this shows they allow to preserve the same predictive power, in addition to them leading to more meaningful process patterns filtering out many non-interesting ones (as illustrated in the user-based evaluation). We drew inspiration from the common evaluation used in "deviance mining" [21] to assess the quality of the set of discovered patterns in predicting the outcome of the process without exploiting the user's knowledge. In this setting, discovered patterns are treated as independent features, while the process's outcome is considered the dependent feature. Frequency-based encoding is used to encode independent features. Specifically, we compare the performance of decision trees (DTs) trained on the K patterns obtained from the Pareto front in each extension step to those trained on the *top* K patterns identified by considering every single dimension, as well as those trained on all discovered patterns. To achieve this, all the K non-dominated patterns in the extension step i^{th} were used as foundational patterns to be extended in iteration $i + 1^{th}$. As interest functions, we maximize the outcome (information-gain-based) and frequency functions and minimize the case distance function. During the pattern discovery procedure, we only considered the training set to prevent potential bias or information leakage in the evaluation.

Datasets. We analyzed the three most commonly used event logs in outcome prediction literature, namely *BPIC2012*, *BPIC2011*, and *Production*, by leveraging preprocessed and labeled logs from prior research [24]. We used all case-related attributes for calculating the case distance function. For the NCR dataset, we divided the survival time into three classes with equal frequency based on the experts' knowledge for this evaluation.

Results. Fig. 5 presents the results of the 5-fold cross-validation (i.e., the average F1-score with minimum and maximum obtained values). The DT trained on the patterns obtained from the Pareto front outperformed or it is as accurate as its counterparts. The only configuration that does better in some cases is the all-patterns configuration, which involves a much higher number of patterns. Indeed, on average, the ratio between the size of the feature set obtained from the Pareto front and the size of the feature set obtained in all-patterns configuration is 47.5%. This result shows that using Pareto optimal solutions, we combine the best of multiple criteria and manage to retain discriminative information with a smaller number of patterns than all possible ones. Another interesting finding is that the results of the DT trained on patterns from the Pareto front consistently rank among the best ones, while results of DTs trained on patterns obtained from single dimensions show a stronger dependency on the dataset.

When comparing single-interest measures, the case distance obtained the worst performance in most of the tested datasets. The *outcome measure* outperformed all single measures in 5 out of 10 studied event logs (BPIC11_2, BPIC11_4, BPIC12_1, BPIC12_2, BPIC12_3), while the *frequency interest* outperformed the other single measures in 3 event logs (BPIC11_3, NCR, Production). This suggests that there might be a relationship between the characteristics of the event log and the predictive power of single-interest dimensions.

5.3   Discussion

The quantitative evaluation indicates that using the Pareto front leads to comparable or better prediction performance than the ones achieved by using single measures, and with much fewer patterns than using all possible ones. Using the Pareto front also has the additional advantage that less effort is required than selecting a threshold for a specific metric. Furthermore, the developed approach provides a flexible means for the user to define the desired pattern characteristics. Note that the quantitative evaluation also shows that the proposed method has the potential to be used in a fully automated setting.

However, a surprising observation is that extending the process patterns often does not improve the prediction results, except for a slight improvement in performance after the first extension in the NCR dataset. This may be due to an overlap between the pattern obtained from the $(i+1)^{th}$ iteration and the foundational patterns in the ith iteration. Considering all patterns in the Pareto front as foundational patterns for being extended in the next iteration may have led to overlap that increases the dimension of the problem without adding much new information. One direction for future research would be to minimize the overlap between patterns obtained from each iteration.

On the other hand, the results of the user-based evaluation demonstrate the usefulness of the IMPresseD framework in discovering process patterns for supporting outcome-oriented process pattern detection. The Pareto front selection of patterns allows users to reach their desired pattern without exploring many non-interesting patterns. Furthermore, the designed visualization dashboard provides effective support to the human analyst in exploring and interpreting the patterns. We would like to point out that, to the best of our knowledge, no other process pattern discovery tool provides these functionalities. However, this evaluation has some threats to validity. First, being based on a use case, these results cannot be generalized to different contexts. Furthermore, only two experts were involved in verifying the discovered patterns. To mitigate these threats, we provide a prototype of the tool to enable other researchers to replicate our results and apply the approach to other case studies. Furthermore, a comprehensive survey involving more experts from different perspectives, such as data scientists and oncologists, is planned to evaluate the proposed method on a wider scale.

6   Conclusion and Future Work

The paper presented the IMPresseD framework, designed to derive interesting and easily interpretable process patterns for the end users. The framework is iterative and interactive and allows the user to select the most interesting patterns to expand further. The paper also discussed a concrete analysis goal of deriving process patterns affecting the process outcome, which is a complex problem that requires considering different aspects. The paper evaluated the proposed approach using a real-life case study in healthcare and in a completely automated setting using publicly available event logs. Overall, the paper contributes to the process pattern discovery literature by introducing a framework that takes into account a multi-dimensional notion of interest and by demonstrating its effectiveness through empirical evaluations. In future work, to further evaluate and enhance the efficacy of our proposed framework, we intend to conduct a comprehensive survey that draws on a wider range of expert knowledge and opinions. This survey will allow us to gather valuable feedback on the usefulness of our framework and explore potential avenues for future research. Additionally, we intend to explore additional extension operators to discover more complex patterns, as well as introduce constraints on the pattern extension in a fully-automated setting.

References

1. Acheli, M., Grigori, D., Weidlich, M.: Discovering and analyzing contextual behavioral patterns from event logs. IEEE Transactions on Knowledge and Data Engineering **34**(12), 5708–5721 (2021)
2. Martin Atzmueller, Stefan Bloemheuvel, and Benjamin Kloepper. A framework for human-centered exploration of complex event log graphs. In International Conference on Discovery Science, pages 335–350, 2019
3. Elisabetta Benevento, Davide Aloini, and Wil MP van der Aalst. How can interactive process discovery address data quality issues in real business settings? evidence from a case study in healthcare. Journal of Biomedical Informatics, 2022

4. Benevento, E., Dixit, P.M., Sani, M.F., Aloini, D., van der Aalst, W.M.P.: Evaluating the Effectiveness of Interactive Process Discovery in Healthcare: A Case Study. In: Di Francescomarino, C., Dijkman, R., Zdun, U. (eds.) BPM 2019. LNBIP, vol. 362, pp. 508–519. Springer, Cham (2019). https://doi.org/10.1007/978-3-030-37453-2_41

5. S. Borzsony, D. Kossmann, and K. Stocker. The skyline operator. In Proceedings 17th International Conference on Data Engineering, pages 421–430, 2001

6. RP Jagadeesh Chandra Bose and Wil MP Van der Aalst. Abstractions in process mining: A taxonomy of patterns. In International Conference on Business Process Management, pages 159–175, 2009

7. RP Jagadeesh Chandra Bose and Wil MP van der Aalst. Trace clustering based on conserved patterns: Towards achieving better process models. In International Conference on Business Process Management, pages 170–181, 2009

8. Cheung, Y., Jia, H.: Categorical-and-numerical-attribute data clustering based on a unified similarity metric without knowing cluster number. Pattern Recognition **46**(8), 2228–2238 (2013)

9. Diamantini, C., Genga, L., Potena, D.: Behavioral process mining for unstructured processes. Journal of Intelligent Information Systems **47**(1), 5–32 (2016). https://doi.org/10.1007/s10844-016-0394-7

10. Diamantini, C., Genga, L., Potena, D., van der Aalst, W.: Building instance graphs for highly variable processes. Expert Systems with Applications **59**, 101–118 (2016)

11. Dirk Fahland. Multi-dimensional process analysis. In Business Process Management: 20th International Conference, BPM 2022, pages 27–33. Springer, 2022

12. Fang, W., Zhang, Q., Sun, J., Xiaojun, W.: Mining high quality patterns using multi-objective evolutionary algorithm. IEEE Transactions on Knowledge and Data Engineering **34**(8), 3883–3898 (2020)

13. Christian W Günther and Wil MP Van Der Aalst. Fuzzy mining-adaptive process simplification based on multi-perspective metrics. In International conference on business process management, pages 328–343, 2007

14. Huang, Z., Xudong, L., Duan, H.: On mining clinical pathway patterns from medical behaviors. Artificial intelligence in medicine **56**(1), 35–50 (2012)

15. Hwang, S.-Y., Wei, C.-P., Yang, W.-S.: Discovery of temporal patterns from process instances. Computers in industry **53**(3), 345–364 (2004)

16. Maikel Leemans and Wil MP van der Aalst. Discovery of frequent episodes in event logs. In International symposium on data-driven process discovery and analysis, pages 1–31, 2014

17. Sander JJ Leemans, Sebastiaan J van Zelst, and Xixi Lu. Partial-order-based process mining: a survey and outlook. Knowledge and Information Systems, 65(1), 1–29, 2023

18. Xixi Lu, Dirk Fahland, Robert Andrews, Suriadi Suriadi, Moe T Wynn, Arthur HM ter Hofstede, and Wil MP van der Aalst. Semi-supervised log pattern detection and exploration using event concurrence and contextual information. In OTM Confederated International Conferences "On the Move to Meaningful Internet Systems", pages 154–174, 2017

19. Xixi Lu, Dirk Fahland, and Wil MP van der Aalst. Conformance checking based on partially ordered event data. In Business Process Management Workshops: BPM 2014 International Workshops, pages 75–88, 2015

20. Felix Mannhardt and Niek Tax. Unsupervised event abstraction using pattern abstraction and local process models. arXiv preprint arXiv:1704.03520, 2017

21. Hoang Nguyen, Marlon Dumas, Marcello La Rosa, Fabrizio Maria Maggi, and Suriadi Suriadi. Mining business process deviance: a quest for accuracy. In OTM Confederated International Conferences "On the Move to Meaningful Internet Systems", pages 436–445, 2014
22. Niek Tax, Benjamin Dalmas, Natalia Sidorova, Wil MP van der Aalst, and Sylvie Norre. Interest-driven discovery of local process models. Information Systems, 77:105–117, 2018
23. Niek Tax, Natalia Sidorova, Reinder Haakma, and Wil MP van der Aalst. Mining local process models. Journal of Innovation in Digital Ecosystems, 3(2), 2016
24. Irene Teinemaa, Marlon Dumas, Marcello La Rosa, and Fabrizio Maria Maggi. Outcome-oriented predictive process monitoring: Review and benchmark. ACM Transactions on Knowledge Discovery from Data, 13(2), 1–57, 2019
25. Aika Terada, David duVerle, and Koji Tsuda. Significant pattern mining with confounding variables. In Pacific-Asia Conference on Knowledge Discovery and Data Mining, pages 277–289, 2016
26. Mozhgan Vazifehdoostirani, Laura Genga, and Remco Dijkman. Encoding high-level control-flow construct information for process outcome prediction. In 2022 4th International Conference on Process Mining, pages 48–55, 2022

Management

Increasing RPA Adoption: An Experiment on Countermeasures for Status Quo Bias

Marie-E. Godefroid[1]([✉]) [iD], Ralf Plattfaut[2] [iD], and Björn Niehaves[3] [iD]

[1] Chair of Information Systems, University of Siegen, Siegen, Germany
marie-elisabeth.godefroid@web.de
[2] Process Innovation and Automation Lab, South Westphalia University of Applied Sciences, Soest, Germany
[3] Chair of Digital Public, University of Bremen, Bremen, Germany

Abstract. This study presents an online experiment to analyse two measures (success stories and additional information) to overcome a potential status quo bias towards adopting robotic process automation (RPA) in the nonprofit sector using 150 participants and two treatments. Data is analysed using PLS-SEM. Our findings indicate that the adoption of RPA technologies in the nonprofit sector is indeed influenced by status quo bias. Moreover, the treatment of success stories might help to overcome some aspects of this bias. Future research should focus on the application of RPA in the nonprofit sector, deepening our understanding of cognitive biases and technology adoption, and testing further potential countermeasures. Our findings should inform organisations that develop a communication strategy within their RPA implementation efforts. This study is one of the first efforts to close the gap of missing RPA studies in the nonprofit sector identified in literature reviews. Moreover, it contributes to a deeper understanding of cognitive biases and technology adoption.

Keywords: Robotic Process Automation · Adoption · Status Quo Bias · Countermeasures

1 Introduction

Robotic Process Automation (RPA) technologies are a major trend in the Business Process Management (BPM) discipline. They have received growing attention in academia [1–3] and practice [4]. RPA describes software or so called "bots" that allow the automation of business process via the graphical user interface (GUI) of underlying core systems with little or no technical skill [3]. It is widely seen as easy to apply in business process improvement efforts and, as such, as a means to digitally transform business processes [1, 2, 5]. While there is striking evidence of the success of RPA implementations in the private sector [e.g., 6, 7], there are only a few accounts of RPA adoption in the public or nonprofit sector [with notable exceptions such as 8]. In fact, missing insights on RPA usage in the nonprofit sector were identified as one research gap in a recent literature review [3]. Reasons for this apparently low adoption of RPA technologies in the nonprofit

© The Author(s), under exclusive license to Springer Nature Switzerland AG 2023
C. Di Francescomarino et al. (Eds.): BPM 2023, LNCS 14159, pp. 323–340, 2023.
https://doi.org/10.1007/978-3-031-41620-0_19

or social sector could be found in lacking methodological support for adoption. Syed et al. call for more research on the issues pertaining to RPA use [2]. Similarly, Plattfaut et al. [9] identify the importance of change management and the active persuasion of stakeholders to use RPA. Naturally, nonprofits' decision structures, skills available or budgetary constraints [10] could also be major influences, but are not the focus in this research effort.

The literature on the adoption and use of technology identifies two main drivers: The expected performance of a technology and the expected effort of using the technology heavily influence adoption and use. This influence was shown in various theoretical models and empirical studies [11–13]. However, this theoretical perspective fails to explain the low adoption and use of RPA technologies in the nonprofit sector. Prior research on RPA unequivocally praised the low efforts of implementation (ad effort expectancy) and the high value of RPA implementations (ad performance expectancy) [1–3].

One prominent avenue to explain the low adoption and use of technologies is status quo bias (SQB), but countermeasures have not yet been tested in information systems (IS) research. SQB describes the effect that individuals typically have a biased preference for the current state of affairs (e.g., for the currently used technology) and, thus, are reluctant to change (e.g., switch to a new technology) [14]. SQB thus impedes organisations from implementing RPA, requiring dedicated change management efforts [9]. But if individuals do not accept changes because of a biased perception of RPA compared to the status quo, valuable improvements and innovations cannot bear fruit. Therefore, information technology (IT) and IS practitioners and researchers are vested in employing valid countermeasures to SQB in the case of technology adoption. But despite a wealth of knowledge that affirms SQB's presence in different situations ranging from the introduction of an enterprise resource planning system [14] to that of a health cloud [15] to date, no study has examined measures to counter SQB in IT in general or RPA in specific.

In other disciplines, studies testing countermeasures are also scarce. Still, two strategies have proven successful, which we also test here: Lorenc et al. (2013) tested the effects of an intervention with additional information material on energy tariff switching behaviour. Shealy et al. (2019) succeeded in getting civil engineers to conceptualise more sustainable buildings with success stories. In this study, we test both ideas of using additional information and success stories to counter SQB towards adopting robotic process automation (RPA) in the nonprofit sector. More specifically, we aim to answer two research questions:

- RQ1: Does SQB affect the technology acceptance of RPA?
- RQ2: Can more information and/or the reference to prior success stories help counter SQB regarding RPA and increase intention to use?

To this end, we conduct an online experiment [16] with 150 participants. Our findings indicate that adoption is indeed affected by SQB and that success stories might help overcome some aspects of this bias.

2 Background

2.1 Robotic Process Automation

RPA is a comparably new technology to automate business processes. While a plethora of different definitions exist [as discussed in, e.g., 1–3], we follow the integrative definition of Plattfaut and Borghoff and understand RPA as *"a technology that allows the development of (multiple) computer programs (i.e., bots) that automate rules-based business processes through the use of GUIs"* [3]. This definition indicates that RPA enables the automation of processes supported by information technology but executed manually. To this end, RPA mimics end-user behaviour through the GUIs. As such, it can be understood as a non-invasive technology as it does not require any underlying systems or infrastructure change.

The literature agrees that RPA comes with comparably low implementation effort. Prior studies highlight that "RPA development is both simple and rapid" [3]. Implementing RPA bots compared to more traditional forms of process automation is "relatively easier and cheaper" [2]. This effect is partly because tech-savvy business people can implement RPA as RPA requires more process and subject matter expertise and fewer IT programming skills [6, 17, 18].

Moreover, the literature also highlights the advantages of RPA regarding its performance [19]. Processes can be executed faster, more reliably, and potentially 24/7 [1]. Case studies especially highlight the great advantages of cost-savings compared to the low investment costs [6]. As such, compared to other more traditional means of process automation, the performance benefits of RPA are high. Hence, RPA is increasingly included in organisations' BPM efforts [17].

2.2 Technology Acceptance and Adoption

Much research has examined technology acceptance and adoption to explain the user acceptance of new IT such as RPA. The starting point for this research was the theory of reasoned action that postulated that an individual's attitudes and subjective norms influence their behavioural intention and, subsequently, their behaviour [20]. The theory of planned behaviour later complemented that with the third construct of perceived behavioural control [21]. This theory offered the theoretical basis for examining users' technology acceptance and adoption behaviour. The most prominent models developed to examine the behavioural intention to use a technology were the technology acceptance model (TAM) and the unified theory of acceptance and use of technology (UTAUT) [11, 12, 22, 23].

Across technology acceptance models, two constructs are prominent: (1) Performance expectancy describes how beneficial an individual perceives the system regarding their job performance [12]. Similarly, perceived usefulness measures how far an individual perceives a new system to enhance their job performance [22]. (2) Effort expectancy describes how easily an individual perceives the use of a system [12]. Similarly, perceived ease of use measures how far an individual believes using a new system is free from effort [22]. We focus on these constructs in our study as researchers have established that these are most influential in determining individual technology acceptance and use behaviour [24]. Furthermore, the relevance of these constructs has also been proven for RPA already [13].

2.3 Status Quo Bias (SQB)

Status quo bias describes a biased preference for a current solution or way of doing things [25]. The literature distinguishes three explanation approaches for this preference: 1) Cognitive misperception or loss aversion refers to the "non-rational" preference of individuals to remain with the status quo because potential losses in the context of change are perceived as unrealistically large [26]. 2) Rational decision-making captures the perception of "rational" aspects like net benefits and costs (uncertainty and transition costs) [14]. 3) Psychological commitment comprises other "non-rational" influences like sunk costs, control, and social norms. In their initial publication, Samuelson and Zeckhauser [27] did not explicitly delineate these rational and non-rational explanation approaches. We thus need to consider them in combination to one another.

In the IS domain, SQB has elicited particular interest in the context of system adoption. Kim and Kankanhalli (2009) studied SQB in the technology acceptance context. They developed an integrative framework combining SQB theory, technology acceptance literature, and the equity implementation model. They then successfully tested parts of this model in the context of an ERP introduction [14]. For the specific case of RPA in the social sector, not all SQB constructs are relevant. As the literature review by Lee and Joshi [28] demonstrates, prior studies typically selected only those SQB constructs relevant to their context. Across the literature, one of the main advantages proposed for RPA is that it is cheap and does not require substantial changes to the processes for automation [2]. Thus, several SQB constructs – especially those related to costs – are less relevant for this context. RPA works on the system interfaces as a human would; therefore, it requires no substantial change to current processes. In the simplest case, tasks that a human did beforehand are now executed the same way with RPA. Nonetheless, individuals could fear that they will lose privileges or other benefits of the current way of working [14] – in the most extreme case, they could fear that their job would also be automated one day. Therefore, loss aversion could be relevant in this context. However, uncertainty and transition costs become irrelevant with a technology that requires no large investments and is easy to implement. Regarding psychological commitment, sunk costs are irrelevant in this context as RPA typically replaces a manual process. Social norms, however, remain relevant as the behaviour of colleagues and superiors could still influence the acceptance of RPA. Finally, control could also be an issue, as introducing RPA could cause insecurities. Therefore, this research focuses on loss aversion, net benefits, social norms, and control as the relevant constructs and thus covers all three explanation approaches to SQB.

Research has discussed several countermeasures for SQB [25]. Two countermeasures stand out both because of their suitability for RPA and because of their empirical grounding (albeit outside the domain of technology acceptance and use): 1) Providing additional information: Lorenc et al. [29] tested the effectiveness of an intervention to motivate energy tariff switching. They conducted two interviews with 150 individuals. In the first interview, they provided information on energy tariffs. Similarly, other researchers recommend providing individuals with additional information or resources to counter SQB: Several researchers recommend enabling users to "feel in control," e.g., via additional resources [30, 31]. For example, it can be more information in general [32, 33], more information about alternative options [29], or specific information on the change [34, 35]. 2) Telling success stories: Shealy et al. (2019) tested the effect of best practice examples (or success stories) on getting engineers to conceptualise more sustainable buildings. Other researchers have also commented on the use of success stories, for example, demonstrations, lighthouse projects or references to successful companies [35–37]. Based on this, we created treatments to test the influence of additional information and success stories on SQB and the adoption of RPA technologies (see Table 1).

3   Hypotheses

To effectively test the influence of the two selected countermeasures (giving individuals additional information and telling them success stories), we combined SQB with technology acceptance theory. We present our hypotheses derived from prior literature on SQB and technology acceptance in the following. We build a conceptual model of the construct relationships shown in Fig. 1 based on three overarching hypotheses.

H1. Based on prior results in the technology acceptance literature, we expect performance and effort expectancy to influence the behavioural intention to adopt [12, 13]. We thus hypothesize *(H1.1) a positive influence of PE on BI to use RPA* and *(H1.2) a positive influence of EE on BI to use RPA.*

H2. Kim and Kankanhalli (2009) proposed a direct influence of SQB constructs on technology acceptance constructs. Due to the specific nature of our context, we focus on loss aversion, net benefits, social norms, and control as relevant constructs for SQB (see above). In building our hypotheses, we rely on several authors who have used these concepts [15, 35, 43, 44]. Li et al. [43] find a positive influence of loss aversion on user resistance, which we would expect to be inverted for behavioural intention. Therefore, we expect a negative influence of loss aversion on the direct determinants. We thus hypothesize *(H2.1) a negative influence of LA on PE* and *(H2.2) a negative influence of LA on EE.* Zhang et al. [44] find a positive influence of perceived benefits on behavioural intention. Kim [35] finds a negative influence of perceived value on user resistance, which we expect to be inverted for the direct determinants of behavioural intention. We assume *(H2.3) a positive influence of NB on PE* and *(H2.4) a positive influence of NB on EE.* Hsieh [15] finds a positive influence of social norms on the behavioural intention to use health clouds. Similarly, Polites and Karahanna [45] also find a positive influence of social norms on behavioural intention. In other words, the positive affirmation of people who are important to the individual should increase the intention to use a system. As Venkatesh et al. (2003) conceptualised performance expectancy and effort expectancy as direct determinants of behavioural intention, we also expect a positive influence of social norms. We thus hypothesize *(H2.5) a positive influence of SN on PE* and *(H2.6) a positive influence of SN on EE.* Hsieh [15] also finds a positive influence of control on the behavioural intention to use health clouds. He refers to the effect that the more resources and knowledge an individual has, the higher their intention to use a new system

Table 1. Constructs derived from literature.

	Constructs	Definition	Source
Tech. Acc	Performance expectancy (PE)	Performance expectancy describes how beneficial an individual perceives the system regarding their job performance	[12]
	Effort expectancy (EE)	Effort expectancy describes how easily an individual perceives the use of a system	[12]
	Behavioural intention (BI)	Behavioural intention describes the behavioural intention of the individual to use the system	[12]
SQB	*Cognitive misperception:* Loss aversion (LA)	Loss aversion describes that individuals prefer to avoid potential losses even when these evenly match with potential gains	[38]
	Rational decision-making: Net benefits (NB)	Net benefits describe the perceived benefits relative to the costs of a change	[14]
	Psychological commitment: Social norms (SN) Control (CO)	Social norms describe the level of influence the individual attributes to the opinions of others	[39]
		Control describes the level of control an individual has regarding a change. They achieve this through resources or capabilities to deal with the new way of doing things	[31]
Treatment	Additional information (AI)	Participants receive additional information on RPA for nonprofit organisations in a textual format adapted from an online source [40]	[29]
	Success story (SU)	Participants receive a success story on RPA for nonprofit organisations in a textual format adapted from an online source [41]	[42]

is. Hence, we argue for *(H2.7) a positive influence of CO on PE* and *(H2.8) a positive influence of CO on EE.*

H3. In line with prior research [29, 42], we tested two countermeasures and expected a significant influence of our two treatments on the SQB constructs. We assume a positive influence on those constructs we hypothesised to influence the intention to use positively and the reverse. *(H3.1) A negative influence of AI on LA; (H3.2) a positive influence of AI on NB and (H3.3) a positive influence of AI on SN and (H3.4) a positive influence of*

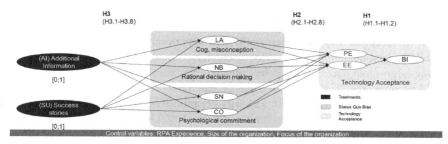

Fig. 1. Research Model.

AI on CO; (H3.5) a negative influence of SU on LA; (H3.6) a positive influence of SU on NB; (H3.7) a positive influence of SU on SN and (H3.8) a positive influence of SU on CO.

4 Research Method

To assess our research model, we employed an online experiment [16]. For analysis, we used structural equation modelling (SEM) using partial least squares (PLS) [46, 47]. PLS-SEM is a well-accepted method in IS research with advantages in case of single-item measurement and lower sample sizes [48].

4.1 Measurement

We used the items suggested by Venkatesh et al. (2003) to measure performance and effort expectancy. For the measurement model of SQB, we relied on Kim and Kankanhalli (2009). As these have not published the items for the SQB constructs, we relied on prior literature for the measurement items [14, 31, 35, 38, 39]. We adopted these in the context of automated processes with RPA. Following prior literature, we used a seven-point Likert scale (1 = strongly disagree, 7 = strongly agree) to assess all measurement items [35]. We conducted two pre-tests with ten employees from nonprofit organisations. The first pre-test indicated conceptual issues regarding the control construct, which led us to adjust the wording. The second pre-test showed no issues.

To be as close to a real-life situation as possible, we adopted both treatments from actual texts available on the internet. For the additional information treatment, we used a text by Tarulata Champawat posted under the title 'How RPA Benefits Nonprofit Organisations?' on Infobeans in February 2020 [40]. For the success story treatment, we used a text by the RPA provider UiPath describing their successful collaboration with the New York Foundling, a US-based non-governmental organisation [41]. We removed pictures and edited the texts minimally to fit into our survey.[1]

---

[1] See the appendix for the survey questions used. The full survey incl. Explanatory texts and treatment material can be requested from the authors.

4.2 Survey Administration and Data Inspection

In total, we recruited 150 participants affiliated with nonprofit organisations from Prolific and via personal connections. Prior research has shown Prolific, an online platform enabling large-scale data collection by connecting researchers and survey participants, to yield adequate results for cognitive bias research [49]. We recruited participants that indicated a concrete affiliation with a charity organisation. This broader focus allowed us to test our treatments in the broader ecosystem of NGOs. Support for new technologies does not only need to come from within the organisation, but a favourable view across stakeholders is vital, especially for organisations with such high transparency and accountability requirements [50, 51].

We administered the online experiment via Unipark, an online survey tool, and randomly assigned participants to four treatment groups. The first treatment group received the text with additional information on RPA and the success story. Participants in groups two and three received only one treatment, and those in group four received no treatment. We checked the resulting dataset for quality measures, e.g., if participants answered attention checks correctly or answered questions in unrealistic times, but no issues appeared. We also recruited several NGO employees to check for significant differences regarding answer times or patterns but found none [52]. No values were missing, as all questions were mandatory, and we did not provide a non-answer option.

5 Results

5.1 Measurement Model Analysis

Following prior literature, we employed PLS-SEM to analyse the hypothesised causal relationships between the constructs. PLS-SEM has been employed for UTAUT [12] and SQB [39]. Moreover, as we had a small sample size and employed non-normally distributed variables (i.e., the treatments), it seemed especially suitable (Hair et al. 2019). We created the outer or measurement model using reflective constructs only following prior researchers' assumption of a causal priority from the construct to the indicators [53]. To test for internal consistency reliability, we used Cronbach's alpha [54] and, for convergent reliability, the average variance extracted (AVE).

We estimated discriminant validity using the Fornell–Larcker criterion [55]. We conducted all analyses using Smart PLS 3.3.3 [56] and used thresholds in line with Hair et al. [46]. Tests for construct reliability and validity were positive for all constructs (see Table 2). All values were above the threshold of 0.7 for Cronbach's alpha and thus considered high [54]. High internal consistency reliability indicates that the items validly measure their corresponding constructs. The same applies to the Fornell-Larcker criterion. The correlations between all constructs were lower than the square root of the AVE, which supports convergent and discriminant validity [55] (Table 3).

Table 2. Fornell-Larcker Criterion

	Cronbach's α	AVE	PE	EE	BI	LA	NB	CO	SN	AI	SU
Performance expectancy (PE)	.858	.706	.840								
Effort expectancy (EE)	.870	.720	.705	.848							
Behavioral intention (BI)	.946	.903	.698	.607	.950						
Loss aversion (LA)	.827	.622	.199	.141	.149	.789					
Net benefits (NB)	.861	.783	.764	.694	.581	.150	.885				
Control (CO)	.817	.646	.691	.765	.622	.140	.715	.803			
Social Norms (SN)	.884	.812	.366	.375	.191	.177	.379	.397	.901		
Add. information (AI)			.001	.011	-.039	-.014	.040	-.011	.009	1.0	
Success story (SU)			.042	.096	.082	.103	.193	.057	.035	0.0	1.0

Table 3. Factor loadings for constructs

Construct	Item	Loading	Construct	Item	Loading
Performance ex-pectancy	pe01	.888	Loss aversion	la01	.812
	pe02	.868		la02	.602
	pe03	.901		la03	.949
	pe04	.686		la04	.752
Effort expectancy	ee01	.871	Net benefits	nb01	.880
	ee02	.833		nb02	.906
	ee03	.831		nb03	.869
	ee04	.859	Social norms	sn01	.914
Behavioural inten-tion	bi01	.948		sn02	.886
	bi02	.942		sn03	.902
	bi03	.961	Control	co01	.806
				co02	.745
				co03	.822
				co04	.838

A look at the factor loadings (see Table 4) confirmed all factor loadings were above the required threshold of 0.7. The only items that were slightly below that threshold were pe04 and la02. We adapted these items from prior literature [12, 35], and they did not appear as an issue during the pre-test. As prior literature consistently relied on the same three items [12, 35, 43, 45, 57], we decided to keep the constructs and items as there was no other way to ensure content validity. To test for common method bias, we employed the standard collinearity test. The variance inflation factor of all constructs with a random variable was below 3.3, indicating that common method bias does not apply here [58, 59].

5.2   Structural Model Analysis

We employed both the general PLS algorithm and bootstrapping to assess our model. For the parameters of the PLS algorithm, we followed Hair et al. [46] and the agreed-upon standards. We used a path weighting scheme with 300 iterations and a stop criterion of 10^{-7}. To test the significance of our model, we employed bootstrapping using 5,000 iterations of randomly drawn subsamples and the parameters as indicated above [46]. As a result, we identified seven significant influences (see Fig. 2), which we present in the following according to our hypotheses.

Table 4 presents the path coefficients we assessed to test our hypotheses. With this, we analysed significance based on the p-values of the path coefficients. In full support of H1, we see a very strong significant positive effect of performance expectancy on behavioural intention to use RPA (H1.1+). We also see a significant effect of effort expectancy on behavioural intention (H1.2+). We expected this result as several studies have tested UTAUT across domains [57].

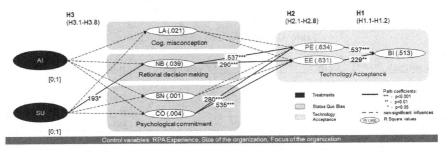

Fig. 2. Results of PLS-SEM analysis

We only find partial support for H2. We see a highly significant positive influence of net benefits on performance expectancy (H2.3+) and on effort expectancy (H2.4+). It stands to reason that a positive perception of the benefits of RPA also reflects in a positive expected performance and a high perception of usage ease. We also find a significant effect of control on performance expectancy and effort expectancy (H2.7+, H2.8+). This finding is also logical as individuals that feel more in control of the technology will be confident about their performance and effort expectancy. However, we find no significant effect of loss aversion on either performance or effort expectancy (H2.1−, H2.2−). Thus, in our case, participants did not fear the loss of the current way of working. This effect could be because the first processes to be automated are rather repetitive and administrative tasks [2], and thus nothing they would miss. Similarly, we find no significant effect of social norms on performance or effort expectancy (H2.5−, H2.6−). This finding implies that, at least in our context for RPA, the opinion of colleagues and friends does not affect the perception of the technology. We discuss possible reasons and avenues for further research in the next section.

Table 4. Path coefficients (*** = p-value < .001, ** = p-value < .01, * = p-value < .05)

	PE	EE	BI	LA	NB	SN	CO
Performance expectancy (PE)			.537***				
Effort expectancy (EE)			.229**				
Behavioural intention (BI)							
Loss aversion (LA)	.016	.074					
Net benefits (NB)	.537***	.290***					
Control (CO)	.280***	.535***					
Social Norms (SN)	.047	.052					
Additional information (AI)				−.014	.040	.009	−.035
Success story (SU)				.103	.193*	.035	.057

In contrast to our expectations derived from prior literature, we only find weak support for H3. We see a weak effect of the success story on one SQB construct and no significant effects of the additional information treatment. The success story treatment positively influenced the perception of net benefits slightly (H3.6). This finding is in line with prior literature that has recommended success stories as a tool to influence rational decision-making [35–37, 42], but we would have expected a stronger effect. Interestingly, we cannot confirm that simply giving individuals additional information about the technology and the implicated change is a valid countermeasure. Thus, at least for our specific context of RPA adoption, this recommendation from the literature [29–35, 43] does not seem to apply.

6 Discussion

Exploring the effects of SQB, we find significant influences of SQB on adopting RPA but only weak or even no support for the countermeasures we tested. We combined a selection of constructs to measure SQB from the literature [14] with technology acceptance constructs to assess two things: Firstly, if the SQB constructs affect technology acceptance of RPA, and second if more information and the reference to previous success stories help counter SQB regarding RPA and increase the behavioural intention to use. Our results indicate that only the SQB constructs net benefits and control affected the direct determinants of technology adoption. This finding highlights the relevance of re-testing established phenomena for "lightweight" technologies. Our hypothesis were based on results of studies with "heavyweight" systems – the first studies on SQB in IS focused on ERP systems for example [25]. This context change apparently reduces the effects of SQB constructs. Secondly, our results show the importance of testing countermeasures. At least in the context of RPA in the social sector, simply giving people more information does not suffice, and success stories only have a slight influence. Even though our results did not fully confirm our hypotheses derived from the literature, we contribute to both theory and practice with three main aspects:

Firstly, we contribute to the body of knowledge on RPA. Different literature reviews have highlighted that more insights into the adoption processes of RPA, especially in the nonprofit sector, are needed [9]. Our online experiment highlights and confirms that RPA adoption depends on perceptions regarding performance and effort. Moreover, people might be biased towards the status quo. However, as our results show, only a part of the constructs derived from measuring the influence of SQB on the adoption of ERP systems is relevant [14]. Our findings imply that the effects of other biases on adopting RPA and similar "lightweight" technologies also need further examination – a relevant insight considering the large number of biases currently employed in IS research [60].

Secondly, we contribute to the theory of technology adoption and cognitive biases. Researchers have found SQB to influence technology decisions; we deliver an additional puzzle piece to understand SQB and its effect and influences as detailed as possible. We modelled SQB with a selected range of concepts from the proposed full set by Kim and Kankanhalli [14]. Prior studies like Kim (2011) focusing only on cost aspects have already alluded to the fact that the full model might not fit all contexts. Our example confirms that the full range of concepts does not work in all contexts. We found significant effects only for the two constructs, net benefits and control. We found no effect for loss aversion and social influence. An interesting find as loss aversion has already been found to influence IS adoption behaviour, e.g., regarding two-factor authentication [61] and social influence is one of the key determinants of technology adoption models like UTAUT [12]. This contradiction implies that the adoption of RPA in the social sector differs from other technologies and contexts. Thus, we must challenge the generalizability of cognitive bias and adoption research so far.

Thirdly, we contribute a more nuanced understanding of countermeasures to SQB. When individuals who are supposed to adopt IT are affected by SQB, technology is either not introduced or not accepted. As such, it cannot unfold its value. Therefore, finding effective countermeasures is important for the IS community – both researchers and practitioners. We managed to test two potential countermeasures, even though our treatments only had a limited effect. There are three main potential reasons for the limited effect of our countermeasures: (1) The first option, which is unlikely regarding the substantial evidence in prior literature, is that our participants did not suffer from SQB concerning RPA. (2) The second option is that, in line with media richness theory [62], our delivery method was inadequate for the technology's complexity. But then, it is still interesting that internet articles, as they exist today alone, are insufficient to sway people's opinions. (3) The third option is that not all often-cited countermeasures work against SQB. If that is the case, researchers should stop recommending them, and practitioners should focus their resources elsewhere.

Moreover, our research also has important implications for practice. Firstly, we provide guidance to IT decision-makers and suppliers specialising in the NGO sector on what to consider when introducing RPA. Secondly, our insights are relevant to managers looking to improve their organisation in a general way. SQB is not a phenomenon unique to IS; thus, it might also impede other changes. Our findings highlight that countermeasures require careful consideration. Thirdly, our work is important as it highlights the transferability of IS research in the private sector to the social sector. Where practitioners often struggle with concepts clearly designed for profit-seeking organisations, our findings demonstrate that the technology acceptance models can also be used in this sector. Nonetheless, we had to consciously tailor the SQB model from prior literature [14]. Thus, further tests of the theories of our domain against a sector that is unique in the motivation of its workforce, the IT infrastructure and skills available and its budgetary constraints [10] are necessary.

We designed our research most carefully. Nonetheless, there are still some limitations, which we cluster around five aspects. Firstly, and most prominent is surely the limited number of participants. With 150 participants, the number of participants is sufficient to give a strong indication, not a final result. Nonetheless, we believe that our results highlight the importance of researching countermeasures to SQB further. Secondly, even though we synthesised our measurement model from prior literature, as we chose to select only part of the concepts due to our context, our choice might not have exhausted the full possibility of relevant concepts. Thirdly, other studies combined SQB with user resistance. In our research effort, we, therefore, expected an inverted effect for intention to use. However, intention to use and user resistance might be more dissimilar than we thought, explaining the lack of significant effects of most SQB constructs. Fourthly, the use of prolific participants, for whom we could not verify the exact nature of their nonprofit affiliation or the type of nonprofit work they were involved with, instead of participants recruited directly might have skewed results despite prior positive results with Prolific in cognitive bias research [49]. Finally, following media richness theory, the text format we chose might not have been sufficient to relay the complex level of information required to understand RPA [62]. Similarly, different texts, e.g., a success story written by a nonprofit and not a software provider, might have had different effects.

Our results and limitations open up four avenues for further research. First, future research could conduct further studies in the nonprofit sector to understand the impact of RPA in more detail. Here, scholars could also try to identify other cognitive biases at play (potentially also other than SQB only). Research could also explicitly try to address our limitations, especially regarding the use of an online panel. Second, future research could also work on the constructs to measure SQB and apply this to other technologies. Our research model was the first attempt to practically measure SQB as conceptualised by Kim and Kankanhalli [14] for the context of RPA in the social sector. Therefore, it is essential to examine our measurement instruments and the hypothesised effects again in other contexts. This further research could also look at user resistance as a dependent variable. Fully understanding the interplay of the different constructs could bring more precision to efforts to design countermeasures. For example, researchers might find that not all constructs selected by Kim and Kankanhalli [14] are relevant for SQB in a specific context; this would also explain the findings of Lee and Joshi [28]. Third, future research should test more countermeasures against the SQB towards RPA usage. Based on established measurement models, it is possible to empirically test the effect of countermeasures and our examples highlight the need for systematic tests. The list of countermeasures in the literature [25] implies there are still many more countermeasures to be tested. Researchers should aim to test more of the countermeasures typically proposed for cognitive biases to ensure the significance of their advice. It may be that other countermeasures are more important in the communication strategies of organisations implementing RPA. Fourth, scholars should further investigate the two countermeasures we tested and the effect of the potential SQB countermeasure of additional experience we identified. The two examples from prior literature only worked to a limited extent – at least in a simple text format. That could mean testing with a video format or, if Covid-19 allows it, a treatment delivered via personal interaction to transport a higher level of complexity following media-richness theory [62].

Appendix

Table 5. Measurement Items used to measure UTAUT and SQB constructs

	Constructs	Items	Source
UTAUT	Performance expectancy (PE)	I would find processes automated with RPA useful in my job Using processes automated with RPA enables me to accomplish tasks more quickly Using processes automated with RPA increases my productivity If I use processes automated with RPA, I will increase my chances of getting a raise	[12]
	Effort expectancy (EE)	My interaction with processes automated with RPA would be clear and understandable It would be easy for me to become skillful at using processes automated with RPA I would find processes automated with RPA easy to use or interact with Learning to use processes automated with RPA is easy for me	
	Behavioral intention (BI)	I intend to use processes automated with RPA in the next 6 months I predict I would use processes automated with RPA in the next 6 months I plan to use processes automated with RPA in the next 6 months	
SQB	Net benefits (NB)	Considering the time and effort that I have to spend, the change to the new way of working with processes automated with RPA is worthwhile Considering the loss that I incur, the change to the new way of working with processes automated with RPA is of good value Considering the hassle that I have to experience, the change to the new way of working with processes automated with RPA is beneficial to me	[14]

(*continued*)

Table 5. (*continued*)

Constructs	Items	Source
Social norms (SN)	I would use processes automated with RPA if people who influence my behavior think I should I would use processes automated with RPA if people who are important to me think I should I would use processes automated with RPA if people whose opinions I value want me to do so	[39]
Control (CO)	I personally have what it takes to deal with the situations caused by processes automated with RPA I have the resources I need to successfully use processes automated with RPA I have the knowledge necessary to use processes automated with RPA I am confident that I will be able to use processes automated with RPA without any problems	[31]

References

1. Santos, F., Pereira, R., Vasconcelos, J.B.: Toward robotic process automation implementation: an end-to-end perspective. Bus. Process Manag. J. (2020). https://doi.org/10.1108/BPMJ-12-2018-0380
2. Syed, R., et al.: Robotic process automation: Contemporary themes and challenges. Comput. Ind. (2020). https://doi.org/10.1016/j.compind.2019.103162
3. Plattfaut, R., Borghoff, V.: Robotic process automation – a literature-based research agenda. J. Inf. Syst. (2022). https://doi.org/10.2308/ISYS-2020-033
4. Kregel, I., Koch, J., Plattfaut, R.: Beyond the hype: robotic process automation's public perception over time. J. Organ. Comput. Electron. Commer. (2021). https://doi.org/10.1080/10919392.2021.1911586
5. van der Aalst, W.M.P., Bichler, M., Heinzl, A.: Robotic process automation. Bus. Inf. Syst. Eng. **60**(4), 269–272 (2018). https://doi.org/10.1007/s12599-018-0542-4
6. Lacity, M., Willcocks, L.P.: Robotic process automation at telefónica O$_2$. MIS Q. Exec. **15**, 21–35 (2016)
7. Asatiani, A., Penttinen, E.: Turning robotic process automation into commercial success – Case OpusCapita. J. Inf. Technol. Teach. Cases (2016). https://doi./.1057/jittc.2016.5
8. Lacity, M., Khan, S., Carmel, E.: Employing U.S. Military families to provide business process outsourcing services: a case study of impact sourcing and reshoring. CAIS (2016). https://doi.org/10.17705/1CAIS.03909
9. Plattfaut, R., Borghoff, V., Godefroid, M., Koch, J., Trampler, M., Coners, A.: The critical success factors for robotic process automation. Comput. Ind. (2022). https://doi.org/10.1016/j.compind.2022.103646
10. Merkel, C., Farooq, U., Lu X., C.G., Rosson, M.B., Carroll, J.M.: Managing technology use and learning in nonprofit community organizations: Methodological challenges and opportunities. In: Proceedings of the 2007 Symposium on Computer Human Interaction for the Management of Information Technology (CHIMIT '07), New York, USA (2007)

11. Davis, F.D., Bagozzi, R.P., Warshaw, P.R.: User acceptance of computer technology: a comparison of two theoretical models. Manage. Sci. **35**, 982–1003 (1989)
12. Venkatesh, M.: Davis: user acceptance of information technology: toward a unified view. MIS Q. **27**, 425–478 (2003)
13. Wewerka, J., Dax, S., Reichert, M.: A user acceptance model for robotic process automation. In: 2020 IEEE 24th International Enterprise Distributed Object Computing Conference (EDOC), Eindhoven, Netherlands, pp. 97–106. IEEE (2020). https://doi.org/10.1109/EDO C49727.2020.00021
14. Kim, H.W., Kankanhalli, A.: Investigating user resistance to information systems implementation: A status quo bias perspective. MIS Q. **33**, 567–582 (2009)
15. Hsieh, P.-J.: Healthcare professionals' use of health clouds: Integrating technology acceptance and status quo bias perspectives. Int. J. Med. Informatics **84**, 512–523 (2015)
16. Fink, L.: Why and how online experiments can benefit information systems research. JAIS (2022). https://doi.org/10.17705/1jais.00787
17. Plattfaut, R.: Robotic process automation - process optimization on steroids? In: Helmut Krcmar, Jane Fedorowicz, Wai Fong Boh, Jan Marco Leimeister, Sunil Wattal (eds.) Proceedings of the 40th International Conference on Information Systems, ICIS 2019, Munich, Germany, December 15–18, 2019. Association for Information Systems (2019)
18. Hallikainen, P., Bekkhus, R., Pan, S.: How OpusCapita used internal RPA capabilities to offer services to clients. MIS Q. Exec. **17**, 41–52 (2018)
19. François, P.A., Borghoff, V., Plattfaut, R., Janiesch, C.: Why companies use RPA: A critical reflection of goals. In: Di Ciccio, C., Dijkman, R., Del Río Ortega, A., Rinderle-Ma, S. (eds.) Business process management, vol. 13420. Lecture Notes in Computer Science, pp. 399–417. Springer International Publishing, Cham (2022)
20. Fishbein, M., Ajzen, I.: Belief, attitude, intention, and behavior: An introduction to theory and research. Addison-Wesley, Reading, MA (1975)
21. Ajzen, I.: The theory of planned behavior. Organ. Behav. Hum. Decis. Process. (1991). https://doi.org/10.1016/0749-5978(91)90020-T
22. Davis, F.D.: A technology acceptance model for empirically testing new end-user information systems: Theory and results. Doctoral thesis, Massachusetts Institute of Technology (1985)
23. Williams, M.D., Rana, N.P., Dwivedi, Y.K.: The unified theory of acceptance and use of technology (UTAUT): A literature review. Journal of Ent Info Management **28**, 443–488 (2015)
24. Ma, Q., Liu, L.: The technology acceptance model: A meta-analysis of empirical findings. Journal of Organizational and End User Computing (2004). https://doi.org/10.4018/joeuc.2004010104
25. Godefroid, M.-E., Plattfaut, R., Niehaves, B.: How to measure the status quo bias? A review of current literature. Manag Rev Q (2022). https://doi.org/10.1007/s11301-022-00283-8
26. Kahneman, D., Tversky, A.: Prospect theory: An analysis of decision under risk. Econometrica **47**, 263–292 (1979)
27. Samuelson, W., Zeckhauser, R.: Status Quo Bias in Decision Making. J. Risk Uncertain. **1**, 7–59 (1988)
28. Lee, K., Joshi, K.: Examining the use of status quo bias perspective in IS research: Need for re-conceptualizing and incorporating biases. Inf. Syst. J. **27**, 733–752 (2017)
29. Lorenc, A., Pedro, L., Badesha, B., Dize, C., Fernow, I., Dias, L.: Tackling fuel poverty through facilitating energy tariff switching: a participatory action research study in vulnerable groups. Public Health **127**, 894–901 (2013)
30. Hsieh, P.-J., Lai, H.-M., Ye, Y.-S.: Patients' acceptance and resistance toward the health cloud: An integration of technology acceptance and status quo bias perspectives. In: Pacific Asia Conference on Information Systems (PACIS) Proceedings, Chengdu, China (2014)

31. Zhang, K.Z., Gong, X., Zhao, S.J., Lee, M.K.: Are you afraid of transiting from web to mobile payment? The bias and moderating role of inertia. In: Pacific Asia Conference on Information Systems (PACIS) Proceedings, Chiayi City, Taiwan (2016)
32. Merriman, K.K., Sen, S., Felo, A.J., Litzky, B.E.: Employees and sustainability: The role of incentives. J. Manag. Psychol. **31**, 820–836 (2016)
33. Weiler, S., Marheinecke, H., Matt, C., Hess, T.: Trapped in the status quo? Cognitive misperceptions' effects on users' resistance to mandatory usage. In: Pacific Asia Conference on Information Systems (PACIS) Proceedings, Xi'an, China (2019)
34. Henkel, C., Seidler, A.-R., Kranz, J., Fiedler, M.: How to nudge pro-environmental behaviour: An experimental study. In: European Conference on Information Systems (ECIS) Proceedings, Stockholm & Uppsala, Sweden (2019)
35. Kim, H.-W.: The effects of switching costs on user resistance to enterprise systems implementation. IEEE Trans. Eng. Manage. **58**, 471–482 (2011)
36. Linnerud, K., Toney, P., Simonsen, M., Holden, E.: Does change in ownership affect community attitudes toward renewable energy projects? Evidence of a status quo bias. Energy Policy **131**, 1–8 (2019)
37. Bekir, I., Doss, F.: Status quo bias and attitude towards risk: An experimental investigation. Manage Decis Econ **41**, 827–838 (2020)
38. Li, Z., Cheng, Y.: From free to fee: Exploring the antecedents of consumer intention to switch to paid online content. J. Electron. Commer. Res. **15**, 281–299 (2014)
39. Hu, T., Poston, R.S., Kettinger, W.J.: Nonadopters of online social network services: Network services: Is it easy to have fun yet? Communications of the Association for Information Systems **29** (2011)
40. Champawat, T.: How RPA benefits non-profit organizations? https://www.infobeans.com/rpa-benefits-nonprofit-organizations (2020). Accessed 22 April 2021
41. UiPath: Nonprofit, the new york foundling, saves 100,000 hours in manual work annually with UiPath. https://www.uipath.com/resources/automation-case-studies/new-york-foundling-ngo-rpa (2020). Accessed 22 April 2021
42. Shealy, T., Klotz, L., Weber, E.U., Johnson, E.J., Greenspan Bell, R.: Bringing Choice Architecture to Architecture and Engineering Decisions: How the Redesign of Rating Systems Can Improve Sustainability. Journal of Management in Engineering **35** (2019)
43. Li, J., Liu, M., Liu, X.: Why do employees resist knowledge management systems? An empirical study from the status quo bias and inertia perspectives. Comput. Hum. Behav. **65**, 189–200 (2016)
44. Zhang, X., Guo, X., Wu, Y., Lai, K., Vogel, D.: Exploring the inhibitors of online health service use intention: A status quo bias perspective. Information & Management **54**, 987–997 (2017)
45. Polites, K.: Shackled to the status quo: The inhibiting effects of incumbent system habit, switching costs, and inertia on new system acceptance. MIS Q. **36**, 21–42 (2012)
46. Hair, J.F., Hult, G.T.M., Ringle, C.M., Sarstedt, M.: A primer on partial least squares structural equations modeling (PLS-SEM), 2nd edn. SAGE, Thousand Oaks (2017)
47. Hair, J.F., Risher, J.J., Sarstedt, M., Ringle, C.M.: When to use and how to report the results of PLS-SEM. Eur. Bus. Rev. **31**, 2–24 (2019)
48. Petter, S.: "Haters gonna hate": PLS and information systems research. ACM SIGMIS Database: the DATABASE for Advances in Information Systems **49**, 10–13 (2018)
49. Bahreini, A.F., Cavusoglu, H., and Cenfetelli, R.: Role of feedback in improving novice users' security performance using construal level and valance framing. In: International Conference on Information Systems (ICIS) Proceedings, India (2020)
50. Schmitz, H.P., Raggo, P., Bruno-van Vijfeijken, T.: Accountability of transnational NGOs: Aspirations vs. practice. Nonprofit and Voluntary Sector Quarterly (2011). https://doi.org/10.1177/0899764011431165

51. Burkart, C., Wakolbinger, T., Toyasaki, F.: Funds allocation in NPOs: The role of administrative cost ratios. CEJOR **26**, 307–330 (2018)

52. Smith, S.M., Roster, C.A., Golden, L.L., Albaum, G.S.: A multi-group analysis of online survey respondent data quality: Comparing a regular USA consumer panel to MTurk samples. J. Bus. Res. **69**, 3139–3148 (2016)

53. Diamantopoulos, A., Winklhofer, H.M.: Index construction with formative indicators: An alternative to scale development. J. Mark. Res. **38**, 269–277 (2001)

54. Hinton, P.R.: SPSS explained. Routledge, London, New York (2008)

55. Fornell, C., Larcker, D.F.: Evaluating structural equation models with unobservable variables and measurement error. Journal of Marketing Research **18** (1981)

56. Ringle, C.M., Wende, S., Becker, J.-M.: SmartPLS 3. www.smartpls.com (2015)

57. Venkatesh, T.: Xu: Consumer acceptance and use of information technology: Extending the unified theory of acceptance and use of technology. MIS Q. **36**, 157–178 (2012)

58. Kock, N.: Common Method Bias in PLS-SEM. International Journal of e-Collaboration (2015). https://doi.org/10.4018/ijec.2015100101

59. Kock, N., Lynn, G.: Lateral Collinearity and Misleading Results in Variance-Based SEM: An Illustration and Recommendations. JAIS (2012). https://doi.org/10.17705/1jais.00302

60. Godefroid, M., Zeuge, A., Oschinsky, F., Plattfaut, R., Niehaves, B.: Cognitive Biases in IS Research: A Framework Based on a Systematic Literature Review. In: Pacific Asia Conference on Information Systems (PACIS) Proceedings, Dubai (2021)

61. Pratama, A.R., Firmansyah, F.M.: Until you have something to lose! Loss aversion and two-factor authentication adoption. Applied Computing and Informatics (2021). https://doi.org/10.1108/ACI-12-2020-0156

62. Ishii, K., Lyons, M.M., Carr, S.A.: Revisiting media richness theory for today and future. Human Behavior and Emerging Technologies **1**, 124–131 (2019)

Stochastic-Aware Comparative Process Mining in Healthcare

Tabib Ibne Mazhar[1], Asad Tariq[1], Sander J. J. Leemans[1]([✉]) [iD],
Kanika Goel[2] [iD], Moe T. Wynn[2] [iD], and Andrew Staib[3] [iD]

[1] RWTH Aachen, Aachen, Germany
s.leemans@bpm.rwth-aachen.de
[2] Queensland University of Technology, Brisbane, Australia
[3] Princess Alexandra Hospital, Brisbane, Australia

Abstract. Evidence-based innovations are critical in optimising the delivery of healthcare services. Process mining aims to provide healthcare stakeholders with insights, derived from historical data recorded in hospital information systems, to optimise healthcare processes. Healthcare processes are well-known for their complexity and control-flow variations are inherent in patient pathways undertaken by different patient cohorts. Comparative process mining can reveal insights from studying the differences between healthcare processes to better understand best-practice patient pathways. In this paper, we take a design science approach to redefine an existing method for process comparison (PCM). Where PCM considers predominantly the control-flow perspective, we extend this method with the stochastic perspective, that is, how likely a particular pathway is for certain patient cohorts, to obtain the Probabilistic Process Comparison Method (P^2CM). Furthermore, we further automate the method. Concretely, we introduce new, stochastic-aware, methods for sub-dividing process behaviour into cohorts based on trace attributes or other trace features, methods for focusing the comparative analysis on specific pairs of interesting cohorts, and provide a new method for in-depth comparison of process differences. The approach is evaluated using three real-life healthcare datasets, of which one case study is conducted with a domain expert from an Australian hospital.

Keywords: process mining · healthcare · comparative process mining

1 Introduction

Healthcare is a field that is confronted with widespread challenges, which require process improvement to be an integral part of the system. Data-informed innovations are important to make healthcare better and efficient [1–3]. New methods can assist healthcare organisations to rapidly adapt their processes to changing needs. Healthcare organisations around the world recognise the need to continually put efforts to improve their clinical as well as administrative processes.

© The Author(s), under exclusive license to Springer Nature Switzerland AG 2023
C. Di Francescomarino et al. (Eds.): BPM 2023, LNCS 14159, pp. 341–358, 2023.
https://doi.org/10.1007/978-3-031-41620-0_20

Healthcare organisations rely heavily on hospital information systems which support clinical and administrative processes, and record executed process steps in process execution data [4]. This data consists of sequences of process steps (activities) executed for patients, hospital stays, etc. (cases).

Process mining is a family of techniques and methods, which can assist in answering questions that are crucial to improving processes in healthcare organisations. One area of process mining focuses on comparing groups of cases (*cohorts*) of a process. In such comparative process analysis, processing of different cohorts is compared, which may lead to insights into the process-based differences between cohorts - if processing is expected to be similar, e.g. leading to the identification of best practices, and into process-based similarities - where differences are expected [5,6]. The insights from such comparative process analysis can then be leveraged to optimise the processes involved. When comparing processes, several perspectives can be identified: the *control-flow* perspective entails the activities that can be performed in a process and their organisation into pathways, while the *stochastic* perspective describes how likely activities, pathways and behaviour in processes are [7]. In comparative process mining analysis, both perspectives may be beneficial: the control flow perspective may indicate that, for instance, a rework loop is possible, however without knowledge of the stochastic perspective that will indicate how likely that rework loop is, it remains unclear what the impact on the process of the rework loop is. A little-executed rework may be part of normal operating procedure, while an often-executed loop may pose a threat to process performance. Thus, a comparison of both perspectives may be beneficial in process comparison to optimise optimisation efforts [1].

Several techniques have been proposed to compare different parts of a process with one another, however applying them effectively in practice requires highly similar processes [5]: benefits have been shown to be derivable from the same (or, supposedly similar) process being executed in different settings. In some literature, such a setting of highly similar processes was known, for instance, comparing fulfilment processes in different geographic regions [8] and building permit processes in different municipalities [9]. To compare two processes with one another, several techniques can be applied [10,11]. However, if a single process is to be considered, a sub-division into variants (or, in log terms, *cohorts*) is necessary first. Several techniques have been proposed to identify cohorts from event logs [6,12].

To assist with applying the combination of these techniques, in [5] a generic method was proposed, the Process Comparison Methodology (PCM). However, as we detail in Sect. 2, PCM does not consider the stochastic perspective, and is highly manual with little automated support. As such, there is no method that takes an event log file as an input and identifies cohorts as output, along with visualisations of similarities and differences between the cohorts.

Given this gap, our problem statement is that we would like to have a method with which analysts can compare sub-processes for stochastic processing differences. In this paper, we use a design science [13] approach to extend the PCM method with stochastic awareness, operationalise PCM in a systematic manner, and provide further guidance on how the techniques can be applied. We refer to this new method as the Probabilistic Process Comparison Method (P^{2}CM). We evaluate our updated method twofold: using two real-life data sets, we validate the applicability of the method and using a case study, we validate the usefulness of the method in practice.

The remainder of this paper is organised as follows: Sect. 2 discusses related work. Section 3 details the research design. Section 4 introduces the updated method, Sect. 5 discusses the evaluative case studies, and Sect. 6 concludes the paper.

2 Related Work

In this section, we discuss related literature and derive our design objectives.

Process Mining in Healthcare. While PCM can be applied to datasets from any domain, in this paper we focus on healthcare processes as these processes typically consist of many variants, and as domain experts are keen to understand how the different patient cohorts pass through a hospital. Healthcare processes are characterised as complex and inherent to significant variations [14]. These variations can be a result of the differences in which the patient pathways proceed in a hospital. Process mining has the potential of uncovering details related to the execution of processes and has been used in healthcare. The potential has been explained in literature reviews [15] and a research agenda paper that highlights various opportunities and challenges [2]. In [16], the authors reviewed a pool of articles to understand how process mining has been applied to clinical pathways. The papers were classified in three categories, (i) discovery of actual execution pathways, (ii) analyse variants of execution pathways, and (iii) improve execution pathways.

As noted, one of the key areas of use of process mining is variant exploration. In [1,17], the authors used process mining to understand the similarities and differences between practices of different hospitals, but this comparison was done manually. Identifying differences between groups of pathway executions using process variant analysis can help to identify areas of potential improvement. Specific challenges related to process variant analysis exist. For example, comparing processes from a resource perspective, checking for compliance, and finding adverse events were mentioned in [2,18]. Despite growing interest in comparing healthcare processes, [2] identified the need for algorithms and methods that provide detailed explanations on the differences between process "variants" as a key challenge. This brings forth our first design objective:

DO1: A method that allows comparative analysis of process-based differences in cohorts of a single process.

Comparative and Stochastic Analysis. To compare multiple event logs with one another, a cross-comparison method has been proposed that first discovers a process model for each event log, and measures the differences between the model and each other event log in a cross-product setting [9]. This method, used in

PCM [5] as well, is susceptible to the trade-offs that are present in process discovery, and consider the stochastic perspective only partially, as the discovered models only consider the control-flow perspective. To compare two event logs with one another, approaches have been proposed based on transition systems [10,11] and fingerprints [19]. Furthermore, [9,20] both cross-compare two event logs: [20] by means of quality measures and [9] by means of deviations. Furthermore, in [21], predicted future process models can be compared. However, these techniques do not consider the stochastic perspective explicitly, which is essential to spot e.g. that exceptional behaviour is much more likely in one part than in the other, and assume that the two to-be compared event logs are known. Therefore, for our method, we specified the following design objective:

DO2: A method that compares the stochastic behaviour of two processes.

Process Comparison Methodology (PCM). PCM [5] has been proposed as a method to support comparative process mining. PCM consists of 5 consecutive phases: (1) in the first phase, the data must be extracted from information systems and pre-processed into the XES event log format [22]. Furthermore, in this step a trace attribute is selected to divide the event log (the α attribute). (2) in the second phase, the event log is divided into sub-logs, and an initial selection of these sub-logs is made, such that this selection will enable the answering of business questions and satisfy the goal of the comparative analysis. (3) in the third phase, suitable pairs of sub-logs (cohorts) are selected for comparison. (4) in the fourth phase, the selected pairs of sub-logs are compared to obtain detailed process-based differences. (5) the fifth phase involves reporting the relevant and impactful differences to the process owner.

In [5], the PCM method was applied to a non-healthcare case study, using semi-automated techniques and visualisations for phases 2, 3 and 4. However, most of the mentioned techniques utilised in PCM [5] only take the control flow – which steps are executed – into account, but only implicitly and unpredictably considering the stochastic perspective – how likely pathways are. Furthermore, considering the phases in detail, the alpha-attribute is chosen in phase 1, but little guidance is provided on how this attribute can be chosen, and data-supported automation that may aid analysts is limited. These details resulted in the following design objective:

DO3: A method that combines guidance for users with automated recommendations derived from data.

3   Research Design

We adopt a design science approach [23] and follow the six phases as described in [24].

(1) Problem Identification. Prior literature conveys that comparative process mining is important, in general, and in the healthcare sector in particular, to visualise the similarities and differences between processes. There is a need to develop comparative process mining techniques and methodological guidance to assist healthcare organisations in identifying potential improvements.

(2) Definition of Design Objectives. The overall objective is to propose a process comparison method that takes an event log file as an input, groups similar cohorts together, and provides visualisation of similarities and differences among the cohorts. Three design objectives (DOs) motivated by the related work in 2 are as follows:

DO1 A method that allows comparative analysis of process-based differences in cohorts of a single process;

DO2 A method that compares the stochastic behaviour of two processes;

DO3 A method that combines guidance for users with automated recommendations derived from data.

(3) Design and Development. The design objectives identified in Step 2 were used to design and develop the P^2CM method. The proposed method leverages the overall structure of PCM and extends several steps with stochastic awareness, provides more automated techniques, and provides more guidance for users of the method. As such, the P^2CM method can be seen as an enhanced version of PCM. The six steps of the P^2CM method are detailed in Sect. 4.

(4) Demonstration. To apply the P^2CM method in practice, we implement scripts for the new alpha attribute selection technique and the new comparative process visualisation algorithm.

(5) Evaluation. To evaluate the P^2CM method, we use the evaluation framework presented by [25]. Two ex-post evaluation strategies are used. First, an experimental controlled experiment [23] is conducted by applying the P^2CM method to two real-life publicly available event logs with the objective of assessing the applicability of the method. Second, we perform an observational case study [23] with the emergency department of a healthcare organisation - the Princess Alexandra Hospital, Brisbane, Australia. The objective was to assess the usefulness of the method, i.e., whether the method we propose can be used to unearth meaningful insights for stakeholders. The findings are presented in Sect. 5. The ex ante evaluation [25] of these design objectives - to, for instance, validate their applicability in practice or their appropriateness with focus groups – is not within the scope of this paper, but would be an interesting area of further research.

(6) Communication. This manuscript is a means of sharing the new P^2CM method. All introduced techniques have been implemented, and their source code is publicly available.

4   Artifact: P^2CM

In this section, we introduce our new method, the Probabilistic Process Comparison Method (P^2CM), which extends and instantiates the PCM framework. As by DO3, we aim to automate the steps as much as possible, we slightly change the overall structure of the PCM method, described in Sect. 2: we denote the selection of the alpha attribute in its own phase, as this step can be automated. Figure 1 provides a visual overview of P^2CM. Furthermore, we change the following phases

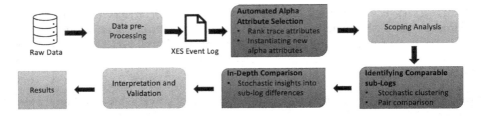

Fig. 1. P²CM. The darker steps are new or different from PCM.

to explicitly consider the stochastic perspective (DO2) and further automate them (DO3): we introduce the new phase 2 (selecting the alpha attribute), we change phase 4 (identifying comparable sub-logs) and change phase 5 (in-depth comparison).

4.1   Assisted Alpha Attribute Selection

The alpha attribute plays a significant role in distinguishing between process variants. However, identifying a suitable alpha attribute in an event log requires domain expertise and a good understanding of the process. To assist the analyst while minimising domain expert input, we rank the trace attributes in an event log by feature importance as a guide to the user. We propose two machine learning techniques, one based on unsupervised learning (ID-K), which groups similar data points without any dependency on a target variable and the other on supervised learning (ID-R), which combines decision trees for classification based on a target variable.. Each method returns a graph of the relative importance of each trace attribute in an event log. Both techniques may indicate the importance of an attribute; an attribute indicated by both provides an even stronger indication.

ID-K: k-Means Clustering. Clustering groups data into clusters based on their similarity in certain features. For our analysis, we start with an XES event log, and consider the trace attributes. Numeric, boolean and categorical features are considered, while unique identifiers, timestamps and free-text comments are not considered. These latter categories are inherently unsuitable as alpha attributes, as they do not sub-divide the traces of the log into clearly defined and understandable sub-logs. To transform data into a format that machine learning algorithms can process, factorisation is used for categorical and boolean attributes[1], which transforms this data into enumerated or categorical values.

Once the data has been transformed, the next step is determining the optimal number of clusters through the Elbow method [26], which allows a user to select the appropriate number of clusters by visualising the within-cluster sum of squares (WCSS) [27]. The inflection point or "elbow" in the plot, where

---

[1] https://pandas.pydata.org/docs/reference/api/pandas.factorize.html.

increasing the number of clusters would not significantly lower the WCSS, shows a levelling out of the inter-cluster variability. For instance, Fig. 3 shows the elbow graph for one of our evaluations. In this graph, the elbow is at number of clusters = 2, as after that there is no significant decrease in the WCSS.

Then, the k-means clustering algorithm is applied to the selected trace attributes. The contribution of each attribute in segregating the traces into different groups is studied and a plot of the relevant importance of each feature is returned. This is done by calculating the significance of each feature for each cluster based on the magnitude of the weight of the feature in the centroid vector.

ID-R: Random Forest Classifier. In a supervised learning setting, we assess the influence of each trace attribute on a target variable. Here, we consider the length of a trace as the target variable, while the training features are the attributes from the log data. We want the outcome to factor in the effect of length of a trace as the count of activities carried out for a case may have interesting reasons, so we extract the count of activities per trace and add it as the target feature in the data set. We choose to utilise an ensemble classifier, specifically the Random Forest classifier, which combines multiple decision trees to enhance the model's overall performance [28]. It operates by training multiple decision trees on randomly selected subsets of the data and then averaging the predictions of all the trees to make a final prediction. This method is robust to high variance and outliers.

In a random forest classifier, the importance of attributes is calculated using the mean decrease impurity (MDI) method [28], which calculates the total reduction of Gini impurity that each attribute provides across all the trees in the forest. The attribute importance is then determined by averaging the reduction in impurity across all trees that use the attribute.

4.2   Identifying Comparable Sub-Logs

The identification of comparable sub-logs is the next phase of the analysis. Sometimes the alpha attribute may have hundreds or thousands of values and comparing them one-on-one creates $n * (n - 1)/2$ comparisons. To reduce the potential number of comparisons, and thereby limit domain expert involvement (DO3), we introduce a new method, consisting of ranking, filtering and clustering. The method follows several steps. First, it ranks the values of the alpha attributes based on their count and takes the top-most frequent ones, based on a user-provided parameter. Second, we reduce the n^2 comparison space by clustering sub-logs based on their similarities with other sub-logs.

In order to take the stochastic perspective into account, we use the Earth Movers' Stochastic Conformance Checking (EMSC) [7] to obtain a sub-log vs sub-log similarity score table. We used the stochastic perspective instead of the control flow perspective to show the likelihood of following a pathway by similar patient cohorts (DO2). Given two sub-logs, EMSC will compute a score that is 1 if the two logs have the same stochastic behaviour, and 0 if the stochastic behaviour of the two sub-logs is completely different.

In the table of similarity scores, each sub-log has a pairwise similarity score between 0–1 with every other sub-log. After creating this table, we need to find the optimum number of clusters, for which we apply the Elbow method. Next, clustering is applied to the vectors of EMSC scores, to identify clusters of sub-logs.

Then, the sub-logs to compare are to be chosen, which is a step that inherently requires domain expertise. Nevertheless, the clustering provides guidance in two ways: pairs of sub-logs in different clusters are likely to differ in stochastic behaviour, while pairs of sub-logs are less likely to differ in stochastic behaviour. The former can be used to study the differences between process cohorts – e.g. to perform auditing –, while the latter can be used to study commonalities – e.g. to identify best practices.

4.3 In-Depth Comparison

Using the techniques of Sects. 4.1 and 4.2, we get the alpha attributes and the sub-logs that we want to compare. In this section, we present a new visualisation technique of in-depth comparison between two sub-logs which we call the Visual Process Comparator (VPC). VPC takes a log (the complete, not-subdivided log), L and two sub-logs of that log, L_1 and L_2. At first, a directly follows graph (DFG) is made from L. In a DFG, every node represents an activity and the edges describe the relationship between the activities [29]. To increase understandability, and to avoid clutter and spaghetti models, we first filter the edges: given a percentage parameter set by the analyst, we remove all edges that are below that threshold. If the log is not very complex, we can choose a threshold of 0. Then we check how many of those connections are present in L_1 and L_2 and use only those. Second, we visualise the differences between L_1 and L_2 on this filtered DFG.

We denote $L_1(a \to b)$ and $L_2(a \to b)$ as the frequency of the DFG edge from a to b in L_1 and L_2 respectively. For a particular node a, $\sum_{(a,b') \in DFG} L_1(a, b')$ and $\sum_{(a,b') \in DFG} L_2(a, b')$ denote the summation of all the edge frequencies out going from that particular node for L_1 and L_2 respectively.

The below formula is used for showing the relative frequency difference D:

$$D_{(a \to b)} = \frac{L_1(a \to b)}{\sum_{(a,c') \in DFG} L_1(a, c')} - \frac{L_2(a \to b)}{\sum_{(a,c') \in DFG} L_2(a, c')}$$

To indicate the importance of an edge, we scale the width according to its relative appearance in both sub-logs varying between a width of 0.5 when the edge is present in neither L_1 or L_2 to 1.5 when the edge is the only outgoing edge of that node. The below formula is used for width calculation,

$$width_{(a \to b)} = \frac{L_1(a \to b)}{\sum_{(a,c') \in DFG} L_1(a, c')} + \frac{L_2(a \to b)}{\sum_{(a,c') \in DFG} L_2(a, c')} + 0.5$$

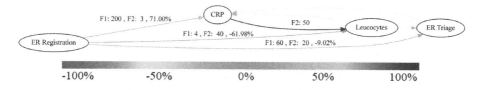

Fig. 2. Example of the Visual Process Comparator.

The edges are coloured so that the user can see the differences instantly. We use the HSV colour scale for the graph. The colour ranges from red (0°, 73%, 96%) to blue (200°, 73%, 96%) and all the colours in between based on D (see Fig. 2). Grey colour indicates the edges are neither in L1 nor L2.

Besides that, the frequency of each edge of each sub-log (denoted as F1 and F2) and the relative frequency differences in percentage between two sub-logs, are also shown on each edge.

For instance, consider Fig. 2. The thin grey edges indicate they are only present in the L but not in L_1 and L_2 and the green edge indicates near 0 relative difference. CRP → Leucocytes edge is deep blue as it only present in L_2. ER Registration → CRP has a teal colour as the relative difference is 71% and ER Registration → Leucocytes has relative difference of -61.98% and the colour is orange here.

5   Evaluation

In this section, we describe the twofold evaluation we performed to verify the applicability and usefulness of the method and its implemented tool support.

5.1   Applicability

As a first evaluation, we assess the applicability of P^2CM by applying it to two real-life healthcare data sets that are publicly available. The aim is to illustrate that P^2CM can be applied to real-life event logs and may lead to insights into the differences in (stochastic) process behaviour of cohorts within a single process with minimal domain experts' input.

Sepsis. Sepsis, a condition characterised by the body's harmful response to infection, is a frequent cause of severe illness and death worldwide [30]. The data setfootnotehttps://data.4tu.nl/articles/dataset/Sepsis_Cases_-_Event_Log/ 12707639 consists of 1 050 patient cases recorded between 2013 and 2015 and includes diagnostic test results, patient demographic information and organisational information. We apply P^2CM to understand the diagnostic journey of sepsis patients and identify factors that may affect patient outcomes. *Data pre-*

processing. The event log has 5 distinct release types, i.e. patient discharges.

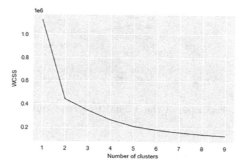

Fig. 3. The Elbow method on our Sepsis analysis.

Table 1. Alpha attribute influence.

Sepsis			MIMIC			PAH		
Feature	ID-R	ID-K	Feature	ID-K	ID-R	Feature	ID-K	ID-R
Age	0.23	0.76	icd_title	0.51	0.15	Time on Ramp	0.10	0.13
Diagnose	0.33	0.30	chiefcomplaint	0.08	0.16	Primary Diagnosis	0.86	0.08
SIRSCritHeartRate	0.03	0.16	acuity	0.44	0.01	Location after Triage	0.44	0.07
SIRSCritTachypnea	0.04	0.16	heartrate	0.07	0.15	Consultation Type	0.03	0.06
Release_type	0.04	0.08	temperature	0.01	0.14	Departure Destination	0.11	0.05

After applying a filter to exclude cases that lacked any release activity, as these cases were deemed incomplete and lacked a definitive end activity, a total of 777 cases remained. We added a new trace attribute to the event log denoting the release type. The event log contains trace attributes such as Case ID, Age, Transition, Organization, Activity Count, Diagnose, and Diagnostic Tests. After removing the non-contributing trace attributes (case identifier, comments and timestamps, see Sect. 4.1), we have 26 trace attributes for our analysis. We removed the cases with missing values for these attributes and this filtered event log has 729 cases.

Assisted Alpha Attribute Selection. We applied the ID-K and ID-R methods to select the alpha attribute from the selected 26 trace attributes and got their respective relevant importance of each feature as output, using the output of the Elbow method in Fig. 3). The relevant importance of the top 5 attributes of both methods is shown in Table 1. It can be observed that ID-K and ID-R both returned Age and Diagnose as the most important features and for our analysis, we have considered both Age and Diagnose as candidate alpha attributes.

Scoping analysis. In the selected candidate alpha attributes, Diagnose has more than 100 distinct values, and we take the top ten most frequent ones choose C,B,E,H,G,D,K,R,Q and S, as a result we have 10 sub-logs; one for each selected diagnosis (the diagnoses are anonymised; knowledge of them is not necessary for P^{2}CM). For Age, we partition the values into 10-year periods, resulting in eight sub-logs and they are 0–20, 20–30, 30–40, 40–50, 50–60, 60–70, 70–80, 80–90.

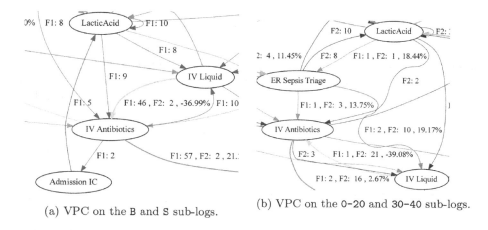

(a) VPC on the B and S sub-logs.

(b) VPC on the 0-20 and 30-40 sub-logs.

Fig. 4. VPC on two Sepsis sub-logs.

Identifying comparable sub-logs. In this step, we apply our new method (see Sect. 4.2) to obtain the log vs log comparison scores. We find the number of optimal cluster is 3 by using the elbow method on the scores. After that, we use k-means clustering to find {C,B,E,H,D,K and, R} in cluster 0, {S} in cluster 1 and {G,Q} in cluster 2.

After applying VPC (Sect. 4.2) on the **age** attribute sub-logs, obtaining the log vs log comparison scores and using the Elbow method (Sect. 3), we get 2 clusters. Then using k-means clustering, we find 0–20, 20–30 in cluster 0 and 30–40, 40–50, 50–60, 60–70, 70–80, 80–90 in cluster 1.

In the next step, we perform pairwise comparison within clusters to find the similarities and between clusters to find key differences. We choose the sub-logs B vs S from **Diagnose** and 0–20 vs 30–40 from **Age** for the in-depth comparison. *In-depth comparison.* Next, we compare the sub-logs using VPC, with a filtering parameter of 40%.

When comparing the sub-logs of B and S (see Fig. 4a), we instantly notice a significant number of bright red edges, which indicates that B has a lot more edges. The edges IV Liquid → IV Antibiotics and LacticAcid Triage → Admission NC (not shown) have relative differences -36.99% and -52.81% which means for sepsis S patients' treatment, this paths play an important role.

Analysing the **Age** attribute of sub-log 0–20 vs 30–40 (see Fig. 4b), we see that around 50% of the edges are only present in the sub-log 30–40. IV Liquid → IV Antibiotics has a percentage difference of -39.08 percent which indicates that for patients' age between 40–50, this path is more important than any other paths. When the sub-logs were compared using VPC, it became clear that the treatment paths for sepsis patients varied significantly and that IV fluids were preferable to IV antibiotics for patients aged 40–50. Overall, P^2CM provided insights into the differences between sub-logs of healthcare pathways with minimal domain expert input.

MIMIC. MIMIC-IV-ED is a database of emergency department (ED) admissions at the Beth Israel Deaconess Medical Center between 2011 and 2019, which contains vital signs, triage information, medication reconciliation, medication administration and discharge diagnoses of around 425 000 ED stays. The ED is a resource limited environment and human care is rationed to provide the best possible patient care [31]. MIMICEL is an event log derived from MIMIC [31].

Data pre-processing. We selected the event attributes `temperature`, `heartrate`, `resprate`, `o2sat`, `sbp`, `dbp`, `pain`, `acuity`, `chiefcomplaint` and `icd_title` and lifted them to the trace level. Other attributes that had a high percentage of null values or were identifiers, timestamps or comments were dropped; 10 trace attributes were used further. Out of the initial 448 972 cases, 436 737 cases remained after filtering traces with missing values. *Assisted Alpha Attribute*

Selection. We utilised both ID-K and ID-R; the top five results are shown in Table 1. The two most important features from ID-K are `icd_title` and `Acuity`, and for ID-R are `icd_title` and `chiefcomplaint`; we selected the common one `icd_title` as the alpha attribute.

Scoping Analysis. Here, we find comparable sub-logs based on the `icd_title` attribute, which has more than a thousand values, and it is not possible to compare these all one-on-one. So we select the top ten most frequent values and generate ten event logs by filtering the event log based on these values. *Identifying*

Comparable Sub-Logs. Then we create ten event logs from the MIMICEL event log based on these ten values. By using our proposed technique for identifying comparable sub-logs, we obtain the log vs log comparison scores. The Elbow method indicates using 3 clusters. We then apply k-means clustering to find `Pneumonia, unspecified organism`, `Altered mental status unspecified`, `Fever, unspecified` in cluster 0, `ALCOHOL ABUSE-UNSPEC`, `Alcohol abuse with intoxication unspecified` in cluster 1 and `Unspecified abdominal pain`, `CHEST PAIN NOS`, `Chest pain unspecified`, `ABDOMINAL PAIN OTHER SPECIED`, `HEADACHE` in cluster 2.

Next, we compare two DFGs - they are `ALCOHOL ABUSE-UNSPEC` vs `Alcohol abuse with intoxication, unspecified` and `HEADACHE` vs `Altered mental status, unspecified`. As `ALCOHOL ABUSE-UNSPEC` and `Alcohol abuse with intoxication, unspecified` are in the same log with very similar diagnosis name, we are interested in understanding the main differences between them. `HEADACHE` and `Altered mental status, unspecified` are in two different clusters, and thus we also like to observe the main differences between them.

In-depth Comparison.
To the selected pairs, we apply the VPC with a filtering parameter of 80%. When we look at `ALCOHOL ABUSE-UNSPEC` vs `Alcohol abuse with intoxication, unspecified` in cluster 1. As the names suggest, there should not be too many differences between them, and the graph validates our intuition. All the edges are green except `Triage in the ED` → `Vital sign check`, -43.29%. This shows that, when there is intoxication involved, more patients are sent for `Vital sign check`.

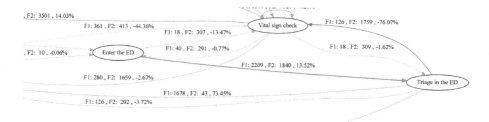

Fig. 5. VPC on `HEADACHE` vs `Altered mental status, unspecified` comparison.

Lastly, `Headache` vs `Altered mental status, unspecified` Fig. 5 shows considerable differences in edges. Specially, `Triage in the ED → Vital sign check`, -76.03% indicates `Vital sign check` is an important step when treating patients with `Altered mental status, unspecified`. Overall, using P²CM we were able to study process-based differences with minimal domain expert input.

5.2 Usefulness

The second evaluation entails an application of P²CM in a case study, performed at the emergency department of the Princess Alexandra Hospital in Brisbane, Australia. The corresponding data set contains ED pathways in 2019–2021. *Data pre-processing.* The data set was converted to XES, after which the activities related to bed management were removed to focus the analysis. Furthermore, cases with data-type mismatches were removed. The remaining log had 2 329 846 events, 134 846 traces and 48 activities.

Assisted Alpha Attribute Selection. In our study on alpha attribute extraction, we utilised our exclusion criteria to select categorical attributes that were likely to be useful alpha attributes. Our methods ID-K and ID-R both on the pre-selected attributes revealed that `Time on Ramp` and `Primary Diagnosis Snomed Code` were the top 2 most important attributes for segregating the traces into sub event-logs. The relative importance of the top 5 attributes of both methods is shown in Table 1. From the alpha attribute selection, we get `Time on Ramp` and `Primary Diagnosis Snomed Code` (Primary Diagnosis) as the alpha attributes.

Scoping Analysis. We binned the `Time on Ramp` in six parts based on their frequency, while attempting to keep the bins balanced in their number of traces. From the `Primary Diagnosis` feature, we chose the top 10 diagnoses based on their frequency, this included `Chest Pain`, `Mental Health Problem`, `Abdominal Pain`, `Viral Illness`, `Syncope`, `Back Pain`, `NSTEMI - Non-ST Segment Elevation MI`, `Headache`, `Cellulitis` and `Alcohol Intoxication`.

Identifying Comparable Sub-logs. The clustering of sub-logs, with 3 clusters identified, was according to expectation: the bins of `Time on Ramp` that were close in value were clustered together. We go through the same process for

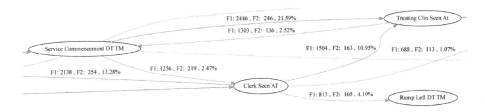

Fig. 6. VPC on `Chest Pain` and `NSTEMI - Non-ST segment elevation MI`.

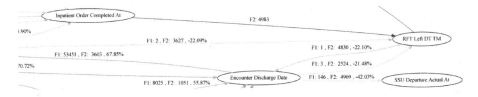

Fig. 7. VPC on time on ramp: (0, 1.0] vs (1.0,105.0] minutes.

`Primary Diagnosis` and obtain 3 clusters. Based on the clustering, the domain expert identified two pairs of potential interest: (1) `mental health problem` vs `viral illness`, as an example of within-cluster differences, and (2) `chest pain` vs `NSTEMI - Non-ST segment elevation MI`, as these are medically closely related, but still showed as being in different clusters. Furthermore, we decided to compare sub-logs based on the `time on ramp` attribute clustering, taking (0,1.0] vs (1.0,105.0] minutes as representatives of two different clusters (3).

In-depth Comparison and Interpretation and Validation. We apply the VPC to these three pairs of sub-logs. For the first pair, we compared `mental health problem` and `viral illness`. Some of the procedural differences highlighted in the visualisation were expected by stakeholders, such as the edge from `Triaged at` to `Treat Nrs`, and `Service Commencement` to `Edip date` as for many mental health problems, neither Emergency Department treatment nor admission occurs. These patients are rapidly transferred to the Emergency Mental Health Unit. Other differences were `Triaged at` to `Clerk seen at`, which stakeholders indicated may be a missing recording step in the process. For the second pair, we compared `Chest Pain` with `NSTEMI - Non-ST segment elevation MI`, shown in Fig. 6. Again, several differences were expected, `Triaged at` → `Clerk Seen`, which indicates that the administrative step in the middle is often skipped for urgent cases. However, the `Service Commencement` to `Treating Clin Seen` edge was a new insight to the expert, while `Service Commencement` to `Edip Date` may again indicate a recording issue. For the third pair, we compared time on ramp being less than one minute (F1) vs 1 to 105 min (F2), shown in Fig. 7. The first observation is that the colours indicate large process-based stochastic differences, as several edges are of teal and yellow colours. These findings suggest both differences that can be explained by differences in the nature of the clinical presentations and their journey through the ED as well as differences related to data recording processes, which may be important in performance reporting.

From our discussion, it is clear that PVC as part of P^2CM, based on the alpha attributes `Primary Diagnosis` and `Time on Ramp`, can be effective in showing the stochastic process-based differences between different patient groups.

5.3   Discussion

P^2CM presented in this paper takes a single event log as input and provides results between cohorts within that event log. P^2CM hence allows comparative analysis of process-based differences (DO1). P^2CM is also a stochastic-aware technique (DO2). The key steps of alpha attribute selection and identifying comparable sub-logs and in-depth comparison, which are stochastic to a larger extent than the original PCM, make P^2CM stochastic-aware. In the future, P^2CM could be extended in the data pre-processing and scoping analysis steps with explicit consideration of the stochastic perspective. Furthermore, P^2CM provides automated techniques that guide users into choosing alpha attributes and comparable sub-logs, and visualises stochastic differences between processes to guide analysts in finding differences or commonalities between cohorts of a process, thus taking a step towards satisfaction of guiding users with automated recommendations (DO3). In the future, it would be interesting to extend automation by refining the comparable sub-logs identification step using heuristics or machine learning to guide analysts further towards potentially notable differences.

The techniques introduced in this paper as part of P^2CM use concepts from existing stochastic-aware techniques, but differ in key ways. In [6], an event log is split along trace attribute values to find their values with the largest influence on stochastic behaviour. Our approach extends it with a full method, non-categorical attributes and a visualisation of the actual differences. Finally, we provide several automated techniques that assist analysts in selecting or creating one or more alpha attributes. Suitable pairs of sub-logs to compare are selected in phase 3, however [5] emphasises the need to use similar processes. We extend the method with a stochastic-aware approach that guides analysts in choosing *similar and dissimilar pairs* of sub-logs for comparison. Furthermore, our approach does not require the discovery of process models to perform this selection, which inherently involves certain well-known trade-offs [32]. The process comparison methods in phase 4 of [5] and literature do not focus on the stochastic perspective. We provide a new process discovery technique/visualisation that highlights differences in stochastic behaviour between two sub-logs.

The experiments can be reproduced using the scripts available at https:// github.com/asadTariq666/BPM-Alpha-Attribute-Selection. The Sepsis data is publicly available, while the MIMIC data is semi-publicly available [31]. For legal/privacy reasons, the data of the Princess Alexandra Hospital cannot be shared.

Several limitations of this work are noted. As the author team applied P^2CM themselves, it was not possible to evaluate the ease of use of P^2CM objectively:

future research is needed to assess this aspect. Second, the experiment covered treatment (Sepsis) and emergency care (MIMIC, PAH), and we see no factors that would prevent applying P^2CM to other areas of healthcare. In order to generalise the application of P^2CM to other domains, it is important that such cases *have* attributes, or, more in general, sub-processes that can sensibly be compared with one another. In case these sub-processes are already known to domain experts, P^2CM might be applicable partially (identifying comparable sub-logs & VPC). Another limitation is that the ordinal encoding used for categorical and boolean attributes may impose an order that can impact k-means clustering. Future research could explore alternative encoding methods like word2vec or one-hot encoding to preserve semantic meaning without introducing implicit order.

6   Conclusion

In health processes, optimisation ideas may be derivable from the comparison of similar but differing processes. In this paper, we applied a design science approach to introduce a method, the Probabilistic Process Comparison Method (P^2CM), to satisfy the design objectives of (i) allowing for comparative analysis of process-based differences in cohorts of a single process, (ii) considering the stochastic perspective of behaviour, and (iii) guiding users with automated recommendations derived from data. We showed that P^2CM adheres to (i) and (iii), while (ii) is satisfied by the combination of the techniques used in P^2CM: in all changed steps, the stochastic perspective is taken into account: most insights obtained were of a stochastic nature, and would have been missed by techniques unaware of the stochastic perspective. Following open-science principles, method and analysis techniques are available to the community and two publicly accessible datasets were used to ensure the reproducibility of our findings.

As further future work, the concepts of process cubes [12] may be applied to expand P^2CM to use the structure between attributes to further reason about (hierarchical) relations between attributes, and guide users towards sub-processes with notable differences.

References

1. Suriadi, S., Mans, R.S., Wynn, M.T., Partington, A., Karnon, J.: Measuring patient flow variations: a cross-organisational process mining approach. In: Ouyang, C., Jung, J.-Y. (eds.) AP-BPM 2014. LNBIP, vol. 181, pp. 43–58. Springer, Cham (2014). https://doi.org/10.1007/978-3-319-08222-6_4
2. Munoz-Gama, J., Martin, N., et al.: Process mining for healthcare: characteristics and challenges. JBI **127**, 103994 (2022)
3. Duma, D., Aringhieri, R.: real-time resource allocation in the emergency department: a case study. Omega **117**, 102844 (2023)

4. Andrews, R., Goel, K., Corry, P., Burdett, R., Wynn, M.T., Callow, D.: Process data analytics for hospital case-mix planning. JBI **129**, 104056 (2022)
5. Syamsiyah, A., et al.: Business process comparison: a methodology and case study. In: Abramowicz, W. (ed.) BIS 2017. LNBIP, vol. 288, pp. 253–267. Springer, Cham (2017). https://doi.org/10.1007/978-3-319-59336-4_18
6. Leemans, S.J.J., Shabaninejad, S., Goel, K., Khosravi, H., Sadiq, S., Wynn, M.T.: Identifying cohorts: recommending drill-downs based on differences in behaviour for process mining. In: Dobbie, G., Frank, U., Kappel, G., Liddle, S.W., Mayr, H.C. (eds.) ER 2020. LNCS, vol. 12400, pp. 92–102. Springer, Cham (2020). https://doi.org/10.1007/978-3-030-62522-1_7
7. Leemans, S.J.J., van der Aalst, W.M.P., Brockhoff, T., Polyvyanyy, A.: Stochastic process mining: Earth movers' stochastic conformance. Inf. Syst. **102**, 101724 (2021)
8. van Eck, M.L., Lu, X., Leemans, S.J.J., van der Aalst, W.M.P.: PM ˆ2 : a process mining project methodology. In: CAiSE, vol. 9097 of LNCS (2015)
9. Buijs, J.C.A.M., Reijers, H.A.: Comparing business process variants using models and event logs. In: BPMDS, vol. 175 of LNBIP, pp. 154–168 (2014)
10. Bolt, A., de Leoni, M., van der Aalst, W.M.P.: A visual approach to spot statistically-significant differences in event logs based on process metrics. In: Nurcan, S., Soffer, P., Bajec, M., Eder, J. (eds.) CAiSE 2016. LNCS, vol. 9694, pp. 151–166. Springer, Cham (2016). https://doi.org/10.1007/978-3-319-39696-5_10
11. Bolt, A., de Leoni, M., van der Aalst, W.M.P.: Process variant comparison: Using event logs to detect differences in behavior and business rules. Inf. Syst. **74**, 53–66 (2018)
12. Bolt, A., van der Aalst, W.M.P.: Multidimensional process mining using process cubes. In: Gaaloul, K., Schmidt, R., Nurcan, S., Guerreiro, S., Ma, Q. (eds.) CAISE 2015. LNBIP, vol. 214, pp. 102–116. Springer, Cham (2015). https://doi.org/10.1007/978-3-319-19237-6_7
13. Hevner, A.R., March, S.T., Park, J., Ram, S.: Design science in information systems research. MISQ **28**(1), 6 (2004)
14. Homayounfar, P.: Process mining challenges in hospital information systems. In: FedCSIS, pp. 1135–1140. IEEE (2012)
15. Erdogan, T.G., Tarhan, A.: Systematic mapping of process mining studies in healthcare. IEEE Access **6**, 24543–24567 (2018)
16. Yang, W., Su, Q.: Process mining for clinical pathway: literature review and future directions. In: ICSSSM, pp. 1–5. IEEE (2014)
17. Partington, A., Wynn, M.T., Suriadi, S., Ouyang, C., Karnon, J.: Process mining for clinical processes: a comparative analysis of four Australian hospitals. ACM Trans. Manag. Inf. Syst. **5**(4), 19:1–19:18 (2015)
18. Caron, F., Vanthienen, J., Vanhaecht, K., Van Limbergen, E., Deweerdt, J., Baesens, B.: A process mining-based investigation of adverse events in care processes. HIMJ **43**(1), 16–25 (2014)
19. Taymouri, F., La Rosa, M., Carmona, J.: Business process variant analysis based on mutual fingerprints of event logs. In: Dustdar, S., Yu, E., Salinesi, C., Rieu, D., Pant, V. (eds.) CAiSE 2020. LNCS, vol. 12127, pp. 299–318. Springer, Cham (2020). https://doi.org/10.1007/978-3-030-49435-3_19
20. Buijs, J.C.A.M., van Dongen, B.F., van der Aalst, W.M.P.: Towards cross-organizational process mining in collections of process models and their executions. In: BPM Workshops, pp. 2–13 (2011)

21. De Smedt, J., Yeshchenko, A., Polyvyanyy, A., De Weerdt, J., Mendling, J.: Process model forecasting using time series analysis of event sequence data. In: Ghose, A., Horkoff, J., Silva Souza, V.E., Parsons, J., Evermann, J. (eds.) ER 2021. LNCS, vol. 13011, pp. 47–61. Springer, Cham (2021). https://doi.org/10.1007/978-3-030-89022-3_5

22. Acampora, G., Vitiello, A., Stefano, B.N.D., van der Aalst, W.M.P., Günther, C.W., Verbeek, E.: IEEE 1849: The XES standard. IEEE Comput. Intell. Mag. **12**(2), 4–8 (2017)

23. Hevner, A., Chatterjee, S., Hevner, A., Chatterjee, S.: Design science research in information systems. In: DRISTP, pp. 9–22 (2010)

24. Peffers, K., Tuunanen, T., Rothenberger, M.A., Chatterjee, S.: A design science research methodology for information systems research. MISQ **24**(3), 45–77 (2007)

25. Venable, J., Pries-Heje, J., Baskerville, R.: A comprehensive framework for evaluation in design science research. In: Peffers, K., Rothenberger, M., Kuechler, B. (eds.) DESRIST 2012. LNCS, vol. 7286, pp. 423–438. Springer, Heidelberg (2012). https://doi.org/10.1007/978-3-642-29863-9_31

26. Marutho, D., Hendra Handaka, S., Wijaya, E., Muljono.: The determination of cluster number at k-mean using elbow method and purity evaluation on headline news. In: ATIC, pp. 533–538 (2018)

27. Ketchen, D.J., Shook, C.L.: The application of cluster analysis in strategic management research: an analysis and critique. SMJ **17**(6), 441–458 (1996)

28. Breiman, L.: Random forests. Mach. Learn. **45**(1), 5–32 (2001)

29. Van Der Aalst, W.M.: A practitioner's guide to process mining: limitations of the directly-follows graph (2019)

30. Reinhart, K., Goolsby, T., et al.: Recognizing sepsis as a global health priority-a who resolution. NEJM **377**(5), 414–417 (2017)

31. Wei, J., He, Z., Ouyang, C., Moreira, C.: Mimicel: mimic-iv event log for emergency department. In: PhysioNet (2022)

32. Buijs, J.C.A.M., van Dongen, B.F., van der Aalst, W.M.P.: On the role of fitness, precision, generalization and simplicity in process discovery. In: Meersman, R. (ed.) OTM 2012. LNCS, vol. 7565, pp. 305–322. Springer, Heidelberg (2012). https://doi.org/10.1007/978-3-642-33606-5_19

On the Cognitive Effects of Abstraction and Fragmentation in Modularized Process Models

Clemens Schreiber[1]([✉]), Amine Abbad-Andaloussi[2], and Barbara Weber[2]

[1] Karlsruhe Institute of Technology, Karlsruhe, Germany
`clemens.schreiber@kit.edu`
[2] University of St. Gallen, St. Gallen, Switzerland
{`amine.abbad-andaloussi,barbara.weber`}`@unisg.ch`

Abstract. Process models support a variety of tasks, which can be organized differently. Notably one can discern local tasks focusing on a single part of a model and global tasks requiring an overview of several parts. These two task types are assumed to affect users' understanding of processes differently especially if the processes are decomposed into many interlinked and self-contained models through modularization. Local tasks can benefit from abstraction as they enable information hiding, while global tasks can be impeded by fragmentation caused by the split attention effect. Following a task-centric approach, we substantiate this hypothesis by investigating the cognitive effects of abstraction and fragmentation in modularization. Therein, we focus particularly on horizontal modularization and study users' cognitive load, comprehension and behavior when solving local and global tasks. Our findings confirm that, compared to abstraction, fragmentation hinders users' comprehension of the model and raises their cognitive load. Additionally, users exhibit different search and integration behaviors when performing local and global tasks. The outcome of this work motivates the shift from artifact-centric to task-centric empirical studies, raises the need for approaches to mitigate the effect of fragmentation and explores different alternatives to achieve this goal.

Keywords: Process model understandability · modularization · cognitive load theory · eye tracking

1 Introduction

Process models serve as a mediator for digital transformation as they support process enactment but also facilitate the communication between stakeholders

First author supported by the Karlsruhe House of Young Scientists Research Travel Grant. Second author supported by the International Postdoctoral Fellowship (IPF) Grant (Number: 1031574) from the University of St. Gallen, Switzerland.

© The Author(s), under exclusive license to Springer Nature Switzerland AG 2023
C. Di Francescomarino et al. (Eds.): BPM 2023, LNCS 14159, pp. 359–376, 2023.
https://doi.org/10.1007/978-3-031-41620-0_21

and promote change in business processes [31]. Different process modeling initiatives exist to support these different activities (e.g., SAP Signavio Process Transformation Suite[1]). Yet, most of the research and tool development in the Business Process Management (BPM) field are focused on the model artifacts themselves (i.e., the graphical representation of processes), with limited interest in the task-perspective [25], i.e., what types of tasks a process model is supposed to support. This perspective is crucial for identifying what kind of information and support are needed in a particular context. In this paper we address this gap by investigating two recurrent tasks encountered in practice: *local* and *global* tasks. The former is typically focused on a local part of the model (e.g., to comprehend or maintain a particular sub-process or model fragment), while the latter covers global aspects requiring an overview of several parts of the model (e.g., cross-cutting concerns spanning over several sub-processes or process fragments). We study these tasks in the light of *horizontal modularization*, i.e., a particular branch of modularization where a system is decomposed into interlinked modules with no strict assumptions on their hierarchy [23,40]. This approach allows modeling flexible and knowledge intensive processes [3,17]. Moreover, it is commonly used to support case-based process execution [3,17] and has shown a positive impact on model comprehension [40].

While one should expect that modularization generally supports model comprehension, existing empirical studies suggest that it does not always make the execution of tasks easier [32,40,43]. Notably, it was hypothesized that modularization enables abstraction as it supports information hiding and pattern recognition, while it causes fragmentation since users are required to split their attention between different fragments to find relevant information [43]. Hence, local tasks, which are constrained to a single module might be supported by abstraction and become easier, whereas global tasks, which refer to several modules might be hindered by fragmentation and become more difficult to perform.

To investigate this hypothesis we conduct an empirical study. Therein, we first confirm the effects of abstraction and fragmentation on the understandability of modularized process models. Secondly, we delve into users' behavior when engaging with modularized models to compare the behavioral traits characterizing the solving of local tasks and global tasks.

Our empirical study is based on eye-tracking. Following the fragment-based modeling approach proposed in [17], we design a horizontally modularized process model and a set of local and global tasks. Then, we test the effects of abstraction and fragmentation on both model comprehension and users' cognitive load. Therein, we use a multi-modal approach covering the typical model comprehension and *subjective* cognitive load measures used in the literature [14,43] – but also advanced *objective* cognitive load measures derived from eye-tracking. Doing so, we provide a multi-perspective empirical account allowing us to draw more robust conclusions on the effects of abstraction and fragmentation. Our findings confirm that the task type is a crucial factor with a significant impact on model comprehension and users' cognitive load. As for the behavioral investigation,

---

[1] See https://www.signavio.com/products/process-transformation-suite/.

we look at eye-tracking measures reflecting the search complexity and cognitive integration effort exhibited by the users when solving local and global tasks. Our analysis shows that global tasks are associated with complex search and higher cognitive integration due to the effect of fragmentation. Based on these insights, we propose several solutions to compensate for this effect and thus support users when solving global tasks on modularized process models.

Overall, this work provides an empirical affirmation for the cognitive effects associated with the solving of global and local tasks. Moreover, our research model demonstrates how to investigate users' comprehension, cognitive load and behavior while reading process models using a wide array of multi-modal measures. From a more applied perspective, the outcome of this work raises the need for incorporating the task type in model comprehension frameworks, developing novel mechanisms to support global tasks and providing new foundations for managerial decisions on task distribution. In the remainder, Sect. 2 provides the background and related work. Section 3 describes our empirical study design. Sections 4 and 5 present and discuss the findings of our study respectively. Section 6 concludes the paper.

2   Background and Related Work

This section describes the theoretical underpinnings related to this work from the process modeling (cf. Sections 2.1-2.3) and cognitive (cf. Section 2.4) perspectives.

2.1   Modularization in Process Modeling

Modularization denotes the process of decomposing a system into interlinked modules with self-contained properties [29]. Authors in the process modeling literature [23,40] have discerned three types of modularization: *vertical, horizontal* and *orthogonal*. Vertical modularization decomposes the process into sub-processes using a hierarchical structure [23,40]. Horizontal modularization, in turn, aims at partitioning the process into several interconnected fragments with no strict assumptions on their hierarchy [23,40]. Lastly, orthogonal modularization divides the process along cross-cutting concerns (i.e., privacy, security) spanning over several elements of the model [23,40].

Since horizontal modularization allows separating process modules regardless of their hierarchical relationship and cross-cutting concerns, this approach appears to be more general and thus well suited to investigate the effects of abstraction and fragmentation on users' comprehension and cognitive load. While a general positive impact of horizontal modularization can be assumed based on [40], we investigate how horizontal modularization affects comprehension tasks focused on single modules (local) versus comprehension tasks focused on multiple modules (global). We use the fragment-based approach [17] (cf., Sect. 2.2) as a representative for this type of modularization.

2.2 Fragment-based Process Modeling Approach

The fragment-based modeling approach [17] is based on Business Process Modeling and Notation (BPMN). It aims at supporting the case-based modeling paradigm [3], which focuses on the dynamical adjustments of processes during execution and emphasizes the role of data as a driving force. At design time, process modules are modeled as separate process fragments, interlinked with data objects. As the linkage between the fragments does not necessarily imply any hierarchical relationship among them, the fragment-based modeling approach can be seen as an instance of horizontal modularization (cf. Section 2.1). At run time, the fragments can be dynamically combined in any required order. The data objects can be defined as input conditions for the activities in the fragments and can thereby pose restrictions on the fragments. A data object belongs to a specific data class and is assigned to a state, which is mutable based on the execution of an activity. The set and order of possible states are depicted by a labeled transition system called lifecycle.

Figure 1 provides an example of the fragment-based modeling approach, as it was used in our experiment. The two fragments "F1_Offload_Container" and "F2_Analyse_Shake_Event" are connected based on the data object "Sensor". As soon as the activity "Activate accelerometer sensor" in the first fragment is executed the state of the sensor changes to "idle" and the second fragment can be executed. When the event "Shake event" occurs the activity "Check intensity of the shake" needs to be executed and the state of the sensor changes to "signalling shake". It is important to notice that the second fragment only depends on a single data-object, i.e., the sensor. This means that as soon as the sensor reaches the state "idle" the fragment can be executed multiple times, independently from any other fragment. The sensor lifecycle "LC1_Sensor" additionally shows that the state of the data-object can alternate between "idle" and "signalling shake".

Similar to the fragment-based approach, other techniques have been proposed in the literature. Therein, events were instead used to link the process fragments (cf. overview in [22]) or other modeling languages such as (colored) Petri nets were extended to achieve the same outcome [13]. We decided to focus on the fragment-based approach [17] since it is based on BPMN (i.e., the de-facto modeling language in the community) and provides in particular a clear execution semantic, based on data input and output conditions for the activities. In this way, the modeling approach provides a higher expressiveness and can depict real-world behavior more accurately.

2.3 The Effects of Abstraction and Fragmentation in Modularized Process Models

The use of modularization to support the comprehension of business processes has been subject to many empirical investigations (overview in [43]). Yet, based on Zugal's Literature review [43], there is no consensus on whether modularization supports or hinders model comprehension as existing empirical studies yielded inconsistent and therefore inconclusive results. The opposing effects of

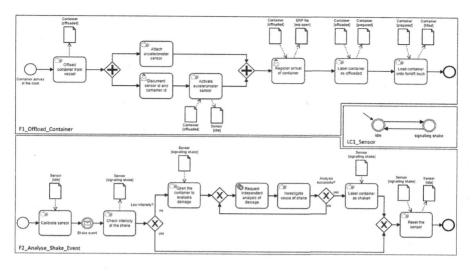

Fig. 1. Example of a horizontally modularized process model following the fragment-based approach. A better resolution is available at github.com/../figures [1].

abstraction and *fragmentation* have been raised in the context of vertical modularization to explain these inconclusive results [32,36,43]. Similarly, they are likely to occur in other modularization types (e.g., horizontal modularization), due to the spatial separation of modules.

Abstraction emerges from the ability of modularization to divide a large process into composable units (or fragments) and thus allowing users to focus their attention on the task-relevant fragments only while hiding the irrelevant ones. Following the insights in [30], users are presumably better at performing comprehension tasks when their attention is focused on the task-relevant parts in the model. Additionally, the recognition of patterns becomes much easier within the individual fragments compared to the overall process model, especially if their size remains within manageable limits. As pointed out in [27] large models are difficult to understand and thus should be decomposed into smaller fragments. Hence, with abstraction, modularization is expected to support model comprehension and lower users' cognitive load. Conversely, fragmentation is caused by the need to repeatedly switch attention between the different fragments, which is likely to cause the *split attention effect* [35,43]. This effect occurs when information is distributed across several locations, requiring the reader to repeatedly switch their attention between these locations, which can be distracting and more cognitively demanding. This effect is likely to impede model comprehension and raise users' cognitive load. Additionally, compared to abstraction, fragmentation is likely to affect users' behavior differently. Indeed, the attention switching and the underlying attention split effect would make the search for relevant information more complex and would raise the need to constantly integrate the information coming from the separated locations (i.e., fragments).

The effects of abstraction and fragmentation can be captured using *local* and *global* tasks. Therein, the local tasks requiring information within a single fragment can benefit from abstraction, whereas the global tasks, in which the relevant information is distributed over several fragments would be impeded by the effect of fragmentation. Accordingly, users are expected to get more challenged when solving global tasks compared to local ones. This hypothesis takes root in the cognitive fit theory [37], positing that the fit between the task at hand and the problem representation (i.e., the artifact presented to the user e.g., a modularized process model) affects the problem-solving procedure [12,37]. Since the fit is lower when solving global tasks (compared to the local ones), the problem-solving procedure would be more complex. Such complexity is likely to affect the model comprehension as well as users' cognitive load and behavior.

Our work extends existing research on the effects of abstraction and fragmentation [32,43] by (1) switching the focus from vertical to horizontal modularization, (2) using a wider array of measures capturing both users' comprehension and cognitive load to confirm these effects from different perspectives and (3) delving into users' behaviour to investigate the traits characterizing their engagement with local and global tasks. The measures used to confirm the effects of abstraction and fragmentation and to study the underlying users' behavior are introduced in the following section (cf. Sect 2.4).

2.4   Investigating Users' Comprehension, Cognitive Load and Behavior

The following paragraphs introduce the concepts and measures serving to (1) investigate the effects of abstraction and fragmentation on *model comprehension* and *cognitive load* as well as (2) assess users' behavior in terms of *search* and *cognitive integration* when solving local and global tasks.

Comprehension. Comprehension denotes the process of interpreting an artifact (e.g., text or process model) and building a mental representation of it [21]. In the literature, comprehension measures have been used as key dependent variables to study the effect of different factors (e.g., presentation medium, model complexity, the naming of activities, task type) on users' understanding of the model (cf. overview in [14]). Therein, low *comprehension accuracy* (i.e., the correctness of provided answers) and low *comprehension efficiency* (i.e., time needed to solve a task) indicate challenges in understanding the model at hand.

Cognitive Load. The cognitive load theory posits that when humans approach the full capacity of their working memory, they become more challenged with the task at hand, which affects their performance negatively and raises the risk of committing errors or taking wrong decisions [10,28]. In the context of process model comprehension, this would translate to difficulties understanding the given model and a higher risk of misinterpreting the encoded behavior.

Cognitive load can be captured *subjectively*, through the means of introspection, or *objectively* through changes in humans' behavior and cognitive states [10, 18, 39]. The subjective assessment of *perceived difficulty* is a measure that has been shown to differentiate different levels of cognitive load [34]. It is typically administrated as a questionnaire based on Likert scales following each task [34] (e.g., a *5-points Likert scale:* [0: "very easy", 4: "very difficult"]).

Eye-tracking provides substantial means to study humans' cognitive load more objectively [10, 18, 39]. Additionally, unlike the perceived difficulty measure which can be administrated only after a task is completed, the measures derived from eye-tracking can be collected continuously along the entire task, providing in turn, measurements with finer granularity in time and potentially more precise insights. In eye-tracking, the eye-mind hypothesis associates *fixations*[2] on the stimulus with mental processing under the assumption that what is fixated by the eyes is being simultaneously processed by the mind [20]. Accordingly, the *average fixation duration* has been used in several studies as an indicator of cognitive load such that the longer the fixation duration, the higher the user's cognitive load will be (e.g., [5, 38]). Another prominent cognitive load measure is the *low/high index of pupil activity* (LHIPA) [11]. This measure is based on the pupillary features of the eyes. It separates the low pupil oscillations frequencies from the higher ones to derive a measure that correlates negatively with cognitive load, i.e., low value refers to high cognitive load [11].

Beside these cognitive load measures, the aforementioned comprehension measures have been also proposed to capture cognitive load in the literature [10].

Search. Search denotes the process of identifying task-relevant information among a set of irrelevant information [42, 43]. In the process modeling literature, investigations of users' search behavior yield, for instance, interesting insights about their engagement with different hybrid process model representations [4], extending imperative process models with linked rules [38] or enriching declarative process models with textual annotations and guided simulations [6]. The duration of fixations has been used, in this vein, to discern different types of mental processes [15, 38]. Notably, in [38], *fixations with duration* ≤ 250 ms were associated with *information screening* behavior. Accordingly, increased numbers of this type of fixations imply a more pronounced search behavior. Another prominent search measure is the *scan-path precision* used in [30]. This measure calculates the ratio of fixations on the task-relevant parts of the model to the total number of fixations. Herein, a high ratio would indicate a simpler search, more focused on the task-relevant parts, whereas a low ratio would imply a more difficult search impeded by too many distracting fixations on non-relevant parts.

Cognitive Integration. Cognitive integration denotes the consolidation of information from different locations [8]. In the process modeling literature, visual

---

[2] A *fixation* refers to the time interval when the eye remains relatively stationary at a specific position within a visual field [18].

associations derived from eye-tracking have been suggested to indicate cognitive integration [8]. In turn, the *average returns to relevant-regions* (or look-backs as defined in [18]) measures how often on average a relevant region (i.e., an area on the screen, which provides relevant information for a comprehension task) has been repeatedly fixated. It can therefore provide insights on visual associations made when recalling and integrating different pieces of task-relevant information located within a single fragment or spread over several fragments.

Overall, through our different measures of model comprehension and cognitive load, we aim at confirming that compared to the local tasks (benefiting from abstraction), the global tasks (impeded by fragmentation) yield lower comprehension accuracy and efficiency as well as higher perceived difficulty, higher average fixation duration, and lower LHIPA. Also, using our behavioral measures we aim at showing, that global tasks are associated with complex search and higher demand for cognitive integration than local tasks. Based the aferomentioned theoretical underpinnings, our measures are particularly suited to investigate the difference in local and global task solving. Their selection is also supported by existing cognitive theories and their successful application in previous experiments.

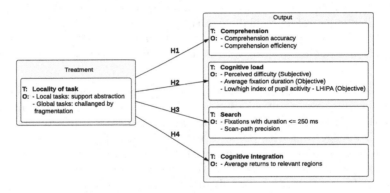

Fig. 2. Research model for hypothesis testing. T - Theoretical construct, O - Operationalization of construct.

3   Research Method

To investigate the effects of abstraction and fragmentation, we have conducted an eye-tracking study following the guidelines in [2]. Sections 3.1 –3.3 provide an overview of the study design, the study execution, and the data analysis procedures respectively.

3.1   Study Design

Research Model. Based on the theoretical background introduced in Sect. 2, our empirical study aims at investigating the impact of abstraction and fragmentation on users' comprehension and cognitive load as well as users' behavior in terms of search and cognitive integration. Our research model is depicted in Fig. 2. As *treatment*, we manipulate the *locality of task* factor, which we separate into two-factor levels with *local tasks* addressing control-flow aspects located within a single process fragment, and *global tasks* addressing control-flow aspects located within two process fragments. As explained in Sect. 2.3, local tasks benefit from the effect of abstraction, whereas global tasks are challenged by the effect of fragmentation. The task locality factor is expected to impact *comprehension, cognitive load, search* and *cognitive integration*, which denote the theoretical constructs at the *output* side. As shown in Fig. 2, we operationalize these theoretical constructs using the measures introduced in Sect. 2.4. Following our research model, we formulate the following hypotheses:

- H_1: Global tasks yield lower model comprehension than local tasks in horizontal modularization.
- H_2: Global tasks yield a higher cognitive load than local tasks in horizontal modularization.
- H_3: Global tasks yield more complex search than local tasks in horizontal modularization.
- H_4: Global tasks yield a higher cognitive integration than local tasks in horizontal modularization.

Material. The material used for the experiment comprises a process model, based on the fragment-based modeling approach proposed in [17], and a set of local and global tasks capturing abstraction and fragmentation effects respectively. The model depicts a logistic process and consists of six different fragments. Each fragment refers to a particular part of the onward carriage process of a container after it arrived at a port, which involves the unloading, scanning, temporally storing and loading of a container, as well as its registration and monitoring. The fragments are connected based on three different data objects, i.e., container, enterprise resource planning file, and sensor. The possible state changes of each data object are additionally visualized by three respective life cycles. To avoid confounding factors [41] the design of the layout and the modeling constructs are carefully aligned with the guidelines proposed by [7], which will be addressed in the following.

All process fragments are of similar size (six to eight unique activities) and complexity. Each fragment contains two to four gateways, which are either exclusive (XOR) or parallel (AND). Since domain knowledge can also be a confounding factor [41], the activities are labeled in layman's terms and specific details of a real logistic process are left out.

In total, the material includes eight different tasks, which have to be solved by every participant. Each task consists of a question, which is posed as a true or false statement. The statements address one of the following control-flow patterns: sequential, concurrent, exclusive, or repetition. Each statement refers to two particular activities, which can either be found in a single fragment (local task) or in two separate fragments (global task). For each control-flow pattern,

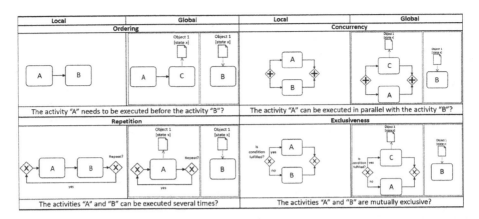

Fig. 3. Difference between local (left) and global (right) tasks and how they are modeled by the fragment-based modeling approach. A better resolution is available at github.com/../figures [1].

a participant has to solve one local and on global task. Figure 3 provides an overview of all eight tasks, the control-flow patterns used in the statements and how they are modeled for the local and the global task type respectively. Additionally, Fig. 3 shows that the used tasks were formulated following a pre-defined template. For each control-flow pattern, the associated local and global tasks were phrased in the same way. Also, each pair of local and global tasks have the same answer (either both true or both false). With this formulation, we ensure that no task-related confounding factors are affecting our design since the only difference between the local and global tasks is the locality of the activities, i.e., either within the same fragment or spanning over two fragments. The experiment material is available online at github.com/../material [1].

Figure 4 depicts a screenshot of the user interface used during the experiment to navigate through the tasks. The process fragments and life cycles are provided in different files, which can be accessed through the file explorer shown on the left side of the screen. Note that only one single file can be viewed at a time. The question is shown at the very top of the screen.

Participants. The eye-tracking study covers 46 participants. The experiment sessions were conducted at four different locations: at the University of St. Gallen (22), at the Karlsruhe Institute of Technology (17), at the research institute Forschungszentrum für Informatik FZI in Karlsruhe (4), and at the IT consulting company Promatis in Karlsruhe (3), within a time frame of 4 weeks. The participants were between 20–50 years old with a majority in the age range of 20–30 years (63%). For the professional background, there were three main groups: researchers at PhD level or higher (22), students (17), and participants with an industry or IT-admin background (7). The participants had different levels of expertise in BPMN. To ensure that they could take part in the exper-

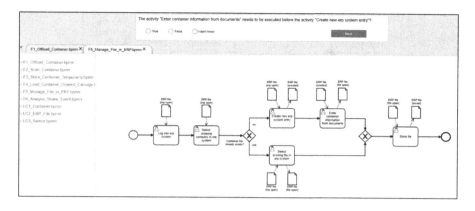

Fig. 4. User interface with an example of an experiment question.

iment, we have conducted a uniform familiarization and tested their knowledge with a quiz as it will be explained in Sect. 3.2. Additionally, to avoid confounding factors related to differences in the participants' expertise, we have used a within-subject design. Therein, every participant was exposed to each factor level multiple times, i.e., four local and four global tasks. Although the participants' expertise might still impact the outcome of each task, based on the within-subject design both treatments are equally impacted and it is therefore possible to draw valid conclusions regarding their distinctiveness. A complete overview of the participants' demographic data is available at github.com/../demograhics [1].

3.2   Experiment Procedure

Each participant was invited to an individual experiment session, which lasted about one hour. The experiment was conducted in a controlled lab environment. The procedure for each session is depicted in Fig. 5. The session starts with familiarizing the participant with BPMN and the fragment-based modeling approach. The familiarization is followed by a short quiz, consisting of two comprehension tasks, to test whether the fragment-based modeling approach is understood. During the quiz, the participant could ask additional questions and receive feedback regarding the solution. The actual experiment starts with two test tasks, which allow the participants to familiarize themselves with the user interface and to get to know the procedure. These tasks are not used for the data analysis later on. This approach is chosen to avoid any bias in the data due to the unfamiliarity of the participants with the experiment setup. To compensate for a potential learning effect, the tasks are randomized and presented in different orders to the participants. For each task, the participant has to answer a comprehension question and two additional follow-up questions. In the first follow-up question, the participant is asked to verbally justify how the task was solved. In the second follow-up question, the participant is asked to provide a self-assessment of how difficult the task was.

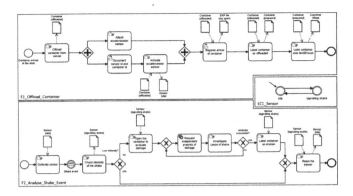

Fig. 5. The different steps during each experiment session.

3.3 Data Analysis

Based on the collected data, the measures introduced in Sect. 2.4 are calculated. From 46 participants, we obtained 184 data points per factor level (i.e., local tasks or global tasks). To avoid interdependence between the data points coming from each individual, a mean value was calculated for the four tasks capturing each factor level for each participant. This resulted in 46 paired data samples. Due to technical issues with the eye-tracker, the data was further reduced to 44 paired data samples for the fixation-based measures and to 43 paired data samples for the pupil-based ones. In the first two cases the brightness in the room was too high, such that the eye-tracker could not detect the participants' gazes correctly. In the third case, the participant kept moving their head which affected the measurement of their pupil dilation. The remaining data was used to compute the descriptive and inferential statistics in order to investigate our hypotheses (cf. Table 1). We used the non-parametric Wilcoxon Signed-Rank test (single-tailed) for the inferential statistics since it is adequate for comparing paired data samples and does not require the data to be normally distributed. Our analysis approach is documented through the used Python notebooks available online at github.com/../analysis [1].

4 Findings

This section presents the findings of our empirical study. Section 4.1 reports the results of the comprehension and cognitive load analysis, while Sect. 4.2 reports the results of the behavioral analysis.

4.1 Comprehension and Cognitive Load Analysis

Based on the results in Table 1, *comprehension accuracy* (in the range [0:incorrect, 1:correct]) was significantly (p< .001) lower for global tasks (*M=0.739*) than for local tasks (*M=0.978*). Likewise, *comprehension efficiency* (measured in seconds) was significantly (p< .001) lower for global tasks (*M=117.593 s*)

Table 1. Comprehension, cognitive load and behavioral analyses. N: number of observations (cf. Section 3.3), M: calculated mean. A *p-value* < .05 means that the pairwise comparison results are significant.

Hypothesis/Construct	Measure	Descriptive			Inferential
		N	Local M	Global M	*p*-value
Comprehension and Cognitive Load Analysis					
H_1/Comprehension	Comprehension accuracy	46	0.978	0.739	< .001
	Comprehension efficiency	46	54.097	117.593	< .001
H_2/Cognitive load	Perceived difficulty	46	0.62	1.891	< .001
	Average fixation duration	44	194.323	200.966	< .001
	LHIPA	43	1.185	0.815	< .001
Behavioral Analysis					
H_3/Search	Fixations with duration $\leq$ 250 ms	44	142.153	298.883	< .001
	Scan path precision	44	0.111	0.077	< .001
H_4/Cognitive Integration	Av. returns to relevant regions	44	4.108	7.844	< .001

compared to the local ones *(M=54.097 s)*. Based on the background presented in Sect. 2.4, these two measures indicate that global tasks yield reduced model comprehension compared to local tasks, which confirms *Hypothesis H_1*.

With regards to cognitive load, the subjective assessment of *perceived difficulty* (measured in the range [0: "very easy", 4: "very difficult"]) was significantly (p< .001) higher for global tasks *(M=1.891)* than for local tasks *(M=0.62)*. Similarly, the *average fixation duration* was significantly (p< .001) higher for global tasks *(M=200.966)* compared to local tasks *(M=194.323)*. Finally, the *LHIPA* score was significantly (p< .001) lower for global *(M=0.815)* than for local *(M=1.185)* tasks. The trends of these measures (cf. Section 2.4) show that when solving global tasks users experience increased cognitive load, compared to local tasks, which confirms *Hypothesis H_2*.

4.2   Behavioral Analysis

Based on Table 1, the *fixations with duration* $\leq$ 250ms are significantly higher (p< .001) for global *(M=298.883)* than for local tasks *(M=142.153)*. This indicates that users' exhibited more information-screening behavior when performing global tasks (cf. Section 2.4). Additionally, the *scan path precision* is significantly lower (p< .001) for global tasks *(M=0.077)* than for local tasks *(M=0.111)*. This indicates that for global tasks users had to look and scan through a higher number of irrelevant process model elements before finding the relevant ones (cf. Section 2.4). These insights hint towards a complex search when dealing with global tasks, which confirm *Hypothesis H_3*.

With regards to cognitive integration, the *average returns to relevant regions* is significantly higher (p< .001) for global tasks *(M=7.844)* compared to the local ones *(M=4.108)*, showing, in turn, more demand for recalling and integrating information, spread over different fragments, when engaging with global tasks (cf. Section 2.4). This finding confirms *Hypothesis H_4*.

5   Discussion

Based on the inferential statistics reported in Sect. 4, we can confirm that the task type has an impact on users' comprehension and cognitive load when engaging with horizontally modularized process models. This validates the existing suppositions in the literature [32,43], i.e., while local tasks require less cognitive effort, due to the effect of abstraction, global tasks require more cognitive effort, due to the effect of fragmentation. Additionally, our behavioral analysis suggests that the cognitive processes underlying global and local tasks differ, as global tasks are associated with a more complex search and require more cognitive integration than local tasks. Since horizontal modularization allows for a very flexible decomposition of process models into modules, regardless of their hierarchical relationships and cross-cutting concerns, it can be assumed that the gained insights generally apply to other modularization approaches as well, which are more restrictive in terms of decomposition.

Following these insights, we suggest that researchers should consider the task type as an integral part of their research models aiming at explaining the comprehension of process models. This proposal finds also support in other literature, demonstrating the impacts of different task types on the understandability of software artifacts [12,33]. The framework of Mandelburger and Mendling [25] can be seen as a prominent example in this direction as it motivates the shift from artifact-centric to task-centric research. This way, the comprehension of process models would not only be justified by the model proprieties (e.g., size, density, connectivity) but also through the characteristics of the task at hand.

From a practitioner's perspective, it should be recognized that not only design choices impact the comprehensibility of a modular system, but also how specific tasks are defined and distributed among development teams. Tasks should, if possible, be confined to specific self-contained modules instead of multiple modules at once. This is indeed in line with existing software engineering practice and the general idea of microservice architecture [19]. At the same time it should be clear that in general, tasks cannot always be confined to a single module. This holds, in particular, true for cross-cutting concerns, such as security and privacy. In this case, the restriction of tasks to specific services or modules hinders the comprehensibility of the overall system and therefore the ability to solve cross-cutting concerns [19]. To overcome this issue our findings motivate the need to develop support for search and cognitive integration to compensate for some of the negative effects of fragmentation.

One possibility to support the search as well as the cognitive integration of information is to use modeling tools interlinking modules to improve their navigation. Yet, empirical evidence is missing, on how such a navigation affects model comprehension [24]. Another approach to avoid the searching and integration issues is to use simulators to evaluate the executions allowed under different scenarios as proposed in [26]. Therein, for certain global tasks users do not necessarily need to search and cognitively integrate information across several modules, since the simulation can provide information regarding global process behavior. To further facilitate the integration of information, one could also provide a general overview or mapping of the modules, e.g., in the form of a model landscape. This approach has also been applied for fragment-based process models and empirical tests show indeed a benefit in terms of process model comprehensibility [16]. Additionally, to particularly facilitate the search for information, one could use natural language processing based on the lexical similarity between task description and the textual corpus of the software artifact. A similar approach has been proposed for feature location in software systems [9].

All in all, future BPM tools should incorporate the characteristics of the user task to augment their support. Based on our empirical findings, for global tasks, this support should particularly focus on reducing the search and cognitive integration effort underlying this type of tasks.

Threats to Validity. Our work can be subject to some validity threats. Although the experiment sessions are conducted in a unified manner based on an experiment protocol and in a controlled environment, one cannot eliminate the possibility of the existence of confounding factors, which might have influenced the outcome of the study and therefore its internal validity. However, several measures are taken to mitigate this risk. To avoid learning effects, the comprehension tasks were presented in a random order to each participant. These tasks also refer to different process fragments or different combinations of fragments. Moreover, each of the used six fragments is only relevant for two out of eight tasks. The relatively high number of subjects for an eye-tracking study (46) further supports the internal validity of the experiment. Each participant received a uniform introduction to BPMN and to the fragment-based modeling approach, thus ensuring that all participants have a similar basic knowledge to participate in the study. A further concern is whether the identified findings could be applied to other process modeling languages. This risk to external validity is again mitigated to some extent by the design of our comprehension tasks addressing different control-flow patterns. Hence, the results are less dependent on particular model structures. Also, it is worthwhile to mention that these control-flow patterns are typically found in process models [43].

6   Conclusion and Future Work

The conducted eye-tracking study confirms that, in horizontal modularization, global tasks impede the model comprehension and raise users' cognitive load in comparison to local tasks. This can be explained by the two opposing effects of abstraction and fragmentation, caused by modularization. Additionally, based on our behavior analysis, it becomes, in particular, apparent that global tasks require complex search and a higher cognitive integration effort, due to fragmentation. To overcome this issue we propose several possible tool augmentations and motivate the need to attend more attention to the task perspective.

In the future, we are planning to investigate tasks addressing the data-flow perspective in fragmented process models. Additionally, we will extend our behavioral investigation to identify other traits characterizing the solving of the covered tasks. Moreover, we can move our analysis of cognitive load to a more fine-grained level to obtain more detailed insights into the particular modeling constructs challenging users in different tasks. Lastly, we can also investigate the effects of the proposed augmented tooling on solving global tasks.

Data Availability Statement. As pointed out in the relevant sections of the paper, the experiment and analysis material are available on GitHub [1].

References

1. Data repository. https://github.com/aminobest/BPM2023TaskType
2. Emperical standard guidelines for experiments. https://github.com/acmsigsoft/EmpiricalStandards/blob/master/docs/Experiments.md
3. Van der Aalst, W.M., Weske, M., Grünbauer, D.: Case handling: a new paradigm for business process support. Data Knowl. Eng. **53**(2), 129–162 (2005)
4. Abbad-Andaloussi, A., Burattin, A., Slaats, T., Kindler, E., Weber, B.: On the declarative paradigm in hybrid business process representations: a conceptual framework and a systematic literature study. Inf. Syst. **91**, 101505 (2020)
5. Abbad-Andaloussi, A., Sorg, T., Weber, B.: Estimating developers' cognitive load at a fine-grained level using eye-tracking measures. In: Proceedings of IEEE/ACM International Conference on Program Comprehension, pp. 111–121 (2022)
6. Abbad-Andaloussi, A., Zerbato, F., Burattin, A., Slaats, T., Hildebrandt, T.T., Weber, B.: Exploring how users engage with hybrid process artifacts based on declarative process models: a behavioral analysis based on eye-tracking and think-aloud. Softw. Syst. Model. **20**, 1437–1464 (2021)
7. Becker, J., Rosemann, M., von Uthmann, C.: Guidelines of business process modeling. In: van der Aalst, W., Desel, J., Oberweis, A. (eds.) Business Process Management. LNCS, vol. 1806, pp. 30–49. Springer, Heidelberg (2000). https://doi.org/10.1007/3-540-45594-9_3
8. Bera, P., Soffer, P., Parsons, J.: Using eye tracking to expose cognitive processes in understanding conceptual models. MIS Q. **43**(4), 1105–1126 (2019)
9. Biggers, L.R., Bocovich, C., Capshaw, R., Eddy, B.P., Etzkorn, L.H., Kraft, N.A.: Configuring latent dirichlet allocation based feature location. Empirical Softw. Eng. **19**, 465–500 (2014)
10. Chen, F., Zhou, J., Wang, Y., Yu, K., Arshad, S.Z., Khawaji, A., Conway, D.: Robust Multimodal Cognitive Load Measurement. HIS, Springer, Cham (2016). https://doi.org/10.1007/978-3-319-31700-7
11. Duchowski, A.T., Krejtz, K., Gehrer, N.A., Bafna, T., Bækgaard, P.: The low/high index of pupillary activity. In: CHI Conference on Human Factors in Computing Systems, pp. 1–12 (2020)
12. Dunn, C., Grabski, S.: An investigation of localization as an element of cognitive fit in accounting model representations. Decis. Sci. **32**(1), 55–94 (2001)
13. Fettke, P., Reisig, W.: Modelling service-oriented systems and cloud services with HERAKLIT. In: Zirpins, C., et al. (eds.) ESOCC 2020. CCIS, vol. 1360, pp. 77–89. Springer, Cham (2021). https://doi.org/10.1007/978-3-030-71906-7_7
14. Figl, K.: Comprehension of procedural visual business process models: a literature review. Bus. Inf. Syst. Eng. **59**, 41–67 (2017)

15. Glöckner, A., Herbold, A.K.: Information processing in decisions under risk: evidence for compensatory strategies based on automatic processes. MPI collective goods preprint (2008/42) (2008)
16. Gonzalez-Lopez, F., Pufahl, L., Munoz-Gama, J., Herskovic, V., Sepúlveda, M.: Case model landscapes: toward an improved representation of knowledge-intensive processes using the fCM-language. Softw. Syst. Model. **20**(5), 1353–1377 (2021). https://doi.org/10.1007/s10270-021-00885-y
17. Hewelt, M., Weske, M.: A hybrid approach for flexible case modeling and execution. In: La Rosa, M., Loos, P., Pastor, O. (eds.) BPM 2016. LNBIP, vol. 260, pp. 38–54. Springer, Cham (2016). https://doi.org/10.1007/978-3-319-45468-9_3
18. Holmqvist, K., Nyström, M., Andersson, R., Dewhurst, R., Jarodzka, H., van de Weijer, J.: Eye Tracking: a Comprehensive Guide to Methods and Measures. OUP Oxford (2011)
19. Jamshidi, P., Pahl, C., Mendonça, N.C., Lewis, J., Tilkov, S.: Microservices: the journey so far and challenges ahead. IEEE Softw. **35**(3), 24–35 (2018)
20. Just, M.A., Carpenter, P.A.: A theory of reading: from eye fixations to comprehension. Psychol. Rev. **87**(4), 329 (1980)
21. Keselman, A., Slaughter, L., Patel, V.L.: Toward a framework for understanding lay public's comprehension of disaster and bioterrorism information. J. Biomed. Inf. **38**(4), 331–344 (2005)
22. Krumeich, J., Weis, B., Werth, D., Loos, P.: Event-driven business process management: where are we now?: A comprehensive synthesis and analysis of literature. Bus. Process Manage. J. **20**(4), 615–633 (2014)
23. La Rosa, M., Wohed, P., Mendling, J., Ter Hofstede, A.H., Reijers, H.A., van der Aalst, W.M.: Managing process model complexity via abstract syntax modifications. IEEE Trans. Ind. Inf. **7**(4), 614–629 (2011)
24. Lübke, D., Ahrens, M.: Towards an experiment for analyzing subprocess navigation in bpmn tooling (2022)
25. Mandelburger, M.M., Mendling, J.: Cognitive diagram understanding and task performance in systems analysis and design. MIS Q. **45**(4), 2101–2157 (2021)
26. Marquard, M., Shahzad, M., Slaats, T.: Web-based modelling and collaborative simulation of declarative processes. In: Motahari-Nezhad, H.R., Recker, J., Weidlich, M. (eds.) BPM 2015. LNCS, vol. 9253, pp. 209–225. Springer, Cham (2015). https://doi.org/10.1007/978-3-319-23063-4_15
27. Mendling, J., Reijers, H.A., van der Aalst, W.M.: Seven process modeling guidelines (7pmg). Inf. Softw. Technol. **52**(2), 127–136 (2010)
28. Paas, F., Tuovinen, J.E., Tabbers, H., Van Gerven, P.W.: Cognitive load measurement as a means to advance cognitive load theory. Educational psychologist (2003)
29. Parnas, D.L.: On the criteria to be used in decomposing systems into modules. Commun. ACM **15**(12), 1053–1058 (1972)
30. Petrusel, R., Mendling, J., Reijers, H.A.: How visual cognition influences process model comprehension. Decis. Support Syst. **96**, 1–16 (2017)
31. Recker, J.C., Lukyanenko, R., Jabbari Sabegh, M., Samuel, B., Castellanos, A.: From representation to mediation: a new agenda for conceptual modeling research in a digital world. MIS Q. Manage. Inf. Syst. **45**(1), 269–300 (2021)
32. Reijers, H., Mendling, J.: Modularity in process models: review and effects. In: Dumas, M., Reichert, M., Shan, M.-C. (eds.) BPM 2008. LNCS, vol. 5240, pp. 20–35. Springer, Heidelberg (2008). https://doi.org/10.1007/978-3-540-85758-7_5

33. Ritchi, H., Jans, M.J., Mendling, J., Reijers, H.: The influence of business process representation on performance of different task types. J. Inf. Syst. **34**(1), 167–194 (2019)
34. Sweller, J., Ayres, P., Slava, K.: Cognitive load theory. Springer, New York (2011). https://doi.org/10.1007/978-1-4419-8126-4
35. Sweller, J., Chandler, P.: Why some material is difficult to learn. Cogn. Instruction **12**(3), 185–233 (1994)
36. Turetken, O., Rompen, T., Vanderfeesten, I., Dikici, A., van Moll, J.: The effect of modularity representation and presentation medium on the understandability of business process models in BPMN. In: La Rosa, M., Loos, P., Pastor, O. (eds.) BPM 2016. LNCS, vol. 9850, pp. 289–307. Springer, Cham (2016). https://doi.org/10.1007/978-3-319-45348-4_17
37. Vessey, I.: Cognitive fit: a theory-based analysis of the graphs versus tables literature. Decis. Sci. **22**(2), 219–240 (1991)
38. Wang, W., Chen, T., Indulska, M., Sadiq, S., Weber, B.: Business process and rule integration approaches-an empirical analysis of model understanding. Inf. Syst. **104**, 101901 (2022)
39. Weber, B., Fischer, T., Riedl, R.: Brain and autonomic nervous system activity measurement in software engineering: a systematic literature review. J. Syst. Softw. **178**, 110946 (2021)
40. Winter, M., Pryss, R., Probst, T., Baß, J., Reichert, M.: Measuring the cognitive complexity in the comprehension of modular process models. IEEE Trans. Cogn. Dev. Syst. **14**(1), 164–180 (2022)
41. Wohlin, C., Runeson, P., Höst, M., Ohlsson, M.C., Regnell, B., Wesslén, A.: Experimentation in Software Engineering. Springer, Berlin (2012). https://doi.org/10.1007/978-3-642-29044-2
42. Wolfe, J.M.: Guided search 2.0 a revised model of visual search. Psychon. Bull. Rev. **1**, 202–238 (1994)
43. Zugal, S.: Applying cognitive psychology for improving the creation, understanding and maintenance of business process models. Ph.D. thesis, University of Innsbruck (2013)

Not Here, But There: Human Resource Allocation Patterns

Kanika Goel[1] , Tobias Fehrer[2,3] , Maximilian Röglinger[2,3] ,
and Moe T. Wynn[1]

[1] Queensland University of Technology, Brisbane, Australia
{k.goel,m.wynn}@qut.edu.au
[2] University of Bayreuth, Bayreuth, Germany
{tobias.fehrer,maximilian.roeglinger}@fim-rc.de
[3] Branch Business & Information Systems Engineering of the Fraunhofer FIT,
Bayreuth, Germany

Abstract. The digital age entails challenges that pressure organisations to redesign their business processes for improved performance. A significant aspect of this effort is the appropriate assignment of human resources – or people – to tasks. Despite the importance, there is a lack of structured guidance on allocating people to tasks considering various performance considerations such as time, cost, quality and flexibility. This paper presents 15 human resource allocation patterns organised into five categories: resource capability, utilisation, reorganisation, productivity and collaboration. The pattern collection is designed to offer guidance on diverse strategies for human resource allocation, focusing on process redesign for performance improvement from a resource perspective. The research was conducted in a two-phase approach. In the first phase, a literature review was conducted to identify existing resource patterns and practices, synthesised into an initial catalogue of human resource allocation patterns. In the second phase, this catalogue was evaluated through expert interviews with ten practitioners. The patterns provide a repository of knowledge guiding academics and practitioners on different ways a person can be assigned to a task for improved process efficiency. These patterns form a strong foundation for future research in the area of human-centred business process redesign.

Keywords: Process improvement · Allocation patterns · Human resource · BPM

1 Introduction

Organisations have faced and continue to face pressure to adapt to trends in an increasingly hyper-connected and fast-moving world [6]. A changing workforce, the shortage of skilled workers and increasing automation significantly impact the work organisation. The need for efficient and effective work allocation to people is more important than ever. A McKinsey survey [5] finds 83% of

© The Author(s), under exclusive license to Springer Nature Switzerland AG 2023
C. Di Francescomarino et al. (Eds.): BPM 2023, LNCS 14159, pp. 377–394, 2023.
https://doi.org/10.1007/978-3-031-41620-0_22

executives believe that strategic use of people can drive value and help deliver higher returns to stakeholders. Similarly, research shows that assigning the right people to the right task can improve process efficiency, income, and customer satisfaction [3,10,33].

Business Process Management (BPM) is a boundary-spanning field of research that supports organisations in operating effectively and efficiently through the discovery, execution, analysis, and redesign of business processes [11]. Business Process Redesign (BPR) (not to be confused with business process re-engineering) is the key activity in BPM for identifying process improvement opportunities. Processes can be analysed and improved from the control-flow, data, and resource perspective [11]. The resource perspective concerns all actors (human and non-human) involved in a business process [11]. The allocation of people has been acknowledged as a significant problem in BPM research [3,4,23]. It is, therefore, imperative that human resource considerations are taken into account during BPR.

Recognising the importance of resource allocation on the performance and efficiency of a process [4], an increasing amount of research is being conducted in this field. One of the earliest works by Russell et al. is on workflow resource patterns [27] that capture how resources (both human and non-human) can be used and represented in workflow systems. Other studies have focused on developing distinct resource allocation methods by leveraging techniques such as machine learning (e.g., [28,31]) and Markov models (e.g., [17]). While prior literature on potential resource allocation mechanisms does exist, it is fragmented. There is a lack of understanding and support for considering human resource allocation in BPR. Given people's continuing important role in executing a process, this study investigates how 'best' to assign people to process tasks. Therefore, we formulate our research question as follows: *How can human resources be best allocated to tasks when redesigning business processes?*

We propose a human resource allocation patterns (HRAPs) catalogue to address this research question. Patterns capture best practices and assumptions from the field and have been used as a common tool to provide methodological support for BPR [15,34]. Patterns have the advantage of being specific to a problem in hand but also generic enough to address future problems [15]. They provide a simple entry point into BPR as they are easy to understand, describe clear redesign ideas, and, hence, can foster creative thinking about BPR options in generic and specific situations [18]. We employ a two-phase research approach to synthesise these patterns. First, we collate and analyse existing literature to develop HRAPs. Next, we evaluate and refine them through expert interviews with BPM professionals with BPR experience. We also evaluate the patterns' impact on process efficiency in terms of time, cost, quality, and flexibility [11]. These 15 patterns provide scholars and practitioners with a repository of knowledge about the different aspects to consider when allocating human resources to tasks in a process.

The remainder of the paper is structured as follows. Section 2 provides background on redesign patterns and related research in people assignment in BPM.

Section 3 introduces the study's methodological setup while Sect. 4 presents the refined set of HRAPs. Section 5 provides an overview of expert interviews conducted for refinement and evaluation purposes and discusses changes to the refined set of patterns. In Sect. 6, we conclude by highlighting research contributions, limitations and avenues for further research.

2 Related Work

BPM is the continuous cycle of discovering, executing, analysing, redesigning, and monitoring business processes [11]. BPR is considered the most value-adding phase in the BPM lifecycle [11,25,34]. Consequently, the BPM discipline seeks principles, methods, techniques, and tools to support this phase [30,34,37].

People are essential elements of enacting business processes, and their roles, capabilities, and interactions play a crucial role in successfully redesigning processes. Much more than machines, people have heterogeneous strengths and preferences that, if correctly leveraged, can positively impact process performance. A key challenge in BPR is identifying the right people with the right skills and knowledge to perform tasks [4]. Best practices in the form of BPR patterns provide the inspiration for creating improvement options. A pattern can be described as a general solution to solve specific problems by reusing experience instead of rediscovering it [15, p. 2]. BPR patterns (commonly also referred to as redesign heuristics) "suggest particular changes to an existing process to influence its operation in certain ways" ([18, p. 193]) and are often documented as text, sometimes augmented by illustrations and implementation examples. BPR patterns are rarely invented but rather observed from the field and compiled for utilisation [34]. Several BPR pattern collections have been published in the past, focusing on general business processes [25,29] as well as specific domains such as customer-centric service design [14], or healthcare [20].

Specific resource-related BPR patterns also exist. For example, Russell et al. [27] present both human and non-human resource patterns in the context of workflow management systems. Reijers and Mansar [25] present a subclass of BPR patterns, which focus on resource view. Zellner [34] presents a set of IT resource-related patterns. However, none of the papers purely focus on human resources. Just as business process performance is a multidimensional construct, redesigns affect the process in a multidimensional way. Selecting performance objectives from high-level yet sometimes opposing dimensions like quality, cost, time, flexibility, or customer/employee satisfaction guides redesign initiatives [11,14].

Several works have been conducted on resource allocation in business processes, recognising the significance of resources in business process improvement. In a literature review, Zhao and Zhao [36] highlighted cross-organisational mining, dynamic resource allocation, and role complexity analysis as significant research directions. Arias et al. [4] proposed a recommendation framework for resource allocation based on process mining at the sub-process level, which considers six dimensions, frequency, performance, quality, cost, expertise, and workload. Zhao et al. [35] presented a resource allocation model that minimises execution time while meeting cost and resource availability constraints for higher process performance. Erasmus et al. [12] used the Fleishman taxonomy for human abilities to specify the abilities required for different tasks and allocate appropriate resources to tasks during run-time. Pika et al. [24] extracted resource behaviour from event logs and quantified outcomes using regression analysis. Kim et al. [16] proposed a framework to derive features that capture people's experience from event log data and used random forest and XG Boost for classification. Yang et al. [32] developed an approach to learn execution contexts from event logs by defining categorisation rules based on case, activity, and time and using a decision tree to learn these rules. In the area of human resource allocation problems, a growing body of research focuses on optimising resource allocation to minimise costs or maximise profits [7]. Various techniques, including assignment methods proposed by Pentico et al. [22], have been developed to address this optimisation problem. These techniques, such as heuristics and branch and bound algorithms, aim to find exact solutions [7]. However, it is important to note that this literature primarily emphasises the development of optimisation techniques rather than providing comprehensive guidance on resource allocation for process improvement.

The literature on human resource allocation in business processes lacks a comprehensive method for improving business process performance in terms of time, cost, quality, and flexibility. Current approaches vary and address specific aspects, such as profiling resources or distributing resources during run-time, but a holistic approach is missing. This study aims to bridge this gap by offering guidance on resource-based personnel allocation in process redesign.

3   Research Design

Our research method included two phases. In the first phase, we developed the initial catalogue of 15 HRAPs through a narrative literature review. In the second phase, we interviewed ten process improvement experts to evaluate and refine the initial catalogue.

3.1   Development Phase

Since there are already some documented examples of BPR patterns in the literature, we started our collection with a narrative literature review adopting guidelines proposed by vom Brocke et al. [8]. We reviewed the proceedings of the BPM conferences of the last 15 years and the databases ABI INFORM, EBSCOhost, ScienceDirect, JSTOR, and Google Scholar. The keywords used were "process redesign patterns", "process redesign heuristics", "process re-design patterns", "process re-design heuristics", "process improvement patterns", "process improvement heuristics", "resource redesign patterns", "resource redesign heuristics", "resource re-design patterns", "resource re-design heuristics", ("resource allocation" AND "patterns" OR "heuristics"), "resource optimisation patterns", and "resource optimization patterns". These keywords were looked for in titles and keywords of papers, resulting in a pool of 848 papers. The titles and abstracts of these papers were reviewed to find papers that address patterns in a BPR context. In addition, a forward and backward search of seminal papers was conducted to collect representative literature [8], resulting in 39 articles.

63 patterns were identified from the remaining papers. These patterns were assessed in more detail using four criteria consistent with the study's motivation. They are: (1) the BPR pattern relates to human resources, i.e., people, (2) the BPR pattern and its use can be evaluated during the (re)design of the process, (3) the BPR pattern should lead to a quantifiable increase in the performance of the process in one or more dimensions of process performance, and (4) the implementation of the BPR pattern should not entail major changes for the organisation (e.g., value proposition, structural changes to the organisation). A systematic application of these criteria yielded a total of 24 patterns. For example, the system-determined work queue content pattern [27], which allows a system to determine the order in which work should be done, can not be used for human resources and hence was not considered. However, the history-based distribution pattern [27] was kept as it advocates the use of performance history before assigning a task, and this concept can be used for human resources. The resulting 24 patterns were: order assignment, flexible assignment, split responsibility, customer teams, numerical involvement, case manager, extra resources, specialist/generalist, empower (all found in [25]), direct distribution, role-based distribution, separation of duties, case handling, authorization, retain familiar, capability-based distribution, history-based distribution, organisational distribution, random allocation, round-robin allocation, shortest queue, delegation, escalation, and additional resources (all found in [27]).

These 24 patterns were reviewed further to see if they could be grouped into a higher-order pattern and augmented with other literature on human resource allocation patterns. This resulted in the initial catalogue of 15 patterns. For example, we merged the specialist pattern [25] and the capability-based distribution patterns [27] into one pattern referred to as expertise-based task assignment. Similarly, shortest queue, round-robin allocation, and random allocation were merged into workload-based task assignment.

3.2   Refinement Phase

Next, interviews were conducted with ten process improvement experts. Purposive sampling [21] was used to recruit experts with several years of experience and varying areas of practice (see Table 1) to enable a more general assessment based on a well-founded judgement. The interviews were semi-structured in nature with the overall objective to gauge the perceived usefulness, pervasiveness, and comprehensiveness of the HRAPs and to estimate their potential effect on process performance.

Each interview was administered by one researcher. In accordance with [19], the interviews started by providing the participants with an overview of the study and the approach to derive these patterns. Each pattern was individually assessed, starting with explaining its basic idea and a simple application example. We presented each pattern individually to the experts together with a brief description. After clarifying possible comprehension questions about the pattern at hand, the experts assessed its usefulness and pervasiveness. Ratings were based on a four-point scale, allowing for simple dichotomies. Next, using qualitative statements, the experts evaluated the potential impact of applying the patterns on performance dimensions (cost, time, quality, and flexibility). For example, they provided assessments such as "very high impact on cost" or "the pattern may slightly improve throughput time." The experts had the opportunity to readjust their previous ratings as the interview progressed to subsequent items. The researcher also recorded any additional expert comments for further analysis. The interview data were analysed using a hybrid approach [13]. Expert ratings were aggregated per pattern to calculate the mean and standard deviation in a deductive approach. Performance assessments and relevant indications from existing literature were consolidated for each pattern. In an inductive approach, the comments were analysed to gather new insights. We report on the survey in Sect. 5.

4   Human Resource Allocation Patterns (HRAP)

We present 15 HRAPs derived from the literature synthesis and expert interviews. The patterns have been grouped across five categories[1] (see Fig. 1):

(1) Capability: human resource allocation based on their ability to complete tasks;
(2) Utilisation: human resource allocation considering their effective use;
(3) Re-organisation: human resource allocation to individual tasks or a process based on strategic and tactical decisions taken by the organisation;
(4) Productivity: human resource allocation based on their efficiency evidenced through historical data; and
(5) Collaboration: human resource allocation based on their interactions with other members within the organisation.

Each HRAP is presented in a structured format consisting of a *title*; a *description*, explaining the core idea for process redesign; an *example*, indicating where/how the process pattern is applied in use; the estimated *impact*, indicating the change in business process performance with regard to the dimensions of time, cost, quality, and flexibility; and *implementation*, describing the key aspects required to implement the pattern in a business process. The patterns are applicable at both process and instance (single and multiple) levels. The following subsections present the HRAPs per category.

---

[1] Four category names have been adapted from Pika et al. [24], while the reorganisation category has been introduced based on the literature synthesis. [24] did not propose any patterns; we have only adapted the category names (e.g., capability).

Capability	Utilisation	Reorganisation	Productivity	Collaboration
o Expertise-based task assignment o Role-based task assignment o Preference-based task assignment	o Workload-based task assignment o Constraint-based task assignment	o Increased resource assignment o Empower resources o Task delegation o Case manager assignment	o Performance-based task assignment o Experience-based task assignment o Quality-based task assignment o Cost-based task assignment	o Teamwork-based assignment o Department-based assignment

Fig. 1. Overview of human resource allocation patterns in five categories.

4.1   Human Resource Capability

The patterns in this category allocate people based on their expertise, role, and preference.

Expertise-Based Task Assignment. Assign tasks based on the unique skills of the person(s) involved. Expertise is defined as the specialised skills possessed by a resource.

Example: To examine a patient who reported a cardiac issue, a doctor with expertise in cardiology is assigned the task.

Impact: This BPR pattern will lead to a high-quality outcome in less time, albeit at a higher cost.

Implementation: The pattern requires prior knowledge of the skills needed to complete the task, including specialised skills of personnel. Skills needed for the task are matched with the skills of available resources, and a person with the required expertise is then assigned to the task.

Role-Based Task Assignment. Assign tasks based on the role of the resource involved in a process.

Example: Upon arriving at the emergency department, the patient is directed to see any available nurse (based on role) for an initial assessment.

Impact: This pattern will increase flexibility and reduce time.

Implementation: The pattern requires knowledge of the roles to complete tasks within the process. When assigning tasks, the required role for the task is matched with the corresponding role within the organisation and then assigned to personnel within that role.

Preference-Based Task Assignment. Assign a task to a person based on the person's preference. Preference is defined as a set of activities that the person has an inclination towards and hence may have been executed more often along with higher execution efficiency by a person.

Example: In Scrum, preference-based task assignment allows team members to choose tasks based on their preferences and past efficiency. For instance, if a team member prefers doing programming tasks, they can choose to work on those tasks.

Impact: This pattern will result in high-quality outcomes in less time and cost.

Implementation: The pattern requires knowledge of the requirements of the process and record keeping of individual preferences of team members. Team members' preferences can be inferred from their past involvement in certain tasks or directly provided by them. During the assignment, tasks and preferences of team members are matched and assigned to those with the corresponding preferences.

4.2   Human Resource Utilisation

This category assigns tasks to people based on workload and execution constraints within a process.

Workload-Based Task Assignment. Assign tasks to people based on their workload, which refers to the number of task instances started but not yet completed by a person.

Example: A team of customer service representatives receives a high volume of incoming calls. The workload-based task assignment pattern assigns new incoming calls to the representative with the least amount of calls in progress.

Impact: The pattern might lead to better resource utilisation, task completion rates, and reduced idle time.

Implementation: The implementation of workload-based task assignments requires the tracking of tasks and their completion status for team members, along with a system to assign tasks based on each person's workload.

Constraint-Based Task Assignment. Assign tasks to resources based on constraints associated with the execution of tasks within the business process.

Example: An organisation's travel process includes approving travel requests and reconciling financial reports. During resource assignment, the resource assigned to these two tasks should be different to ensure that the separation of duties constraint adheres.

Impact: This pattern will result in an outcome of high quality and ensures compliance. However, flexibility, creativity and throughput time could suffer from strict adherence to constraints.

Implementation: The implementation involves identifying the constraints associated with the execution of tasks, defining the rules for assigning tasks based on those constraints, and incorporating those rules into the task assignment process.

4.3 Human Resource Re-Organisation

This category involves allocating people based on the need for resource expansion, task delegation, empowering people to make decisions, and assigning an accountable person to the process.

Increased Resource Assignment. Allocate more people to a task in a process for strategic reasons, such as speeding up the process or reacting to trends.

Example: When launching a new product, increase the number of customer service agents available to handle customer inquiries and support requests to ensure timely and efficient service and enhance the customer experience.

Impact: The pattern will result in a high-quality outcome in less time, albeit at a higher cost.

Implementation: For increased resource assignment, details related to expertise, preference, role, workload, productivity, the collaboration of resources, and constraints in process execution need to be known. Based on the needs of the process, additional resources will be matched and allocated to the process.

Empower Resources. Grant decision-making authority to people rather than seeking approval from a supervisor.

Example: To streamline the insurance claim process, resources are empowered to make decisions in place of middle management, reducing approval bottlenecks.

Impact: This pattern will result in less time and cost.

Implementation: To empower resources, gather information about their capability, productivity, collaboration, and utilisation. Based on this information, identify resources capable of making decisions and provide them with the necessary authority. Communicate the reasons behind the decision to empower resources and the expectations and constraints associated with the decision-making authority.

Task Delegation. Task delegation is when a person who was originally assigned a task passes it on to another person based on their position in the organisational hierarchy.

Example: The customer complaint manager delegates two complaints to one team member and three to another.

Impact: This pattern will result in an outcome in less time but may decrease the quality.

Implementation: Prior knowledge related to the privileges of people within the organisation is required. People can then be granted the authority to delegate tasks to team members based on their position. Ensure transparency to those with delegation powers on the capabilities and utilisation of delegates for good delegation decisions to be made.

Case Manager Assignment. Assign a person as a manager for a case who is responsible and accountable for all decisions taken during the process execution.

Example: An organisation assigns a new resource as a recruitment process manager responsible for decision-making related to process flow, resource allocation, data handling, and other relevant decisions.

Impact: This pattern will result in high-quality outcomes in less time and cost. Ensures clear accountability and responsibility for the process, while it can create a bottleneck if the case manager is not readily available.

Implementation: Involves identifying the process to be managed, selecting a suitable resource as the case manager, providing the necessary authority and resources to carry out the responsibilities, and ensuring effective communication and collaboration among the resources involved.

4.4   Human Resource Productivity

The patterns in this category allocate people based on past performance, experience, quality, and cost.

Performance-Based Task Assignment. Assign tasks based on a person's past performance, measured by execution time and successful outcomes.

Example: An organisation assigns a complex data analysis task to a resource who has previously demonstrated expertise in that area and consistently produced accurate and timely results.

Impact: This pattern will result in a high-quality outcome in less time and cost.

Implementation: Performance metrics must be defined and measured (in particular time taken to complete activities) for different tasks. Based on these metrics, persons can be assigned tasks that they have performed well in the past.

Experience-Based Task Assignment. Assign a task to a person based on their experience, measured by the number of times they have executed a work item, been involved in a case, and interacted with others.

Example: An organisation assigns a senior manager role in data and analytics to a person with at least 10 project involvements, five project leadership experiences, and team management experience of at least 20 people.

Impact: This pattern will result in high-quality outcomes in less time.

Implementation: Experience metrics for people must be available. At the time of execution, the experience required for the task will be matched with existing data and to select an appropriate person.

Quality-Based Task Assignment. Assign a task based on prior internal or external customer feedback or quality metrics.

Example: An organisation evaluates the performance of its customer service representatives based on customer feedback and assigns high-performing representatives to handle complex customer complaints.

Impact: This pattern will result in high-quality outcomes.

Implementation: For quality-based task assignments, details related to the customer evaluation feedback for resources need to be known. At the time of allocation, people with the best quality feedback will be chosen for the task.

Cost-Based Task Assignment. Assign tasks based on the cost of a resource, defined as the person's per-hour cost when executing a task.

Example: An organisation assigns a data entry task to persons that charge a lower hourly rate for the same quality of work.

Impact: This pattern will result in lower cost. *Implementation:* For cost-based

task assignments, data related to the cost per hour of employees within an organisation must be known. A resource with an appropriate cost will be assigned for a particular task.

4.5   Human Resource Collaboration

This category of patterns allocates people based on interactions within a team and with different functional units within an organisation.

Teamwork-Based Assignment. Assign a task to a person based on their experience working with other resources, which is measured by factors such as the time taken for handovers, number of interactions, and diversity of experience with different people.

Example: The review risk task in a loan application process requires two resources to work together. The task is hence allocated to resources A and B as they have evidenced working well together in the past.

Impact: This pattern will result in a high-quality outcome in less time and cost.

Implementation: For a teamwork-based assignment, a prior understanding of the interaction of a resource with other resources needs to be known. Based on that understanding, appropriate resources will be allocated.

Department-Based Assignment. Assign tasks to people based on their inter-actions with other departments to involve multiple departments or limit involvement.

Example: The review loan application task has been assigned to a resource from the finance and human resources departments, as they have shared responsibility.

Impact: This pattern will result in a high-quality outcome in less time and cost.

Implementation: For department-based assignments, prior information related to different departments, people in those departments, their skills, and the time involved in handovers may be required. Based on this information and the objective of the process, the appropriate resources would be allocated to the tasks of the process.

5   Evaluation

Ten BPR experts were interviewed to share their opinions. Table 1 provides an overview of the interviewees. Table 2 presents the results obtained from expert feedback on the perceived usefulness and pervasiveness of the HRAPs. Similarly, expert comments related to certain patterns were analysed and three main types have been identified: (a) Comments reinforcing the value of a pattern and its rank among others, which served as a consistency check for the quantified ratings of perceived usefulness and pervasiveness. (b) Comments seeking clarification of a pattern's description, which were utilised to improve the wording and naming of the patterns. (c) Comments sharing practical examples or further experiences related to pattern usage, which were incorporated into the pattern descriptions in Sect. 4. The results show that all patterns are perceived to be at least useful: expertise-based task assignment, case manager assignment, and increased resource assignment were considered very useful. In a similar vein, the perceived pervasiveness of the patterns was also agreed by experts, with expertise-based task assignment, workload-based task assignment, role-based task assignment, and cost-based task assignment being used often. For example, "we often use expertise-based assignment. It is very useful" (INT8).

The discussions, especially on the patterns with high standard deviations, yielded interesting results. Preference-based task assignment often scored lower where experts used the context of high-volume processes and higher where the experts demonstrated affinity to knowledge-intensive processes. Experts overall emphasise the value of the selected prioritisation in case handling for the patterns in the productivity category, especially experience-based task assignment and quality-based task assignment. Yet, on performance-based task assignment, three experts noted that it could lead to disorder, resource overload and imbalanced workloads. Although "cost-based task assignment is highly prevalent" (INT1), four experts commented that while the pattern may result in lower cost, it may compromise the quality of outcome if the people do not have adequate skills to perform the task (e.g., "Cost-based task assignment is used often but we need to ensure that quality and time involved remain good" (INT3)) This indicates that

Table 1. Overview of interview participants.

INT	Role	Exp.	Industry	Employees
1	Head of Organisation and BPM	12 yrs.	Real Estate	500
2	Head of Business Process Excellence	20 yrs.	Construction Industry	32,000
3	Business Analyst	6 yrs.	Academic Consulting	250
4	Business Performance Improvement Consultant	4 yrs.	BPM Consulting	100
5	Business Analyst	5 yrs.	Academic Consulting	250
6	Technical Business Analyst	4 yrs.	IT services & consulting	13000
7	General Manager	20 yrs.	Public Services Research and Innovation	12
8	Head of Business Services	20 yrs.	Insurance	159,000
9	Senior Business Analyst	5 yrs.	Finance and Insurance	13000
10	Senior Business Analyst	6 yrs.	Banking	700

Table 2. Evaluation results across the HRAPs.

Category	Pattern	p. usefulness[1]		p. pervasiveness[2]	
		Mean	Std.Dev	Mean	Std.Dev
Capability	Expertise-based task assignment	2.6	0.49	2.7	0.46
	Role-based task assignment	2.4	0.66	2.7	0.64
	Preference-based task assignment	2.1	0.83	2.0	0.63
Utilisation	Workload-based task assignment	2.5	0.50	2.7	0.46
	Constraint-based task assignment	2.5	0.50	2.5	0.50
Re-organisation	Increased resource assignment	2.6	0.66	2.3	1.00
	Empower resources	2.5	0.50	2.1	0.70
	Task delegation	2.5	0.67	2.6	0.66
	Case manager assignment	2.8	0.40	2.6	0.49
Productivity	Performance-based task assignment	2.1	0.83	1.7	0.46
	Experience-based task assignment	2.1	0.70	2.3	0.64
	Quality-based task assignment	2.5	0.67	2.1	0.83
	Cost-based task assignment	2.5	0.50	2.7	0.46
Collaboration	Teamwork-based assignment	2.4	0.66	2.3	0.64
	Department-based assignment	2.1	0.70	2.5	0.50

[1] Responses from each interviewee for perceived usefulness were mapped as follows: not useful (0), somewhat useful (1), useful (2), and very useful (3).
[2] Responses from each interviewee for perceived pervasiveness were mapped as follows: not at all (0), rarely (1), sometimes (2), and often (3).

in some situations, patterns might need to be used in pairs to allocate resources optimally: one could use both the empower resources and experiencebased task assignment patterns together to identify and empower experienced persons to handle activity without the involvement of a supervisor. Another interesting revelation from interviews was how patterns were viewed based on the nature of the processes within an organisation. In large-scale processes designed to roll out in several markets or regions, role-based task assignment was deemed more relevant, while it was considered less relevant for smaller-scale processes. This conveys that factors such as process standardisation, knowledge and creativity involved in processes, and others may make certain HRAPs more relevant for a process. Furthermore, through inductive analysis, new insights were gained from further statements. Besides overall statements like "this collection of patterns is useful"(INT1), themes were commented on that we already outline in the paper: The call for process improvement methods that consider the human with its distinct features and needs. "BPM has an operations view, not a people development view"(INT2). "Mapping HR positions to roles is challenging in practice"(INT3). In addition, new people-specific considerations were raised in the interviews, which have less impact on the classic process performance dimensions but may have longer-term implications for the organisation. Two experts commented on the need to assign tasks such that they contribute to building the capabilities of an employee, e.g., "The notion of knowledge sharing/development should be integrated into human-centric BPM"(INT1) and "The long-term perspective of employee development is usually not considered as part of BPM"(INT2). Another expert suggested that avoiding monotonous work over long periods of time is necessary for people to maintain their psychological well-being. "Themes like personal traits, mental well-being, the intrinsic joy of work, and personal development are worth considering"(INT3). These comments communicate the significance of considering up-skilling and well-being when allocating tasks to people. Organisations are facing a paradigm shift in the management of employees from simple recruiting and terminating to managing the well-being and retention of employees [2]. Up-skilling employees are recognised as a major factor for retention [26] as is meaningful work [9]. Being able to do diverse tasks and feeling a sense of purpose is also associated with employee mental health and well-being [1]. Henceforth, The interviews suggest a new human resource allocation theme that should be further explored in future research: employee well-being. Moreover, the gravity of considering various factors that may affect the efficiency of people and their ultimate impact on process efficiency is also indicated. Henceforth, identifying the key indicators of resource efficiency is another area of future investigation.

6   Conclusion and Future Work

While digital processes continuously evolve for efficiency and effectiveness, the significance of humans remains undeniable as a crucial factor for exceptional customer experiences. Therefore, it is imperative to reimagine resource allocation considerations such that people can be assigned tasks in a value-creating manner. In this paper, we present 15 human resource allocation patterns that provide guidance on how to allocate people best. The patterns were derived from the synthesis of literature followed by expert evaluations. Our evaluation conveys the perceived usefulness and pervasiveness of these HRAPs and their impact on process performance. The patterns provide a repository of knowledge on the different aspects to consider when allocating people. They also provide guidance on the data that needs to be maintained to make these allocation decisions at design time. Thus, this BPR pattern catalogue can serve as a common language or a reference point for academics and practitioners.

We acknowledge the limitations associated with this work. The patterns were deduced from a narrative literature synthesis within the BPM discipline. Unlike systematic literature reviews, narrative reviews are known to carry risks of bias and subjectivity, lack replicability, and may feature inconsistent quality appraisal. Nevertheless, our review was designed to encompass a wide breadth of literature, all critically examined with the explicit objective of unearthing patterns related to human-centric process improvement. The findings are not presented as a narrative interpretation of the literature, but instead, we have elected to list the patterns. This list represents our efforts to cluster and merge patterns, thus minimising subjective interpretation and enhancing the clarity and applicability of the results. To mitigate bias, we evaluate the patterns with process improvement experts. It should be highlighted that even though our pattern collection is derived from a comprehensive review of literature and expert discussions, it isn't deemed to be all-encompassing. Certain constraints tied to the expert survey, such as the utilization of a purposive sample, could potentially skew the results. There may also be bias emanating from the researcher-led interview approach. Furthermore, gauging the value of patterns via single-rated items may not fully capture the intricacy of usefulness. To counter these shortcomings, we have incorporated relevant expert comments to supply further contextual information.

The theme "mental health and well-being" did emerge from expert interviews, and further work is required to investigate the significance of this category and any new HRAPs that may arise. We also recognise that the design and level of detail of the fields collected may not be fully comprehensive. For example, the authors are interested in exploring suitable techniques to automate the recognition and application of the patterns in further research. To this end, research should also investigate how preferences and past experiences can be processed and presented for decision making. Beyond that, the work presented in this manuscript opens various avenues for future research. Future researchers might conduct surveys to evaluate the completeness of the patterns in the field and both collect and conduct case studies for pattern application in the field. In addition, techniques can be developed to execute the rules described in the patterns, which can be integrated into tools such as workflow management systems or business process management systems. Finally, research is also advocated to identify and harness the key factors that lead to value-creating human work in processes.

References

1. Achor, S., Kellerman, G.R., Reece, A., Robichaux, A.: America's loneliest workers, according to research. HBR **19**, 13–28 (2018)
2. Agarwal, V., Mathiyazhagan, K., Malhotra, S., Saikouk, T.: Analysis of challenges in sustainable human resource management due to disruptions by industry 4.0: an emerging economy perspective. IJMr **43**(2), 513–541 (2022). https://doi.org/10.1108/IJM-03-2021-0192
3. Arias, M., Munoz-Gama, J., Sepúlveda, M., Miranda, J.C.: Human resource allocation or recommendation based on multi-factor criteria in on-demand and batch scenarios. EJIE **12**(3), 364–404 (2018). https://doi.org/10.1504/EJIE.2018.092009
4. Arias, M., Rojas, E., Munoz-Gama, J., Sepúlveda, M.: A framework for recommending resource allocation based on process mining. In: Reichert, M., Reijers, H.A. (eds.) BPM 2015. LNBIP, vol. 256, pp. 458–470. Springer, Cham (2016). https://doi.org/10.1007/978-3-319-42887-1_37
5. Atsmon, Y.: How nimble resource allocation can double your company's value. McKinsey Special Collection Resource Allocation, pp. 1–3 (2016). https://www.mckinsey.com/capabilities/strategy-and-corporate-finance/our-insights/how-nimble-resource-allocation-can-double-your-companys-value
6. Beverungen, D., et al.: Seven paradoxes of business process management in a hyperconnected world. Bus. Inf. Syst. Eng. **63**(2), 145–156 (2020). https://doi.org/10.1007/s12599-020-00646-z
7. Bouajaja, S., Dridi, N.: A survey on human resource allocation problem and its applications. Oper. Res. Int. J. **17**(2), 339–369 (2016). https://doi.org/10.1007/s12351-016-0247-8
8. vom Brocke, J., Simons, A., Riemer, K., Niehaves, B., Plattfaut, R., Cleven, A.: Standing on the Shoulders of giants: challenges and recommendations of literature search in information systems research. CAIS **37** (2015). https://doi.org/10.17705/1cais.03709
9. Chitra, S., Vanitha, A.: Effectiveness of employee upskilling program: a study on private insurance industry in Chennai. SDMIMD JoM **13**, 27–34 (2022). https://doi.org/10.18311/sdmimd/2022/27922
10. Djedovic, A., Karabegovic, A., Avdagic, Z., Omanovic, S.: Innovative approach in modeling business processes with a focus on improving the allocation of human resources. Math. Probl. Eng. **2018** (2018). https://doi.org/10.1155/2018/9838560
11. Dumas, M., La Rosa, M., Mendling, J., Reijers, H.A.: Fundamentals of Business Process Management, 2nd edn. Springer, Cham (2018). https://doi.org/10.1007/978-3-662-56509-4
12. Erasmus, J., Vanderfeesten, I., Traganos, K., Jie-A-Looi, X., Kleingeld, A., Grefen, P.: A method to enable ability-based human resource allocation in business process management systems. In: Buchmann, R.A., Karagiannis, D., Kirikova, M. (eds.) PoEM 2018. LNBIP, vol. 335, pp. 37–52. Springer, Cham (2018). https://doi.org/10.1007/978-3-030-02302-7_3
13. Fereday, J., Muir-Cochrane, E.: Demonstrating rigor using thematic analysis: a hybrid approach of inductive and deductive coding and theme development. Int J Qual Methods **5**(1), 80–92 (2006). https://doi.org/10.1177/160940690600500107
14. Frank, L., Poll, R., Röglinger, M., Lea, R.: Design heuristics for customer-centric business processes. BPMJ **26**(6), 1283–1305 (2020). https://doi.org/10.1108/BPMJ-06-2019-0257

15. Gamma, E., Helm, R., Johnson, R., Vlissidies, J.: Design Patterns: Elements of Reusable Object-Oriented Software, 4th edn. Addison-Wesley, Boston (1995)
16. Kim, J., Comuzzi, M., Dumas, M., Maggi, F.M., Teinemaa, I.: Encoding resource experience for predictive process monitoring. DSS **153**, 113669 (2022). https://doi.org/10.1016/j.dss.2021.113669
17. Koschmider, A., Yingbo, L., Schuster, T.: Role assignment in business process models. In: Daniel, F., Barkaoui, K., Dustdar, S. (eds.) BPM 2011. LNBIP, vol. 99, pp. 37–49. Springer, Heidelberg (2012). https://doi.org/10.1007/978-3-642-28108-2_4
18. Limam Mansar, S., Reijers, H.A.: Best practices in business process redesign: use and impact. BPMJ **13**(2), 193–213 (2007). https://doi.org/10.1108/14637150710740455
19. Myers, M.D., Newman, M.: The qualitative interview in is research: examining the craft. Inf. Organ. **17**(1), 2–26 (2007)
20. Netjes, M., Mans, R.S., Reijers, H.A., van der Aalst, W.M.P., Vanwersch, R.J.B.: BPR best practices for the healthcare domain. In: Rinderle-Ma, S., Sadiq, S., Leymann, F. (eds.) BPM 2009. LNBIP, vol. 43, pp. 605–616. Springer, Heidelberg (2010). https://doi.org/10.1007/978-3-642-12186-9_58
21. Patton, M.Q.: Qualitative Research & Evaluation Methods: Integrating Theory and Practice. Sage Publications, Thousand Oaks (2014)
22. Pentico, D.W.: Assignment problems: a golden anniversary survey. Eur. J. Oper. Res. **176**(2), 774–793 (2007). https://doi.org/10.1016/j.ejor.2005.09.014
23. Pereira, J.L., Varajão, J., Uahi, R.: A new approach for improving work distribution in business processes supported by BPMS. BPMJ **26**, 1643–1660 (2020)
24. Pika, A., Leyer, M., Wynn, M.T., Fidge, C.J., ter Hofstede, A.H.M., van der Aalst, W.M.P.: Mining resource profiles from event logs. ACM TMIS **8**(1), 1–30 (2017)
25. Reijers, H.A., Limam Mansar, S.: Best practices in business process redesign: an overview and qualitative evaluation of successful redesign heuristics. Omega **33**(4), 283–306 (2005). https://doi.org/10.1016/j.omega.2004.04.012
26. Rombaut, E., Guerry, M.A.: The effectiveness of employee retention through an uplift modeling approach. Int. J. Manpow. **41**(8), 1199–1220 (2020)
27. Russell, N., van der Aalst, W.M.P., ter Hofstede, A.H.M., Edmond, D.: Workflow resource patterns: identification, representation and tool support. In: Pastor, O., Falcão e Cunha, J. (eds.) CAiSE 2005. LNCS, vol. 3520, pp. 216–232. Springer, Heidelberg (2005). https://doi.org/10.1007/11431855_16
28. Talib, R., Volz, B., Jablonski, S.: Agent assignment for process management: agent performance evaluation framework. In: ICDM, pp. 1005–1012. IEEE (2010)
29. van der Aalst, W.: Re-engineering knock-out processes. DSS **30**(4), 451–468 (2001). https://doi.org/10.1016/S0167-9236(00)00136-6
30. vom Brocke, J., et al.: Process science: the interdisciplinary study of continuous change. SSRN (2021). https://doi.org/10.2139/ssrn.3916817
31. Xu, R., Liu, X., Xie, Y., Yuan, D., Yang, Y.: A gaussian fields based mining method for semi-automating staff assignment in workflow application. In: ICSS, pp. 178–182 (2014). https://doi.org/10.1145/2600821.2600843
32. Yang, J., Ouyang, C., ter Hofstede, A.H.M., van der Aalst, W.M.P.: No Time to dice: learning execution contexts from event logs for resource-oriented process mining. In: Di Ciccio, C., Dijkman, R., Del Río Ortega, A., Rinderle-Ma, S. (eds.) Business Process Management. BPM 2022. Lecture Notes in Computer Science, vol. 13420, pp. 163–180. Springer, Cham (2022). https://doi.org/10.1007/978-3-031-16103-2_13

33. Yeon, M.S., Lee, Y.K., Pham, D.L., Kim, K.P.: Experimental verification on human-centric network-based resource allocation approaches for process-aware information systems. IEEE Access **10**, 23342–23354 (2022). https://doi.org/10.1109/ACCESS.2022.3152778

34. Zellner, G.: Towards a framework for identifying business process redesign patterns. BPMJ **19**(4), 600–623 (2013). https://doi.org/10.1108/BPMJ-Mar-2012-0020

35. Zhao, W., Yang, L., Liu, H., Wu, R.: The optimization of resource allocation based on process mining. In: Huang, D.-S., Han, K. (eds.) ICIC 2015. LNCS (LNAI), vol. 9227, pp. 341–353. Springer, Cham (2015). https://doi.org/10.1007/978-3-319-22053-6_38

36. Zhao, W., Zhao, X.: Process mining from the organizational perspective. In: Wen, Z., Li, T. (eds.) Foundations of Intelligent Systems. AISC, vol. 277, pp. 701–708. Springer, Heidelberg (2014). https://doi.org/10.1007/978-3-642-54924-3_66

37. Zuhaira, B., Ahmad, N.: Business process modeling, implementation, analysis, and management: the case of business process management tools. BPMJ **27**(1), 145–183 (2021). https://doi.org/10.1108/BPMJ-06-2018-0168

A Novel Multi-perspective Trace Clustering Technique for IoT-Enhanced Processes: A Case Study in Smart Manufacturing

Yannis Bertrand$^{(\boxtimes)}$ ⬤, Jochen De Weerdt ⬤, and Estefanía Serral ⬤

Research Centre for Information Systems Engineering (LIRIS), KU Leuven,
Warmoesberg 26, 1000 Brussels, Belgium
{yannis.bertrand,jochen.deweerdt,estefania.serralasensio}@kuleuven.be

Abstract. *IoT-enhanced business processes (BPs)* are processes supported by Internet of Things (IoT) technology, such as sensors capable of monitoring the physical environment where processes are executed. Although the execution of BPs is typically recorded in event logs, IoT-enhanced BPs also generate IoT data that contain vital contextual information. Such BPs are typically found in manufacturing contexts, where, for instance, temperature sensors can provide valuable insights into the storage conditions of sensitive raw materials. However, the potential of this *IoT-enhanced process mining (PM)* has not been fully explored. In this paper, we propose TROPIC, an approach for multi-perspective trace clustering that considers three key perspectives: the control-flow perspective, the trace attribute data perspective and the time series sensor data perspective. We demonstrate the efficacy of our approach in a real-world manufacturing use case. The evaluation of the resulting clusters revealed that integrating the three different perspectives enabled the detection of process variants and anomalous instances that would have been missed using any one of the perspectives in isolation.

Keywords: Process mining · Internet of Things · Trace clustering · IoT-enhanced process mining

1 Introduction

Currently, the use of Internet of Things (IoT) devices in organisations is becoming increasingly common, providing support to their business processes (BPs), known as *IoT-enhanced BPs* [16,36]. The execution of BP activities is usually recorded in event logs, which can be analysed to gain insights into the BP and identify opportunities for improvement. When BPs are augmented with IoT devices, these devices can also provide critical contextual information. One of the main domains where IoT-enhanced BPs are found is smart manufacturing. In these BPs, sensors can track time series (TS) data on various process parameters, such as, for example, flow, temperature, and pressure, which can aid in

© The Author(s), under exclusive license to Springer Nature Switzerland AG 2023
C. Di Francescomarino et al. (Eds.): BPM 2023, LNCS 14159, pp. 395–412, 2023.
https://doi.org/10.1007/978-3-031-41620-0_23

predicting process outcomes and automating tasks. However, due to the unique characteristics of IoT data, such as granularity and storage independent of the process system [4], it is necessary to develop new PM techniques designed specifically for them. This emerging field of *IoT-enhanced process mining (PM)* is still in its early stages [4], with only limited research being already done, focusing primarily on decision mining using IoT data [2,32].

In this paper, we propose TROPIC (TRace attributes, cOntrol-flow Plus Iot Clustering), a novel approach for multiperspective trace clustering that is capable of integrating the TS sensor data perspective, in addition to the control-flow and trace attribute data perspectives. By integrating these different perspectives, multi-perspective trace clustering can effectively identify process variants and anomalous process executions in smart manufacturing that may not be apparent from analysing the control-flow or another single perspective alone. Knowing these variants can, in turn, help organisations identify and propagate best practises to enhance process efficiency and increase the likelihood of positive process outcomes. To demonstrate the effectiveness of our approach, we apply it to a real-life manufacturing process and provide a detailed evaluation of the results. This case study highlights the potential of our approach to analyse and improve IoT-enhanced BPs.

The remainder of the paper is organised as follows. First, Sect. 2 provides an overview of previous research in multi-perspective PM, IoT-enhanced PM, and trace clustering. In Sect. 3, we present TROPIC, our two-level approach for multi-perspective trace clustering, and apply it to the manufacturing process in question in Sect. 4. The experimental results are discussed in Sect. 5, before concluding the paper in Sect. 6 with final remarks and suggestions for future work.

2   Background

2.1   Trace Clustering

Trace clustering is a technique used to group similar process instances, for instance, based on their shared sequential activity patterns. Traditionally, trace clustering has been used to improve process discovery by splitting the event log into sublogs consisting of instances that share comparable activity sequences, and mining a model of each sublog separately. This approach produces simpler and better fitting models that describe different process variants [5,9,13]. However, more recently, trace clustering has been applied to other goals, such as concept drift detection and process evolution analysis [19] and outlier detection [11]. Although improving process discovery results can typically rely only on the control-flow perspective, other objectives can greatly benefit from incorporating context information in clustering.

According to [8], three main categories of trace clustering approaches have been proposed: distance-based, feature-based, and model-based. Distance-based approaches directly cluster traces based on the distances between traces as sequences of activities, using distance metrics such as the Hamming distance,

Levenshtein distance, Damerau-Levenshtein distance, and geodesic distance. Feature-based techniques, on the other hand, derive features from the traces, such as scalars, graphs and embeddings and cluster based on the feature values. Finally, model-based techniques aim to create clusters of traces that produce the best process models [9], optimising criteria such as model fitness. These three approaches have their advantages and disadvantages depending on the nature of the data and the intended application. Choosing the appropriate approach is critical to the effectiveness of the trace clustering process.

2.2 Multi-perspective Process Mining

Multi-perspective PM refers to process analysis techniques that take more than one process perspective into account, e.g., the control-flow and data attributes. The following perspectives are listed in [22] lists the following perspectives:

- Control-flow perspective: Activity ordering in each process instance;
- Resource perspective: Human and non-human resources executing tasks;
- Data perspective: Trace and event attributes;
- Time perspective: Activity duration, throughput time, business rules, etc.;
- Function: Granularity of the activities of the process.

Multi-perspective techniques have been proposed for various types of PMs, such as multi-perspective process discovery [18,24] and multi-perspective conformance checking [14,23]. In trace clustering, a multi-perspective approach is proposed in [15], where a distance metric is presented to compare traces based on the control-flow perspective, the resource perspective, and the data perspective. The (possibly weighted) average of these metrics is computed and used as a pairwise multi-perspective distance measure to perform hierarchical clustering.

However, extending such a technique to TS data can be challenging, as TS typically need to be characterised by many features. For example, [12] reviewed the proposed TS characteristics in the literature and identified a list of approximately 7,700 characteristics to fully represent the TS data. Therefore, proceeding in one step, inputting TS features in a feature vector or including them in an average as in [15], would likely result in either TS features dominating over other perspectives or require very carefully selecting TS features beforehand. This problem grows dramatically when considering multivariate TS, which are very common in manufacturing. To address this issue, we propose a two-step approach that is more versatile than the simple average of distances computed over multiple perspectives.

2.3 IoT-Enhanced PM

Event Log Derivation. The existing literature on IoT-enhanced PM has primarily focused on deriving high-level events of the process from low-level IoT data to create event logs. Subsequently, traditional PM techniques have been employed to analyse these event logs and discover control-flow models of the processes. Several techniques have been proposed specifically for manufacturing processes. In [35], a four-step framework is presented to generate event logs from industrial IoT data, including data preprocessing, clustering of low-level data, classification to derive events from clusters, and creation of the final event log. Also, focussing on industrial applications, [34] propose to transform raw IoT data into an XES event log using complex event processing and event detection and refinement techniques. The authors present another approach in [33] to detect activities interactively from sensor data based on visualisation and exploratory analysis. In [37], a domain-specific language is developed to extract event logs from IoT data by specifying the case and activity identifiers.

Process Contextualisation. Next to event log derivation, some context-aware techniques have also been investigated, e.g., IoT data-aware process discovery [2,20], sensor TS-aware decision mining [32], and IoT-aware conformance checking [28]. In a manufacturing context, [32] outlines an approach to derive decision rule patterns from TS sensor data by automatically featurising the sensor data and training a decision tree to learn the rules. A different problem is addressed by [28], who present an approach for IoT-enhanced deviation detection. In their paper, they argue that traditional conformance checking cannot take into account data that changes over time independently of the events of the process (i.e., TS data). They subsequently proposed a method to detect patterns in the TS data directly.

3   Methodology

TROPIC involves a two-step clustering process (see Fig. 1) currently tailored to the setting of smart manufacturing, typically characterised by highly structured processes around which sensor data are collected in the form of TS. Indeed, in such manufacturing BPs, sensor data and process activities are usually correlated, with process activities leaving recognisable patterns in the sensor data and certain sensor data values triggering the execution of certain process activities. In the clustering process of TROPIC, process instances are first clustered *separately* according to three perspectives: the *control-flow*, *trace attribute data* and *TS sensor data* perspectives. In this step, each perspective is considered independently, providing a detailed view of each aspect of the process. Then, the distances to each centroid in each clustering are computed and used as features for a second clustering step, which takes into account all three perspectives together. This results in a multi-perspective clustering that groups instances based on their unique combinations of control-flow, trace attributes and TS sensor data, providing a comprehensive view on the process.

Next, we explain the approach applied for each single-perspective clustering, followed by the multi-perspective clustering.

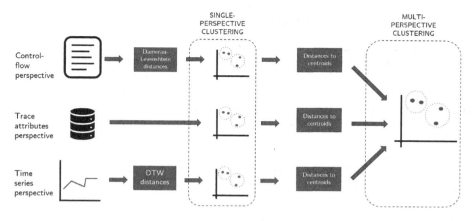

Fig. 1. Overview of TROPIC, our proposed approach.

3.1 Control-Flow Perspective

As mentioned in Sect. 2, three main categories of trace clustering have been proposed: distance-based, feature-based, and model-based approaches. Our approach follows the former by using the Damerau-Levenshtein (DL) distance. The DL distance is a string metric used to compute the edit distance between two strings, which is the minimum number of single-character edits (i.e., insertions, deletions, substitutions, and transpositions) required to transform one string into the other. It extends the Levenshtein distance by also including transpositions of characters. The DL distance between strings A and B, denoted DL(A,B), is computed as follows:

$$
DL(A, B) = \begin{cases} \max(|A|, |B|) & \text{if } \min(|A|, |B|) = 0 \\ \min \begin{cases} DL(A_{1..i-1}, B) + 1 \\ DL(A, B_{1..j-1}) + 1 \\ DL(A_{1..i-1}, B_{1..j-1}) + \delta_{a_i, b_j} \\ DL(A_{1..i-2}b_i, A_{1..j-2}a_j) + 1 \end{cases} & \text{otherwise} \end{cases} \tag{1}
$$

where $|A|$ denotes the length of string A, a_i denotes the i-th character of string A, and δ_{a_i, b_j} is the Kronecker delta function, which is equal to 1 if $a_i = b_j$, and 0 otherwise. The last term in the minimum function corresponds to transposition, and is only included if $i, j > 1$ and $a_{i-1} = b_j$ and $b_{j-1} = a_i$.

Due to the strictly ordered nature of control-flow data in many manufacturing processes, other trace clustering paradigms are usually less suitable. Additionally, activities are often logged at a fairly low level of granularity, making model-based techniques less appropriate. It is worth noting that manufacturing processes tend to be more structured in nature, and thus may not require more complex trace clustering techniques designed for less structured processes.

3.2   Trace Attribute Data Perspective

Trace attributes are usually numerical, categorical, or ordinal features that can be clustered using traditional clustering techniques. Common clustering techniques include hierarchical techniques [38], distance-based techniques, such as K-means [21] or K-medoids [27], model-based techniques, such as self-organising maps [17]; and density-based techniques such as DBSCAN [10]. TROPIC uses K-means, as a generic technique for mixed-type input features, which is most often the case in smart manufacturing. Moreover, its simplicity makes it easily understandable for non-experts. However, depending on the specific process, other techniques could be applied as well; for a general discussion of clustering techniques, see [31].

3.3   Time Series Sensor Data Perspective

In TS analysis, [1] distinguishes three categories of techniques to cluster whole TS: distance-based features, using measures such as Euclidean or dynamic time warping (DTW) distance [30]; structure-based features, which characterise the whole TS; and shape-based features, created by searching for common motifs.

We use DTW distance, which allows a direct comparison of whole TS and is suitable for TS that are expected to share a common general structure as is the case in most manufacturing processes but can differ in length and speed (i.e. certain subsequences can last longer in one TS than in the other). Intuitively, it corresponds to the distance remaining between two series after eliminating timing differences, i.e., correcting for varying activity duration. It relies on the computation of a warping function mapping time points from two series together to minimise the distance between the two series. More specifically, given two series $A = a_1, a_2, ..., a_i, ..., a_n$ and $B = b_1, b_2, ..., b_j, ..., b_m$, with distance $d_{i,j} = ||a_i - b_j||$, DTW aims at finding an optimal mapping function $F = c_1, c_2, ..., c_k, ..., c_l$ such that the total distance $E(F) = \sum_{k=1}^{l} d(c(k)) \cdot w(k)$ is minimised:

$$DTW(A, B) = \min_F \left[\frac{\sum_{k=1}^{l} d(c(k)) \cdot w(k)}{\sum_{k=1}^{l} w(k)} \right] \tag{2}$$

where $w(k)$ is a weight coefficient for the elements of the mapping function.

Applying this for each pair of batches yields a distance matrix which can be used as input for clustering techniques like K-medoids or hierarchical clustering.

3.4   Multi-perspective Clustering

Once process instances are clustered separately in each perspective, the results are combined by clustering them together. Single-perspective clusters can be represented in different ways, such as using their labels as categorical features or computing distances to the centroids. We follow the latter approach, which retains more information for multi-perspective clustering.

Moreover, perspectives can be weighted to adjust their contribution to the multi-perspective clustering. For example, control-flow can be given more weight to ensure it has sufficient influence on the final clustering. Weights can also be used to account for differences in the number of clusters generated by each perspective, where more clusters may result in more features and a greater impact on the final clustering.

4   A Case Study in Smart Manufacturing

4.1   Use Case

Process Description. We applied TROPIC to a real use case at a partner company active in the production of chemical products. Their production process can be summarised in four main steps:

1. Preparing raw material and loading the tank;
2. Mixing the raw material in the tank;
3. Circulating the product through filters to remove impurities;
4. Bottling and packing the finished product.

Sometimes, the quality of the product is not high enough after filtering, i.e., some characteristics of the product do not meet the specifications. In this case, an adjustment is applied by loading additional raw materials into the tank and repeating steps two and three, resulting in the high-level production process depicted in Fig. 2.

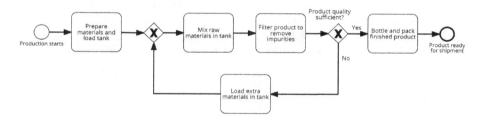

Fig. 2. High-level model of the process analysed in the experiment.

This seemingly simple process has to be executed with extreme precision and care as the slightest presence of impurities in the finished product greatly diminishes its quality. This is why the company is interested in analysing production logs and TS sensor data together to find out variation in process execution.

Data. Two main data sources are used in this use case: 1) logs from the production system, which contain the sequences of activities executed for each process instance and trace attributes and 2) TS data from sensors tracking the flow of the product in the four tanks and in the pipes leading through the filters every second. The data span a period from October 2020 to April 2022, representing 161 complete process instances and 199.4 million rows of sensor data.

Data Preprocessing. First, relevant TS pump circulation flow data was extracted for each batch. The data were resampled to one measurement per minute for smoothing and to reduce their length (the raw TS for the longest batch counted more than one million measurements before resampling), and some missing values due to the storage format were imputed. Finally, all data were normalised.

4.2   Clustering and Evaluation Approach

Multi-perspective Trace Clustering. We applied our two-step multi-perspective trace clustering approach to the obtained data. For the *control-flow perspective*, we followed a distance-based approach by computing the DL distance between the event sequences for each pair of batches and using the resulting distance matrix as input for the K-medoids. The number of clusters was set to five by plotting inertia and following the elbow method. The clusters contained 28, 23, 52, 22 and 36 instances, respectively. Secondly, regarding the *trace attributes perspective*, we applied the K-means algorithm with K = 5 (based on the elbow method). This yielded clusters of 23, 48, 41, 25 and 24 instances. Note that the attributes "tank open time" and "time in tank" are considered trace attributes as they measure batch quality and not timeliness. Third, we applied a distance-based TS clustering approach for the *TS sensor data perspective*, computing the DTW distance between the TS of each pair of batches to obtain a TS distance matrix used as input for K-medoids, with K = 6 (based on the elbow method), which formed clusters of sizes 9, 44, 59, 20, 21 and 8. Finally, to perform *multi-perspective clustering*, we computed the distances to centroids for each single-perspective clustering. Then we weighed the clusterings to take into account the different values of K in each clustering and applied K-means to all distances to centroids together, with K = 4 based on the elbow method. When K-means were applied, centroids initialisation was optimised to speed up convergence of the clustering by sampling centroids based on marginal inertia decrease, while when K-medoids were applied, medoids were randomly initialised.

Clustering Evaluation. The evaluation of clustering results is a challenging task that often depends on the specific domain and task at hand. A range of metrics are available to score clusterings based on intrinsic properties, such as the Davies-Bouldin (DB) score [6], which measures the similarity of clusters to their respective most similar cluster (lower value is better), or the Silhouette index [29], which compares the similarity between an instance and instances in its cluster with the similarity between this instance and instances in other clusters (higher value is better). Other metrics compare clusterings with known classes in the data or other clusterings, such as the Rand index [26], entropy, or purity. However, it is worth noting that better-formed clusters may not necessarily be more useful in practise, hence obtaining external validation from experts is critical to evaluate clustering results.

In our case study, we compared the clusters obtained from the multi-perspective approach with those derived from single-perspective clustering, using

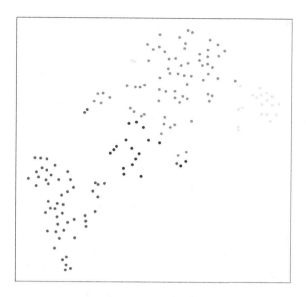

Fig. 3. Visualisation of multi-perspective clustering with t-SNE (cluster 1 = purple, cluster 2 = blue, cluster 3 = green, cluster 4 = yellow). (Color figure online)

both metrics and expert feedback. We computed silhouette indexes and DB scores for each clustering to assess the quality of the clusters in each approach. We also computed adjusted Rand indexes (ARI; where a higher value indicates higher similarity) and entropy scores (where a lower value indicates higher similarity) to determine the degree of similarity between the clusterings and to identify which perspective has the most influence on multi-perspective clustering. To validate our clustering results, we presented them to a senior process engineer at our partner company. Specifically, we showed the engineer the centroids of each multi-perspective cluster, as well as an overview of each cluster, including a directly-follows graph (DFG) for the control-flow, the mean or mode of trace attributes, and the DTW barrycenter average (DBA) [25] for the TS perspective, which is a method to compute the average of several TS taking into account potential time shifts.

4.3   Results

Multi-perspective clustering with K = 4 resulted in clusters of sizes 18, 53, 69, and 21 (see Fig. 3). In the remainder of this section, we provide visualisations of the clusters and report the values of the metrics and the interpretation and evaluation of the clusters by the process expert for each perspective.

Clustering Quality Assessment and Visualisation. The Silhouette score and the DB index are reported in Table 1. As can be seen, multi-perspective clustering has better scores than other clusterings for both metrics. Trace attributes clustering has the worst scores, while control-flow and TS clusterings have similar values.

Table 1. Internal validation metrics for each clustering.

Metric	Multi-perspective	Control-flow	Trace attributes	Time series
Silhouette index	0.2516	0.0331	-0.0168	0.0284
DB score	1.3845	3.0526	4.5942	3.2061

Table 2 reports the cluster similarity metrics. Both entropy and ARI show that multi-perspective and control-flow clusterings have the highest similarity, i.e., they most often group the same instances together. On the other hand, trace attribute data clustering has high entropy and low ARI for all other clusterings, indicating that it forms very different clusters than the other perspectives.

We visualised the multi-perspective clusters by modelling the DFGs of their control-flows (see Figs. 4–5, where high-level steps from Fig. 2 are highlighted), computing the mean and the mode of their attributes (see Table 3) and plotting the DBAs of their TS data (see Figs. 6–7). DFGs and DBAs were used and are put forward in this paper as they can provide intuitive visualisations of the control-flow and the TS data of many instances of a process at once, enabling business experts to quickly understand and analyse whole clusters. Note that while all the results of the multi-perspective clustering are shown, only particularly interesting results are displayed for the other clusterings, and that activity labels as well as some trace attribute values were anonymised on request of the company.

Expert-Based Validation. When showing the multi-perspective clusters, the process expert categorised them as follows. Cluster 3, the largest cluster and the ones with the fewest distinctive characteristics, was identified as representing the standard execution of the process. Cluster 2 typically included traces with fewer adjustment activities and a lower material adjustments attribute than those in the other clusters, as shown in Fig. 4b and Table 3. In contrast, cluster 1 represented batches that required more adjustment activities and have a higher value for the material adjustments attribute (see Fig. 4a and Table 3) than batches in the other clusters. Having more adjustments also caused the filtering step to last longer, which can also be seen in the TS data by comparing Figs. 6a and 6b (filtering being characterised by long periods with a stable flow). Finally, cluster 4 included traces with missing activities that were necessary for proper process execution. These instances were identified as anomalies caused by improper logging of these activities.

4.4   Comparison of the Clusterings

In general, single-perspective clusters are more difficult to interpret than multi-perspective clusters. While control-flow clustering also groups together batches

Table 2. Pairwise similarity metrics values.

	Multi-perspective	Control-flow	Trace attributes	Time series
Multi-perspective entropy	0	0.8661	1.1585	0.9749
Multi-perspective ARI	1	0.2552	0.0301	0.0886
Control-flow entropy	1.1806	0	1.4154	1.3783
Control-flow ARI	0.2552	1	0.0340	0.0359
Trace attributes entropy	1.4790	1.4213	0	1.4663
Trace attributes ARI	0.0301	0.0340	1	0.0141
Time series entropy	1.2930	1.3817	1.4638	0
Time series ARI	0.0886	0.0359	0.0141	1

that required more adjustments, no cluster groups instances with fewer adjustments as neatly as multi-perspective cluster 2 (see Figs. 5a–5b). It is particularly difficult to recognise consistent patterns across perspectives in data clusters, while TS clusters succeed to some extent in grouping together instances with similar control-flows. Next to this, the most difficult perspective to interpret in all clusterings seems to be the TS perspective, where DBAs have difficulty capturing typical TS shapes, partly due to the presence of batches with missing data. This being said, DBAs based on TS clustering (see Fig. 7) seem more distinct and more easily interpretable.

5 Discussion

TROPIC successfully integrates TS sensor data in multi-perspective trace clustering, resulting in clusters that consider different process perspectives. The two-step structure makes it easy to disentangle the different perspectives, adjust their importance, and compare them. In our manufacturing use case, comparing multi-perspective trace clustering with single-perspective clustering showed that by leveraging underlying relationships between different perspectives, multi-perspective trace clustering could outperform single-perspective clustering even in their own perspective. For instance, multi-perspective trace clustering grouped instances with few adjustments better than control-flow clustering, as other perspectives helped recognise these instances.

In addition, the process expert found multi-perspective clusters more meaningful from a business point of view, as they identified variants and anomalies. This insight could help the company investigate the differences between clusters 1 and 2 to reduce the number of necessary adjustments in the future.

Furthermore, some anomalous process instances were detected in the use case, although we did not apply any anomaly detection technique. This observation highlights the potential of multi-perspective anomaly detection using TROPIC by applying outlier detection to the distances to centroids.

In addition, the choice of K for K-means and K-medoids clustering could have a great impact on the results of clustering at both stages. In this paper,

Table 3. Mean or mode of the trace attributes for each cluster of all clusterings (standard deviations between brackets).

Cluster	Materials	Adj. materials	Tank open time	Solvent	Time in tank	Tank
Multi perspective 1	5.7222 (5.3776)	3.8889 (3.2519)	6392.9444 (4161.9357)	A /	381671.8889 (223893.4571)	T3 /
Multi perspective 2	7.2453 (2.0466)	0.0755 (0.3848)	7552.4151 (2760.7288)	B /	261959.6038 (60081.1298)	T4 /
Multi perspective 3	7.0 (1.4349)	2.7681 (1.5449)	8689.6377 (2785.0268)	B /	298360.3478 (84948.1399)	T1 /
Multi perspective 4	7.2381 (1.9211)	2.8095 (1.4703)	10599.1429 (2919.2145)	B /	260751.6667 (97850.3688)	T4 /
Control-flow 1	7.8929 (3.6952)	1.1071 (2.3308)	7877.0 (1878.0679)	B /	270904.6786 (69923.7843)	T4 /
Control-flow 2	6.2609 (2.4349)	3.9565 (2.5132)	8857.0 (3929.4461)	A /	325021.1739 (118121.8841)	T1 /
Control-flow 3	7.1538 (1.9742)	2.0769 (1.4799)	8587.3077 (3264.1794)	B /	301473.25 (108380.1455)	T4 /
Control-flow 4	5.8636 (2.1447)	2.2273 (1.3778)	9019.5909 (3878.9293)	A /	283188.4545 (118979.1435)	T4 /
Control-flow 5	7.1111 (1.6695)	1.25 (1.9911)	7452.2222 (2711.0095)	A /	273584.1389 (124958.7138)	T1 /
Trace attributes 1	8.4348 (3.4089)	4.6957 (2.6187)	9494.7391 (3112.0703)	B /	291919.2609 (94352.3756)	T4 /
Trace attributes 2	7.2083 (1.688)	1.1667 (1.3262)	8225.375 (2075.6587)	B /	257210.7083 (65348.9021)	T4 /
Trace attributes 3	6.7317 (2.3667)	1.3415 (1.5266)	7048.2683 (3339.7471)	A /	282512.2683 (73998.1907)	T1 /
Trace attributes 4	5.84 (2.5113)	2.52 (1.8735)	7396.36 (3202.4594)	A /	354616.76 (180006.1447)	T3 /
Trace attributes 5	6.6667 (2.1196)	1.75 (1.7508)	10434.7083 (3427.8528)	B /	304496.7917 (128730.3709)	T4 /
TS 1	5.6667 (2.2913)	2.2222 (1.8559)	9548.1111 (4358.0709)	Other /	217580.6667 (38620.8273)	T4 /
TS 2	6.6136 (2.4133)	1.75 (1.6999)	7262.3182 (2992.4131)	A /	355988.5455 (123518.9904)	T3 /
TS 3	7.1356 (2.5492)	2.0169 (2.4878)	8272.678 (2755.9548)	B /	260759.0508 (97720.4561)	T4 /
TS 4	7.7 (1.8382)	1.9 (1.8035)	9000.9 (2929.1435)	B /	267001.65 (68154.5069)	T4 /
TS 5	6.7143 (2.1941)	2.5714 (2.0874)	9366.0 (4099.0399)	B /	312286.619 (120953.309)	T2 /
TS 6	8.0 (3.4641)	2.0 (2.3299)	8406.375 (2185.4744)	B /	239005.5 (24086.8173)	T4 /

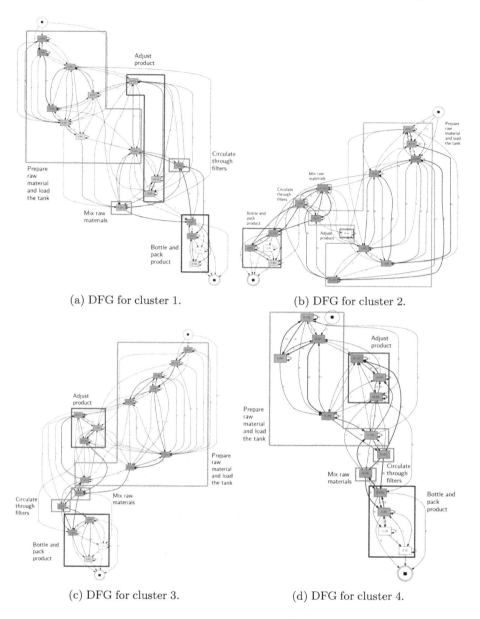

(a) DFG for cluster 1.

(b) DFG for cluster 2.

(c) DFG for cluster 3.

(d) DFG for cluster 4.

Fig. 4. DFGs for each cluster of the multi-perspective clustering.

the popular elbow method was used and yielded good results, as the clusters formed were insightful from a business perspective. Future work could investigate more complex methods to determine the value of K, e.g., based on stability or separation as in [7].

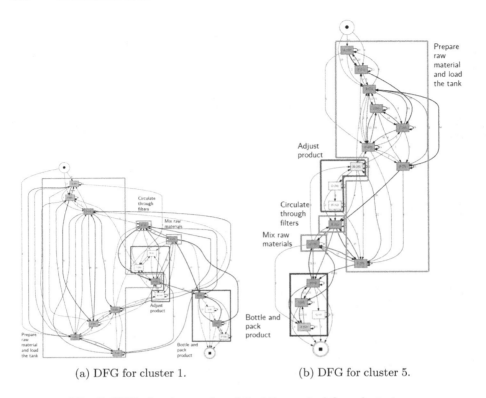

(a) DFG for cluster 1.   (b) DFG for cluster 5.

Fig. 5. DFGs for clusters 1 and 5 of the control-flow clustering.

However, ARI and entropy indicated that the control-flow perspective produced a clustering more similar to the other perspectives. This result suggests that the control-flow perspective might be more important than other perspectives in the multi-perspective trace clustering. Weighting the perspectives could rebalance their contributions, but as all perspectives are correlated, weighting may not fundamentally change the clustering in the use case.

Finally, although we focused on three specific perspectives in this paper, we believe our approach could be extended to consider other perspectives. For example, a similar approach to that applied to the TS data obtained from IoT sensors could be applied to other processes that evolve over time, such as process performance. Such a different perspective could serve as a substitute for one of the current three dimensions, or the approach could easily be adapted to a higher dimensionality, allowing for several other perspectives to be included, such as the resource perspective.

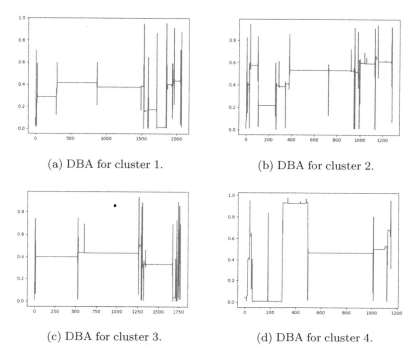

(a) DBA for cluster 1. (b) DBA for cluster 2.

(c) DBA for cluster 3. (d) DBA for cluster 4.

Fig. 6. DBAs for each cluster of the multi-perspective clustering.

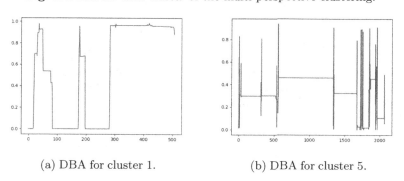

(a) DBA for cluster 1. (b) DBA for cluster 5.

Fig. 7. DBAs for clusters 1 and 5 of the TS clustering.

6 Conclusion

In this paper, we presented a novel approach for multi-perspective trace clustering of manufacturing processes that considers three perspectives: control-flow,
trace attributes, and TS sensor data. This approach can reveal process variants
that are homogeneous across all three perspectives simultaneously. We evaluated
the approach in a real-life use case of a smart manufacturing process, where it

revealed meaningful clusters and anomalous instances for a specific IoT-enhanced BP, both actionable insights to improve process design and execution.

In future work, we plan to extend this approach in various ways. One possibility is to propose a generalisation to n arbitrary perspectives. We could also consider including event attributes and incorporating TS data at the event level. Furthermore, we could explore other clustering techniques for the multi-perspective clustering if our approach were to be used for more flexible types of processes, such as ensemble clustering methods or soft clustering techniques. Finally, we find that integrating contextual information in the log in the form of events, as suggested in [3], is an interesting alternative approach.

Acknowledgement. This research was supported by the Flemish Fund for Scientific Research (FWO) with grant number G0B6922N.

References

1. Aghabozorgi, S., Shirkhorshidi, A.S., Wah, T.Y.: Time-series clustering-a decade review. Inf. Syst. **53**, 16–38 (2015)
2. Banham, A., Leemans, S.J., Wynn, M.T., Andrews, R., Laupland, K.B., Shinners, L.: xPM: enhancing exogenous data visibility. Artif. Intell. Med. **133**, 102409 (2022)
3. Bertrand, Y., De Weerdt, J., Serral, E.: A bridging model for process mining and IoT. In: Munoz-Gama, J., Lu, X. (eds.) ICPM 2021. LNBIP, vol. 433, pp. 98–110. Springer, Cham (2022). https://doi.org/10.1007/978-3-030-98581-3_8
4. Bertrand, Y., De Weerdt, J., Serral, E.: Assessing the suitability of traditional event log standards for IoT-enhanced event logs. In: In: Cabanillas, C., Garmann-Johnsen, N.F., Koschmider, A. (eds.) Business Process Management Workshops. BPM 2022. Lecture Notes in Business Information Processing, vol. 460, pp. 63–75. Springer, Cham (2023). https://doi.org/10.1007/978-3-031-25383-6_6
5. Bose, R.J.C., Van der Aalst, W.M.: Context aware trace clustering: towards improving process mining results. In: Proceedings of the 2009 SIAM International Conference on Data Mining, pp. 401–412. SIAM (2009)
6. Davies, D.L., Bouldin, D.W.: A cluster separation measure. IEEE Trans. Pattern Anal. Mach. Intell. **2**, 224–227 (1979)
7. De Koninck, P., De Weerdt, J.: Similarity-based approaches for determining the number of trace clusters in process discovery. In: Koutny, M., Kleijn, J., Penczek, W. (eds.) Transactions on Petri Nets and Other Models of Concurrency XII. LNCS, vol. 10470, pp. 19–42. Springer, Heidelberg (2017). https://doi.org/10.1007/978-3-662-55862-1_2
8. De Weerdt, J.: Trace clustering (2019)
9. De Weerdt, J., Vanden Broucke, S., Vanthienen, J., Baesens, B.: Active trace clustering for improved process discovery. IEEE TKDE **25**(12), 2708–2720 (2013)
10. Ester, M., Kriegel, H.P., Sander, J., Xu, X., et al.: A density-based algorithm for discovering clusters in large spatial databases with noise. In: kdd, vol. 96, pp. 226–231 (1996)
11. Fani Sani, M., van Zelst, S.J., van der Aalst, W.M.P.: Applying sequence mining for outlier detection in process mining. In: Panetto, H., Debruyne, C., Proper, H.A., Ardagna, C.A., Roman, D., Meersman, R. (eds.) OTM 2018. LNCS, vol. 11230, pp. 98–116. Springer, Cham (2018). https://doi.org/10.1007/978-3-030-02671-4_6

12. Fulcher, B.D., Little, M.A., Jones, N.S.: Highly comparative time-series analysis: the empirical structure of time series and their methods. J. R. Soc. Interface **10**(83), 20130048 (2013)
13. Greco, G., Guzzo, A., Pontieri, L., Saccà, D.: Mining expressive process models by clustering workflow traces. In: Dai, H., Srikant, R., Zhang, C. (eds.) PAKDD 2004. LNCS (LNAI), vol. 3056, pp. 52–62. Springer, Heidelberg (2004). https://doi.org/10.1007/978-3-540-24775-3_8
14. Grüger, J., Geyer, T., Kuhn, M., Braun, S.A., Bergmann, R.: Verifying guideline compliance in clinical treatment using multi-perspective conformance checking: a case study. In: ICPM Workshops, pp. 301–313 (2021)
15. Jablonski, S., Röglinger, M., Schönig, S., Wyrtki, K.M.: Multi-perspective clustering of process execution traces. EMISAJ **14**, 2 (2019)
16. Janiesch, C., et al.: The internet of things meets business process management: a manifesto. IEEE Systems, Man, Cybern. Mag. **6**(4), 34–44 (2020)
17. Kohonen, T.: The self-organizing map. Proc. IEEE **78**(9), 1464–1480 (1990)
18. Leno, V., Dumas, M., Maggi, F.M., La Rosa, M.: Multi-perspective process model discovery for robotic process automation. In: CAiSE 2018, vol. 2114, pp. 37–45 (2018)
19. Luengo, D., Sepúlveda, M.: Applying clustering in process mining to find different versions of a business process that changes over time. In: Daniel, F., Barkaoui, K., Dustdar, S. (eds.) BPM 2011. LNBIP, vol. 99, pp. 153–158. Springer, Heidelberg (2012). https://doi.org/10.1007/978-3-642-28108-2_15
20. Lull, J.J., et al.: Exploration with process mining on how temperature change affects hospital emergency departments. In: Leemans, S., Leopold, H. (eds.) ICPM 2020. LNBIP, vol. 406, pp. 368–379. Springer, Cham (2021). https://doi.org/10.1007/978-3-030-72693-5_28
21. MacQueen, J.: Classification and analysis of multivariate observations. In: 5th Berkeley Symposium on Mathematical Statistics and Probability, pp. 281–297. University of California, Los Angeles (1967)
22. Mannhardt, F.: Multi-perspective process mining. In: BPM (Dissertation/Demos/Industry), pp. 41–45 (2018)
23. Mannhardt, F., De Leoni, M., Reijers, H.A., Van Der Aalst, W.M.: Balanced multi-perspective checking of process conformance. Computing **98**, 407–437 (2016)
24. Mannhardt, F., de Leoni, M., Reijers, H.A., van der Aalst, W.M.P.: Data-driven process discovery - revealing conditional infrequent behavior from event logs. In: Dubois, E., Pohl, K. (eds.) CAiSE 2017. LNCS, vol. 10253, pp. 545–560. Springer, Cham (2017). https://doi.org/10.1007/978-3-319-59536-8_34
25. Petitjean, F., Ketterlin, A., Gançarski, P.: A global averaging method for dynamic time warping, with applications to clustering. Pattern Recogn. **44**(3), 678–693 (2011)
26. Rand, W.M.: Objective criteria for the evaluation of clustering methods. J. Am. Stat. Assoc. **66**(336), 846–850 (1971)
27. Rdusseeun, L., Kaufman, P.: Clustering by means of medoids. In: Proceedings of the Statistical Data Analysis Based on the L1 Norm Conference, vol. 31 (1987)
28. Rodriguez-Fernandez, V., Trzcionkowska, A., Gonzalez-Pardo, A., Brzychczy, E., Nalepa, G.J., Camacho, D.: Conformance checking for time-series-aware processes. IEEE TII **17**(2), 871–881 (2020)
29. Rousseeuw, P.J.: Silhouettes: a graphical aid to the interpretation and validation of cluster analysis. J. Comput. Appl. Math. **20**, 53–65 (1987)
30. Sakoe, H., Chiba, S.: Dynamic programming algorithm optimization for spoken word recognition. IEEE TASSP **26**(1), 43–49 (1978)

31. Saxena, A., et al.: A review of clustering techniques and developments. Neurocomputing **267**, 664–681 (2017)
32. Scheibel, B., Rinderle-Ma, S.: Online decision mining and monitoring in process-aware information systems. In: Ralyte, J., Chakravarthy, S., Mohania, M., Jeusfeld, M.A., Karlapalem, K. (eds.) Conceptual Modeling. ER 2022. Lecture Notes in Computer Science, vol. 13607, pp. 271–280. Springer, Cham (2022). https://doi.org/10.1007/978-3-031-17995-2_19
33. Seiger, R., Franceschetti, M., Weber, B.: An interactive method for detection of process activity executions from IoT data. Future Internet **15**(2), 77 (2023)
34. Seiger, R., Zerbato, F., Burattin, A., García-Bañuelos, L., Weber, B.: Towards IoT-driven process event log generation for conformance checking in smart factories. In: EDOCW 2020, pp. 20–26. IEEE (2020)
35. Trzcionkowska, A., Brzychczy, E.: Practical aspects of event logs creation for industrial process modelling. MAPE **1**(1), 77–83 (2018)
36. Valderas, P., Torres, V., Serral, E.: Modelling and executing IoT-enhanced business processes through BPMN and microservices. J. Syst. Softw. **184**, 111139 (2022)
37. Valencia Parra, Á., Ramos Gutiérrez, B., Varela Vaca, Á.J., Gómez López, M.T., García Bernal, A.: Enabling process mining in aircraft manufactures: extracting event logs and discovering processes from complex data. In: BPM2019IF (2019)
38. Ward, J.H., Jr.: Hierarchical grouping to optimize an objective function. J. Am. Stat. Assoc. **58**(301), 236–244 (1963)

The Impact of Process Complexity on Process Performance: A Study Using Event Log Data

Maxim Vidgof[1]([✉])[ID], Bastian Wurm[2][ID], and Jan Mendling[1,3,4][ID]

[1] Wirtschaftsuniversität Wien, Welthandelsplatz 1, 1020 Vienna, Austria
maxim.vidgof@wu.ac.at
[2] Ludwig-Maximilians-Universität München, LMU Munich School of Management,
Ludwigstrasse 28, 80539 Munich, Germany
bastian.wurm@lmu.de
[3] Humboldt-Universität zu Berlin, Unter den Linden 6, 10099 Berlin, Germany
jan.mendling@hu-berlin.de
[4] Weizenbaum Institute, Hardenbergstraße 32, 10623 Berlin, Germany

Abstract. Complexity is an important characteristic of any business process. The key assumption of much research in Business Process Management is that process complexity has a negative impact on process performance. So far, behavioral studies have measured complexity based on the perception of process stakeholders. The aim of this study is to investigate if such a connection can be supported based on the analysis of event log data. To do so, we employ a set of 38 metrics that capture different dimensions of process complexity. We use these metrics to build various regression models that explain process performance in terms of throughput time. We find that process complexity as captured in event logs explains the throughput time of process executions to a considerable extent, with the respective R-squared reaching up to 0.96. Our study offers implications for empirical research on process performance and can serve as a toolbox for practitioners.

Keywords: Process complexity · Process performance · Throughput time

1 Introduction

Business processes management (BPM) provides various analysis techniques for improving the performance of business processes (see, for example, [11]). Several of these techniques support the identification of root causes behind performance issues of a process. Some studies have pointed to the connection between process complexity as a root cause of bad process performance. More specifically, it has

Jan Mendling: The research by Jan Mendling was supported by the Einstein Foundation Berlin under grant EPP-2019-524 and by the German Federal Ministry of Education and Research under grant 16DII133.

© The Author(s), under exclusive license to Springer Nature Switzerland AG 2023
C. Di Francescomarino et al. (Eds.): BPM 2023, LNCS 14159, pp. 413–429, 2023.
https://doi.org/10.1007/978-3-031-41620-0_24

been established that standardized business processes are connected with better process performance [23] and outsourcing success [32]. For this reason, high business process complexity is often a motivation for business process redesign initiatives [14], but also a challenge for standardization efforts [28].

However, these studies largely build on perceptual measures, which entails at least three key issues. First, such measures require specific attention in order to meet potential validity concerns [17]. Second, perceptual differences exist along the organizational hierarchy. The so-called hierarchical erosion effect states that perceptions become less favourable towards the lower levels of the hierarchy [12]. Third, a study based on perceptual measures is often restricted to an observation at only a single point in time. All of this raises the question to which extent a more precise investigation of the connection between process complexity and performance is possible.

In this paper, we address this research problem. To this end, we utilize available event log dataset to a) calculate complexity measures and b) throughput time as a performance measure over different time windows. The connection between complexity and performance is then investigated by means of statistical regression. Our results suggest that process complexity is closely connected to throughput time, but also dependent on idiosyncratic factors. We discuss implications of this finding for research and practice.

The remainder of the paper is structured as follows. Section 2 introduces complexity metrics and the related concepts. Section 3 presents our approach for the calculation of process complexity, throughput time, and the creation of our statistical models. Section 4 presents our results and showcases the best statistical models. Section 5 provides the discussion of the results and points to avenues for future research. Finally, Sect. 6 concludes with a summary.

2   Background

In this section, we discuss the background against which we position our work. First, we summarize related work on process complexity. Second, we outline research on process performance and the role that complexity plays for it.

2.1   Process Complexity

In BPM research, process complexity has often been approached from a process model perspective. Most notably is the work by Mendling on the relationship between process model complexity and error probability [19,21,22]. Recently, various metrics for complexity based on event logs have been defined, partially inspired by work in neighboring disciplines, such as organization science. These measures can be used to quantify different aspects of business processes complexity that are visible from event log data. The various measures can be organized in five categories as presented in Table 1.

The first category encompasses measures pertaining to the *size* of a given event log. These measures count properties of an event log, such as the number of *events*, *sequences*, and minimum, average, and maximum *sequence length* [15].

Table 1. Complexity Measures for Business Processes based on Event Logs (adapted from [1])

Category	Measure	Reference
Size	Number of Events	[15]
	Number of Event Types	[15]
	Number of Sequences	[15]
	Minimum, Average, Maximum Sequence Length	[29]
Variation	Number of Acyclic Paths in Transition Matrix	[25]
	Number of Ties in Transition Matrix	[16]
	Lempel-Ziv Complexity	[24]
	Number and Percentage of Unique Sequences	[29]
Distance	Average Affinity	[15]
	Deviation from Random	[24]
	Average Edit Distance	[24]
Simple Entropy	Sequence entropy	[1]
	Variant entropy	[1]
	Normalized sequence entropy	[1]
	Normalized variant entropy	[1]
Enriched Entropy	Enriched Sequence entropy	[31]
	Enriched Variant entropy	[31]
	Enriched Normalized sequence entropy	[31]
	Enriched Normalized variant entropy	[31]

The second category contains measures capturing the variation of process behavior as documented in the event log. Many of the measures in this category build on a transition matrix that is derived based on the directly-followed relations as captured in the event log [1]. Pentland et al. [25] operationalize process complexity as the number of *acyclic paths* provided by the transition matrix. Closely related to this is the measure by Hærem et al. [16], who measure complexity as the number of *ties*, i.e. directly-follows relations, over all distinct *sequences*. Further measures that depict variation are Pentland's [24] approach to compress an event log based on the Lempel-Ziv algorithm as well as the absolute and relative number of *unique sequences* [29] contained in an event log.

The third category includes measures that are based on different notions of distance [1]. Günther [15] suggests a measure of *affinity* of two event sequences, capturing the extent to which directly-follow relations of the sequences overlap. His *average affinity* measure calculates the mean of the pair-wise affinity over all sequences in the event log [15]. This measure is similar to Pentland's [24] *deviation from random* of the transition matrix. Pentland [24] further proposes average *edit distance* between event sequences based on optimal matching [4].

The fourth category of measures builds on graph entropy and has been recently proposed by Augusto et al. [1]. They distinguish between measures for sequence and variant entropy of an event log. Additionally, they suggest that each of the measures can be normalized to take a value between 0 and 1. We refer to these measures as simple entropy measures.

Fifth, the measures by [1] have been extended beyond the control flow to incorporate data variety [30]. In contrast to the simple entropy measures, we refer to this class of measures as *enriched entropy* measures.

Several of these measures have been applied to study an increasing breadth of research problems. The above named study by Augusto et al. [1], for example, investigated the influence of process complexity on the quality of process models derived from event log data. They find that process complexity is negatively correlated with the quality of discovered process models. Thus, the more complex the event log, the poorer will be the model discovered by process mining algorithms. Importantly, different discovery algorithms are more sensitive to certain complexity measures than others [1].

There are also some behavioral studies that examine how process complexity changes over time. Pentland et al. [25] simulate how process complexity changes over time. They find that organizational processes undergo different phases of process complexity. At the initiation of their simulation, processes exhibit low levels of complexity. After several iterations of the simulation, process complexity suddenly sharply increases, leading to *bursts* of complexity. Afterwards, complexity again decreases resulting in limited but ongoing variation in the process. Further, Wurm et al. [34] investigate process complexity in the Purchase-to-Pay and Order-to-Cash processes of a multinational enterprise. While they find that process complexity changes continuously, they do not find any indication for sudden bursts of complexity in the examined processes.

Importantly, both studies [25,34] rest on measures that are not precise [1]. As shown in [1], corner cases can be identified that illustrate that the measures used tend to overestimate the actual complexity of a process.

2.2 Process Performance

The literature suggests a clear link between process complexity and process performance. Empirical studies indicate that the standardization of business processes ultimately leads to better process performance [23] and outsourcing success [32]. In particular, Münstermann et al. [23] have found that process standardization is positively associated with different process performance dimensions, such as process time, cost, and quality. By means of standardization, organizations aim to reduce process complexity, i.e. the number of ways that a process can be performed [26,35].

At a second look, however, the relationship between process complexity and process performance is not that clear-cut. Detailed findings by Münstermann et al. [23] show that the effect of standardization is conditional to the industry and type of process in question. Specifically, they find that process standardization only significantly influences process performance in the service industry and for companies that can be classified as analyzers [23].

Furthermore, studies that measure the complexity of processes and their corresponding performance rely on perceptual measures that can cause several important validity issues. First, there are important validity concerns that need to be taken into account when developing perceptual measures for organizational and process performance [17]. Second, there are perceptual differences that need to be considered when interpreting results from perceptual measures. For example, the hierarchical erosion effect [12] describes that perceptions at lower levels of an organization's hierarchy tend to be less favourable. Similarly, Pentland [24] shows that process stakeholders' perception and actual enactment of process variation diverge substantially. Third, the use of perceptional measures to determine changes in properties of business processes is inefficient. In order to assess the effect of an improvement initiative on process performance, one would have to survey process stakeholders again and again. For example, to assess the success of a standardization initiative, a company would have to survey process stakeholders at least twice: prior to the initiative and after the initiative. Thus, studies based on perceptual measures are often restricted to data that is collected at a single point in time and only provide a static perspective of processes and their performance.

In light of these limitations, several authors have proposed to use process mining to move from opinion-based to evidence-based measures for business processes [2,3,13,27]. In this regard, the studies by [2,3] are the first to define measurability of process performance indicators and evaluate process redesign best practices based on event logs, respectively.

In the following, we develop an evidence-based and time-sensitive measure for process complexity based on the recent work by [1]. As this measure is based on graph entropy it is precise and allows researchers and process managers to quantify process complexity in a comprehensive way at any given point in time. In addition, it allows to continuously monitor how process complexity changes over time. We further evaluate the measure by applying it to a set of event logs from the Business Process Intelligence (BPI) Challenge allowing us to closely examine the relationship between process complexity and process performance.

3   Approach

In this section, we describe our approach. First, we introduce the notion of forgetting that allows us to weight events in the event log differently based on their time of occurrence. Then, we prepare the dataset by splitting it into time periods, performing complexity and performance measurements as well as filtering out the outliers. Then, we automatically build regression models and systematically reduce the number of variables in them. All computations were performed on a laptop with Intel®Core ^{TM} i7-8565U CPU @ 4.60 GHz x 4 and 16 GB of DDR4 RAM, Linux kernel 4.15.0-88-generic 64-bit version, Python version 3.8.10 and R version 4.0.3. The code for complexity and performance measurement as well as the data are available on GitHub[1,2].

3.1 Forgetting

An important concern for prediction models is the potential evolution of the data-generation mechanism over time [18, p.525]. The available measures described in Sect. 2.1, such as sequence entropy and normalized sequence entropy, rely on counting the events distributed over different partitions of an automaton. However, they weight all events equally. This can lead to undesirably high influence of older complex execution paths on current complexity measurements.

Here, we consider the idea of *forgetting*, which means the events that happen earlier should add less to the sequence entropy than more recent ones. To this end, we assign a weight to each event based on its timestamp, or, to be more precise, based on the time difference between each event and the most recent event in the log. Thus, the older the event, the more it will be discounted.

There are two ways of doing so. The first, naïve way, is calculating the weight linearly as in Formula 1. Thus, this method is called *linear forgetting*. *Sequence entropy with linear forgetting* can be then computed similarly to the original *sequence entropy* by summing up the weights of the events instead of counting them.

$$w_l(e) = 1 - \frac{ts_{max} - ts(e)}{ts_{max} - ts_{min}} \tag{1}$$

While linear forgetting provides a first glimpse of how forgetting can be incorporated into sequence entropy, it has a number of problems, all of which are connected to the weight assignment. First, the weight of the earliest observed event is 0, meaning the contribution of this event to process complexity is disregarded. This is an inadequate solution. Second, it implies a linear nature of forgetting itself, which does not reflect reality closely enough.

Thus, we introduce a more advanced method – *exponential forgetting*. It is similar to the first method, the only difference being a slightly more complex weighting Formula 2.

$$w_e(e) = exp(-k\frac{ts_{max} - ts(e)}{ts_{max} - ts_{min}}) \tag{2}$$

With such weighting, the weight of the most recent event is 1 and earlier events have decreasing weights that never reach 0. In addition, the forgetting coefficient $k > 0$ is introduced. It enables further control over the contribution of the older events. The larger the coefficient, the less the weight of the event. The coefficient is considered to be 1 by default and in this paper we proceed with this default value. If it is desired to decrease the weight of older events even more, a larger coefficient $k > 1$ can be set. In the opposite case, one should use $0 < k < 1$.

---

[1] https://github.com/MaxVidgof/process-complexity.
[2] https://github.com/MaxVidgof/complexity-data.

3.2 Data Preparation

Our dataset comprises 14 publicly available real-life event logs from Business Process Intelligence Challenge (BPIC) [5–10]. In order to use them for statistical analysis, we apply the following procedure. First, we split each event log into time periods. Then, we measure process performance and complexity for each period. Afterwards, we create a merged dataset with all event logs and add an *industry* label to specify which industry the process belongs to. Finally, we remove the outliers. In this section, we describe these steps in more detail.

Time Periods. We start by splitting the event logs into time periods. First, we extract the minimum and maximum timestamp of the log events. We then set the month of the earliest event to be the starting period and the month of the last event to be the end period. Afterwards, we split the event log into months using an *intersecting* filter, as outlined in [34]. The filter assigns all traces to a period that started before or during the period and ended during or after the period. In other words, it selects all active traces in a given period. The choice of this filter further entails that a trace can be assigned to multiple periods.

Theoretically, it is possible to choose any granularity level at this point. I.e., we could choose shorter time periods like weeks or even days, but also longer ones like years. In any case, the choice of the the time interval mostly depends on case duration, e.g. setting time periods as granular as weeks if a process instance takes half a year on average will only increase the amount of data points without providing any additional value.

Complexity Measurement. For each time period, we measure the complexity of the process, treating traces in every period as separate event logs. We use all metrics defined in Sect. 2.1 and presented in Table 1. Furthermore, we add forgetting to both simple and enriched sequence entropy, as outlined in Sect. 3.1. Note that when calculating entropy metrics with forgetting, only a log partition (one month in our case) is considered, the minimal and maximal timestamp refer to the first and last event in this partition, not in the entire event log. It is also worth noting that from this point on we treat entropy metrics (both simple and enriched) as one group, which will be important in future steps. In addition, we also measure a set of what we called *generic* metrics. These are the metrics that can be measured out of the box by PM4Py[3] and include number of cases, number of activity repetitions, among others. We calculate a total of 38 complexity measurements for each time period.

Performance Measurement. As already discussed, there are various ways to assess process performance, including time and cost dimensions. However, most publicly available event logs lack cost data, as is the case for the event logs we chose for analysis. We thus focus on measuring process time as an indicator for process performance. More specifically, we examine the throughput time of the respective processes. While using cycle time could potentially be more insightful, many event logs do not contain information about starting timestamps of

---

[3] https://pm4py.fit.fraunhofer.de/.

activities, thus making it impossible to calculate cycle time. Throughput time, in contrast, can be easily calculated for any event log. For each time period, we calculate the median throughput time of all traces in that period. While average throughput time might seem a more intuitive measure, median throughput time is more robust.

Combined Dataset. After calculating the measurements for all logs, we combine them in a single dataset. In this dataset, each row consists of the originating event log, time period and corresponding measurements. We also want to control for industry-specific process characteristics that may influence performance but will not be captured by the complexity metrics. Thus, we also introduce the variable *industry* that specifies which industry the event log belongs to according to the SIC division[4]. Following the classification, we assigned BPIC 2011 [6] to *healthcare*, BPIC 2015 [5] and 2018 [10] to *public administration*, BPIC 2017 [7] to *finance*, BPIC 2019 [8] to *manufacturing*, and BPIC 2020 [9] to *education*.

Outlier Removal. The last step of our data preparation procedure is the removal of outlier periods. At the beginning and at the end of each event log, there are periods that contain considerably less traces than the rest of the log. Our assumption is this is a by-product of data extraction. Consider the following case: if it is decided to extract all traces from January until December of year Y, then all the traces that were *ongoing* in this time period will be extracted. However, some of them might have started earlier than January and some also ended later than December. It is indeed better to keep those traces in full rather than trim them (which might result in removing the start or end events and harm process discovery) or remove them entirely (in which case we would not have full information about resource usage). Still, in this case the extracted log will contain events occurring (at least) in years $Y - 1$ and $Y + 1$. For our approach, however, this is critical as this produces periods having not all traces, which, in turn, reduces the overall data quality. Thus, we filter out these periods either based on the event log description or based on the number of cases. Note that we only remove outliers on the level of time periods, not individual traces. The resulting dataset and some descriptive statistics are presented in Table 2.

3.3   Regression Analysis

After data preparation, we can now continue with building statistical models to explain throughput time based on process complexity. We start with two sets of independent variables. One with and one without industry as dummy variable. For each set, we have the following procedure. First, we build the models automatically. Then, we reduce model size in terms of number of variables in two steps such that we are left with simple yet powerful models. Note that we only consider linear combination of independent variables in this work. The remainder of the section describes the procedure in more detail.

---

[4] https://en.wikipedia.org/wiki/Standard_Industrial_Classification.

Table 2. Dataset description.

Source	Industry	Event logs	Periods	Traces	Events	Median throughput time
BPIC 2011 [6]	Healthcare	1	34	1143	150291	333 days
BPIC 2015 [5]	Public adm	5	51	832–1409	44354–59681	38–108 days
BPIC 2017 [7]	Finance	1	13	31509	1202267	19 days
BPIC 2018 [10]	Public adm	1	34	43809	2514266	267 days
BPIC 2019 [8]	Manufacturing	1	13	251734	1595923	64 days
BPIC 2020 [9]	Education	5	34	2099–10500	18246–86581	7–72 days

Independent Variables. We use two sets of independent variables: complexity metrics with and without *industry*. The reason for including *industry* is to account for effects on throughput time that are in the nature of the specific process and do not depend on complexity of its execution sequences. On the other side, however, it is interesting whether process performance can be explained purely in terms of theory-backed complexity measures.

Automated Model Selection. We use automated model selection procedures to select the best regression models based on Akaike Information Criterion (AIC). We use three directions: *forward*, *backward* and *both*. With *forward* selection, we start with a small set of variables (only *industry* if it is used or empty set of variables otherwise) and then add new variables one by one. At each step, the model with the lowest AIC is selected for the further step. The procedure stops if adding more variables does not decrease AIC or if all variables are already included. In the *backward* direction, we start from the model having all variables and remove them, also in a stepwise manner. Using *both* directions, we start from a simple model and then at each step we can either add or remove a variable, depending on what yields the best AIC. As a result, we get 3 models for each of the 2 independent variable setups.

Significant Variables. While the models produced in the previous step tend to have high explanatory power, they include a large number of independent variables, making them difficult to interpret and to use in practice. However, these models can often be further reduced in terms of the used variables. As a first step in this reduction, we remove all non-significant variables, i.e. variables with p-value larger than 0.001, from the models. We remove all variables that are not highly significant in one step. Interestingly, after this, some other variables in the model become less significant as well, thus we repeat the procedure until all the variables in the model are highly significant. In some cases, this procedure allows us to considerably reduce the size of the models, while keeping the explanatory power mostly unchanged. There are, however, cases, where the reduction leads to considerably lower explanatory power.

Minimal Models. Finally, we create *minimal* models. I.e., we further reduce the size of the models, such that at most one independent variable from each of the five categories (size, variation, distance, entropy, generic) is left. When selecting among variables in one category, the one with the lowest p-value is taken. In case two variables have the same p-value, the one that yields higher R-squared if left in the model is selected.

4   Results

In this section, we present our results. We start with automatically generated models that include *industry* as dummy variable. We present summaries of full models, significant models as well as minimized models. Then, we show the best of the minimized models in terms of R-squared. Afterwards, we present models that use only *theoretical* variables following the same structure.

4.1   Full Metrics

We started with a model where throughput time depends on industry and automatically added complexity metrics. The best model was achieved with *backward* selection. It included 23 variables and had R-squared of 0.9566887. Other models were very similar: *forward* selection also produced a model with 23 variables with very close R-squared of 0.9566887; selection in *both* directions produced a slightly smaller model – 18 variables – with still similarly high R-squared of 0.9556369.

Restricting to only significant variables allowed to massively reduce the complexity of the models: to 11 for *forward* and *backward* selection and to 10 for selection in *both* directions. The explanatory power of such models, however, only marginally decreased to roughly 0.94 for these models.

While being rather small, these models still contained some redundancy. They contained multiple variables belonging to the same categories of complexity defined in Sect. 2. In most cases, variables belonging to the same category were highly correlated, which is not surprising given they measure the same aspects of complexity. When removing such redundancy by leaving only one variable per category, we arrived at two models: the one resulting from minimizing the *forward* selection model had 6 variables and R-squared of almost 0.92, and the one resulting from minimizing the *backward* selection model had only 5 variables because the *generic* variables were not among the significant variables, and had R-squared of 0.87. Minimizing the model resulting from selecting in *both* directions resulted in the same model as from *forward* selection and thus was dropped out.

The summary of the models can be viewed in Table 3. The best of the minimal models was the model resulting from *forward* selection. It is presented in more detail in Table 4. Note that while several possible values for the *industry* variable are present (*finance, healthcare, manufacturing* and *public*), only one of them is present in the formula for each observation. The default value is the remaining industry, *education*, thus in case of process in education no industry variable should be considered for the estimation.

Table 3. Summary of regression models built from full set of metrics.

Size	Variable selection method	Number of variables	R-squared
Full	forward	23	0.9554849
Full	backward	23	0.9566887
Full	both	18	0.9556369
Significant	forward	11	0.9489275
Significant	backward	11	0.9442622
Significant	both	10	0.9391087
Minimal	forward	6	0.9195022
Minimal	backward	5	0.8726173

Table 4. Best model with industry as dummy variable.

Variable	Estimate
Intercept	101.96221156
Finance	−199.51131885
Healthcare	280.14409159
Manufacturing	−251.45111110
Public	−124.96774620
Magnitude	0.00950598
Level of detail	3.04980147
Affinity	−195.87822418
Number of activity repetitions in period	−0.00192884
Enriched variant entropy	−0.00079535
R-Squared	0.9195022

4.2 Theoretical Metrics

While the models presented above explain most of the variance in through-put time, they heavily rely on the *industry* variable that accounts for industry-specific process characteristics. However, we see that relying only on *theoretical* variables, i.e. only complexity variables, without any adjustments, still gives valuable results.

First, we can see that some of the automatically generated models are in fact even better than the ones shown in the previous section. Namely, the models achieved with *forward* selection and selection in *both* directions have slightly higher R-squared (0.969 vs 0.955) and at the same time have less variables (19 and 15 vs 23 and 18, respectively). The model generated with *backward* selection is slightly worse, still on par with its counterpart.

Reducing the models to significant variables only gave differing results, some of them are very optimistic. Indeed, the model with *forward* selection could be reduced to only 9 variables while still having R-squared of 0.94. For the other

Table 5. Summary of regression models built from theoretical metrics.

Size	Variable selection method	Number of variables	R-squared
Full	forward	19	0.9696743
Full	backward	26	0.9513178
Full	both	15	0.9698724
Significant	forward	9	0.93589
Significant	backward	18	0.9479194
Significant	both	7	0.8202257
Minimal	forward	2	0.7965621
Minimal	backward	4	0.6385549
Minimal	both	4	0.7294595

Table 6. Best model with theoretical variables only.

Variable	Estimate
(Intercept)	−142.1478
Average trace length	4.8333
Affinity	274.9644
R-Squared	0.7965621

two models, the results are also very good, yet not that impressive. For instance, the model with *backward* selection could be reduced from 26 to 18 variables while maintaining its R-squared of over 0.94, it is still very large. The model with selection in *both* directions could be reduced to 7 variables only, however, at a price of considerably lower R-squared of 0.82.

The models could be reduced even further, with the smallest minimal model containing only 2 variables. However, the explanatory power of such models barely reaches 0.8 in the best case. The models are summarized in Table 5. The best minimal model containing only theoretical variables is presented in more detail in Table 6. Interestingly, the significant model for *forward* selection only contained complexity metrics from 2 categories, thus the minimal model has only 2 variables.

5   Discussion

5.1   Implications

During this work, we have made some interesting observations. First of all, we see that industry alone explains 80% of variance in the dependent variable throughput time. This means that processes in different industries and in different companies are so different in their nature that knowing where the process is executed allows us to infer a lot about its throughput time without considering its complexity at all.

However, adding the complexity dimension on top of the industry allows us to gain even more insightful information, explaining up to 95.6% of variation in throughput time. One can look at it from different perspectives. On the one hand, when having 80% of the output explained by one categorical variable that does not even need any further computation, one can say that all possible additions to it are only marginal and are not worth considering. On the other hand, being able to explain more than 95% of variance in throughput time is a valuable capability that is worth the effort. In addition, there is a compromise solution with the minimal models. One can still achieve remarkable explanatory power using only a handful of metrics: 5 complexity metrics on top of industry can explain roughly 92% of variance.

Despite having such high explanatory power, models containing *industry* as an independent variable have received criticism as the observations used to build them only consider one or two processes per industry and thus are not necessarily representative. In the light of this criticism, we also developed models explaining throughput time solely by the complexity of the corresponding processes. The good news is that these *theoretical* variables successfully managed to compensate the information gained by industry. After all, the industry a process belongs to is not something that directly influences process performance by itself but rather a factor that contributes to how the process is set up, thus it is not surprising that these differences are (at least to some extent) visible in the complexity of the process.

These models with theoretical variables achieve similar results in terms of R-squared while having slightly smaller variable counts. Interestingly, the full models that were generated in the first step have even slightly higher R-squared. This resulted from different starting points: for the models with industry, the lower boundary for model selection was a model already containing the industry because it was considered a baseline. The theoretical models, instead, had a constant as their lower boundary. The interpretation of this is that if we set no boundaries and allow (but do not force) selecting the *industry* variable, it might be the case that the best models will still not contain it, which even better supports our idea of being able to explain process performance using its complexity only.

It is also interesting to look at these models in more details because they are structurally different from the ones including *industry*. In the first step, these two kinds of models are similar in terms of both the number of variables and R-squared. In the second step, where we restrict the models to only containing significant variables, some differences become visible. Theoretical models achieved with *forward* selection and selection in *both* directions have slightly smaller R-squared than their counterparts including *industry*. However, this can be attributed to just having also slightly smaller number of variables. The model achieved with *backward* selection, however, does not fit the pattern. Its full version had a lot of significant variables, thus it could not be reduced much, which also allowed it to keep most of its R-squared. The most interesting differences become visible in the last reduction step, where we only choose one variable per category. Theoretical models contain a more homogeneous sets of variables, i.e. more variables from one category, while some categories can be missing entirely. Thus, reducing them to minimal models yields much smaller (2–4 variables) but also much less powerful (R-squared 0.63–0.79) models. Note that models containing industry cannot have such low values at all as *industry* alone would have R-squared of 0.8.

Up to this point, our goal was to develop the most simple yet powerful explanatory models. However, as we achieved this, we asked ourselves whether we can tweak the models a bit further to gain more explanatory power while not increasing the complexity of the models by much. The first approach we tried was to add interplay between the variables in our models. This, however, was not very fruitful. Adding pairwise interplay terms between all variables in models did not improve R-squared. Adding all possible interplay terms between two but also more variables in the model did increase the R-squared (for instance, we achieved R-squared of 0.96 with only 6 variables for the model including industry), however, such terms are very difficult to explain.

Another approach that we took was clustering the event logs based on their median throughput times and developing separate models for each cluster. With this approach, we could achieve slightly better (or smaller) models in the first step. However, R-squared falls drastically when we try to minimize models.

The last observation that we did, also going in the direction of clustering, is that in the end processes are different and while we can explain a lot of their variance in terms of complexity, there is no one-fits-all solution. Our results and models should be thus considered as toolbox, and the practitioners should analyze which exact variables makes sense in case of their processes.

5.2   Future Work

We see several promising avenues for future research. First, our quantitative analysis of throughput time can be extended in several directions. On the one hand side, future work can use principal component analysis to cluster the independent variables. This would not only reduce model size, but might lead to interesting insights how different theoretical variables can be empirically grouped together. On the other hand side, our analysis presented in this paper can be complemented by the prediction of throughput time. We deem our models a suitable starting point for such an endeavour.

Second, behavioral studies can investigate how process complexity develops over time. For example, how process complexity is reduced and increases in course of business process standardization initiatives. Such a study could focus on the specific actions taken by management and unpack how they influence process performance and overall complexity. We deem such studies particularly fruitful, if they can complement insights from event logs with detailed interviews with key stakeholders, such as managers and process experts.

Third, more generally, there are plenty of opportunities for behavioral business process research due to the increasing availability of digital trace data [13]. Future research can make use of digital trace data from event logs to investigate how business process change over time, contributing to theory on business process change and routine dynamics [20,33].

6   Conclusion

In this paper, we reported on a study in which we empirically examined the link between process complexity and throughput time. Based on 14 event logs and 38 different process complexity metrics, we created various statistical models that explain the throughput time of business processes. Our models are able to explain a large share of the variance in the throughput time, reaching R-squared values of up to 0.96. Our results provide important implications for research on process complexity and process standardization. Practitioners can use our implementation of the different complexity measures to monitor their processes.

References

1. Augusto, A., Mendling, J., Vidgof, M., Wurm, B.: The connection between process complexity of event sequences and models discovered by process mining. Inf. Sci. **598**, 196–215 (2022). https://doi.org/10.1016/j.ins.2022.03.072
2. Cappiello, C., Comuzzi, M., Plebani, P., Fim, M.: Assessing and improving measurability of process performance indicators based on quality of logs. Inf. Syst. **103**, 101874 (2022). https://doi.org/10.1016/j.is.2021.101874
3. Cho, M., Song, M., Comuzzi, M., Yoo, S.: Evaluating the effect of best practices for business process redesign: an evidence-based approach based on process mining techniques. Decis. Support Syst. **104**, 92–103 (2017)
4. Cornwell, B.: Social Sequence Analysis: Methods and Applications, vol. 37. Cambridge University Press, Cambridge (2015)
5. van Dongen, B.B.: BPI challenge 2015 (2015). https://doi.org/10.4121/uuid: 31a308ef-c844-48da-948c-305d167a0ec1, https://data.4tu.nl/collections/BPI_Cha llenge_2015/5065424/1
6. van Dongen, B.: Real-life event logs - hospital log (2011). https://doi. org/10.4121/UUID:D9769F3D-0AB0-4FB8-803B-0D1120FFCF54, https://data.4 u.nl/articles/_/12716513/1
7. van Dongen, B.: BPI challenge 2017 (2017). https://doi.org/10.4121/UUID: 5F3067DF-F10B-45DA-B98B-86AE4C7A310B, https://data.4tu.l/articles/_/126 96884/1
8. van Dongen, B.: BPI challenge 2019 (2019). https://doi.org/10.4121/UUID: D06AFF4B-79F0-45E6-8EC8-E19730C248F1, https://data.4tu.nl/articles/_/1271 5853/1
9. van Dongen, B.: BPI challenge 2020 (2020). https://doi.org/10.4121/UUID: 52FB97D4-4588-43C9-9D04-3604D4613B51, https://data.4tu.nl/collections/_/50 65541/1
10. van Dongen, B., Borchert, F.F.: BPI challenge 2018 (2018). https://doi. org/10.4121/UUID:3301445F-95E8-4FF0-98A4-901F1F204972, https://data.4tu. nl/articles/_/12688355/1

11. Dumas, M., Rosa, M.L., Mendling, J., Reijers, H.A.: Fundamentals of Business Process Management, 2nd edn. Springer, Cham (2018)
12. Gibson, C.B., Birkinshaw, J., McDaniel Sumpter, D., Ambos, T.: The hierarchical erosion effect: a new perspective on perceptual differences and business performance. J. Manage. Stud. **56**(8), 1713–1747 (2019)
13. Grisold, T., Wurm, B., Mendling, J., vom Brocke, J.: Using process mining to support theorizing about change in organizations. In: 53rd Hawaii International Conference on System Sciences, HICSS 2020, Maui, Hawaii, USA, 7-10 January 2020, pp. 1–10. ScholarSpace (2020). http://hdl.handle.net/10125/64417
14. Gunasekaran, A., Nath, B.: The role of information technology in business process reengineering. Int. J. Prod. Econ. **50**(2–3), 91–104 (1997)
15. Günther, C.: Process mining in flexible environments. Ph.D. thesis, Technische Universiteit Eindhoven (2009). https://doi.org/10.6100/IR644335
16. Hærem, T., Pentland, B.T., Miller, K.D.: Task complexity: extending a core concept. Acad. Manag. Rev. **40**(3), 446–460 (2015)
17. Ketokivi, M.A., Schroeder, R.G.: Perceptual measures of performance: fact or fiction? J. Oper. Manag. **22**(3), 247–264 (2004)
18. Kuhn, M., Johnson, K., et al.: Applied Predictive Modeling, vol. 26. Springer, Cham (2013)
19. Mendling, J.: Metrics for Process Models: Empirical Foundations of Verification, Error Prediction, and Guidelines for Correctness. Lecture Notes in Business Information Processing, vol. 6. Springer, Cham (2008)
20. Mendling, J., Berente, N., Seidel, S., Grisold, T.: The philosopher's corner: Puralism and pragmatism in the information systems field: the case of research on business processes and organizational routine. ACM SIGMIS Database: DATABASE Adv. Inf. Syst. **52**(2), 127–140 (2021)
21. Mendling, J., Reijers, H.A., van der Aalst, W.M.: Seven process modeling guidelines (7pmg). Inf. Softw. Technol. **52**(2), 127–136 (2010)
22. Mendling, J., Sánchez-González, L., García, F., La Rosa, M.: Thresholds for error probability measures of business process models. J. Syst. Softw. **85**(5), 1188–1197 (2012)
23. Münstermann, B., Eckhardt, A., Weitzel, T.: The performance impact of business process standardization. Bus. Process. Manag. J. **16**, 29–56 (2010)
24. Pentland, B.T.: Conceptualizing and measuring variety in the execution of organizational work processes. Manage. Sci. **49**(7), 857–870 (2003)
25. Pentland, B.T., Liu, P., Kremser, W., Hærem, T.: The dynamics of drift in digitized processes. MIS Quart. **44**(1) (2020)
26. Pentland, B.T., Mahringer, C.A., Dittrich, K., Feldman, M.S., Wolf, J.R.: Process multiplicity and process dynamics: weaving the space of possible paths. Organ. Theory **1**(3), 2631787720963138 (2020)
27. Recker, J.: Scientific Research in Information Systems: A Beginner's Guide, 2nd edn. Springer, Cham (2021). https://doi.org/10.1007/978-3-030-85436-2
28. Schäfermeyer, M., Rosenkranz, C., Holten, R.: The impact of business process complexity on business process standardization. Bus. Inf. Syst. Eng. **4**(5), 261–270 (2012)
29. van der Aalst, W.M.: Process Mining: Data Science in Action, 2nd edn. Springer, Cham (2016)

30. Vidgof, M., Mendling, J.: Leveraging event data for measuring process complexity. In: Montali, M., Senderovich, A., Weidlich, M. (eds.) Process Mining Workshops - ICPM 2022 International Workshops, Bozen-Bolzano, Italy, 23–28 October 2022, Revised Selected Papers. Lecture Notes in Business Information Processing, vol. 468, pp. 84–95. Springer, Cham (2022). https://doi.org/10.1007/978-3-031-27815-0_7

31. Vidgof, M., Mendling, J.: Leveraging event data for measuring process complexity. In: Montali, M., Senderovich, A., Weidlich, M. (eds.) Process Mining Workshops, pp. 84–95. Springer Nature Switzerland, Cham (2023). https://doi.org/10.1007/978-3-031-27815-0_7

32. Wüllenweber, K., Beimborn, D., Weitzel, T., König, W.: The impact of process standardization on business process outsourcing success. Inf. Syst. Front. **10**(2), 211–224 (2008)

33. Wurm, B., Grisold, T., Mendling, J., vom Brocke, J.: Business process management and routine dynamics. Camb. Handb. Routine Dyn., 513–524 (2021)

34. Wurm, B., Grisold, T., Mendling, J., vom Brocke, J.: Measuring fluctuations of complexity in organizational routines. In: Academy of Management Proceedings, vol. 2021, p. 13388. Academy of Management Briarcliff Manor, NY 10510 (2021)

35. Wurm, B., Schmiedel, T., Mendling, J., Fleig, C.: Development of a measurement scale for business process standardization (2018)

Deviation from Standards and Performance in Knowledge-Intensive Processes: Evidence from the Process of Selling Customized IT Solutions

Mikhail Monashev[1]([✉]) [iD], Michal Krčál[1] [iD], and Jan Mendling[2,3,4] [iD]

[1] Masaryk University, Lipová 41a, Brno 602 00, Czech Republic
mikhail.monashev@mail.muni.cz
[2] Humboldt-Universität zu Berlin, Unter den Linden 6, 10099 Berlin, Germany
jan.mendling@hu-berlin.de
[3] Weizenbaum-Institut e. V, Hardenbergstraße 32, 10623 Berlin, Germany
[4] Wirtschaftsuniversität Wien, Welthandelsplatz 1, 1020 Vienna, Austria

Abstract. Standardization has been shown to be a reliable method of reducing unpredictability and consequently improving the performance of routine processes. However, it is surprising that the literature on knowledge-intensive processes (KiPs) rarely discusses this option or portrays such processes as inherently unsuitable for standardization. This presents the question of whether and to what extent standardization and following standards benefit KiPs. In this paper, we report findings from a case study on the impact of deviations from standards on the sales process of an IT service provider. Each instance of the sales process is a new project which involves a series of tasks characterized by different degrees of knowledge intensity. The findings are based on two data sources: (i) process documentation, and (ii) semi-structured interviews with managers and process participants. We applied the constructivist grounded theory method in the analysis of these materials. Our analysis yielded a series of propositions that characterize the benefits and issues that deviations from standards may bring to KiPs and the circumstances under which they are likely to materialize. Our study implies that deviations from standards mostly undermine the performance of KiPs unless they are initiated internally by process actors when standards are not sufficiently robust.

Keywords: Knowledge-Intensive Processes · Standardization · Communication

1 Introduction

Process standardization via imperative modeling, which prescribes all possible execution paths, is considered an essential management mechanism for improving routine processes with a low degree of unpredictability [24]. Imperative standardization allows companies to decrease the costs and time of process execution and increase the quality of process outputs [14, 15]. However, it is argued that imperative standardization is ineffective for knowledge-intensive processes (KiPs), which are highly unpredictable [12, 24], and therefore often deviate from the expected paths [10].

© The Author(s), under exclusive license to Springer Nature Switzerland AG 2023
C. Di Francescomarino et al. (Eds.): BPM 2023, LNCS 14159, pp. 430–446, 2023.
https://doi.org/10.1007/978-3-031-41620-0_25

For this reason, efforts to standardize in KiPs mostly focus on setting boundaries for the process execution by declaring possible procedures and rules that need to be followed when executing them [18]. The body of literature dedicated to standardizing KiPs [9, 13, 18] focuses mainly on the design aspects of declarative modeling languages. Nevertheless, little is known about the various aspects of applying declarative standards in practice.

The aim of this paper is to expand the current body of literature on standardization in KiPs by looking at: (i) how declarative standards are followed in practice, (ii) what makes process actors deviate from pre-defined rules, and (iii) what are the consequences of these deviations for process performance. To do so, we conducted a qualitative case study and derived propositions from the collected data following the constructivist grounded theory method [2]. We explored deviations from standards in the sales process of an IT service provider by obtaining information from the process documentation and conducting a series of semi-structured interviews with management and participants in two specific instances of the sales process. Our findings reveal both positive and negative consequences of deviating from standards in the execution of KiPs and the circumstances under which these consequences are likely to materialize.

The paper is structured as follows. Section 2 reviews the literature on process standardization and KiPs. Section 3 describes the procedures for the data collection and data analysis. Section 4 provides background on the sales process and sales projects. Section 5 deduces propositions from the collected evidence. Section 6 relates the findings of our study to the literature and outlines practical implications. In Sect. 7 we present our conclusions.

2 Theoretical Background

This section first summarizes previous findings on the positive effects of standardization on various aspects of process performance and drivers of and barriers to standardization. It goes on to characterize KiPs in distinguishing them from routine processes and outlines the benefits standardization might bring to KiPs.

2.1 Process Standardization

Making the different process instances as uniform as possible [15] is expected to yield lower costs of process execution, improved collaboration, decreased process throughput time, improved quality and control, or process automation. Similar to Schäfermayer et al. [16], we understand standardization as the "unification of business processes and the underlying actions within an organization." Standardization aims to "enable handoffs across process boundaries in terms of information, and to improve collaboration and develop comparative measures of process performance" [3, p. 102].

Most empirical studies on process standardization have focused either on drivers and influencing factors of standardization or on the impact of standardization on business or process performance that has been identified in theory. Process standardization has been statistically shown to positively impact process performance in terms of time, cost, and quality [14].

Many studies that have focused on process standardization have been conceptual. For instance, Wurm and Mendling [23] proposed a theoretical model of process standardization that identified that standardization could be increased by: (i) developing a formalized process manual and process training, and (ii) constraining IT. Both factors could be positively influenced by a culture of uncertainty avoidance and collectivism. Conceptual studies underline that the positive effects of standardization might be explained by its ability to reduce uncertainty and unpredictability in process execution.

Despite the positive effects that standardization may bring to companies, the opportunities for standardization might be limited by a number of factors. For instance, process complexity, which is a frequently cited characteristic of KiPs [1, 4], has been shown to hamper standardization [16]. It is therefore necessary to examine how KiPs are conceptualized in the literature.

2.2 Knowledge-Intensive Processes

In contrast to routine processes, KiPs involve social interactions and depend on unexpected events, which means these processes are highly unpredictable [4, 6]. Szelagowski [19] defines three types of KiPs classified in terms of the increasing degree of unpredictability associated with the execution of them: (i) structured processes with ad-hoc exceptions, (ii) unstructured processes with pre-defined elements, and (iii) unstructured processes. In the latter two types of processes, the unpredictability is so high that it becomes impossible to pre-define the control flow of the KiPs in advance, and thus the process structure emerges during the execution of them [4, 10]. Therefore, the unpredictability associated with KiPs presents a challenge to ensuring high process performance [8].

The approach to addressing this challenge in the literature is twofold. On the one hand, it is claimed that the challenge of unpredictability cannot be addressed by standardization [11, 12], as process execution would inevitably deviate from the expected paths [10]. As an alternative to standardizing KiPs, various publications recommend empowering workers involved in the process execution by giving them decision-making autonomy, setting high requirements for creativity and innovation that allow for dealing with unexpected situations, and improving communication and collaboration between process stakeholders [1, 4, 6, 18].

On the other hand, there have been attempts to solve the challenge of unpredictability by modifying the approach to standardization to the needs of the KiPs and developing declarative modeling notations [9, 13, 18]. In contrast to pre-defining all the possible execution paths, as is traditionally done in imperative approaches [20], declarative approaches prescribe rules and constraints that set boundaries for the process execution as well as possible procedures that may be executed to achieve process goals, but without specifying the exact order of these procedures. Although publications on declarative process modeling outline the approach to standardizing KiPs, to the best of our knowledge, they do not investigate the practical aspects of applying these standards, including the consequences of deviating from the standards and the circumstances under which these deviations occur.

3 Methodology

We studied standardization and knowledge intensity in sales processes, which are specifically interesting in this context, because they are considered an example of KiPs [7]. We collected our data in a medium-sized Czech IT company, which delivers customized IT solutions for its clients (hereafter referred to as the Company). The project-oriented culture of the Company provides employees with substantial opportunities for autonomy and self-organization, in order to address the volatile nature of the business.

We studied the process documentation and conducted a series of semi-structured interviews to obtain an insight into two sales projects, Projects A and B, which represent instances of the sales process and vary in multiple aspects, as described in Sect. 4. We then applied the constructivist grounded theory method [2] to derive theoretical propositions from the collected evidence and validated these propositions by means of semi-structured interviews with some of the initial respondents and external experts.

3.1 Data Sources

To obtain an in-depth insight into the sales process of the Company, we relied on two data sources: (i) process documentation, and (ii) semi-structured interviews with the CEO and with participants in Projects A and B.

Process Documentation. The Company gave us access to the Confluence[1] pages that describe the standard flow of the sales process and contain methodological guidelines on executing specific sub-processes within the process. Examining the process documentation served as the input for developing a description of the standardized process flow (see Sect. 4.1) and helped us structure the interviews.

Semi-structured Interviews. We conducted two entry interviews with the CEO of the Company, who had previously held the post of Head of Sales and currently remains involved in the sales process. In the first interview, we asked the CEO to describe the sales process, while in the second we asked him to identify recent sales projects of different sizes that could serve as input for our analysis. The CEO evaluated Projects A and B as those best suited to our criteria.

We obtained insights into Projects A and B by interviewing the project participants as specified in Table 1. The participants were asked to describe the project they were involved in, reflect on their role, and identify what went well and what could have been done better. Moreover, R1, R2, and R7 were asked to reflect on the degree of knowledge intensity and standardization associated with each sub-process of the sales process, as, unlike other respondents, they were involved in all of the sub-processes.

All the interviews were semi-structured. Following Scheibelhofer's [17] recommendations, we started the interview with broad, open-ended questions, making notes while the interviewee addressed the questions, and then asking follow-up questions based on these notes.

---

[1] https://www.atlassian.com/software/confluence.

Table 1. Summary of interview data collected

ID	Position	Role in Project A	Role in Project B
R1	CEO	Business Representative	Formal Acceptance
R2	Bid Manager, Sales	Bid Manager	N/A
R3	Business Analyst, Analysis	Business Analyst	N/A
R4	Head of Project Management Office	Project Manager	N/A
R5	Lead of JavaScript Area	Solution Architect	N/A
R6	Quality Assurance (QA) Engineer	N/A	Solution Architect
R7	Head of Sales	N/A	Bid Manager

In total, we conducted nine semi-structured interviews. The interview duration varied between 20 and 59 min. The two initial interviews with the CEO were conducted on-site, while the remaining seven interviews were done via online calls. We recorded the interviews and sent anonymized transcripts to the interviewees for validation. R2 and R4 provided clarifying notes to the transcripts of their interviews.

3.2 Data Analysis

This study applied the constructivist grounded theory method to analyze the collected data. We followed a three-stage process of initial, focused, and theoretical coding, as per recommendations by Charmaz [2]. The initial coding was done line-by-line to restrict analyst-imposed interpretations and allow codes to emerge from the data.

We integrated the initial codes into five focused codes: (i) "process performance", (ii) "characteristics of standardization", (iii) "problems in execution", (iv) "things that went well in execution", and (v) "deviation from standards". The latter focused code, which describes the degree to which the execution of specific sub-processes in Projects A and B was guided by standard procedures, for example, included initial codes such as "not being good at following standard procedures" and "not following standardized methodology."

In the theoretical coding, we attempted to define how the focused codes related to each other. When we related the focused codes "deviation from standards" and "process performance", we found that deviations from standards had happened in both projects but had led to different results, as the performance of Project A was characterized as *low*, while the performance of Project B was *high*. Based on the principles of the constructivist grounded theory method [2], we followed up on this surprising finding and revised the focused codes, focusing on deviations from standards.

After this revision, we came up with the following focused codes: (i) "factors that had led to a deviation from standards (e.g., communication, robustness)", (ii) "circumstances under which a deviation had happened (e.g., maturity of relationships with the client, solution complexity)", and (iii) "drivers of standardization" (i.e., external, internal). In the subsequent theoretical coding, we related these revised focused codes and deduced four theoretical propositions.

We validated the propositions by conducting semi-structured interviews, in which the respondents were asked to relate the propositions to their past experiences. First, we conducted the internal validation with R1 and R7, who are directly involved in the sales process and therefore have insight into most sales projects executed in the Company. Then we conducted four semi-structured validation interviews with workers involved in project-oriented KiPs, but not affiliated with the Company. The profiles of experts who participated in the internal and external validations are provided in Table 2.

Table 2. Expert profiles

ID	Process	Experience (years)	Company size
R1	Sales (in the Company)	10+	Medium
R7	Sales (in the Company)	5+	Medium
E1	Software deployment	10+	Large
E2	Software development	10+	Medium
E3	Supply chain management	5+	Large
E4	Software testing	5	Large

4 Background on the Sales Process and the Sales Projects

This section introduces the standard flow of the sales process and categorizes each of its five sub-processes in terms of their embedded unpredictability (and thus, knowledge intensity) based on Szelagowski's [19] framework. Moreover, the section presents two specific process instances – Projects A and B. The section concludes by reflecting on how the process was followed in each of the Projects.

4.1 Standard Flow of the Sales Process

The sales process consists of five sub-processes, as presented in Fig. 1. The process starts when a new opportunity is identified in the market. At this point, the Head of Sales either assigns the lead to one of the three bid managers (BM) or to himself, thereby taking on the role of the BM. Once the BM has been assigned, he/she continues with qualifying the opportunity.

Qualify Opportunity. When qualifying the opportunity, the BM collects the data necessary to address the questions in a standardized checklist. The process documentation specifies that the sub-process is considered complete only if all the questions on the checklist have been addressed. However, the procedure of data collection is not standardized and emerges based on the judgment of the BM. Therefore, we characterized the "Qualify Opportunity" sub-process as *unstructured with pre-defined fragments*.

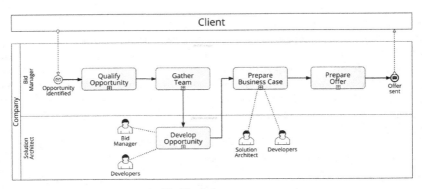

Fig. 1. Sales process

Gather Team. The process continues with gathering the presales team. The allocation of people is initiated by the BM at a weekly traffic meeting. After the traffic meeting, the team leaders of the requested people check their availability and either approve the BM's request or suggest other people who could take up the roles. Therefore, the process documentation clearly prescribes a control flow for the "Gather Team" sub-process. At the same time, the standard does not prevent the BM from going to the traffic meeting with a list of pre-selected candidates for the team. Therefore, we characterized the "Gather Team" sub-process as *structured with ad-hoc exceptions.*

Develop Opportunity. During the opportunity development, the BM focuses on gathering additional data via intensive communication with the client. In parallel, the Solution Architect (SA) and developers progress with the design of the solution.

As with the "Qualify Opportunity" sub-process, the work of the BM is guided by a standardized checklist. At the same time, the work of the SA was characterized by R5 as follows:

I don't have [any documents I can consult]. It is basically only [my] knowledge of an [software] application and [my] imagination.

This confirms that process execution is reliant on previous knowledge and expertise, as well as on the creativity of the workers involved in its execution, which signals the high knowledge intensity of the process.

Although the process documentation does not prescribe the specific activities that need to be executed during the "Develop Opportunity" sub-process, it limits the scope of possible activities by clarifying the following point:

The presales team works on identifying the information necessary for the creation of a business case.

Based on the above-mentioned features, we concluded that the "Develop Opportunity" sub-process is *unstructured with pre-defined fragments.*

Prepare Business Case. The goal of the "Prepare Business Case" sub-process is to complete a standardized Excel sheet with the financial and workload estimates. As illustrated by the comment from R4 below, the estimates are done on the basis of the previous experience and knowledge of workers:

> I assess all these risks by literally thinking about the project.

Therefore, we characterized the "Prepare Business Case" sub-process as *unstructured with pre-defined fragments*.

Prepare Offer. In the final stage of the sales process, the BM prepares an offer. At this stage, outputs from previous steps are combined into a pitch presentation. The offer is then sent to the client. The process is successfully concluded when the client accepts the offer.

The standard procedure prescribes a control flow for the "Prepare Offer" sub-process and also the standardized scope of documentation that needs to be prepared. Moreover, new documentation is prepared based on existing documentation from past projects. However, the project is associated with ad hoc exceptions, as reported by the R1:

> The offer … is usually accompanied by mandatory documents … There's very basic stuff that is used in individual cases. But it doesn't really need any sort of major experience or input or insight.

On the basis of the above-mentioned observations, we characterized the "Prepare Offer" sub-process as *structured with ad-hoc exceptions*.

To sum up, the sales process in the Company consists of sub-processes that are either *structured with ad-hoc exceptions* or *unstructured with pre-defined fragments*. Therefore, the theoretical propositions presented in Sect. 5 would only hold for these types of processes and do not hold for purely structured or purely unstructured types, therefore covering most KiPs encountered in practice.

4.2 Sales Projects

To see how the sales process is executed in practice, we looked at two specific instances of it, represented by Projects A and B. Table 3 compares the projects in terms of six factors that were reflected in the semi-structured interviews related to both projects, including performance and deviation from standards.

Project A. Client A, an investment brokerage firm, was a new client. In Project A, the Company was co-developing a new software application for Client A together with a third party. The Company was responsible for the front-end development, while the third party was responsible for the back-end development. The project started in December 2021, when the Company won the tender initiated by Client A. It was paused in October 2022 because of the delay in back-end development. The project restarted in January 2023 and was completed in March 2023 after the client accepted the offer.

When reflecting on Project A's performance, which took sixteen months to complete, R1 reported that the project was associated with "ridiculous overheads" and took "too much time", therefore, we evaluated the project performance as *low*.

The R2 reported that the execution of the sales process in Project A had followed the standards for all sub-processes apart from the "Develop Opportunity" sub-process, in

Table 3. Comparison of the sales projects

Factor	Project A	Project B
Client	New	Long-term
Type of project initiation	Reactive	Proactive
Third parties involved	1	0
Duration	16 months (including a 3-month freeze)	2 months
Performance	Low	High
Deviation from standards	Yes	Yes

which the presales team "started the analysis phase without having a signed contract", which is against the rules specified in the process documentation. R3 and R5, who were involved in the analysis, confirmed the non-standard character of the "Develop Opportunity" sub-process.

Project B. Client B, a postal service firm, was a long-term client. The Company built a web portal for the client through a series of consecutive projects. Project B started in January 2023, when R7 approached Client B with a proposal to automate the quality assurance of the web portal, which Client B agreed to consider. After specifying the offer and negotiating the details, the contract was signed in February 2023, thereby marking the completion of Project B.

When reflecting on Project B's performance, R1 referred to the project as "successful". In particular, he made the following comment:

I think that [Project B] went well. We were able to handle the risk properly, we were able to make the pitch correctly.

The execution of the sales process in Project B also deviated from the process documentation. The deviations happened in the execution of the "Prepare Business Case" sub-process, as reported by R7.

To sum up, although Projects A and B varied in terms of process performance, the execution of both projects deviated from standards. This contradiction attracted our attention and guided the further analysis. We wanted to ascertain the reasons why deviation from standards was associated with different levels of process performance and the circumstances under which deviations occurred in both projects.

5 Findings

This section describes within-case and cross-case analyses of Projects A and B and deduces propositions regarding (i) associations between deviations from standards and process performance and (ii) necessary conditions enabling these associations. The section concludes with the results of an internal and external validation of the derived propositions.

5.1   Deviation from Standards and Low Process Performance (Project A)

The performance of reactively initiated Project A, whose execution deviated from the rules prescribed in the documentation, was *low* (see Sect. 4.2). The within-case analysis of Project A revealed the reasons for the deviation, which are described in more detail below.

Externally Driven Deviation. We first looked at the characteristics of the deviation from standards that occurred in Project A. Our data showed that the deviation from standards was driven by Client A with the Company having to adjust to the client's demands, which is supported by the comment below from R3:

> Client A was cooperating with UX directly without user stories … At some point, we were losing the continuity and the Company offered my service to Client A to do the documents for the project, but Client A didn't accept this offer, they were doing their own documents. So, my role in the team was to check their documents [for] gaps and continuity between [back end and front end].

We labeled this deviation from standards initiated by an external process stakeholder (e.g., a client) an "externally driven deviation" (EDD). We then looked at the dynamics of the execution of Project A and identified two factors that changed between the pre-freeze phase of Project A (Phase 1), which lasted ten months (from December 2021 to October 2022), and the post-freeze phase (Phase 2), which lasted three months (from January to March 2023). Our analysis showed that the quality of communication and unpredictability changed dramatically between the two phases.

Communication. In Phase 1, R2, R3, and R5 highlighted issues related to communication with Client A. They reported feeling that representatives of Client A lacked a common opinion and were trying to find a consensus during the joint meeting rather than beforehand. R3 summarized the situation as follows:

> [Client A representatives] didn't communicate with each other. Marketing would make a user story, but [Back-end Development] wouldn't read it.

In Phase 2, R1 had to intervene in communication with Client A and began to negotiate with the client:

> We went up front and we said: "Hey guys, we really don't need this project. So, if we are going to, like, walk away from it, neither side will be happy, because you won't get your application, we won't get our money, but ultimately, we don't really care. So, let's really work hard together to, you know, find this mutually agreeable solution."

Consequently, Project A finalized shortly after overcoming the communication barriers, resulting in the signing of the contract.

Unpredictability. Another factor that evidently changed over the course of Project A was the unpredictability. On the basis of our analysis, we characterized Phase 1 as highly unpredictable, whereas Phase 2 was associated with less unpredictability. As stated in Sect. 4.2, Project A was the first time Client A and the Company had interacted. Therefore, the initial maturity of the relationships between the Company and the client was low, which subsequently improved during the project execution, as confirmed by R1 and R2.

The second factor that affected the unpredictability was the dependence on the third party in the solution development. In fact, the project was frozen due to delays in the back-end development on the third-party side. In Phase 2, the dependency on the third party was removed, since the back end had been delivered, which allowed the Company to proceed with specifying requirements for the front end. R3 and R5, who were directly involved in the task, reported that this situation led to more predictability and certainty.

To conclude, the situation of Project A was characterized by an EDD, low communication quality and high unpredictability, which undermined the process performance in Phase 1. Once it had improved in Phase 2, the project was promptly completed.

5.2 Deviation from Standards and High Process Performance (Project B)

We conducted a within-case analysis of proactively initiated Project B, in which the deviation from standards was associated with *high* process performance. The analysis allowed us to identify reasons and get a better comprehension of the circumstances of the situation.

Internally Driven Deviation. We started by looking at the characteristics of the deviation from standards that occurred in Project B. Our data showed that the deviation from standards in the "Prepare Business Case" sub-process was initiated by R7, who did not use the Excel template but did a rapid estimate, as indicated by the following comment:

> I knew there was a large difference [between the overhead costs and the price of the project], so I didn't do the business case.

Therefore, we refer to the type of deviation in Project B as an "internally driven deviation" (IDD), which we conceptualize as a deviation from standards initiated by a process actor.

Robustness of Standards. When we asked R7 to elaborate on the reasons he did not execute the "Prepare Business Case" sub-process to the full extent, his response was as follows:

> Preparing a business case for a time-and-material project is something that feels unnecessary, because there is the same difference in margin. There is still a margin, and it doesn't matter how many man-days you sell.

This led us to speculate that the process documentation did not prescribe the best approach to dealing with the situation in this specific context (i.e., in the time-and-material type of project), and therefore, using Endsley's [5] terminology, we characterized the process documentation as a case of *low robustness*, i.e., insufficiently able to handle current and potential situations.

5.3 Cross-Case Analysis and Deviations from Standards in Knowledge-Intensive Processes

In the cross-case analysis, our aim was to identify and compare factors that were revealed in the within-case analysis of Projects A and B. Table 4 summarizes the results of the comparison.

Table 4. Comparison of the sales projects

Factor	Project A	Project B
Externally driven deviation	Present	No evidence found
Quality of communication	Low in Phase 1 Moderate in Phase 2	High
Unpredictability	High in Phase 1 Moderate in Phase 2	Low
Internally driven deviation	No evidence found	Present
Robustness of standards	Low	No evidence found

After revisiting transcripts of interviews with the members of the presales teams in both projects, we found no evidence for the occurrence of IDD in Project A and for an EDD in Project B. Moreover, we did not find any evidence to characterize the robustness of the process documentation for Project A, in which the Company was dealing with a different type of project (i.e., a fixed time and fixed price).

In contrast to Project A, the unpredictability in Project B was *low*, as the Company was not dependent on third parties, it initiated the project proactively by approaching the client with a pre-defined proposal, and it was able to predict the behavior of Client B based on previous interactions. Simultaneously, the communication quality was reportedly high, as confirmed by R1, who highlighted that a state of mutual understanding and ultimate transparency between the Company and long-term Client B was reached quickly and effortlessly. R6 and R7 were of the same opinion as R1 regarding the high quality of communication in Project B.

Based on the results of the within-case and cross-case analyses, we created a series of propositions regarding the necessary but insufficient conditions that lead to either a negative association between an EDD and process performance or a positive association between an IDD and process performance.

In Project A, where an EDD was negatively associated with process performance, two such conditions were: high unpredictability and low quality of communication. In formal terms:

Proposition 1. High unpredictability is a necessary but not sufficient condition for a negative association between an externally driven deviation and process performance to occur.

Proposition 2. Low communication quality is a necessary but not sufficient condition for a negative association between an externally driven deviation and process performance to occur.

In Project B, where an IDD was positively associated with process performance, two necessary conditions were: low unpredictability and low robustness of standards. In formal terms:

Proposition 3. Low unpredictability is a necessary but not sufficient condition for a positive association between an internally driven deviation and process performance to occur.

Proposition 4. Low robustness of standards is a necessary but not sufficient condition for a positive association between an internally driven deviation and process performance to occur.

5.4 Results of Validation

The six derived propositions were validated internally with R1 and R7 and externally with four external experts (E1 to E4; see Sect. 3.2). The results of the validation are presented in Table 5.

Table 5. Results of validation

Proposition	R1	R7	E1	E2	E3	E4	Valid?
1	NE	S	NE	NS	NE	NE	No
2	S	S	NE	S	NE	S	Yes
3	S	S	NE	S	S	S	Yes
4	S	S	S	S	S	S	Yes

Note: S – supported, NS – not supported, CS – conditionally supported, NE – no evidence

First, we would like to stress that most experts confirmed the occurrence of a negative association between an EDD and process performance and a positive association between an IDD and process performance across instances of the sales process in the Company and in other KiPs executed in medium- and large-sized companies. Additionally, R7 and E2 highlighted that the association between an IDD and process performance might also be negative, although this does not contradict our theory, as we did not encounter any cases in which such a situation occurred.

The validation of the necessary conditions for a negative association between an EDD and process performance (Propositions 1 and 3) allowed us to confirm the role of quality communication and reconsider the role of unpredictability. As became evident from interviews with R1, E1, E2, and E4, unpredictability increases as a result of an EDD and does not lead to its emergence, as we initially assumed.

The validation of Propositions 3 and 4 related to the necessary conditions for a positive association between an IDD and process performance allowed us to consider both propositions *valid*, since most of the experts found supporting examples in their experience for both propositions.

6 Discussion

In this section, we relate the valid propositions to the existing literature and outline the practical implications of our study.

Proposition 2. Our findings showed that low quality of communication is a necessary condition for a negative association between an EDD and process performance to occur. This proposition is relevant to KiPs, as, due to their interactive nature, these processes are highly reliant on effective communication [4], which is not the case for routine processes.

The association between communication and standardization was studied by Wüllenweber et al. [22], whose findings indicated that process standardization positively affects intra-organizational communication. Our findings bring a new perspective to these relationships, where intra-organizational communication of low quality might lead to a deviation from standards, which consequently increases the unpredictability in process execution that these standards aim to reduce, thereby undermining process performance. Therefore, our findings contribute to the Business Process Management (BPM) literature by expanding the rationale underlying the relationships between intra-organizational communication and standardization.

From a practical standpoint, it is evident that to avoid increased unpredictability and consequent reduction in process performance, it is necessary to have a high quality of communication from the initial phases of process interactions. This especially concerns situations in which companies need to interact with external stakeholders who they do not know (e.g., new clients).

Proposition 3. According to Proposition 3, low unpredictability is a necessary but not sufficient condition for a positive association between an IDD and process performance to occur. This proposition is relevant to KiPs, since it outlines the role of unpredictability, which is a commonly reported characteristic of these processes [4, 6, 11].

As described in Sect. 2.2, unpredictability presents challenges to the performance of KiPs. As we did not find an empirical confirmation for the benefits of deviating from standards in the context of high unpredictability, our findings extend the current understanding of how this challenge might be addressed by showing that standardization and following standards are beneficial for reducing unpredictability and improving performance in KiPs.

From a practical perspective, deviating from standards in situations of high unpredictability is not recommended, as it might lead to unknown risks and, in the end, undermine the process performance.

Proposition 4. According to Proposition 4, low robustness of standards is a necessary but not sufficient condition for a positive association between an IDD and process performance to occur. Although this proposition is not directly relevant to KiPs, it brings an important contribution by outlining the overlap between BPM and Human-Automation Interaction (HAI) research. The concept of robustness is commonly known in HAI research [5] and is argued to be one of the factors influencing trust in automation, which, in turn, is associated with process performance [21].

As indicated by an interview with R7, the situation of distrust of standards is common in the Company and is explained by the low robustness of standards in certain situations. Therefore, our data allows us to conclude that the theoretical mechanism explaining interactions between robustness, trust, and performance discussed in HAI research in the context of automation is also applicable in the context of standardization, and therefore BPM.

As for the practical implications, we would recommend that companies do not encourage IDDs even in situations of low unpredictability, but rather update the standards, so that they incorporate the previous experience of process actors. This would lead to the increased effectiveness of standards. One way to achieve this is to discuss any deviations from standards in lessons learned sessions after the end of the process iteration and adjust standards based on the results of these discussions.

In general, we conclude that deviations from standards in the execution of KiPs are undesirable, since they increase unpredictability and consequently undermine process performance. Companies are recommended to increase the quality of communication with external process stakeholders and keep standards up to date, incorporating the growing experience of process actors to minimize deviations from standards.

7   Conclusions

Our primary goal was to investigate the reasons and consequences of deviating from standards in the context of the execution of KiPs. We did this by examining the sales process of an IT service provider (the Company) and conducting a cross-case comparison of two instances of this process, the subjects of which were Projects A and B. The collected evidence revealed: (i) a negative association between an externally driven deviation from standards and process performance, and (ii) a positive association between an internally driven deviation from standards and process performance.

Moreover, we deduced four propositions regarding the necessary conditions for the above associations to occur and considered three of them valid in the results of validation. Relating these three valid propositions to the literature illustrated their relevance for KiPs and allowed us to make certain theoretical contributions. In addition, we put forward a number of recommendations for practitioners on how to avoid deviations from standards.

Nevertheless, our study has a number of limitations related to data collection and validation. First, we only compared two process instances of a single process in a single company. Although the internal validation allowed us to compensate for the insufficiency of the collected evidence, further data collection is required. Following the principles of theoretical sampling [2], we would like to focus our efforts on collecting more evidence on (i) the negative association between an EDD and process performance, (ii) the positive association between an IDD and process performance; and also on the hitherto undiscovered situations of (iii) the negative association between an IDD and process performance, which was mentioned by several experts in the process of validation, and possibly (iv) the positive association between an EDD and process performance.

We plan to continue data collection in the Company to get further insight based on more instances of the sales process. Subsequently, we are considering collecting data on instances of the sales process in a large-sized company. The data collection can then be continued by conducting research on other KiPs across companies of different sizes and operating in various industries.

Another limitation of our study was the validation of the derived propositions. The validation sessions were conducted in the form of semi-structured interviews with a limited number of internal and external experts. In further research, we plan to initiate panel discussions to enable an exchange of opinions between experts, involve bid managers in internal validation, as they would have more detailed insights into specific instances of a sales project, and involve experts from academia in the external validation.

Acknowledgements. This research was supported by Masaryk University internal grant No. MUNI/A/1315/2022, entitled "An Integration of Business Process Management and Knowledge Management."

References

1. Aureli, S., Giampaoli, D., Ciambotti, M., Bontis, N.: Key factors that improve knowledge-intensive business processes which lead to competitive advantage. Bus. Process Manage. J. **25**(1), 126–143 (2019). https://doi.org/10.1108/BPMJ-06-2017-0168
2. Charmaz, K.: Constructing Grounded Theory, 2nd edn. Sage, Los Angeles (2014)
3. Davenport, T.: The coming commoditization of process. Harvard Bus. Rev. **83**(6), 100–108 (2005)
4. Di Ciccio, C., Marrella, A., Russo, A.: Knowledge-Intensive Processes: Characteristics, Requirements and Analysis of Contemporary Approaches. Journal on Data Semantics **4**(1), 29–57 (2014). https://doi.org/10.1007/s13740-014-0038-4
5. Endsley, M.: From here to autonomy: Lessons learned from human–automation research. Hum. Factors. **59**(1), 5–27 (2017). https://doi.org/10.1177/0018720816681350
6. Eppler, M., Seifried, M. Ropnack, A.: Improving knowledge intensive processes through an enterprise knowledge medium. In: Prasad, J (ed.) ACM SIGCPR, pp. 222–230 (1999)
7. Erol, O., Sauser, B., Mansouri, M.: A framework for investigation into extended enterprise resilience. Ent. Inf. Syst. **4**(2), 111–136 (2010). https://doi.org/10.1080/17517570903474304
8. França, J.B.S., Baião, F.A., Santoro, F.M.: Towards characterizing Knowledge Intensive Processes. Proceedings of the IEEE International Conference on Computer Supported Cooperative Work in Design (CSCWD), vol. 16, pp. 497–504, Wuhan, IEEE (2012). https://doi.org/10.1109/CSCWD.2012.6221864
9. Gonzalez-Lopez, F., Pufahl, L., Munoz-Gama, J., Herskovic, V., Sepúlveda, M.: Case model landscapes: toward an improved representation of knowledge-intensive processes using the fCM-language. Softw. Syst. Model. **20**(5), 1353–1377 (2021). https://doi.org/10.1007/s10270-021-00885-y
10. Gromoff, A., Bilinkis, Y., Kazantsev, N.: Business Architecture Flexibility as a Result of Knowledge-Intensive Process Management. Glob. J. Flex. Syst. Manag. **18**(1), 73–86 (2017). https://doi.org/10.1007/s40171-016-0150-4
11. Işik, Ö., Mertens, W., Van den Bergh, J.: Practices of knowledge intensive process management: quantitative insights. Bus. Process Manag. J. **19**(3), 515–534 (2013). https://doi.org/10.1108/14637151311319932

12. Marjanovic, O.: A novel mechanism for business analytics value creation: improvement of knowledge-intensive business processes. J. Knowl. Manag. **26**(1), 17–44 (2022). https://doi.org/10.1108/JKM-09-2020-0669

13. Mertens, S., Gailly, F., Poels, G.: Towards a decision-aware declarative process modeling language for knowledge-intensive processes. Expert Syst. Appl. **87**(1), 316–334 (2017). https://doi.org/10.1016/j.eswa.2017.06.024

14. Münstermann, B., Eckhardt, A., Weitzel, T.: The performance impact of business process standardization. Bus. Process Manage. J. **16**(1), 125–134 (2010). https://doi.org/10.1108/14637151011017930

15. Romero, H.L., Dijkman, R.M., Grefen, P.W.P.J., van Weele, A.J.: Factors that Determine the Extent of Business Process Standardization and the Subsequent Effect on Business Performance. Bus. Inf. Syst. Eng. **57**(4), 261–270 (2015). https://doi.org/10.1007/s12599-015-0386-0

16. Schäfermeyer, M., Rosenkranz, C., Holten, R.: The Impact of Business Process Complexity on Business Process Standardization. Bus Inf Syst Eng **4**, 261–270 (2012). https://doi.org/10.1007/s12599-012-0224-6

17. Scheibelhofer, E.: Combining Narration-Based Interviews with Topical Interviews: Methodological Reflections on Research Practices. Int. J. Soc. Res. Methodol. **11**(5), 403–416 (2008). https://doi.org/10.1080/13645570701401370

18. Sid, I., Reichert, M., Réda Ghomari, A.: Enabling flexible task compositions, orders and granularities for knowledge-intensive business processes. Enterp. Inf. Syst. **13**(3), 376–423 (2019). https://doi.org/10.1080/17517575.2018.1556815

19. Szelągowski, M.: Practical assessment of the nature of business processes. IseB **19**(2), 541–566 (2021). https://doi.org/10.1007/s10257-021-00501-y

20. van der Aalst, W., Weske, M., Grünbauer, D.: Case handling: a new paradigm for business process support. Data Knowl. Eng. **53**(2), 129–162 (2005). https://doi.org/10.1016/j.datak.2004.07.003

21. Wickens, C.D., Li, H., Santamaria, A., Sebok, A., Sarter, N.B.: Stages and Levels of Automation: An Integrated Meta-analysis. Proceedings of the Human Factors and Ergonomics Society Annual Meeting **54**(4), 389–393 (2010). https://doi.org/10.1177/154193121005400425

22. Wüllenweber, K., Beimborn, D., Weitzel, T., König, W.: The impact of process standardization on business process outsourcing success. Inf. Syst. Front. **10**(2), 211–224 (2008)

23. Wurm, B., Mendling, J.: A Theoretical Model for Business Process Standardization. In: Fahland, D., Ghidini, C., Becker, J., Dumas, M. (eds.) BPM 2020. LNBIP, vol. 392, pp. 281–296. Springer, Cham (2020). https://doi.org/10.1007/978-3-030-58638-6_17

24. Zelt, S., Recker, J., Schmiedel, T., vom Brocke, J.: A theory of contingent business process management. Bus. Process Manage. J. **25**(6), 1291–1316 (2019). https://doi.org/10.1108/BPMJ-05-2018-0129

Benevolent Business Processes - Design Guidelines Beyond Transactional Value

Michael Rosemann[✉], Nadine Ostern, Marleen Voss, and Wasana Bandara

Centre for Future Enterprise, Queensland University of Technology, Brisbane, Australia
{m.rosemann,n.ostern,marleen.voss,w.bandara}@qut.edu.au

Abstract. The academic and professional BPM discipline is concentrated on process performance and conformance. However, changing consumer demand patterns, ESG requirements and concepts such as conscious capital have increased the pressure to create more than 'transactional value' from business processes. One such consequence is the requirement to provide trusted processes and the embedded request for benevolence. A benevolent process prioritizes customers' demands over providers' interests. Following a Design Science approach and informed by primary data collected from three large service organisations and related secondary data, this paper presents eight design guidelines, grouped in four pairs, for benevolent processes. These design guidelines conceptualise an entirely new set of process aims and have the potential to initiate BPM research and professional practice exceeding common transactional value propositions.

Keywords: Trust · benevolence · design guidelines · transactional value

1 Beyond Transactional Value

Traditionally, the management of business processes has been driven by the economic paradigm. As a result, process improvement initiatives have predominantly targeted reducing costs and processing time and increasing process quality [1] and a plethora of approaches along the process lifecycle have been developed to support exactly these objectives. However, customers nowadays require more than transactional excellence and developments such as the ESG framework (environmental, social and governance) have led to formal requirements for process analysts and owners to go beyond established metrics of successful process execution. This movement is further amplified by management approaches such as conscious capitalism [2], shared value [3] and the notion of the purpose-led organisation [4]. These approaches postulate that all stakeholders, and not just shareholders need to be considered when designing business processes. In this context, the established focus on process performance and conformance and the related set of skills and tools are still necessary, but no longer sufficient. Rather, new approaches for the design and management of business processes need to be invented to respond to these calls in the academic and professional domain of BPM.

This overall notion of 'doing good' beyond economic incentives is best captured in the concept of benevolence. Benevolence prioritises the well-being of the other party

© The Author(s), under exclusive license to Springer Nature Switzerland AG 2023
C. Di Francescomarino et al. (Eds.): BPM 2023, LNCS 14159, pp. 447–464, 2023.
https://doi.org/10.1007/978-3-031-41620-0_26

and is therefore oppositional to the dominating provider-centric view of BPM which tends to emphasise flawless, economically efficient process execution. Benevolence as a design criterion in a corporate context overall, and in the specific domain of BPM, is in its infancy which makes its operationalisation difficult. This is despite the fact that various studies have demonstrated that benevolence increases customer loyalty [5] and comes with reputational gains [6]. Benevolence also creates an emotional attachment to the firm [7], has positive effects on the long-term sustainability of organisations [8], and promotes relationships [9] and interpersonal trust [10]. Studies have also shown that in specific contexts benevolence matters more than competence [11]. However, while the latter is a typical focus in the allocation of work along a process to staff members, the former has so far not made it as a primary concern into the BPM literature. Research that integrates trust in general and benevolence in particular into business process design has only emerged recently [12]. Related research on customer-centric process design tends to focus on demand-oriented process practices from an economic lens (see approaches such as customer journey mapping) as opposed to an authentic benevolent motivation.

Thus, with a specific focus on an organisation's business processes, we aim to accelerate and guide the design of benevolent processes. For this, we are using design guidelines derived from primary and secondary data to inform the design of benevolent business processes. Therefore, the research question of this paper is *"What are design guidelines that can inform benevolent business processes?*

In order to address this research question, first the existing body of knowledge on benevolence was comprehensively studied and used to derive the key dimensions of benevolence. Following a Design Science approach, and building on this consolidated body of knowledge, we then started to create the desired artefact, i.e., guidelines for the design of benevolent processes. For this, we first interacted with executives from three Australian organisations in the domains of financial and insurance services. These were identified from within members of the Brisbane Trust Alliance, a community of professionals dedicated to advancing the professionalisation of trust management, and also via a white paper we circulated to executives within Australia. This set of primary data allowed us to inductively build a set of design guidelines, which we grouped into four pairs. These guidelines were then further revised and specified in light of complementary secondary data which reported on process-related benevolence across industries and which we identified based on feedback we received from our three interview partners as well as based on the initial literature review.

This paper is structured as follows. Section 2 contextualises the research by summarising the body of knowledge on benevolence and differentiates these from related constructs such as customer centricity, corporate social responsibility and also process patterns. Section 3 describes the research method of our study before Sect. 4, the core of this paper, presents eight benevolent process design guidelines grouped in four pairs. Section 5 discusses our findings and elaborates on those contextual factors that matter for a successful adoption of these guidelines. Finally, Sect. 6 sums up the paper with conclusions, limitations and possible future research directions.

2   Research Context

2.1   Benevolence

The word benevolence comes from the Latin words "bene" and "volens", which means *"wanting the good"* [13]. In the management literature, benevolence is understood as caring for the other party and manifests in actions that go beyond contractual aspects [13, 14]. Benevolence is demonstrated by the giving party being responsive to and considerate of the receiving party's needs in order to enhance the receiving party's well-being [15].

Depending on the social semantics, benevolence is divided into two forms [5]. Altruistic benevolence is *"the extent to which a trustee is believed to want to do good to the trustor, aside from an egocentric motive"* [14, p. 718] and comes without expectation of future gain. Mutualistic benevolence, on the other hand, is defined as *"the degree to which one party is genuinely interested in the other's well-being and seeks joint gain"* [16, p. 36]. This form of benevolence is based on reciprocal utilitarian motives. For example, employees provide their customers with exceptional support in the hope of strengthening their loyalty [5]. The expected gain of the benevolent party, however, is unpredictable, delayed and cannot be guaranteed. Against this background, a benevolent business process has three characteristics: (1) the process constitutes an immediate benefit for the recipient party; (2) the process is an investment for the provider in the short term, (3) the benevolent behaviour in the process is not mandated.

A further differentiation of benevolence can be made based on the social level of the interaction. The micro level refers to benevolent behaviour between two individuals. Livnat (2004) [17] describes three elements (the emotive, the performative, and the cognitive element) that comprise benevolent acts between two stakeholders. The meso level refers to benevolence in the context of an organisation's business processes. Doney and Cannon (1997) [16] determined processes by which industrial buyers demonstrate benevolence and as a result develop trust in a supplier. Finally, the macro level captures regulated benevolence [18] to protect the rights of customers and employees.

Benevolent processes can be distinguished into public and private. Public benevolent processes are known to the customer through advertising, publications or past interactions with the company. For example, a customer might be allowed to exchange a product without the need to provide reasons, and this is explicitly stated in the company's return policy. Private benevolence, on the other hand, is situational and not expected by the customer. Tesla's over-the-air update which extended the possible distance to be travelled for customers close to a hurricane (USA) or bush fire (Australia) was such a not to be expected type of private benevolence.

If the benevolent behaviour is codified in a defined process, script or guideline, it is formal benevolence. Formalisation ensures consistency in the provision of benevolent actions whereas informal benevolence is situational and not prescribed.

The focus of this paper is on mutualistic, formal, micro benevolence only.

2.2 Related Concepts in BPM

2.2.1 Customer Centric Process Design

Customer centricity puts the customer at the centre of all business activities [19]. The focus of customer centric process design is on the needs of customers and their expectations in order to achieve above-average customer satisfaction and loyalty [20, 21]. In the domain of BPM, customer-centricity has been triggered by Lean Management and gave rise to approaches such as moments of truth, design-thinking and customer journey mapping.

While customer centricity is dedicated to the understanding and satisfaction of customer needs via convenient processes (e.g., first contact resolution, choice of channel [19]), benevolence matters in those 'moments of truth' in which empathy matters more than convenience (e.g., the customer's inability to repay a debt within a given period due to a personal hardship). Customer centricity is for most organisations a necessity to remain competitive whereas benevolence is optional, often not even expected by customers [15]. Unlike customer centricity which is the desire to have an attractive market offering, benevolence is driven by the desire to address a customer's circumstances.

2.2.2 CSR Business Processes

Corporate Social Responsibility (CSR) is the notion that companies have an obligation to constituent groups in the society they operate in, beyond shareholders [22]. Despite CSR being discussed over decades, it is still widely debated on facets such as the degree of voluntary behaviour that drives a firm's CSR, which stakeholder groups (customers, employees, communities, etc.) a firm should serve with CSR or how to balance the serving of multiple stakeholder-groups' interests. Companies are expected to report on their CSR efforts and many firms have dedicated CSR teams to design and execute CSR initiatives. So far, only limited research has considered an integrated perspective on BPM and CSR [23].

CSR and benevolence both aim for social-good, are driven by value-creation objectives, and can influence a firm's 'social license to operate' [24]. They both can help organisations to proactively manage corporate social 'irresponsibility' [25]. While more contemporary CSR debates such as 'CSR for Corporate Social Performance' versus 'Obligatory CSR' aimed at reducing harm and increasing benefits at the societal level [26] bring the resemblance between CSR and benevolence very close, benevolence is different from traditional CSR practices in several ways. First, benevolence is purely voluntary and not influenced by any reporting requirements or external law/policy enforcements. Second, benevolence in the context of this paper is targeted as acts by the firm towards one key stakeholder group; customers. And third, unlike CSR, benevolence is conducted by the firm as part of its regular business operations.

2.3 Process Patterns and Design Guidelines

Process patterns are proven solutions to recurring situations and are popular artefacts within Business Process Management. Among others, process patterns have been proposed as a foundation for process-aware systems [27], which inspired related work in data patterns and resource patterns within the process management community.

Patterns have been widely used to categorize process weaknesses as in the seven wastes of lean management, and as broad process improvement patterns [28]. More recently, explorative process patterns [29] have been proposed to identify new revenue opportunities for an existing process. An overview about process patterns can be found at bpmpatterns.org. Process patterns are atomic building blocks for business processes and aim to trigger specific considerations among stakeholders dealing with processes (e.g., identify a weakness, reflect on a re-design suggestion).

In contrast to process patterns, design guidelines refer to higher order principles and are motivated by a design intention – here the design of benevolent processes and the inherent creation of immediate value for a customer beyond the expected transactional value of the business process. Thus, design guidelines are applicable constructs embedding research contributions that help with the creation of artefacts (here: benevolent processes) in practice [30]. Patterns as atomic and often (semi) formalised templates can be a part of a design guideline. In the following we propose a set of process design guidelines unique in their intention; the design of benevolent business processes. Each guideline will be described as follows: (1) an example to provide clarity to the guideline, (2) the context describes the factors leading to the emergence of the benevolent process design guideline; (3) a definition of the benevolent process design guideline; (4) the type of value created for both parties involved (customer and provider), (5) a simple textual description of the pattern inherent to the design guideline; (6) required considerations for the application of the design guideline, and finally, (7) a discussion of the perceived ease of automation of this design guideline.

3   Research Method

The study followed a Design Science Research Paradigm. In the *first,* guideline-building phase, a comprehensive analysis of benevolent business practices was conducted based on a detailed literature review. As the existing body of knowledge on benevolent process design is very limited, we collated 'moments of corporate benevolence' primarily from publicly available resources (e.g., company web pages, media releases, social networks and forums that shared customer experiences). These were further complemented by collecting input from our industry and academic network where we sought for examples and/or elaborations of corporate benevolence people have experienced or witnessed. We extracted over 40 examples, which were exposed to detailed researcher corroborations.

Second, the guideline-building phase continued, while we approached organisations participating in the Brisbane Trust Alliance, a community of 40 + professionals across various industries committed to improving their trust literacy. This community meets on a monthly base since end of 2021, and at each meeting dedicates their discussions towards one trust-related topic. Two organisations (here after referred to as Case 1 and 2) expressed an interest to assess the potential applicability of these eight guidelines in the sense of a naturalistic, ex-ante artefact evaluation [31].

Case 1 is an Australian mutual organisation with more than 1.7 million members and more than 2,200 employees providing road assistance, insurance and banking services to its members. Its motivation to consider benevolent business processes has been the alignment with its value system as a member-based organisation. The Chief Purpose Officer, together with several of the most senior executives of the organisation, considered the applicability of our guidelines and presented the findings for discussion to the Trust Alliance. *Case 2* is an Australian automotive and accessories retailer with more than 300 stores employing more than 14,000 staff across Australia and New Zealand. This retailer sees trust as an opportunity to derive a competitive advantage in an economic environment in which competing on product and price is increasingly challenging. Benevolence is a key feature of trust and the retailer has in particular an interest in deploying services recovery practices, here called compensation. The General Manager, eCommerce and Marketing, of one of the four brands of the organisation, committed to assess our guidelines and together with input from other senior executives shared related views with the Trust Alliance.

The two executives (Chief Purpose Officer of Case 1 and the General Manager, eCommerce and Marketing of Case 2) were involved in the following activities related to building the artefact; (1) individual briefing between at least one of the authors and the executives to contextualize and motivate the research; (2) conceptual application, i.e. the executive contemplated – guideline by guideline – the deployment of the artefact presented; (3) presentation of the applicability of the artefact within each respective organisation (by the executive) to the Brisbane Trust Alliance; and (4) related discussion among all Trust Alliance members (e.g., correctness of application, adequate naming of the guideline, need to refine definition of artefact constructs). Both executives identified for each of these eight guidelines (a) their related practices, (b) new practices they could deploy (pre an economic assessment), and (c) those practices that would be impossible to implement. This discussion helped us to further improve the definition, intention and constraints of our eight design guidelines, provided us with additional practices and helped us to better understand prerequisites for the implementation of benevolent business processes.

In the *third* phase, the focus was on early evaluation of the design guidelines. Here we summarized our initial conceptualization in a white paper that we circulated beyond the members of the Brisbane Trust Alliance leading to an expression of interest by an international insurance and healthcare group of more than 43 million customers and 84,000 employees globally. This became the third case study (*Case 3*) which was used for quasi validation of the design guidelines. The Chief Customer and Strategy Officer from their Melbourne office assessed our eight design guidelines against their more than 100 defined micro-moments in customer facing processes. In an interview, we extracted the feedback and used the insights and examples provided to further revise our definitions and the details for each guideline.

The purpose of the benevolent process design guidelines is to make BPM professionals aware of a design rationale that goes beyond the common focus on transactional value. The assessment of the actual (technical, ethical, legal) feasibility, desirability, and viability of these guidelines is context-specific and beyond the scope of this paper.

4 Benevolent Process Design Guidelines

Figure 1 summarizes the eight benevolent process guidelines as per our joint codification and shows how they are bundled in four pairs, which we call benevolent principles. In the following we will describe each of these guidelines according to the attributes outlined in Sect. 3.

4.1 Be Fair

Benevolence materialises when a business process demonstrates that a customer's well-being matters more than the immediate interests of the organisation. Unfortunately, such practices are not always common. The Australian Royal Commission into Misconduct in the Banking, Superannuation and Financial Services Industry, established in December 2017 (royalcommision.gov.au/banking), as an example, surfaced practices, in which the financial advice provided to banking customers was incentivised more by commissions than suitability of the products for the customer.

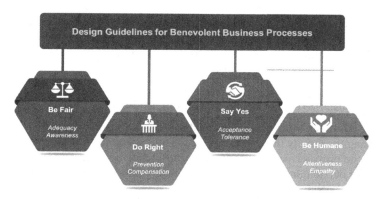

Fig. 1. The eight benevolent process design guidelines grouped in four principles

'Being fair' is relevant when products are complex and it is difficult for the customer to select the right one, when the products are not commonly used or when the products (e.g., due to interest rates changes) change frequently. 'Be fair' captures two guidelines dedicated to making sure customers receive the product or service, right for them. These two guidelines can be differentiated based on the strength to which they ensure fit-for-purpose; is it an action (adequacy) or recommendation (awareness) that ensures this fit?

4.1.1 Adequacy

Example. As part of their customer engagement process, a telecommunication company runs monthly queries to identify customers of mobile data plans who have consistently used 50% or less of their data plans over at least six months. Those customers are then proactively downgraded to the next lower data plan and informed about the decision. Customers have two weeks to respond in case they would like to keep their current data plan, otherwise the downgrade takes place. From then, the customer is charged less each month. In a similar way, some Neo Banks proactively move their customers' funds into higher interest retail banking products unless the customer declines within a specified timeframe. One of our three case partners, the global health insurer, referred to running regular processes during which over-insured customer were proactively approached to discuss and adjust their insurance policies according to their actual requirements (e.g., excluding age-related services no longer needed). A more common form of benevolent processes increases the quality of services (e.g., increased Internet speed or data volume) without price changes.

Context. In sales processes in which a customer has to estimate upfront consumption volumes (e.g., telco data plans, software purchases), some customers will over-estimate their demands meaning they end up in the language of lean management with an over-produced product or service. Benevolence is demonstrated by adjusting this service to an adequate offering that matches the actual demands of the customer once they are understood. Thus, adequacy is a form of down-selling. However, while common down-selling reactively offers more budget-friendly solutions to a hesitant buyer (e.g., after they took out an item of their shopping cart), adequacy is about proactively ensuring that the customer only pays what is needed to get a sufficient product or service. Adequacy works well in subscription processes such as telecommunication or energy plans, cloud solutions, but also retail banking or insurance processes.

Definition. Adequacy is an event-driven activity in which a customer's service is proactively adjusted to the most appropriate, sufficient service.

Value. The customer will receive an immediate monetary benefit (paying less) without compromising the quality of service received. Thus, adequacy is a true net gain. In the short-term, the provider will miss out on the difference between previous and revised pricing model but demonstrates benevolence which might lead to retention [5].

Pattern. Adequacy is triggered by a time-based event (e.g., once the month) which is the result of a report identifying all those customers who under-utilize existing services they subscribed to. The identified customers are then adjusted to the most adequate, and for them more cost-effective, service and notified about it. Customers are offered a period (e.g., five business days) to decline the offer, otherwise it is regarded as accepted.

Consideration. The implementation of the adequacy guideline requires a definition of 'under-utilization', the period for which this under-utilization occurs, and a business rule for how the customer declines the downgrade.

Process Automation. Adequacy is a benevolent process design guideline that can be automated reasonably well. Data analysis can help to identify those customers that under-utilize existing service and trigger a downgrade of the customers' service to the most adequate one. The customer is then notified about this downgrade.

4.1.2 Awareness

Example. A provider of white goods sells its washing machines, dish washers etc. with a standard two-year warranty during which any issues will be fixed free of charge. Four weeks before the warranty expires, the customer is contacted to ensure they are aware of the upcoming expiry date, and that any issues could be addressed within the next four weeks without any cost. Similarly, the global health insurance provider ensures that their customers are aware of free access to wellbeing services such as massages or acupuncture. Such services can be part of a comprehensive bundle of services (e.g., car insurance), but also, like in the case of the road assistance company, be complementary services (e.g., a taxi service in case of a breakdown).

Context. Customers often have access to relevant services free of charge but they are unaware of them. While the adequacy design guideline interprets the non-use of these services as a service oversupply, the awareness design guideline is driven by the intention to make the customer fully aware of potentially relevant services. The services are often time-stamped (like a warranty) so that this pattern is more of a one-off than a regularly executed activity.

Definition. Awareness matters for about-to-expire, in advance paid services such as maintenance, warranty as well as products that complement a primary product.

Value. Making the customer aware of a free-of-charge service means access to a value add without any costs. In return, the provider is required to offer this service and carry related costs, costs that otherwise would not have occurred. Again, the act of benevolence has the potential to lead to increased loyalty and advocacy.

Pattern. Awareness requires either a time-based event that correlates with a service about to expire or a definition of services available to a customer. The customer will have a certain period to request or consume this service.

Consideration. Awareness is a notification process encouraging customers to access complementary services. The impact of the uptake of these services in terms of resourcing and costs needs to be assessed.

Process Automation. Awareness can be automated with current technologies. For this, it needs to be defined when and for what type of service what type of customer group will be notified using what type of communication channel. A nuanced execution of awareness will require artificial intelligence and machine learning (e.g., if only specific customers in defined contexts are about to be notified).

4.2 Do Right

Doing right is about ensuring that harm is prevented from the customer and recognising when an organisation's processes did not deliver to their promise. Consequently, we call these two design guidelines (1) prevention which occurs proactively and early on in customer engaging processes and (2) compensation which takes places reactively and at a late stage in a process.

4.2.1 Prevention

Example. When customers order the same book again on Amazon, they will be asked to re-confirm if they are certain that they like to proceed as the same book has been ordered by this Amazon account before. This prevents the customer from ordering the book by mistake. Prevention might also be an alert highlighting that the purchase about to be initiated is different to previous purchases (e.g., in the case of repetitive refill purchases). Moreover, prevention is deployed in processes in which a provider aims to keep the customer safe. For example, if someone is about to rent an e-scooter after 10pm the provider's app might request that the user enters 'Y E S' on the phone's keyboard to prevent possibly intoxicated customers from using the scooter. Aldi's cashierless shops are using AI-empowered facial recognition to estimate the age of a customer and will not enable sales of age-constrained items (e.g., alcohol) unless the customer provided evidence that they are over the minimum age. Prevention occurs in cars (like Teslas) that stop proactively at a red light or headsets that automatically reduce the volume of music after the user has listened to it on a volume level that could be damaging for too long.

Context. An organisation embeds a preventative step in its sales process to avoid regret or even harm on the customer's side. Such a process might be in contrast to the aim of friction free process execution according to the customer's requirements.

Definition. Prevention is a process activity by a provider with the intention to trigger a consideration of the intended purchase in order to avoid post-order regrets or harm.

Value. The customer benefits from prevention if the intended purchase is made in a hurry or without sufficient consideration (e.g., if the item has been ordered before). The company demonstrates potentially trust-building values that come with the loss of immediate revenue.

Pattern. A purchasing or ongoing consumption process is extended with an additional prevention activity which is context-dependent (e.g., time-dependent, previous purchases).

Consideration. The deployment of the prevention design guideline requires an identification of possible regrets and harm due to the process to be executed. This demands a careful assessment of the process impact. Prevention might be a simple alert that can be over-ridden (e.g., a customer purchases an item in a grocery store that they are allergic to according to their customer record) or a firm decline of a purchase.

Process Automation. Prevention might require the customer contributing some personal information to assess fit of purchase. For example, the automotive retailer warns their customer if they are about to buy a car battery that does not fit their vehicle according to the car registration they provided. The automation of prevention requires an identification of possible regrets such as redundant or incompatible purchases. This requires access to data such as previous transactions or assets owned by the customer.

4.2.2 Compensation

Example. The fast-food chain McDonald's offers the meal ordered for free if the customer has to wait for too long. A similar process is in place for Starbucks. The German Railway has a policy according to which a passenger will get a 25% (50%) discount of the purchased ticket of the train is delayed by 60 (120) minutes. The latter is a typical example for public (announced) benevolent process. Compensation does not always have to be time related. For example, a retailer might proactively offer vouchers in case of non-satisfactory process execution (e.g., an item was not available for collection).

Context. Compensation is the practice of offering a reimbursement to the customer, for example by offering a discount in response to underperformance. Instead of a discount, it might also be a value-add (e.g., adding an additional product or service for free such as access to the airline's lounge in compensation for a disrupted flight).

Definition. The literature discusses compensation as service recovery [32], defined as the process of recovery from an insufficient customer experience with the aim to regain the loyalty of the customer.

Value. Compensation creates customer value by reducing, reversing or even over-compensating for a previous mistake in the customer engagement process. The company has to fund the costs of this service discovery.

Pattern. Compensation comes in a variety of ways ranging from ad-hoc, informal actions to well-specified processes. Whereas the former requires empowerment at the edges so that staff involved can make relevant decisions, only the latter is pattern-like and repeatable. It requires contextual rules for the process execution need (e.g., in case waiting time exceeds x minutes, a reimbursement of $y will be offered).

Consideration. The organisation needs to identify relevant process failure, the extent to which this failure impacts and is actually experienced by the customer, and the type of compensation offered, including the option of staged discounts (see German Railway). It needs to be decided if compensation is well-published, and if so, how it is communicated to the customer. Compensation as a benevolence practice also requires considering potential fraud, and how this can be prevented.

Process Automation. The ease and ways of automating compensation will vary depending on its configuration. Public compensation is a well-defined process with precise business rules which come in the form of ECA (event-condition-action) triplets. Thus, an automated process compensation requires three features. First, it needs to be assessed if the process failure can be captured automatically (e.g., via video analytics, event logs or IoT devices capturing delays or unavailability of products). Second, the condition for process initiation needs to be coded (e.g., threshold, type of customer). Third, the actual form and delivery of automated compensation needs to be defined (e.g., printing voucher, direct funds transfer).

4.3 Say Yes

Saying yes to a customer as a benevolent principle means that all possible attempts are made within a process to fulfill a customer's request. Such requests might be either exceptional (qualitative) enquiries requiring a case-by-case decision (acceptance) in a process, or they are frequent requests that can be anticipated and quantified (tolerance).

4.3.1 Acceptance

Example. An airline received a phone call from one of its members. The caller asked to put her membership on hold as she was undergoing cancer therapy, and as such could not fly during the period of this treatment. The call centre agent looked into the airline's policy and responded that only pregnancy was a valid reason to put a membership on hold, and as such had to reject this request. This incident made it into the media and the airline's CEO later stated that he could not blame the call centre agent as they had acted compliant with the airline's call centre processes. An Australian superannuation provider received a request from a customer who wanted to access income protection as he was about to donate their liver. However, the company's policy stated that only *receiving* an organ qualified for access to income protection. The case also made the news as the customer wanted to donate their liver to their two-year old daughter who was facing a life threating illness. On the pressure of the media, both companies decided to increase the benevolence of their processes and accepted their customers' requirements.

Context. Acceptance is characterized by the specificity of the customer's enquiry, making the definition of upfront rules difficult. The need for benevolent action is most visible when customers approach an organisation with an off-script enquiry (put membership on hold) and the systemically proposed next best action (rejection) is not one with the benevolence required. Constrained by predefined routines, the responding business processes often do not not have the flexibility required to support what seems like the right, kind-hearted action.

Definition. Acceptance is the fulfillment of a customer's enquiry despite it being a service that is typically not available.

Value. The customer experiences value in the form of getting what they ask for whereas the organisation needs to resource and fund the provision of the related service.

Pattern. Despite the fact that events requiring acceptance come in various forms, companies can in some scenarios be proactive and follow formal processes. For example, some banks embed a 'loan holiday' in their mortgage product allowing customers to activate this holiday – a period during which they pause repayments – in exceptional personal circumstances. This is a practice in place by many Canadian banks (e.g., The Royal Bank of Canada, Bank of Montreal). Similarly, restaurants might consider cash payments in a variety of currencies if they are in a tourist-intensive, transitional place (e.g., at an airport).

Consideration. If anticipating and classifying requests is too difficult or simply impossible, organisational governance solutions are needed. This could be in the form of the practice at the Ritz Carlton [33]. This luxurious hotel chain moved from prescribing the actions to be taken by their concierges (e.g., carry luggage) to empowering and funding their front-line decision making. As a result, a concierge can now react situationally to a specific context and has the funds to accept unusual requests (e.g., a request for a special vehicle). Such empowerment as a mechanism of benevolence only works when the staff is sufficiently trained and experienced. An alternative and more scalable governance arrangement is practiced by some organisations in their call centres. When an agent receives an exceptional request but needs to make an immediate decision (as time-consuming escalation to a supervisor would not be seen as 'benevolent enough'), they can reach out to an immediate colleague close by and seek confirmation. If both agents agree, and this is to be documented, the request will be accepted (or declined).

Process Automation. Acceptance is an automated benevolent process only in the case of anticipated requests (e.g., banking customers asking to pause mortgage payments). Here a stand-by process variant is defined that can be activated if needed. Otherwise, access to a database of previous acceptances allows front-line staff to accept customer enquiries based on previous reference cases.

4.3.2 Tolerance

Example. One of the authors of this paper was on a domestic flight in Australia. Their suitcase was 24 kgs, one kilogram above the upper limit of 23 kg. This would typically lead to an extra fee of AUS$85. However, instead of charging or mentioning anything, our colleague noticed how the airline member lifted the suitcase when it was on the scale and checked it in without any request for payment. We don't know if this was a spontaneous action of the staff, or an action according to a defined script of the airline. In any case, it is an example for tolerance – 'saying yes' to a customer's (suit)case that exceeds a defined threshold. Unlike acceptance, tolerance can be quantified and related requests can be anticipated. For example, an organisation might tolerate requests for a free-of-charge repair of an item a few days (or weeks) after the product's warranty has expired. Similarly, late payments of a customer might be tolerated without a penalty or late arrivals (for a booking at a restaurant or a pick-up of a requested item from a retailer) might be tolerated as a sign of benevolence.

Context. Many business processes come with defined, quantified boundaries for their initiation (e.g., years of experience in a recruitment process, weight and size in a logistical process, opening hours of a shop, etc.). By default, exceeding these thresholds means either 'no process at all' or additional costs or other efforts. There are in some cases legal reasons (e.g., age limits for certain products), but in many cases the thresholds are created for internal reasons such as demand control or considering economic requirements.

Definition. Tolerance is the formalised benevolent practice in which customer requests that are going beyond quantified thresholds are still allowed though this will often be a hidden allowance, i.e. in advance unknown to the customer.

Value. Despite not fitting the requirements, the process is conducted for the customer so they can receive the desired service. The tolerating provider of the process carries the negative consequences of allowing such a request (e.g., costs of working hours).

Pattern. Tolerance is a guideline that is applicable in processes with quantified thresholds only. The related process configuration requires a hidden adjustment, i.e., the published tolerance level is different to the actual tolerance level.

Consideration. The provider needs to decide on the tolerance policy. Process mining is here a potential source of valuable datasets, e.g., what percentage of processes fall within what tolerance levels. Furthermore, the costs and other implication of tolerating processes beyond the defined thresholds need to be calculated and assessed.

Process Automation. A company deploying tolerance as a benevolence design guideline needs to define the tolerance spectrum and context factors moderating it. For example, in light of the workload this type of benevolence might be more feasible in low demand periods. Augmented tolerance is the interplay of a system providing context-specific data and human judgement by an empowered staff member.

4.4 Be Humane

The previous six design guidelines relate to the transactional business processes an organisation has with its customers and adds a benevolence lens to the established focus on process performance and conformance. The fourth, and final pair of design guidelines is different. 'Being humane' is about the customer as an individual with a personal life that has its ups and downs. Attentiveness is a set of practices that recognises positive life events whereas empathy is the opposite and is about how the organisation reacts sensitively to negative life events.

4.4.1 Attentiveness

Example. When a customer checks into a hotel on their birthday, there might be a certain chance to be upgraded. The hotel uses the simple data point birthday as an external variable of the check-in process. A builder in Australia has a principle that they purchase a gift for the new owner of the home they built to the value of up to AUS$500. The only prerequisite is that the gift needs to fit into the life of the customer (e.g., an expensive cactus for someone who likes outdoor gardening). Similar practices are deployed by some banks on the day a mortgage is fully paid off or health providers at the end of a successful therapy. Many companies have small gift procedures that are triggered by a customer's milestone date.

Context. Attentiveness goes beyond immediate fulfillment as it is about taking part in the life of a customer. The benevolent act here can be closer (paying off a mortgage) or further removed (wedding gift) from the vicinity of the business.

Definition. Attentiveness is a proactive execution of an action that demonstrates support and celebration for a customer's achievement or positive life event.

Value. The customer experienced value in the form of feeling seen and cared for by an organisation. The organisation needs to allocate resources and funds for the planning and provision of gift and attention given to the customer.

Pattern. Attentiveness is initiated by two types of data points. Internal to the process it is the successful completion of a process that is personally significant to the customer (e.g., paying off a home loan). Externally, it is positive life event (e.g., birth of a child).

Consideration. A prerequisite for being attentive is to have legitimate access to a life event and to comply with regional data protection rules. For example, the European GDPR prevents organisations from even sending Christmas cards to their customers as their addresses have only been collected for the sole purpose of conducting business transactions. Being 'too attentive' can also be seen as an unwanted intrusion by the customer. The Dutch airline KLM had an initiative called 'KLM Surprise' during which they used social media data of their travelling customers to learn more about the motives of their customers. However, when they then surprised their customers during the check-in process (e.g., with baseball caps of the team they are about to watch during their trip to Florida), not all customers appreciated this level of attentiveness.

Process Automation. Attentiveness requires a definition of the start events that matter and how they can be accessed. This could be the simple re-purposing of a process end event (e.g., home loan has been paid off) as a start event for an attentive practice (e.g., sending a gift). As soon as personal data is a trigger of an attentive practice, regional data compliance needs to be carefully considered including gaining customer's consent.

4.4.2 Empathy

Example. The call centre of an Australian insurance company received a call in which a father wanted to cancel the car insurance for his daughter. The young agent in the call centre was trained to retain the customer by exploring rate discounts or other forms of down-selling. However, the caller was not interested – his daughter had passed away. The agent cancelled the insurance immediately and closed the file. But empathy is not just about reacting to specific events. It also materialises in explicitly committing time to conversations with customers. A related benevolent practice has been established by the second largest supermarket chain in The Netherlands, Jumbo. 1.3 million people in The Netherlands are older than 75 years, often lonely and a visit to a supermarket provides a rare, but much valued opportunity for interaction. Therefore, Jumbo opened in the summer of 2019 its first Kletskassa – 'chat checkout'. Here, staff at the checkout are committed to a personal chit-chat with the customers, the opposite of the supermarket checkout process designed for high throughput rates. This practice has been so appreciated by Jumbo's customers, that now more than 200 of its stores have a Kletskassa.

Context. Demonstrating empathy is needed in special, often emotionally or even traumatic circumstances and goes beyond successful fulfillment. Empathy requires additional activities that demonstrate care and do not come with immediate benefits for the provider. Empathy requires taking the perspective of the effected customer and showing feelings such as sympathy or compassion.

Definition. Empathy is the "affective response more appropriate to someone else's situation than one's own" [34]. Showing empathy is the proactive execution of an activity that demonstrates care for the challenging situation of a customer.

Value. Empathy leads to customer value that is non-monetary whereas the providers' efforts to be empathic can often be quantified (see Jumbo's additional checkout).

Pattern. Empathy follows the same pattern like attentiveness, just that it is triggered by negative life events.

Consideration. Showing empathy requires a careful consideration of the circumstances and a classification of what type of action justifies a demonstration of empathy. The most appropriate next call of action, which might be a pre-described one (if then then this) or an undefined choice for the staff in charge. Comparability, appropriateness and the requirement for individualisation needs to be carefully considered.

Process Automation. Many forms of empathy will be beyond a scripted process automation and rely on the emotional intelligence of empowered staff. However, empathy by design is also possible as the dedicated check-out line of Jumbo shows. The automation of an empathic practice might be as simple as having an initiating trigger ('send flowers') as part of a call centre agent's script.

5   Discussion

We presented eight benevolent process design guidelines summarising our study of related practices within three organisations and complementary secondary data. We aimed to consider all the practices we observed but cannot claim completeness.

Beyond the actual benevolent process design guidelines, benevolence needs to be seen from its organisational, technological and economic setting as well as from the viewpoint of the customer's individual considerations. Though the idea of committing to customer's well-being is compelling, it has to be feasible and aligned with the strategic and operational context and the technological potential of the organisation.

As trust accounting does not exist, and gains in loyalty and advocacy are less tangible than the funds benevolent practices require, many organisations tend to remain focused on rating quarterly targets higher than building long-term partnerships. However, a changing environmental climate in which concepts such as shared value, conscious capitalism and the purpose-led organisation are getting increasing attention, is expected to provide fertile ground for more benevolent practices. The formalisation of ESG reporting makes purpose-related reporting the new 'business-as-usual' and trust-building actions are increasingly becoming a primary concern.

The three organisations we worked with and who mapped these design guidelines against their own processes and assessed the applicability of the proposed design guidelines overall, all highlighted the need to revise their customer service charter, i.e., their rules, policies, procedures and service-level agreements as well as work on internal changes in terms of their culture of customer engagement. This will be required to establish a system capable of delivering authentic and scalable benevolent processes.

6   Conclusions, Limitations and Future Research

The academic and professional BPM discipline possesses a comprehensive set of mature and validated approached to derive transactional value via business processes. However, changing customer demands and regulatory conditions require value propositions that go beyond a time-cost-quality ambition. Demonstrating authentic benevolence is one of these values, and this paper takes a first step in providing operational advice for the design of benevolent processes using process design guidelines.

As part of our study, we interviewed executives of three organisations. However, like most companies, their benevolent practices are only at their infancy meaning we needed to rely more on perceptions than actual practices. As such the absence of substantial empirical insights led to compromises in terms of the understanding of each of the benevolent process patterns or the completeness of this list of design guidelines. Moreover, we did not provide a detailed formalisation of each design guideline or assessed boundary conditions, i.e., in which context these guidelines are applicable.

As one of the first contributions towards benevolent process design guidelines, this paper has the potential to initiate *future research directions*. At this stage, the design guidelines are presented as a 'list' of individual options to consider, but there are interrelationships between them. In particular some design guidelines can set pre-conditions for other design guidelines. An empirical investigation of these interrelationships will provide a deeper understanding of their application. The current design guidelines and examples are focused on corporates (large, for-profit organisations), but the concept of benevolent process design guidelines is equally applicable for government agencies, not-for-profit organisations and smaller enterprises. Thus, how these design guidelines can be extended and/or adapted within diverse sectors can be investigated further and complemented with a collection of rich case examples from these other sectors. Finally, knowledge about the actual impact of these mutualistic benevolent process design guidelines in the form of retention, advocacy or trust is limited, but a highly relevant factor before organisations would proceed their implementation.

References

1. Dumas, M., La Rosa, M., Mendling, J., Reijers, H.A.: Fundamentals of Business Process Management, 2nd edn. Springer, Berlin (2018)
2. Mackey, J., Sisodia, R.: Conscious Capitalism: Liberating the Heroic Spirit of Business. Harvard Business Review Press, Boston (2013)
3. Porter, M.E., Kramer, M.R.: Creating shared value. Harvard Business Review, January-February (2011)
4. Quinn, R.E., Thakor, A.V.: Creating a purpose-driven organisation. Harvard Business Review, July-August (2018)
5. Ngyuen, N.: Reinforcing customer loyalty through service employees' competence and benevolence. Serv. Ind. J. **36**, 721–738 (2016)
6. Ngyuen, N.: Competence and benevolence of contact personnel in the perceived corporate reputation: an empirical study in financial services. Corp. Reput. Rev. **12**(4), 345–356 (2010)
7. Colquitt, J.A., Scott, B.A., LePine, J.A.: Trust, trustworthiness and trust propensity: a meta-analytic test of their unique relationship with risk taking and job performance. J. Appl. Psychol. **92**(4), 909–927 (2007)
8. Karakas, F., Sarigollu, E.: Benevolent leadership: conceptualization and construct development. J. Bus. Ethics **108**(4), 537–553 (2010)
9. Podsakoff, P.M., et al.: Transformational leader behaviors and substitutes for leadership as determinants of employee satisfaction commitment, trust, and organizational citizenship behaviors. J. Manag. **22**(2), 259–298 (1996)
10. Mayer, R.C., Davis, J.H.: The effect of the performance appraisal system on trust for management. J. Appl. Psychol. **84**(1), 123–136 (1999)
11. Akbolat, M., et al.: Benevolence or competence which is more important for patient loyalty? J. Int. Health Sci. Manag. **5**(9), 76–84 (2019)
12. Rosemann, M.: Trust-aware process design. In: Hildebrandt, T., van Dongen, B.F., Röglinger, M., Mendling, J. (eds.) Business Process Management. Lecture Notes in Computer Science, vol. 11675, pp. 305–321. Springer, Cham (2019). https://doi.org/10.1007/978-3-030-26619-6_20
13. Mercier, G., Deslandes, G.: Formal and informal benevolence in a profit-oriented context. J. Bus. Ethics **165**(1), 125–143 (2019). https://doi.org/10.1007/s10551-019-04108-9
14. Mayer, R.C., Davis, J.H., Schoorman, F.D.: An integrative model of organizational trust. Acad. Manag. Rev. **20**(3), 709–734 (1995)
15. Lee, D.-J., Sirgy, M.J., Brown, J.R., Bird, M.M.: Importers' benevolence toward their foreign exporter suppliers. J. Acad. Mark. Sci. **32**(1), 32–48 (2004)
16. Doney, P.M., Cannon, J.P.: An examination of the nature of trust in buyer-seller relationships. J. Mark. **61**(2), 35–51 (2007)

17. Livnat, Y.: On the nature of benevolence. J. Soc. Psychol. **35**(2), 304–317 (2004)
18. Tirole, J.: The internal organization of government. Oxf. Econ. Pap. **46**(1), 1–29 (1994)
19. Frank, L., Poll, R., Roeglinger, M., Lea, R.: Design heuristics for customer-centric business processes. Bus. Process Manag. J. **26**(6), 1283–1305 (2020)
20. Galbraith, J.R.: Designing the Customer-Centric Organization. A Guide to Strategy, Structure, and Process. Wiley, Hoboken (2005)
21. Moormann, J., Pavölgyi, E.Z.: Customer-centric business modelling: setting a research Agenda. In: Proceedings of the IEEE International Conference on Business Modeling, Vienna, 15–18 July, pp. 173–179 (2013)
22. Jones, T.M.: Corporate social responsibility revisited, redefined. Calif. Manage. Rev. **22**(3), 59–67 (1980)
23. Breitenbach, F.J.: Application of business process management for corporate sustainability. PhD Thesis, University of Plymouth (2020)
24. Boutilier, R.G.: Frequently asked questions about the social licence to operate. Impact Assess. Proj. Appraisal **32**(4), 263–272 (2014)
25. Mena, S., Rintamäki, J., Fleming, P., Spicer, A.: On the forgetting of corporate irresponsibility. Acad. Manag. Rev. **41**(4), 720–738 (2016)
26. Mitnick, B.M., Windsor, D., Wood, D.J.: CSR: undertheorized or essentially contested? Acad. Manag. Rev. **46**(3), 623–629 (2021)
27. van der Aalst, W.M.P., ter Hofstede, A.H.M., Kiepuszewski, B., Barros, A.P.: Workflow Patterns. Distrib. Parallel Databases **14**, 5–51 (2003)
28. Reijers, H.A., Mansar, S.L.: Best practices in business process redesign: use and impact. Bus. Process. Manag. J. **13**(2), 193–213 (2007)
29. Rosemann, M.: Explorative process design patterns. In: Fahland, D., Ghidini, C., Becker, J., Dumas, M. (eds.) Business Process Management. Lecture Notes in Computer Science, vol. 12168, pp. 349–367. Springer, Cham (2020). https://doi.org/10.1007/978-3-030-58666-9_20
30. Peffers, K., Tuunanen, T., Rothenberger, M.A., Chatterjee, S.: A Design science research methodology for information systems research. J. Manag. Inf. Syst. **3**(24), 45–77 (2007)
31. Venable, J., Pries-Heje, J., Baskerville, R.: A comprehensive framework for evaluation in design science research. In: Peffers, K., Rothenberger, M., Kuechler, B. (eds.) Design Science Research in Information Systems. Advances in Theory and Practice. Lecture Notes in Computer Science, vol. 7286, pp. 423–438. Springer, Heidelberg (2012). https://doi.org/10.1007/978-3-642-29863-9_31
32. Krishna, A., Dangayach, G.S., Rakesh, J.: Service recovery: Literature review and research issues. J. Serv. Sci. Res. **3**, 71–121 (2011)
33. Hall, J.M., Johnson, M.E.: When should a process be art, not science? Harvard Business Review, March 2009
34. Hoffman, M.: Empathy and Moral Development: Implications for Caring and Justice. Cambridge University Press, Cambridge (2001)

PEM4PPM: A Cognitive Perspective on the Process of Process Mining

Elizaveta Sorokina[1]([⊠]) [iD], Pnina Soffer[1] [iD], Irit Hadar[1] [iD], Uri Leron[2], Francesca Zerbato[3] [iD], and Barbara Weber[3] [iD]

[1] University of Haifa, City Campus, Hanamal St. 65, Haifa, Israel
`esorokin@campus.haifa.ac.il`, `{spnina,hadari}@is.haifa.ac.il`
[2] Technion - Israel Institute of Technology, Technion City, 3200003 Haifa, Israel
`uril@technion.ac.il`
[3] Institute of Computer Science, University of St, Gallen, Switzerland
`{francesca.zerbato,barbara.weber}@unisg.ch`

Abstract. During the last decades, process mining (PM) has matured and rapidly increased in its adoption. Making sense of data is a main part of the work of PM analysts, which involves cognitive processes. Recent work has leveraged behavioral data to explain these processes. Still, the process of process mining (PPM) is yet to be well understood and a theoretical foundation for explaining how these processes unfold is missing. This paper attempts to fill this gap by understanding how PPM data can be analyzed in a theory-guided manner and what insights can be gained from this analysis. To investigate these aspects, we analyzed verbal data and interaction traces obtained from analysis sessions with 29 participants performing a PM task. The analysis was based on the Predictive Processing (PP) theory and the derived Prediction Error Minimization (PEM) process, anchored in cognitive science. The results include (1) a theoretical adaptation of the PEM theory to the PPM context, (2) four strategies utilized by PM analysts, identified, and validated based on the adapted theory, and (3) an understanding of the differences in performance between analysts using different strategies and independence of the expertise level and the strategy choice.

Keywords: Process Mining · Predictive Processing · Prediction Error Minimization · Analysis Strategies · Mixed Methods

1 Introduction

Over the past decades, information systems penetrated all areas of people's lives and led to an incredible growth of produced event data in various branches of industry. This growth has led to the development of techniques to analyze such data, such as process mining, which stood out as a valuable discipline for extracting insights from event logs and supporting the discovery, monitoring, and improvement of real processes.

Until now, process mining research has mainly concentrated on the technical perspective, proposing new approaches and tool enhancements [1, 2]. Still, recent studies on

© The Author(s), under exclusive license to Springer Nature Switzerland AG 2023
C. Di Francescomarino et al. (Eds.): BPM 2023, LNCS 14159, pp. 465–481, 2023.
https://doi.org/10.1007/978-3-031-41620-0_27

the "process of process mining" (PPM) have highlighted the importance of the *individual* perspective and focusing on the individual analyst's perception and behavior [3].

The PPM includes different activities such as choosing the data sources, inspecting, cleaning, and preprocessing the data, selecting and implementing process mining algorithms, and interpreting and presenting the results. A main goal of individual analysts during the PPM is to make sense of data. The sense-making process entails an iterative cycle, where attention is focused according to a set goal, leading to the generation of predictions about the world, which are then tested and reconsidered for minimizing the prediction error, i.e., the discrepancy between the predicted and the actual input. These are complicated cognitive processes that may entail high cognitive load [4, 5]. Although recent works have attempted to leverage behavioral data to explain such processes, e.g., by eliciting patterns of observed behavior or analysts' challenges [6–10], these processes are not yet well understood and there is a lack of theory in the process mining (PM) field that can explain how these processes unfold.

This paper attempts to fill this gap by proposing a theory-guided analysis approach based on a theory from a neighboring field. More specifically, previously mentioned characteristics of the PPM point to a close correspondence with the Prediction Error Minimization (PEM) theory, which is a part of the Predictive Processing (PP) theory [11–13]. In this research, we adapted the PEM theory to the context of the PPM, yielding the PEM4PPM model. This model provides a theoretical lens for analyzing the PPM from a cognitive perspective and understanding the analysts' cognitive processes. We applied this model to analyze data recording analysts' interactions with PM tools as well as think-aloud data verbalizing their thinking processes while performing PM tasks during the analysis phase of the PPM. With our analysis, we aim to answer the following two research questions:

- RQ1: How can PPM data be analyzed in a theory-guided manner (specifically, based on the PEM theory)?
- RQ2: What insights can be gained from this analysis?

To address these questions, we employed a mixed-method (qualitative and quantitative) approach [14]. RQ1 was addressed in a qualitative manner. First, we made the theoretical adaptation of the PEM model for the PPM, which resulted in the first version of the PEM4PPM model. Second, data collection and analysis of students' data were performed with the goal to refine the model. As a result of this step, the PEM4PPM model was refined to include additional cognitive steps, and four analysis strategies were identified. Next, the model and the strategies were validated against data stemming from experienced analysts. In addition, an analysis protocol was established. Finally, we integrated our findings regarding the identified strategies from both datasets and formed hypotheses for RQ2. To answer RQ2, we applied a quantitative approach, focused on analyzing the influence of the previously discovered strategies on the analysts' performance and checking the association between the strategies and the level of expertise. The main contributions of this paper are (i) a theoretical adaptation of the PEM to the PPM context, (ii) four strategies utilized by PM analysts, identified and validated based on the adapted theory, and (iii) an understanding of the difference in performance between analysts using different strategies and independence of the expertise level and the strategy choice.

The remainder is organized as follows: Sect. 2 provides background concepts and related work. Section 3 presents the proposed PEM4PPM model. Section 4 describes the research method of this work and Sect. 5 reports on the findings. Section 6 discusses the findings and limitations of our work. Section 7 provides conclusions and outlines future research directions.

2 Related Work

2.1 Human Involvement in the PPM

Vom Brocke et al. [3] proposed a five-level framework for research on PM. Two of the levels, group and individual, focus on the humans involved in PM. While the group level characterizes the effect of PM on people's interaction and work mode, the individual level concentrates on people's perception and behavior. There are additional recently published papers taking an individual perspective and focusing on different aspects related to the behavior of PM analysts. These include analysts' patterns of behavior, goals, and strategies in the initial exploration phase [6], strategies applied by analysts during the analysis stage [7], the development of PM questions [8], the discovery of process improvement opportunities by exploring how analysts use PM [9], and challenges perceived by process analysts during different project phases [10].

This work also takes the individual perspective but extends the body of knowledge through a theory-guided approach. More specifically, this work attempts to unveil and explain the complicated cognitive processes that PM analysts follow to make sense of event data representing real-world behavior from a cognitive perspective [4, 5]. Such a theoretical cognitive lens is still missing.

2.2 The Prediction Error Minimization Theory

Our choice of a conceptual basis for analyzing the PPM from a cognitive perspective is based on the previously outlined similarities (cf. Section 1). PEM conceives the brain as a probabilistic inference system, which attempts to predict the input it receives by constructing models of the possible causes of this input. This system aims to minimize the prediction error, namely, the discrepancy between the predicted and the actual input. If the prediction error is small, then there may be no need to revise the model that gives rise to the prediction. If, on the other hand, the prediction error is large, then it is likely that the model fails to capture the causes of the inputs and, therefore, must be revised. In this case, there are two optional courses of action for reducing the prediction error. The first is to revise one's model of the world until the prediction error is satisfactorily diminished (termed 'perceptual inference') and the second is to change the world (e.g., by manipulations that yield additional inputs) so that it matches the model (termed 'active inference'). Moreover, PEM suggests that predictive models are arranged in a hierarchy. Interpretation of sensory signals is at the lowest level of the hierarchy, while low-level features are grouped into objects at the higher levels. Even further up the hierarchy, these objects are grouped together as parts of larger scenes entailing multiple objects [11, 12]. It is important to note, that PEM is not only a formal apparatus which allows generating precise quantitative models, but it is also a general framework that can be applied in the service of different explanatory programs. Thus, this framework can be enriched with additional assumptions about a specific domain, e.g., PM [15].

3   The Proposed PEM4PPM Model Based on the PEM Theory

In this section, we present the PEM4PPM model, which we developed based on the PEM theory, by adaptation to the PPM as a part of answering RQ1 (cf. Section 4.1).

The PEM process involves receiving input, focusing attention (possibly based on a goal) [16], and then iteratively creating a model and respective predictions, testing predictions, minimizing errors and eventually, when the error is small enough, acting upon the prediction. The initial PEM4PPM model was created in a series of brainstorming sessions, where the bi-directional mapping between PEM steps and PM operations took place, considering what PM operations could accomplish each PEM step and what PEM step could correspond to each PM operation. The model is shown in Fig. 1, which illustrates a diagram of a goal-oriented PPM conceptualized in terms of the PEM theory. Each step or sequence of steps in the diagram can be skipped. Cognitive operations are color-coded according to the PEM steps to which they correspond, and example mining operations appear in italics below PPM steps where relevant.

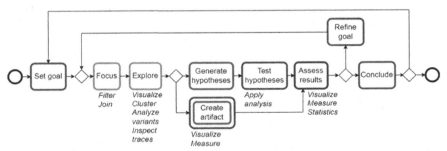

Fig. 1. The PEM4PPM model

According to the model, high-level business goals serve as a starting point for each PM endeavor. They can be later decomposed or refined into more specific goals. The refinement of goals can be done until the goal is concrete enough to be achieved by an available mining operation. Additional goals can be set or refined at any stage during this process. More concretely, when a goal is to be refined, a series of operations are involved. In the beginning, the relevant subset of the data needs to be filtered and organized. Next, this data needs to be explored to find behavior that may be of interest. These both are operations that enable focusing the attention on parts and aspects of the input data. Based on this, concrete hypotheses can be formed regarding the identified behavior, as predictions to be tested. Predictions can also be generated and tested by the creation of specific artifacts (e.g., creating a process model through discovery). Accordingly, available PM techniques are applied for testing hypotheses or creating artifacts. Then, the results are assessed against the goal or the hypothesis for minimizing the prediction error. There are two possible outcomes for this assessment: (1) conclusion – goal achieved; (2) the goal needs to be further refined and this can be repeated iteratively. Next, the results can point to a new goal to be set.

4   Research Methodology

This section presents the main steps of the research method followed in this paper.

4.1   Overview

The objective of this work was to gain an understanding of how to analyze PPM data in a theory-guided manner (based on the PEM model) (cf. RQ1) and to explore what insights can be gained from this analysis (cf. RQ2). For this purpose, a mixed-method research approach combining both qualitative and quantitative methods was applied [14]. Figure 2 provides a high-level depiction of the method used for this research.

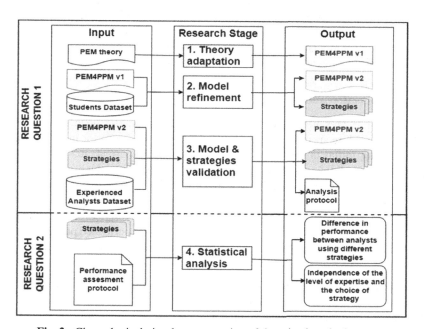

Fig. 2. Chronological visual representation of the mixed method process

To address RQ1 (*"How can PPM data be analyzed in a theory-guided manner (specifically, based on the PEM theory)?"*) we applied a qualitative research approach [17]. Our aims were to adapt the PEM theory as a basis for analyzing the PPM and establish and validate an adapted PEM model and a derived analysis protocol.

First, during Stage 1, we adapted the PEM theory, creating a first version of the PEM4PPM model through brainstorming sessions (as described in Sect. 3).

Second, as a part of Stage 2, we refined the model using behavioral data. For this, we collected data from student participants (Students Dataset) and analyzed it. We decided to use students for model refinement since they were expected to have sufficient knowledge for performing a process mining analysis (all of them took the same "Process Mining" course). This stage resulted in the second version of the model, PEM4PPM v2, which included some additional, refined cognitive steps. Moreover, we used the model to analyze user behavior and derived strategies, as patterns of behavior, applied by the students during their work.

Finally, during Stage 3, we validated the PEM2PPM v2 model discovered in stage 2 and the identified strategies by analyzing the data that were collected from experienced analysts (Experienced Analysts Dataset). Experienced analysts were chosen for this stage since they differ in their expertise and experience from students and therefore could contribute to further validating, and possibly refining our findings. In addition, during this stage, we developed an analysis protocol describing the cognitive steps of the PEM4PPM v2 model.

To address RQ2 (*"What insights can be gained by analyzing the PPM on the basis of PEM model?"*) a quantitative research approach was used. The strategies identified from analyzing the Students and Experienced Analysts Datasets in the context of RQ1, gave the direction for RQ2, i.e., to gain an understanding of whether the difference in performance between analysts using different strategies is significant and whether the level of expertise and the choice of strategy are independent of each other.

4.2 Data Collection and Settings

Overall, the study presented here comprises 29 participants: 16 students and 13 experienced analysts. More specifically, the Students Dataset was used for model refinement and for gaining initial insights (Stage 2), whereas the Experienced Analysts Dataset served for their validation (Stage 3). Despite having been collected separately, these two datasets are comparable since they were collected in the context of two studies of a similar design and set-up but different tasks and participants. More details about the two datasets are provided online: https://doi.org/10.5281/zenodo.8055223, including demographic information, interview protocols and examples of experienced participants' coding for the entire duration of the session, showing the chain of evidence from statements to findings.

Collection of Data from Students. The data collection took place between June and July 2021. The study consisted of individual Zoom[1] video sessions with 16 participants, all first- and second-degree students taking the "Process Mining" course provided by the University of Haifa[2]. All the students agreed to be observed and recorded via Zoom for the entire duration of the study. The study consisted of a process mining task (guided exploration) and a semi-structured interview. Before taking part in the study, students were asked to perform an initial exploration of the log guided by a few questions provided to them as a part of course assignments. This was done to help them avoid possible

---

[1] Zoom https://zoom.us/

[2] The full study design, including participants' recruitment procedure, was approved by the Institutional Ethics Committee (Approval no. 238/21).

technical issues which may be related to various operations such as installing the tool, preprocessing, filtering, or importing/converting the log. In addition, these questions were designed to give students the opportunity to better understand the process, its activities and event attributes, and to familiarize with the tool before participating in the study. Our assumption was that this preliminary stage enabled participants to better concentrate on the provided question and guidelines during the data collection.

For the guided exploration, the students were asked to analyze the Road Traffic Fine Management (RTFM) event log, i.e., a real-life event log representing the process of managing fines for road traffic violations by the Italian police [18]. The log was chosen as it comes with extensive documentation about the process which allowed us to prepare the materials for the study [19]. The participants were asked to answer the following question related to the log "Describe two to three circumstances in which a penalty is added. Is the penalty addition always legitimate, i.e., do people that receive a penalty deserve it?". This question was designed to guide participants in exploring different scenarios but also go deeper into the data and investigate penalty addition. For their analysis, the participants could choose among different tools (Disco, ProM, PM4Py and bupaR), but all the participants chose Fluxicon Disco[3]. During the task, the participants were instructed to verbalize everything they were doing in a think-aloud manner [20]. In this way, we aimed to gain insights into their cognitive processes. The session was supervised by one author, who reminded participants to speak out in case of long sequences of actions happened without explanation. No feedback on their analysis strategies and outcomes was provided to the students during the session and they could conduct the analysis at their own pace.

After the participants completed the process mining task, we interviewed them to improve our understanding of analysis strategies and cognitive factors underlying them. The interviews were semi-structured [21] to allow expanding and gathering of information on relevant aspects emerging during the session.

For each student participant, we collected the recording of the session (screen recording of the analysis conducted in Disco and audio of the think-aloud), the log produced by Disco, i.e., a log that is produced automatically by the tool which records both the debug information and user's actions in the PM tool, and the audio recordings of the interviews. The session lasted up to one hour in most cases, with an average duration of 38 min. The interviews lasted between 8 and 13 min.

For this dataset, the signed participant consent, as approved by the Institutional Review Board (IRB) committee, does not allow for making this data publicly available in its raw form.

Collection of Data from Experienced Analysts. Thirteen experienced analysts served for validation, giving us the opportunity to validate the PEM4PPM v2 model, which we had refined based on the students' data. The data collection took place between May and July 2021, and involved analysts in the professional network of the authors who met the following requirements: (i) having analyzed at least two real-life event logs in the two years prior to the study and (ii) being knowledgeable of at least one among bupaR, Celonis, Disco, ProM and PM4Py, i.e., the PM tools available for the task. These two criteria allowed us to select participants familiar with the PM but having varying levels

---

[3] Fluxicon Disco https://fluxicon.com/disco/

of experience and expertise. For this paper, we selected participants who used Disco to keep this set comparable to the Students Dataset and to reduce the dependency of the validation of the PEM4PPM v2 model and the strategies on the working mode imposed by the tool. The experienced analysts differ in their affiliation, job role and position, and tool knowledge. The data collection followed a similar design and the same materials used with the students: participants were asked to engage in a PM task while verbalizing their thoughts in a think-aloud manner. What differed was the scenario question we asked on the RTFM log: "Can you describe the three most prevalent circumstances in which the fines and the related expenses are not paid (in full) in this process?". Then, as done with the students, we prompted participants to look deeper into reasons for which fines are not paid. In this case, we did not ask participants to engage in an initial exploration phase, as the risks for technical issues and lack of experience with PM analysis were mitigated by the set-up (a remote desktop environment with the study materials and tools) and the participation requirements. Data collection was supervised by one author who prompted participants to speak in case of long silences, but, as for the students, no feedback on their analysis nor its outcome was given.

Finally, we conducted semi-structured interviews asking participants about (i) activities and artifacts, (ii) goals, (iii) strategies, and (iv) challenges of PM analysis. In detail, we asked them to reflect on the task they had just executed but also prompted them to share general anecdotes and work experiences with PM.

For each participant, we collected the same data as for the students: the screen and audio recording of the session, the log produced by Disco and the audio recordings of the interviews.

In this case, the sessions lasted between 20 and 90 min, with an average duration of 43. The interviews lasted between 23 and 43 min, and about 31 min on average. The data is collected as part of the ProMiSE project, for which we plan to share results and anonymized data with the research community upon project completion.

4.3   Data Preparation and Analysis

This section details how the collected data was analyzed to address RQ1 and RQ2 (Stages 2–4) leveraging the PEM4PPM model developed in Stage 1 (cf. Section 3).

Data Preparation: For the data preprocessing, we engaged in the following activities. First, (i) we transcribed the verbal utterances from the audio recordings of the session with the help of Vocalmatic[4] (Students Dataset) and MAXQDA[5] (Experienced Analysts Dataset), which allowed us to keep information about the timestamps of each utterance. Then, (ii) we watched all the screen recordings and reported next to each transcribed utterance a short description of non-verbal behavior observed from the videos. Examples of non-verbal behavior of interest for our analysis are pointing with the cursor to a specific process activity, reading documentation related to the task, filtering the data, etc. Then (iii) we merged and cleaned the logs from Disco removing debug information that did not relate to actions performed by participants in the tool and, thus, was not relevant to

---

[4] Vocalmatic https://vocalmatic.com/
[5] MAXQDA https://www.maxqda.com/

our research. As a final step, (iv) we merged and organized the data from (i), (ii) and (iii) in Excel, using the timestamps produced by the transcription tools and Disco to synchronize the different sources. In this way, we obtained a multimodal dataset [22] for our analysis, which included both verbal utterances and interaction traces of the participants with Disco.

Stage 2: To refine the PEM4PPM model we used the Students Dataset. This stage involved two activities: (i) coding the data according to the steps of the PEM4PPM v1 model to refine and tune the model and (ii) identifying different strategy types, characterizing, and grouping them into categories.

We utilized provisional coding, a first-cycle coding method, to code the Students Dataset [23]. The behavior and thinking process of each participant (as reflected by the verbal utterances and interaction traces) was analyzed individually. Since the steps of the PEM4PPM v1 model were not yet validated during this first analysis iteration, we analyzed the data described in the previous paragraph by combining both deductive and inductive methods. In detail, at the beginning of the analysis, a set of a priori codes, reflecting the steps of the PEM4PPM v1 model, was used. The coding was performed by assigning a code to each multimodal piece of data (think-aloud, screen recordings, and Disco logs), and using one or the other source to disambiguate in case the user cognitive step was not clear. The first author coded all the data in 3 iterations. At the end of each iteration, two other authors checked the codes independently to ensure consistency and assess the reliability of the process. Throughout the analysis, the authors revised and refined the codes collaboratively in several meetings. The final coding was determined through discussions among authors in an iterative consensus building process, similar to [24]. Since a one-to-one mapping of the data to the PEM4PPM v1 steps was not possible in all the cases, two additional steps ("Task understanding" and "Interpret data") emerged as additional steps in the PPM process. Next, they were validated against the theoretical literature [25, 26] and we added them to our model, obtaining PEM4PPM v2.

Then, based on the PEM4PPM v2 model, high-level patterns were identified during the second cycle of coding. For this purpose, theoretical coding (sometimes also called "Selective Coding" or "Conceptual Coding") [27] was utilized. It allowed us to group the first cycle summaries (steps of the PEM4PPM v2 model) into strategies applied by participants during their work. Analyzing commonalities and variabilities among these processes allowed us to identify the repetitive use of certain patterns and strategies that can be associated with different analyst profiles. These strategies were characterized by the presence or absence of the steps "Interpret data", "Generate hypotheses", or "Test hypotheses" in the verbal utterances, since other steps were present in most cases. These steps relate to cognitive operations of "creating prediction" and "testing". As a result of this analysis, we discovered four strategies associated with different analyst profiles. To investigate the possible reasons for applying specific strategies by students the interview data was examined.

Stage 3: To validate the PEM4PPM v2 model and the strategies discovered by analyzing the students' data, we coded and analyzed the data collected from experienced analysts. The coding and analysis of this dataset were performed following a similar approach to the analysis of the students' data described above. However, in this stage, we used the steps of the PEM4PPM v2 model in a deductive manner. The coding was performed by 2 authors in two iteration and, similarly to what was described for the students, we worked collaboratively in the team of authors to reach a consensus. The validation of the PEM4PPM v2 resulted in the analysis protocol describing the cognitive steps of the PEM4PPM v2 model, their explanation, example statement, and detection method. Table 1 presents a part of this protocol. The full analysis protocol can be found in the supplementary materials: https://doi.org/10.5281/zenodo.8055223.

Table 1. Analysis protocol (partial) describing the cognitive steps of the PEM4PPM v2 model. For each quote, we write the student (S) and experienced (E) participant's ID in parenthesis.

Cognitive step	Explanation	Example statement
Task understanding	Understanding the task/problem, understanding the data (e.g., attributes, model, activities), verifying with researcher/stakeholder information regarding the data or task	"Wait, *the question is about* the fine or the additional penalty?" (S7)
Set/ Refine goal	Formulating a goal related to the question	"So, the question is to identify the flows where the payment has not been done completely. That is why now I isolate the flows, the different flows we have…" (E26)
Focus	Concentrating on a specific part/activity of the map	"So, I think I first want to *focus on the cases* in which there was the addition of a penalty, so I'm actually going to filter by this element [activity Add penalty]" (S3)
Explore	Learning about the process	"He has a fine of 20, he has a fine of 22, he pays 22, then he had a penalty of 44. I don't understand why and then he paid another 35." (S8)
Interpret data	Explaining the process based on previously inspected behaviors, information about the process, the participant's previous knowledge and experience, or their combination	"Yeah, this is one third of the cases. Here it says 30% of the cases. We saw on the diagram that it was one third. Oh, yeah. 30% is one third. Um, yeah, one third of the citizens are good [laughing]. I have never driven in Italy, but it doesn't look like people drive following the rules. And I think that the data confirms…" (E3)
Create artifact	Producing deliverable objects driven by a goal	"Hmm. And it will make a difference, probably. Is it possible for me *to make a screenshot* of this and put it aside?" (E3)

(continued)

Table 1. (*continued*)

Cognitive step	Explanation	Example statement
Generate hypotheses	Forming a hypothesis about previously identified behaviour or possible answers for the question	"*I assume that* it can be a difference between Amount for example, paid and Amount counted." (E28)
Test hypotheses	Checking hypotheses using available PM techniques	NA (this step can be identified only by actions in the PM tool performed to test previously generated hypotheses)
Assess results	Evaluating the findings against previously defined hypothesis or goal	"There are actually quite a lot of cases where there was a payment in the dataset, there was no value for the Payment Amount." (E16)
Conclude	Providing a final answer to the question or its part	"So, most of the fines are legitimate. The majority behaves in a way that is legitimate, but there are some cases, like I said, that either they paid and then received a penalty or they sent an appeal or received the appeal and did not receive an answer to it, but received a penalty. These are actually the illegitimate situations." (S5)

Stage 4: Finally, aiming to address RQ2, we performed a combined analysis of both datasets focusing on the elicited strategies. Inspired by the analysis conducted to address RQ1, we formulated hypotheses regarding the difference in performance between analysts using different strategies and the independence of the level of expertise and the choice of strategy. In this stage, we tested such hypotheses. In order to assess the analysts' performance, we developed a gold standard: one author "graded" the answers provided by the participants, then two other authors validated the grades independently, and an iterative consensus building process took place (like the one applied for the coding). The answers were assessed using a grading scale from 0 to 100, where 0 indicates the lowest performance and 100 is the highest. The grades reflected the correctness of the analysts' answers to the questions of the task.

More details about the data analysis and findings, including the analysis protocol, the coding scheme with examples, the protocols for assessing the participants' performance and statistical outputs are available online: https://doi.org/10.5281/zenodo.8055223.

5 Findings

In this section, we present the results of our analysis for RQ1 (cf. Section 3 and Sect. 5.1) and RQ2 (cf. Section 5.2).

5.1 Qualitative Analysis

Validated PEM4PPM model. During Stage 3 of the empirical study (cf. Section 4.1), we validated the PEM4PPM v2 model. The resulting model is shown in Fig. 3, additional steps when compared to the initial model version are highlighted with grey. The step "Task understanding" reflects situations in which participants spent time understanding the task at hand from the documentation available or asking clarification questions to the researcher supervising the data collection. Instead, "Interpret data" implies explaining the process based on previously observed behaviors, information about the process, the participant's experience, or their combination.

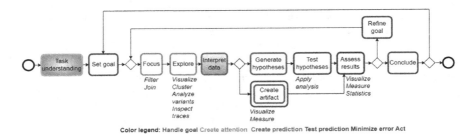

Fig. 3. The PEM4PPM model after validation (PEM4PPM v2)

Identifying and Validating PM Analysts' Strategies. This section presents the four strategies applied by PM analysts. These strategies were identified using the Students Dataset during Stage 2 and validated with the data collected from experienced analysts in Stage 3. Here we present the strategies we identified, with sample participants' quotes from interviews. For each strategy we indicate the count of students (S) and experienced analysts (E) using it in the form of #/29 (S = #, E = #):

1. *No indicated interpret data - No indicated hypothesis - No testing (NNN) strategy* which was observed for 2/29 (S = 2, E = 0) of the participants. The conclusions of the participants were based on their observations from the exploration stage. These participants spent the entire session understanding the data. "*I think I would try to use filters and not switch between cases and maps and try to draw conclusions through filters and this is also a way to work*". (S12)
2. *With interpret data - No indicated hypothesis - No testing (WNN) strategy* which was observed for 12/29 (S = 8, E = 4) of the participants. Instead of basing their conclusions on the testing of hypotheses, these participants made them mainly based on exploration and data interpretation. "The *process was very new, for me, it was a really challenge to understand what's. It was a simple process of course. It was just finding and these things. But it was new for me, so it took a time for me to understand different steps of the process and the process was really strange in some ways.*" (E6)
3. With interpret data - With hypothesis - No testing (WWN) strategy which was observed for 5/29 (S = 2, E = 3) of the participants. Participants formulated hypotheses but for some reason (e.g., insufficient tool knowledge (S8, S16, E28) or question

understanding (S16), were reluctant to follow a trial-and-error approach while being observed (E11)) did not test them. *"Indirectly, knowing that you were looking at the screen while doing this, I think not consciously or unconsciously, I think I was a little bit reluctant on the trial and error."* (E11)

4. *With interpret data - With hypothesis - With testing (WWW)* strategy which was observed for 10/29 (S = 4, E = 6) of the participants. The WWW strategy reflects the PEM4PPM v2 model by containing all its steps. *"So, it was not like you do the hypothesis and then you check it is more like you do a preliminary hypothesis. You check that the hypothesis make sense in a secondary step once the data is already been explored and then you do, you check the hypothesis after that always related with the questions."* (E7)

It is important to note that "No indicated interpret data" or "No indicated hypothesis" means that these cognitive steps were not explicitly expressed during the session by the participants. Although they could be performed implicitly, they are clearly not emphasized. Looking through the data after its classification according to the strategies, we noticed that experienced analysts tend to apply the WWW strategy (46%) more than students (25%).

Moreover, we observed that the answers of participants utilizing the WWW strategy are of higher quality (in terms of correctness and accuracy) than those of participants using other strategies. The students' mean grade was 51, while the experienced analysts' mean grade was 59. These assessments, together with the defined strategies, formed a basis for formulating four hypotheses:

- H1a: analysts (students and experienced) who use the WWW strategy have higher performance (grades) than analysts using other strategies.
- H1b: students who use the WWW strategy have higher performance (grades) than students using other strategies.
- H1c: experienced analysts who use the WWW strategy have higher performance (grades) than experienced analysts using other strategies.
- H1d: the level of expertise (student vs experienced analyst) and the choice of strategy (WWW vs any other strategy) are not independent.

5.2 Quantitative Analysis

The hypotheses formulated based on the findings of RQ1 aimed to investigate the differences in performance between analysts using different strategies and check whether the level of expertise and the choice of strategy are independent of each other. For the analysis, the data regarding the strategy applied by each participant and their performance was organized and exported to an IBM SPSS version 23 environment for statistical analysis.

In order to test hypotheses a-c, we conducted the Mann-Whitney test (also known as the Wilcoxon test for independent samples) [28]. This test is used to compare differences between two independent groups when the dependent variable is either ordinal or continuous, but not normally distributed.

The conclusions resulting from testing hypotheses a-c (Table 2) is that the grades of analysts (in general, and students and experienced analysts separately) using the WWW strategy are significantly higher than the grades of those using other strategies.

Table 2. Results of Mann-Whitney test; students (S), experienced analysts (E).

Hypothesis	N	U	Z	Calculated effect size (r)	One-tailed p-value	Decision for H0 (reject H0 if p-value < α (0.05))
a (S and E)	29	13.5	-3.762	-0.699	.0000185	rejected
b (S)	16	2.5	-2.688	-0.672	.002	rejected
c (E)	13	6	-2.161	-0.599	.0175	rejected

To check whether selecting the WWW over any other strategy is independent of being a student or an experienced analyst (hypothesis d), we conducted the Chi-Squared Test of Independence of Categorical Variables [29]. It is a statistical test used to check whether some categorical variables are statistically independent in a population.

The conclusion from testing hypothesis d is that H0 cannot be rejected ($\chi (1)^2 = 1.421$, p-value $= .233$). Therefore, we concluded that the choice of the WWW strategy over another does not depend on being a student or an experienced analyst.

6   Discussion

In this section, we review our findings, link them to related work, and discuss their impact on future research. Our work proposes the PEM4PPM model, a theoretical foundation for explaining the cognitive processes of individual analysts during the PPM, derived from theory and refined and validated with user behavior data. This approach complements fully data-driven work, e.g., [6–10], and extends the body of knowledge on the individual perspective with a theoretical lens.

In addition, based on the PEM4PPM model we identified four strategies applied by PM analysts. While some analysts performed all operations (WWW strategy), other analysts did not "generate hypothesis" and "test hypothesis" (NNN and WNN strategies). A possible explanation for not investigating the data more in-depth may be that analysts pursuing the NNN and WNN strategies were still in the stage where they collected evidence in the hope to suggest a hypothesis, rather than in the stage of testing a hypothesis they had in mind [30]. The development of strategies aiming to improve performance by realizing all their steps, usually in a specific order, is well renowned in the problem-solving and information-processing field [31, 32]. Analysts' strategies were also the focus of [7]. Unlike this work they organized them by analysis goals (understand, plan, analyze, and evaluate) and based their analysis on post-session interviews, rather than on the participants' real-time think-aloud.

The established analysis protocol together with the developed and validated PEM4PPM model can be used for future research. Moreover, this has implications for tool development or improvement allowing to consider the individual and his/her analysis processes. Since identified strategies differ in terms of performance this can give rise to the development of process guidance for PM analysts. These findings also can have implications for teaching students how to approach process mining. Beyond the expected contributions to the field of process mining, the outcomes of this research may be possible to generalize and adapt for additional data science tasks, which also entail extracting knowledge from data.

Limitations. The findings of this study should be considered in the light of some limitations. First, although participants were constantly encouraged to verbalize their thoughts in a think-aloud manner, it is possible that some thoughts might have remained concealed. We mitigated this risk by triangulating the verbal utterances with the log data. In the future, adopting additional techniques, e.g., eye tracking or EEG [33], might help to fill this gap, since additional fine-grained information about where the analyst focused their attention might be collected. Second, although the data coding was done carefully and iteratively by building consensus among researchers, it is still based on subjective assessment and could be done differently. Yet, we note that this is a common procedure [24], and that such an iterative process, where coding changes along the iterations, intercoder reliability is usually not measured. Third, while the findings reported here have been established, specifically with statistically significant results for the quantitative analysis, the setting of the study is limited, using one dataset and one process mining tool. For addressing our research questions, of how to analyze PPM data and what insights can be gained, this is sufficient. Yet, to draw general conclusions about PPM, additional studies are still needed. At last, one important consideration to make is the confidentiality of the collected data, which, as clarified in Sect. 3, cannot be fully shared. While we acknowledge that this may limit the replicability of the study, the inclusion of representative examples adds transparency and enables the reader to follow the chain of evidence from the raw data to the presented results.

7 Conclusions

In this paper, we have presented the findings of an empirical study investigating how PPM data can be analyzed in a theory-guided manner (specifically, based on the PEM theory) and what insights can be gained from this analysis. We adapted the PEM theory to the context of the PPM, yielding the PEM4PPM model. This model provided a theoretical lens for analyzing the PPM from a cognitive perspective and understanding the analysts' cognitive processes. Applying the PEM4PPM model for the analysis of the verbal utterances and interaction traces obtained from sessions with 29 participants we identified four strategies applied by PM analysts. We characterized these strategies by the sets of cognitive operations involved. Moreover, we investigated whether the difference in performance between analysts using different strategies is significant and whether the level of expertise and the choice of strategy are independent. We found that: (i) the grades of analysts (in general, and students and experienced analysts separately) using the WWW strategy are significantly higher than the grades of those using other strategies; (ii) the level of expertise (student or experienced analyst) and the choice of strategy are independent.

Future Work. We established PEM4PPM as a basis for analyzing PPM data and showed that it can yield benefits. Having established this, the generalizability of the findings may be improved by conducting additional studies with more participants and in different settings e.g., different PM tools and tasks using different event logs. In addition, a better understanding of the thinking processes of PM analysts may be gained by collecting data with technologies such as eye tracking or EEG. Another possible avenue for future research is the development of a support tool, designed in accordance with the PEM4PPM model, to assist PM analysts throughout the PPM.

Acknowledgment. Acknowledgment. This work was partially funded by the Israel Science Foundation under grant agreement 2005/21 and the Swiss National Science Foundation (SNSF) under Grant No.: 200021 197032.

References

1. R'bigui, H., Cho, C.: The state-of-the-art of business process mining challenges. Int. J. Bus. Process Integr. Manage. **8**(4), 285 (2017). https://doi.org/10.1504/IJBPIM.2017.10009731
2. Tiwari, A., Turner, C.J., Majeed, B.: A review of business process mining: state-of-the-art and future trends. Bus. Process. Manag. J. **14**, 5–22 (2008). https://doi.org/10.1108/146371 50810849373
3. vom Brocke, J., Jans, M., Mendling, J., Reijers, H.A.: A five-level framework for research on process mining. Bus. Inf. Syst. Eng. **63**(5), 483–490 (2021). https://doi.org/10.1007/s12599-021-00718-8
4. Sweller, J.: Cognitive load during problem solving: effects on learning. Cogn. Sci. **12**, 257–285 (1988). https://doi.org/10.1207/s15516709cog1202_4
5. Jans, M., Soffer, P., Jouck, T.: Building a valuable event log for process mining: an experimental exploration of a guided process. Enterpr. Inf. Syst. **13**, 601–630 (2019). https://doi.org/10.1080/17517575.2019.1587788
6. Zerbato, F., Soffer, P., Weber, B.: Initial insights into exploratory process mining practices. In: Polyvyanyy, A., Wynn, M.T., Van Looy, A., Reichert, M. (eds.) Business Process Management Forum. LNBIP, vol. 427, pp. 145–161. Springer, Cham (2021). https://doi.org/10.1007/978-3-030-85440-9_9
7. Zerbato, F., Soffer, P., Weber, B.: Process mining practices: evidence from interviews. In: Di Ciccio, C., Dijkman, R., del Río, A., Ortega, S.R.-M. (eds.) Business Process Management, pp. 268–285. Springer International Publishing, Cham (2022). https://doi.org/10.1007/978-3-031-16103-2_19
8. Zerbato, F., Koorn, J.J., Beerepoot, I., Weber, B., Reijers, H.A.: On the origin of questions in process mining projects. In: João, P.A., Almeida, D.K., Guizzardi, G., Montali, M., Maggi, F.M., Fonseca, C.M. (eds.) Enterprise Design, Operations, and Computing, pp. 165–181. Springer International Publishing, Cham (2022). https://doi.org/10.1007/978-3-031-17604-3_10
9. Kubrak, K., Milani, F., Nolte, A.: Process mining for process improvement - an evaluation of analysis practices. In: Guizzardi, R., Ralyté, J., Franch, X. (eds.) Research Challenges in Information Science, pp. 214–230. Springer International Publishing, Cham (2022). https://doi.org/10.1007/978-3-031-05760-1_13
10. Zimmermann, L., Zerbato, F., Weber, B.: Process mining challenges perceived by analysts: an interview study. In: Augusto, A., Gill, A., Bork, D., Nurcan, S., Reinhartz-Berger, I., Schmidt, R. (eds.) Enterprise, Business-Process and Information Systems Modeling, pp. 3–17. Springer International Publishing, Cham (2022). https://doi.org/10.1007/978-3-031-07475-2_1
11. Hohwy, J.: The self-evidencing brain. Noûs **50**, 259–285 (2014). https://doi.org/10.1111/nous.12062

12. Clark, A.: Busting Out: Predictive Brains, Embodied Minds, and the Puzzle of the Evidentiary Veil. Noûs **51**, 727–753 (2016). https://doi.org/10.1111/nous.12140

13. de Bruin, L., Michael, J.: Prediction error minimization: implications for embodied cognition and the extended mind hypothesis. Brain Cogn. **112**, 58–63 (2017). https://doi.org/10.1016/j.bandc.2016.01.009

14. Venkatesh, V., Brown, S.A., Bala, H.: Bridging the qualitative-quantitative divide: guidelines for conducting mixed methods research in information systems. MIS Q. **37**(1), 21–54 (2013). https://doi.org/10.25300/MISQ/2013/37.1.02

15. de Bruin, L., Michael, J.: Prediction error minimization as a framework for social cognition research. Erkenntnis **86**(1), 1–20 (2018). https://doi.org/10.1007/s10670-018-0090-9

16. Williams, D.: Predictive processing and the representation wars. Mind. Mach. **28**(1), 141–172 (2017). https://doi.org/10.1007/s11023-017-9441-6

17. Miles, M.B., Huberman, A.M., Saldana, J.: Qualitative Data Analysis (2019)

18. de Leoni, M., Mannhardt, F.: Road traffic fine management process. Eindhoven University of Technology. Dataset (2015)

19. Mannhardt, F.: Multi-perspective process mining. Ph.D. thesis, Technische Universiteit Eindhoven (2018)

20. Denzin, N.K., Ericsson, K.A., Simon, H.A.: Protocol analysis: verbal reports as data. Contemp. Sociol. **14**, 125 (1985). https://doi.org/10.2307/2070501

21. Myers, M.D., Newman, M.: The qualitative interview in IS research: examining the craft. Inf. Organ. **17**, 2–26 (2007). https://doi.org/10.1016/j.infoandorg.2006.11.001

22. Giannakos, M.N., Sharma, K., Pappas, I.O., Kostakos, V., Velloso, E.: Multimodal data as a means to understand the learning experience. Int. J. Inf. Manage. **48**, 108–119 (2019). https://doi.org/10.1016/j.ijinfomgt.2019.02.003

23. Saldaña, J.: The Coding Manual for Qualitative Researchers. Sage (2021)

24. Recker, J., Safrudin, N., Rosemann, M.: How novices design business processes. Inf. Syst. **37**, 557–573 (2012). https://doi.org/10.1016/j.is.2011.07.001

25. Westbrook, L.: Mental models: a theoretical overview and preliminary study. J. Inf. Sci. **32**, 563–579 (2006). https://doi.org/10.1177/0165551506068134

26. Staggers, N., Norcio, A.F.: Mental models: concepts for human-computer interaction research. Int. J. Man Mach. Stud. **38**, 587–605 (1993). https://doi.org/10.1006/imms.1993.1028

27. Strauss, A., Corbin, J.: Basics of Qualitative Research: Techniques and Procedures for Developing Grounded Theory (1998).https://doi.org/10.1604/9780803959392

28. Nachar, N.: The Mann-Whitney U: a test for assessing whether two independent samples come from the same distribution. Tutor. Quant. Methods Psychol. **4**(1), 13–20 (2008). https://doi.org/10.20982/tqmp.04.1.p013

29. Mendenhall, W., Beaver, B.M., Beaver, R.J.: Mendenhall's brief introduction to probability and statistics. Brooks/Cole (2001). https://doi.org/10.1604/9780534396091

30. Baron, J.: Thinking and Deciding. Cambridge University Press (2006). https://doi.org/10.1017/CBO9780511840265

31. Wallace, A.F.C.: Plans and the structure of Behavior. Am. Anthropol. **62**(6), 1065–1067 (1960). https://doi.org/10.1525/aa.1960.62.6.02a00190

32. Polya, G.: How to Solve It: A New Aspect of Mathematical Method (2004)..https://doi.org/10.1604/9780691119663

33. Dehaene, S., Lau, H., Kouider, S.: What Is Consciousness, and could machines have it? In: von Braun, J., S. Archer, M., Reichberg, G.M., Sánchez Sorondo, M. (eds.) Robotics, AI, and Humanity, pp. 43–56. Springer, Cham (2021). https://doi.org/10.1007/978-3-030-54173-6_4

Author Index

© The Editor(s) (if applicable) and The Author(s), under exclusive license
to Springer Nature Switzerland AG 2023
C. Di Francescomarino et al. (Eds.): BPM 2023, LNCS 14159, pp. 483–484, 2023.
https://doi.org/10.1007/978-3-031-41620-0

Printed in the United States
by Baker & Taylor Publisher Services